Handbook of Research on Implementation and Deployment of IoT Projects in Smart Cities

Krishnan Saravanan
Anna University Chennai - Regional Office Tirunelveli, India

Golden Julie
Anna University, India

Harold Robinson
SCAD College of Engineering and Technology, India

A volume in the Advances in Civil and Industrial Engineering (ACIE) Book Series

Published in the United States of America by
IGI Global
Engineering Science Reference (an imprint of IGI Global)
701 E. Chocolate Avenue
Hershey PA, USA 17033
Tel: 717-533-8845
Fax: 717-533-8661
E-mail: cust@igi-global.com
Web site: http://www.igi-global.com

Library of Congress Cataloging-in-Publication Data

Names: Saravanan, Krishnan, 1982- editor. | Julie, Golden, 1984- editor. | Robinson, Harold, 1980- editor.
Title: Handbook of research on implementation and deployment of IoT projects in smart cities / Krishnan Saravanan, Golden Julie, and Harold Robinson, editors.
Description: Hershey, PA : Engineering Science Reference, [2020] | Includes bibliographical references.
Identifiers: LCCN 2018058996| ISBN 9781522591993 (h/c) | ISBN 9781522592013 (eISBN)
Subjects: LCSH: Smart cities. | Internet of things.
Classification: LCC TD159.4 .H365 2020 | DDC 307.1/160285--dc23 LC record available at https://lccn.loc.gov/2018058996

This book is published in the IGI Global book series Advances in Civil and Industrial Engineering (ACIE) (ISSN: 2326-6139; eISSN: 2326-6155)

British Cataloguing in Publication Data
A Cataloguing in Publication record for this book is available from the British Library.

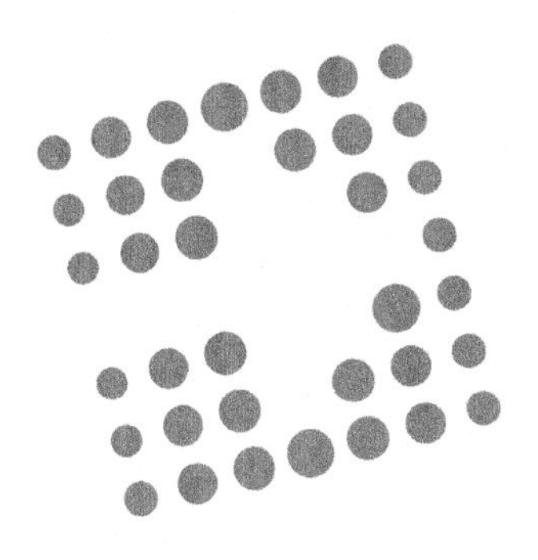

Advances in Civil and Industrial Engineering (ACIE) Book Series

Ioan Constantin Dima
University Valahia of Târgovişte, Romania

ISSN:2326-6139
EISSN:2326-6155

Mission

Private and public sector infrastructures begin to age, or require change in the face of developing technologies, the fields of civil and industrial engineering have become increasingly important as a method to mitigate and manage these changes. As governments and the public at large begin to grapple with climate change and growing populations, civil engineering has become more interdisciplinary and the need for publications that discuss the rapid changes and advancements in the field have become more in-demand. Additionally, private corporations and companies are facing similar changes and challenges, with the pressure for new and innovative methods being placed on those involved in industrial engineering.

The **Advances in Civil and Industrial Engineering (ACIE) Book Series** aims to present research and methodology that will provide solutions and discussions to meet such needs. The latest methodologies, applications, tools, and analysis will be published through the books included in **ACIE** in order to keep the available research in civil and industrial engineering as current and timely as possible.

Coverage

- Hydraulic Engineering
- Ergonomics
- Earthquake engineering
- Optimization Techniques
- Coastal Engineering
- Construction Engineering
- Materials Management
- Urban Engineering
- Quality Engineering
- Transportation Engineering

IGI Global is currently accepting manuscripts for publication within this series. To submit a proposal for a volume in this series, please contact our Acquisition Editors at Acquisitions@igi-global.com or visit: http://www.igi-global.com/publish/.

The Advances in Civil and Industrial Engineering (ACIE) Book Series (ISSN 2326-6139) is published by IGI Global, 701 E. Chocolate Avenue, Hershey, PA 17033-1240, USA, www.igi-global.com. This series is composed of titles available for purchase individually; each title is edited to be contextually exclusive from any other title within the series. For pricing and ordering information please visit http://www.igi-global.com/book-series/advances-civil-industrial-engineering/73673. Postmaster: Send all address changes to above address.

Titles in this Series

For a list of additional titles in this series, please visit: www.igi-global.com/book-series

Recycled Waste Materials in Concrete Construction Emerging Research and Oportunities
Jahangir Mirza (Research Institute of Hydro-Québec, Canada) Mohd Warid Hussin (Universiti Teknologi Malaysia, Malaysia) and Mohamed A. Ismail (Miami College of Henan University, China)
Engineering Science Reference • © 2019 • 169pp • H/C (ISBN: 9781522583257) • US $165.00

Ambient Urbanities as the Intersection Between the IoT and the IoP in Smart Cities
H. Patricia McKenna (AmbientEase, Canada)
Engineering Science Reference • © 2019 • 333pp • H/C (ISBN: 9781522578826) • US $195.00

Optimizing Current Strategies and Applications in Industrial Engineering
Prasanta Sahoo (Jadavpur University, India)
Engineering Science Reference • © 2019 • 382pp • H/C (ISBN: 9781522582236) • US $245.00

Big Data Analytics in Traffic and Transportation Engineering Emerging Research and Opportunities
Sara Moridpour (RMIT University, Australia)
Engineering Science Reference • © 2019 • 188pp • H/C (ISBN: 9781522579434) • US $165.00

Optimization of Design for Better Structural Capacity
Mourad Belgasmia (Setif 1 University, Algeria)
Engineering Science Reference • © 2019 • 283pp • H/C (ISBN: 9781522570592) • US $215.00

Reusable and Sustainable Building Materials in Modern Architecture
Gülşah Koç (Yildiz Technical University, Turkey) and Bryan Christiansen (Global Research Society, LLC, USA)
Engineering Science Reference • © 2019 • 302pp • H/C (ISBN: 9781522569954) • US $195.00

Measuring Maturity in Complex Engineering Projects
João Carlos Araújo da Silva Neto (Magnesita SA, Brazil) Ítalo Coutinho (Saletto Engenharia de Serviços, Brazil) Gustavo Teixeira (PM BASIS, Brazil) and Alexandro Avila de Moura (Paranapanema SA, Brazil)
Engineering Science Reference • © 2019 • 277pp • H/C (ISBN: 9781522558644) • US $245.00

Big Data Analytics for Smart and Connected Cities
Nilanjan Dey (Techno India College of Technology, India) and Sharvari Tamane (Jawaharlal Nehru Engineering College, India)
Engineering Science Reference • © 2019 • 348pp • H/C (ISBN: 9781522562078) • US $225.00

701 East Chocolate Avenue, Hershey, PA 17033, USA
Tel: 717-533-8845 x100 • Fax: 717-533-8661
E-Mail: cust@igi-global.com • www.igi-global.com

List of Contributors

Table of Contents

Detailed Table of Contents

This chapter covers an introduction to internet of things (IoT), various architectures proposed by the researchers, IoT protocols, and interoperability as a major challenge. Millions of devices will be connected to internet in a few years. The major problems in accomplishing the vision of IoT is the incompatibility of devices and standard protocols to support heterogeneity. Interoperability in IoT has been a major problem with the constant increase in various kinds of sensors and manufacturers and lack of a common platform or protocol to all such sensors to share data among each other. As we traverse from physical layer to the higher layers in the IoT stack, each layer has interoperability challenges that have to be addressed uniquely such as network layer interoperability, messaging protocol interoperability, data format interoperability, and semantic interoperability.

Majority of the humanity is living in the cities. Cities have adverse environmental impact. Their environmental footprints need to be reduced. As the world's living conditions deteriorate, the survival of the humanity depends on the precautions taken. These precautions can include sustainable living styles, new technologies, and circular economy principles. Furthermore, climate change caused disasters can have adverse consequences as they can be deadly and as they can result in economic loss. The cities need to be resilient so that disasters adverse consequences can be reduced and the post-disaster phase rescue and recovery processes can be effectively carried out. Circular, smart and connected cities based on the new technologies such as big data, Information and Communication Technology (ICT), and Internet of Things (IoT) can contribute to the cities' sustainability and resilience performance. This chapter aims to investigate the roles of big data, IoT, ICT, as well as circular, connected and smart cities in enhancing sustainability and resilience of the cities. With this aim, based on the literature review, this chapter covers: need for, pillars of and aspects of smart, sustainable, circular, and resilient cities as well as ways for transforming the cities into smart, sustainable, circular, and resilient ones. This chapter can be beneficial to the researchers, academics, construction professionals, and policy makers.Keywords: Disaster; Internet of Things; Information and Communication Technology; Smart Cities; Big Data; Resilience; Building Information Modelling; Circular Cities; Circular Economy

The internet of things (IoT) is the new buzzword in technological corridors with most technology companies announcing a smart device of sorts that runs on internet of things (IoT). Cities around the world are getting "smarter" every day through the implementation of internet of things (IoT) devices. Cities around the world are implementing individual concepts on their way to becoming smart. The services are automated and integrated end to end using internet of things (IoT) devices. The chapter presents an array of internet of things (IoT) applications. Also, cyber physical systems are becoming more vulnerable since the internet of things (IoT) attacks are common and threatening the security and privacy of such systems. The main aim of this chapter is to bring more research in the application aspects of smart internet of things (IoT).

Internet of things (IoT) is the current area of research that allows heterogeneous devices to have a homogeneous connectivity based on the designed and desired application of the user. With the latest development in connectivity via smart phones, there is an exponential increase in users who access internet. However, various applications have already been designed based on the user's requirement. Therefore, this chapter intends to provide a detailed view on applications on IoT. Industrial applications help in monitoring the machinery so that production increases with minimum chaos if any error occurs. Safety helmet for mining based on IoT is used to measure the gas and temperature levels in the coal mines. Garbage management system is used for monitoring and clearing of dust bins. IoT-based domestic applications help users to have a better access over the equipment they use. As a business application, emotion analysis is performed to obtain the customers mood while shopping. Monitoring of crops from a remote location is another application which provides data on the health of the crop.

Internet of things (IoT) is an integral part of modern digital ecosystem. It is available in different forms. Narrowband IoT (NBIoT) is one of the special forms of the IoTs available for deployment. It is popular due to its low power wide area (LPWA) characteristics. For new initiatives such as smart grids and smart cities, a large number of sensors will be deployed and the demand for power is expected to be high for such IoT deployments. NBIoT has the potential to reduce the power and bandwidth required for large IoT projects. In this chapter, different practical aspects of NBIoT deployment have been addressed. The LPWA features of NBIoT can be realized effectively if and only if its deployment is done properly. Due to its large demand, it has been standardized in a very short span of time. However, the 5G deployment of NBIoT will have some new provisions.

The smart cities mission of the Government of India has opened up new pathways for urban redevelopment and transformation. But given the limited resources available with a developing country, a more pragmatic approach would be to first learn from the best international experiences and approaches and then implement those in Indian context. With this view, the chapter examines some of the best practices related with different aspects of a smart city and suggests their relevancy for the development of smart cities in India. The study found that by focusing on the five core areas (i.e., urban mobility and public transport, safety and security of citizens, health and education, water management, and robust IT connectivity and social networking) the concerned authorities in India can successfully achieve their goal of urban redevelopment and transformation with scarce resources. Limitations and scope for future research are discussed in the end.

Smart city technology evolved with the developments in wireless sensor networks (WSN) and the internet of things (IoT). IoT-based waste management is an advanced waste management system offered in smart cities. The practice of monitoring, transporting, and processing of solid waste are included in the waste management. Litter bins play an indispensable role in the waste collection process at the primary level. The process of monitoring litter bins would become difficult for the ones placed at out of reach areas and remotely located sites. Smart litter bin (SLB) is generally embedded with different types of sensors where used for sensing the garbage levels and locating the bins location. Radio frequency identification (RFID), sensors, global positioning systems (GPS), general packet radio service (GPRS) are the components in smart waste management system and are discussed in this chapter. These components were used to monitor the collection, transportation, processing, and dumping. This chapter also focuses on the perception of IoT architecture to upgrade waste management in smart cities.

Smart grid is a new emerging trend as it is more efficient because of its computation process and energy utilization. It provides many benefits due to the communication between the service company and users. Smart grid includes smart metering, which is used to collect the data from the appliances, devices, sensors, etc. and transfer the data to the network, and the data is analyzed for energy consumption. The transmission and distribution of energy; the smart grid architecture; communication technologies like LAN, HAN, and NAN; and security and privacy challenges are discussed in this chapter.

integration. IoT system provides a digital access to waste management. This system uses online smart monitor sensors that monitor the performance of water supply and effluent handling system utilizing a cloud-based platform. This enhances real-time planned performance and increases life-cycle equipment. This technique enhances the synergistic use of resources due to climate mitigation and adaptation for sustainable growth and this technology also uses air quality sensors across the city to collect open data platform for monitoring and reducing primary and secondary pollutants and systematically instruct the pollutant-causing sources to maintain ambient air quality.

Generally, the rate of technological advancement is increasing with time. Specifically, the technologies that are the building blocks of Farming 4.0 are now advancing at a rapid pace never witnessed before. In this chapter, the authors study the advances of major core technologies and their applicability to creating a smart farm system. Special emphasis is laid on cost of the technology; for, expensive technology will still keep small farmers at bay as major population of farmers inherently are new to technology, if not averse. The authors also present the pros and cons of alternatives in each of the subsystems in the smart farm system.

Smart farming is a key to develop sustainable agriculture, involving a wide range of information and communication technologies comprising machinery, equipment, and sensors at different levels. Seawater, which is available in huge volumes across the planet, should find its optimal way through irrigation purposes. On the other hand, underwater wireless sensor networks (UWSNs) finds its way actively in current researches where sensors are deployed for examining discrete activities such as tactical surveillance, ocean monitoring, offshore analysis, and instrument observing. All these activities are based on a radically new type of sensors deployed in ocean for data collection and communication. A lightweight Hydro probe II sensor quantifies the soil moisture and water flow level at an acknowledged wavelength. The freshwater absorption repository system (FARS) is matured based on the mechanics of UWSNs comprised of SBE 39 and pressure sensor for analyzing atmospheric pressure and temperature. This necessitates further exploration of FARS to complement smart farming. Discrete routing protocols have been designed for data collection in both compatible and divergent networks. Clustering is an effective approach to increase energy efficient data transmission, which is crucial for underwater networks. Furthermore, the chapter attempts to facilitate seawater irrigation to the farm lands through reverse osmosis (RO) process. Also, the proposed irrigation pattern exploits residual water from the RO process which is identified to be one among the suitable growing conditions for salicornia seeds and mangrove trees. Ultimately, the cost-effective technology-enabled irrigation methodology suggested offers farm-related services through mobile phones that increase flexibility across the overall smart farming framework.

Preface

Smart City aims at developing a comprehensive eco-system which requires development of institutional, physical, social and economic infrastructure. It deals with High priority pillars and core smart city infrastructure element for India. The main focuses of the chapter present "Core Smart City Infrastructure Elements". This elements are classified ion to Urban Mobility and Public Transport, Safety and Security of Citizens, Health and Education, Water Management and Robust IT Connectivity and Social Networking finally it focus on Limitations and future research directions.

Chapter 1 covers Introduction to Internet of Things (IoT), various architectures proposed by the researchers, IoT Protocols and Interoperability as a major challenge. Millions of devices are expected to be connected to internet within few years. The major problems in accomplishing the vision of IoT are the incompatibility of devices and standard protocols to support heterogeneity. Interoperable environment requires a well-established model that selects the right protocol by providing the message formats and the frameworks. IoT requires adaptive and dynamic models that enable diverse devices to interoperate and communicate. This is possible by the manufacturers of the devices by systematically and incrementally publishing the changes without affecting the advances in the development of applications. Horizontal linking of data through the standards is the key that enables heterogeneous devices from various industries to be on the common platform to improve the end-user experience.

Chapter 2 aims to investigate the role of big data, IoT, ICT, as well as circular, smart and connected cities in enhancing sustainability and resilience of the cities. These new technologies along with circular economy principles will help to improve the quality life of the urban citizens.

In Chapters 3 and 4, Internet of Things (IoT) is the current research area which is of prominence. IoT enables to connect heterogeneous devices enabling them to have a homogeneous type of connectivity based on the application that is desired by the user. With the latest development of connectivity in smart phones via Long Term Evolution (LTE) and Large Scale Convergence (LSC) there is an exponential increase in terms of users who access internet, resulting in globally connected Ethernet. It is expected that almost one to two trillion devices will get interconnected by the end of 2025. Such connectivity will improve the life style of humans providing them better solutions to their requirements. In this juncture, IoT requires a better framework which can connect these devices which are heterogeneous in nature. Further, the applications that are being designed for these devices must also be capable enough in terms of processing, storage and other better capabilities. Applications that are designed based on IoT can be broadly classified as Industrial and Domestic. Industrial applications are these applications which help industry personnel in analyzing the industrial day to day activity such that the production is increased.

Chapter 5 discusses the NBIoT (Narrow Band IoT), which is used in LPWAN areas. NBIoT has the potentials to reduce the power and bandwidth required for large IoT projects. In this chapter, different practical aspects of NBIoT deployment have been addressed. The LPWA features of NBIoT can be realized effectively if and only if its deployment is done properly.

Chapter 6 discusses the The Smart Cities Mission of government of India with different aspects of a Smart City and suggests their relevancy for the development of Smart Cities in India.

Chapters 7, 16, and 20 focus on automatic waste management system using IoT. Chapter 7 describes the solid waste management IoT system implemented in Tirunelveli City Municipal Corporation. Chapter 16 proposed a system which helps to implement smart city for further development in urbanization. Tracking the bins and self-efficient network groups are framed to focuses on handling of large volume of data, cost of the underlying network topology, merging of devices, and speed of the data connectivity. Architecture model are proposed to handle the challenges .The proposed architecture model handles possible challenges in developing smart city and suggestions are given to overcome those challenges. Chapter 20 illustrates the automated IoT waste management system without any human intervention.

The smart Grid (Chapter 8) will overcome all the problems that occur in the power Grid like insecurity, damage in the transmission power line, physical and cyber-attack etc. An IoT based architecture diagram is given below, which has four layers. They are Application layer, Support layer, Network layer and Perception layer. The lowest layer is the perception layer; it has sensors and WSN, etc. The main purpose of this layer is to collect data from the end devices (things in the real world) and transmit it to the upper layer (network layer). The root layer in IoT architecture is network layer, which is connected with the servers, devices and smart things, and it transfers the data from the perception layer to the higher layer.

Chapter 9 deals with blockchain based smart cities IoT. To make the IoT data get communicated with each other and make the devices talk to each other in a coherent way, the device data transactions are made to communicate through the blockchain network. This work focuses on, when adding the vendor specific IoT devices to the public or private block chain the emerging challenges and the possible solutions.

Security issue in the smart city application (Chapter 10) is one of the major issues. Threat detection and behavioural monitoring will be done after analysing the traffic data. Main aim is to analyse and predict real time streaming data to achieve security. Different analytical tools on security will be used to obtain the optimal result in smart cities. Traffic analysis which is treated as a real time stream will be applied in street and traffic lights, transportation and parking space occupancy and soon. Large volume of data that are received from different sensors and cameras will be given as the input in order to analyse traffic in a smart cities. Intelligent traffic congestion control system will be developed in order to analyse the heavy traffic on roadside. Security in IoT is proposed which includes encrypting and decrypting the user request which is further to be processed by the Central Processing hub in order to prevent from unauthorized access.

In smart transport and parking slot reduce the traffic using automated system (Chapter 11). Parking slots are accurately predicted the available free slots are intimated to the driver through speaker. IoT sensors are used to reduce the traffic based on arrival and departure of vehicles. Vehicular ad hoc networks (VANET) provide collusion free traffic in the busy road and also it reduces accidents. Challenges in smart Transport and Parking: In smart vehicle traffic and parking system contain mobility. Due to dynamic nature system should guide the traffic and taking final decision about parking slot is very time effective.

Chapter 12 contributes the design and development of an Automated Online Condition Monitoring System for elevators (AOCMSE) using IoT techniques to avoid failures. The failure rate of elevators can be prevented by sending this detected fault information to the maintenance team.

Chapter 13 developed Early Earthquake Warning (EEW) system using IoT, which can minimize damage of earth quake and saves countless lives. An early warning system can be effectively implemented by the proposed method, along with high-end processors and the IoT (Internet of Things), which has the ability to collect and transfer data over networks with no manual intrusion.

Chapter 14 implemented unified framework to support the “smart planning” of a university environment, intended as a “smart campus”. The main goal is to improve the management, storage and mining of information coming from the university areas and main players.

Chapter 15 discussed the smart IoT meters for traffic control, transport management, managing spare resources like power and water, solid waste management, e-health monitoring, infrastructure management. many IoT based smart devices namely smart garbage bins, automatic parking system, smart electric meters, Supervisory Control and Data Acquisition (SCADA) for water distribution have been devised and used successfully in many cities.

Chapter 17 proposed Farming 4.0, which focussed on the advances of major core technologies and their applicability to creating a Smart Farm System. The pros and cons of alternatives of each of the Smart Farm subsystems are detailed.

Chapter 18 implemented Freshwater Absorption Repository System (FARS) based on the mechanics of UWSNs comprised of SBE 39 and pressure sensor for analyzing atmospheric pressure and temperature. Smart farming is a key to develop sustainable agriculture, involving a wide range of information and communication technologies comprising machinery, equipment, and sensors at different levels.

Chapter 19 deals with smart cities important issue - Air Pollution. Air pollution is a major environmental health problem affecting everyone. It occurs when the environment is contaminated by any chemical, physical or biological agent that modifies the natural characteristics of the atmosphere. In Air Quality Monitoring, Air pollutants are added in the atmosphere from variety of sources that change the composition of atmosphere and affect the biotic environment. Next, it deals with Smart Cities and IoT Enabled Environment Sensors. Followed by SURAT SMART CITY, Surat is a city located on the western part of India in the state of Gujarat. Surat was selected for the Smart City Mission in the first phase itself.

In Chapter 21, palm oil mill effluent (POME) as an agricultural colloidal wastewater is one of the highly polluted industrial wastewater ever. Organic characteristics of the pollutants in POME cause to limitation of capable treatment method for this dark brown effluent as well as its straightly high concentration of non-degradable or low-degradable organic pollutants in very wide range of sizes. he most important part of POME treatment which is a capable agricultural effluent for the fastest, the easiest and the cheapest microbiological treatment. Direct application of municipal wastewater as microbial resources through the aerobic process could be a smart method for meeting environmental discharges in the shortest time ever.

Acknowledgment

The Successful completion of this edited book is made possible with the help and guidance received from various quarters. I would like to avail this opportunity to express my sincere thanks and gratitude to all of them.

I deeply indebted to Almighty god for giving this opportunity. With the presents of God only it's possible.

I extend my deep sense of gratitude to My Son Master H.Jubin, for moral support and encouragement, at all stages, for the successful completion of this Book.

I extend my deep sense of gratitude to my scholars and friends for writing the chapter in time and amenities to complete this book. I sincerely thank my parents and family member for providing necessary support.

Finally, I would like to take this opportunity to specially thank IGI global publisher for his kind help, encouragement and moral support.

Harold Robinson & Golden Julie

My sincere thanks to my family members – Sudha (Wife), Mazhilan Krish (Elder son), Raghvan Krish (Younger son), Dhanalakshmi (Mother) & Krishnan (Father) for their unconditional love and moral support.

Krishnan Saravanan

Chapter 1
Internet of Things:
Architecture, Protocols, and Interoperability as a Major Challenge

Jyoti Ramachandra Desai
PES University, India

ABSTRACT

This chapter covers an introduction to internet of things (IoT), various architectures proposed by the researchers, IoT protocols, and interoperability as a major challenge. Millions of devices will be connected to internet in a few years. The major problems in accomplishing the vision of IoT is the incompatibility of devices and standard protocols to support heterogeneity. Interoperability in IoT has been a major problem with the constant increase in various kinds of sensors and manufacturers and lack of a common platform or protocol to all such sensors to share data among each other. As we traverse from physical layer to the higher layers in the IoT stack, each layer has interoperability challenges that have to be addressed uniquely such as network layer interoperability, messaging protocol interoperability, data format interoperability, and semantic interoperability.

INTRODUCTION

Information technology has profound impact on human's daily life. The internet era has erased the geographical borders and exchange of information is instantly processed, unlike traditional ways which required days or months for larger distance. The internet which is used for interconnecting end users has evolved to a stage which connects the physical objects that communicate with each other and with humans, offering a service. 'Internet-of-Things'(IoT) is used as an umbrella keyword which covers various aspects of spatially distributed devices with embedded identification, sensing and actuation capabilities.

'Internet of Things' evolved in 1980's with coffee vending machine, and the original term is coined by Kevin Ashton, the Executive Director of Auto-ID Labs in MIT in 1999. The concept of 'Internet of Things' first became very popular through the Auto-ID centre in 2003 and in related market analysts publications (S. Madakam et al., 2015). IoT enables different style of living and interacting with the

DOI: 10.4018/978-1-5225-9199-3.ch001

content and services available around us. This is the need of large corporations that benefit them to become more efficient, flexible, reduce errors in order to maintain accuracy. The technological evolution and the internet revolution has made academicians, industrialists and researchers to think about how to connect everything on the planet by transforming the real world objects into smart virtual objects through intelligence, the type of wireless communications to be built into devices, and the need of the internet infrastructure to support constrained devices.

IoT has two words: "Internet" and "Things". *Internet* is interconnection of communication networks that use standard internet protocol(TCP/IP). *Things* are real objects in this physical world. These real objects can be equipped with identifying, sensing, networking and processing capabilities that will allow these objects to communicate with one another which are of similar capabilities to accomplish some objective. *Internet of Things* allows people and things to be connected at *Anytime*, from *Anyplace* and with *Anything* and *Anyone* which implies addressing elements such as convergence, content collections, computing, communication and connectivity. Machine-to-Machine(M2M) solutions which is a subset of IoT suffice the needs of industries which requires minimal direct human intervention to deliver the services using wireless network technology.

The communication industry had seen tremendous growth in the last two decades and IoT is a paradigm shift in Information Technology. IoT is maturing and becoming the most hyped concept in IT world by overcoming the barriers such as size and cost of wireless communication, addressing issue to billions of devices connecting to the internet, power constrained devices etc. As the technology evolves the electronic industries/companies have started building Wi-Fi and cellular wireless connectivity into a wide range of devices. The challenge of addressing to the growing number of physical objects connected to the internet is solved by introducing IPv6 which allows us to assign a communication address to billions of devices. The battery technology has advanced tremendously by using solar recharging concept incorporated into numerous devices. That's how IoT aims at unifying everything in our world under a common infrastructure by giving us control of things around us.

Figure 1. Characteristics of IoT

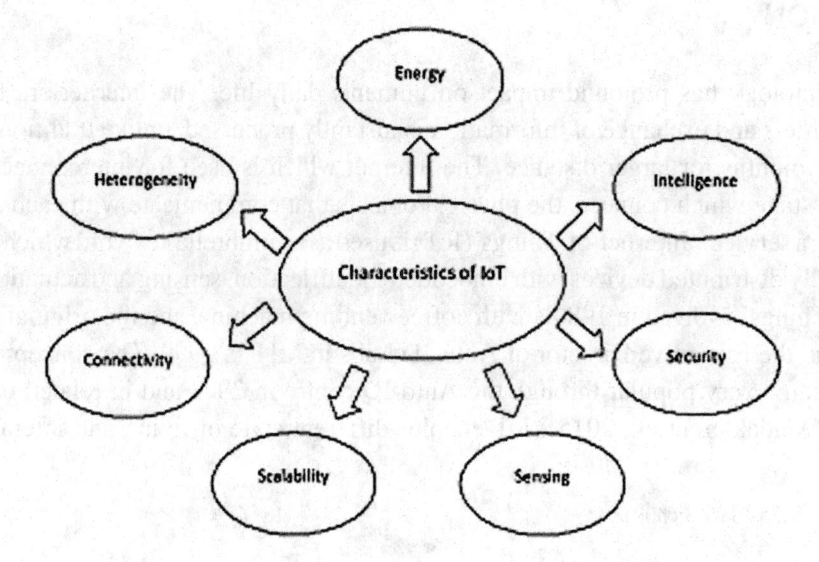

CHARACTERISTICS OF IOT

There are a number of characteristics that are needed in deployment of IoT applications. Some of the key characteristics are shown in Fig.1 and are discussed as follows:

1. **Intelligence**: Things having identities and virtual personalities operating in smart spaces use intelligent interfaces to connect and communicate through the use of intelligent decision making algorithms in software applications to get rapid and appropriate responses based on the information sensed and collected and also considering the patterns of the historic data.
2. **Connectivity**: Development in IoT is happening at the breathtaking pace only due to the appropriate connectivity solutions which enable communication, exchange data and derive actions. Connectivity solutions for IoT can be structured into two high-level categories (Wired and Wireless). Wireless solutions can be further divided into long-range connectivity standards (ex: Low Power Wide Area Network) and short-range connectivity standards(ex: Bluetooth Low Energy).
3. **Scalability**: IoT connects both inanimate and living things, hence the number of devices that will be connected to the internet will be much larger. Managing of these devices and the data generated by these devices needs to be interpreted for application purposes and becomes challenging.
4. **Sensing**: Embedded sensors in the devices will have the sensing capabilities which makes IoT detect or measure the data generated and can report the device about the environment status. Sensing technologies provide the means to create capabilities that reflect a true awareness of the physical world and the people in it. The sensing information is simply the analogue input from the physical world, but it can provide the rich understanding of our complex world.
5. **Heterogeneity**: Heterogeneity is the key characteristics in IoT. IoT aims at interconnecting large number of heterogeneous devices in order to provide advanced applications that improve quality of life by making the human task easier. Heterogeneous devices are used in diverse domains, with various protocols, standards and architecture. Hence, interoperability is seen as major challenge in integrating various representation models and data.
6. **Security**: IoT security is the technology concerned with safeguarding of the devices and networks which are connected to the Internet of Things. As the devices and networks connected to the internet are open for the vulnerabilities, IoT security technologies plays an important role in safeguarding them. According to the analysis made by the researchers with respect to IoT security, the most important technologies in providing security are enlisted as IoT network security, IoT authentication, IoT encryption, IoT PKI, IoT security analytics, IoT API security and so on.
7. **Energy**: Consumers would like to have smart devices which are of low cost, and devices should have small physical size. Many devices operate with small battery and others use a self-energizing energy source. Hence batteries are the critical source in the smart devices. Battery technology is playing an important role in design of the IoT devices. It has been observed that battery technology is doubling in performance approximately every 10 years. There are number of factors that needs to be considered for selecting the suitable battery for the applications, such as physical requirement (size, shape, weight), operating voltage level, duty cycle, service life, environmental conditions (temperature, pressure, humidity), shelf life, safety and reliability and cost (Minoli, 2013).

Figure 2. Three-Layered and Five-Layered Architecture of IoT
(Ala Al-Fuqaha et al.,2015)

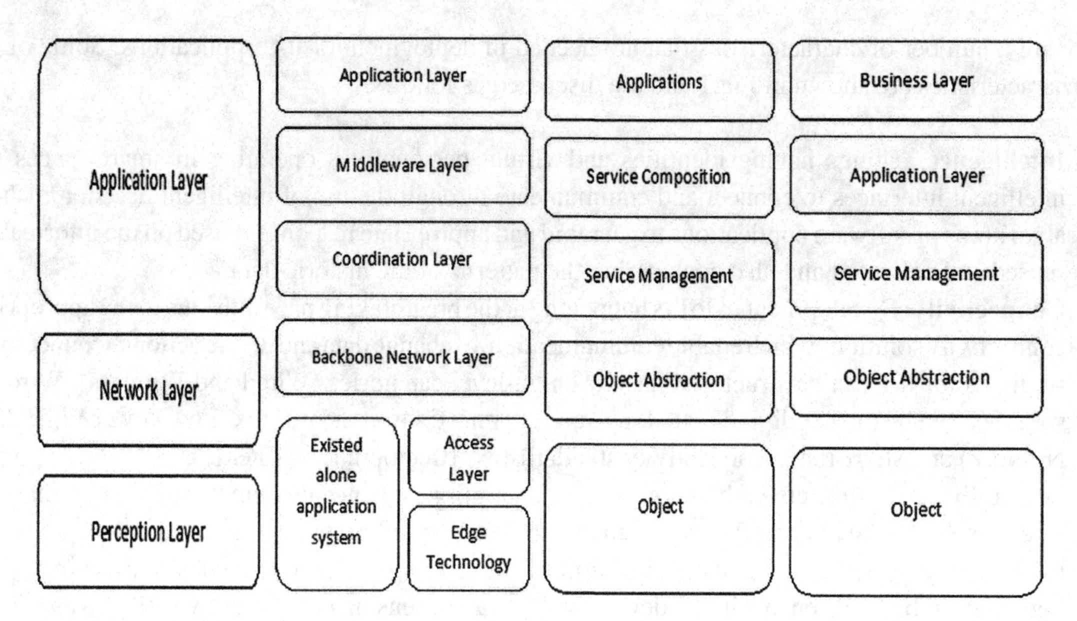

ARCHITECTURE FOR IOT

Architecture is defined as a framework that specify the network's physical components and their functional organization and configuration, its operational principles and procedures, as well as data formats used in its operation. The architecture for IoT has to be chosen carefully considering various factors, such as *specific domains* (like health care, finance, administration etc), *communication models* (Things-to-Application, Things-to-Human, Things-to-Things), *addressing mechanism* etc. A detailed study of centralized and decentralized architecture gives us a fair idea on what are the advantages and the risks of both the types of architecture and which suites the best for IoT. Architecture should be flexibly designed and should also consider security as a pre-requisite where in, it provides privacy and handles the data protection issues if exists. Architectural ecosystem should embed necessary interfaces in all layers where in various parties can contribute and leverage the international standards.

Researchers have proposed different architectures. There is no universally agreed architecture based on consensus. Researchers have proposed Three-Layer and Five-Layer architecture. Researchers(Ala Al-Fuqaha et al.,2015)mentions about these architectures and the explanation is as follows:

In the early stages of research in the field of IoT, three-layer architecture was introduced as shown in Fig. 2(a). Three layers were namely physical, network and application layers.

1. **Physical Layer:** This layer has physical devices which will have sensors and actuators to sense and gather information about the environment.
2. **Network Layer:** This layer connects the smart capable things, network devices and servers. Network layer may also include a gateway and provides interface between sensors and the internet. It helps in processing the sensed data and transmission.

Figure 3. Sensors and Actuators in a device

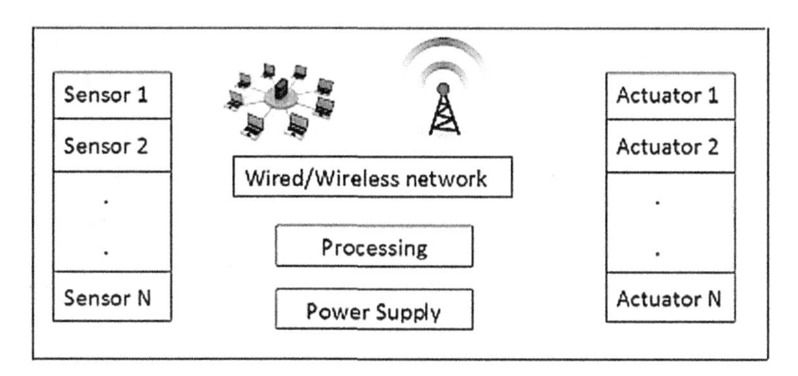

3. **Application Layer:** It's responsible for delivering application specific services to the users by sending commands to real world objects over the internet through mobile applications or web applications.

As researchers focus on finer aspects of IoT, three-layer architecture was not felt sufficient and hence additionally included middle-ware based architecture (Fig.2b), service oriented architecture (Fig.2c) and business layers in five-layer architecture (Fig. 2d).

4. **Objects Layer:** It is same as the physical layer in the three-layer architecture which represents physical devices with various sensors and aiming to collect the data and process the collected information. The devices will have sensors and actuators which has different functionalities. The device could be represented as shown in Fig. 3. The data generated by heterogeneous devices have to be digitized and transferred to the next immediate layer i.e Object Abstraction Layer. Huge amount of data is getting generated and can be said that creation of big data is initiated at this layer.
5. **Object Abstraction Layer:** This layer transfers data from Objects Layer to the service Management Layer. Various technologies used are RFID, GSM, WiFi, Bluetooth Low Energy(BLE),Infrared(IR), Zigbee etc.
6. **Service Management Layer:** This layer allows the IoT application programmers to work with heterogeneous devices based on the unique identifiers and the names of the devices irrespective of the hardware platform. As the name suggests, this layer receives the data, analyses, makes decisions and deliver the required services to the requester.
7. **Business Layer:** The word business here indicates money making. The responsibility of this layer is to build a business model, graphs based on the data received by the Application Layer. This layer receives data from the application layer and make the information knowledgeable and modify into a meaningful service and helps in decision making process.

IOT STACK AND PROTOCOLS

As shown in Fig.4, it depicts some of the open standards applicable to the data link layer. There are various other protocols such as WirelessHART, Z-Wave, DASH7,LoRaWAN,LTE-A etc. Following are

Figure 4. IoT stack and Protocols
(Ala Al-Fuqaha et al.,2015)

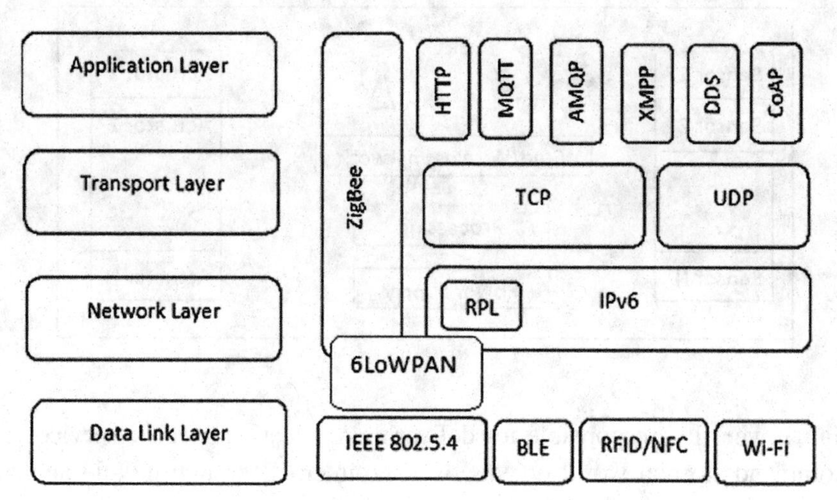

some of the data link layer standards, network layer standards, transport layer standards, session layer standards and application layer standards discussed in detail (Salman et al., 2017).

1. **IEEE 802.15.4:** IEEE 802.15.4 is the most commonly used IoT standard for MAC. This helps in identifying how a particular communication can happen across various nodes in the network. Unlike in traditional networks, the frame formats which are not suitable for low power networking, an extension of IEEE 802.15.4 standard facilitates low power communication. The maximum data rate will be 250 kb/s, output power will be 1mW and the maximum packet size will be 127 bytes. Synchronization and Channel hopping are the key features to realize high reliability, low cost and meet IoT communication requirements.
   a. **Synchronization:** To achieve synchronization, following are the two approaches:
      i. **Acknowledgment Based:** In this mode, nodes are already in communication and they send acknowledgement to maintain connectivity.
      ii. **Frame Based:** In this mode of operation, nodes are not communicating and hence empty frames are exchanges with each other for a pre-specified interval of time.
   b. **Channel Hopping:** IEEE 802.15.4e introduces channel hopping for time slotted access to the wireless medium. A mechanism in which frequency is altered with the help of a pre-determined random sequence that brings frequency diversity and reduces the effect of interference.
2. **Bluetooth Low Energy(BLE):** BLE uses a short range radio signals and this standard has been developed by smartphone makers. BLE is efficient as compared to ZigBee in terms of energy consumption while transmission of the data. The range coverage is about 100 meter and the transmission power range between 0.01mW to 10mW which makes it suitable for IoT applications.
3. **Radio Frequency Identification(RFID):** EPCglobal is an organization responsible for the genesis of EPC and RFID technology and standards. Electronic Product Code(EPC) is a unique identification number stored on RFID tag and commonly used in supply chain management in identifying the items.

There are two main components, RFID tag and RFID reader. RFID tags can be classified as active, passive or assisted passive. Active tags periodically broadcast their identity, passive tags are ideal for devices without batteries and assisted passive tags become active when RFID reader is present.

4. **Wi-Fi:** Wi-Fi is a technology for wireless local area networking based on IEEE 802.11 standards. The IEEE 802.11a/b/g/n specifications have contributed immensely to the wireless networking technology. IEEE 802.11n offers the highest data throughput, but at the cost of high power consumption, so IoT devices might only use IEEE 802.11b or g for power conservation reasons. Although Wi-Fi is adopted within many prototype and current generation IoT devices, as longer-range and lower-power solutions become more widely available, it is likely that Wi-Fi will be superseded by these lower-power alternatives.
5. **6LoWPAN:** 6LoWPAN is an IPv6 adaptation layer standard for low power wireless PAN(LoWPAN) which sits between the MAC layer and the IP network layer. The idea of 6LoWPAN originated to show that the Internet Protocol could be applied to the smallest devices that has limited processing capabilities. LoWPAN is characterized as lossy, low power, low bit-rate, short range with many nodes saving energy with long sleep periods. 6LoWPAN is a network layer encapsulation protocol which shows how IPV6 over Low Power Wireless Personal Area Network efficiently transports IPV6 packets over IEEE 802.15.4 links by encapsulating long IPV6 headers in small MAC frames (<128 byte length). It also implements header compression mechanism which decreases the transmission overhead. Types of headers in frames transported by LoWPAN with 00,01,10,11 indicates Not a LoWPAN frame, LoWPAN IPv6 addressing header, LoWPAN mesh header and LoWPAN fragmentation header respectively.
6. **Routing Protocol for Low Power and Lossy Networks(RPL):** Low power and Lossy Networks(LLN) are a class of networks in which both the routers and the interconnect are constrained. Existing routing protocols fail in meeting the goals of the IoT applications. As discussed (Minoli,2013), RPL implementation will support LLN applications with the necessary functionalities. RPL has a feature of dissemination of information over the dynamically formed network topology which is very essential in IoT. RPL uses Directed Acyclic Graph(DAG) and Destination-oriented DAG(DODAG) as shown in Fig. 5. A node rank defines a node's relative position within a DODAG with respect to the DODAG root. Each node in the network shares a DODAG Information Object(DIO) which gradually propagates across the network and forms the entire DODAG structure. Based on the information in the DIOs, the node chooses parents that minimize path cost to the DODAG root. Therefore, RPL computes the optimal path by building up a graph of the nodes in the network based on dynamic metrics and constraints like minimizing energy consumption or latency.
7. **IPv6:** At the Internet Layer, devices are identified by IP addresses. IPv6 is typically used for IoT applications over legacy IPv4 addressing. IPv4 is limited to 32-bit addresses, which only provide around 4.3 billion addresses in total, which is less than. the current number of IoT devices that are connected, while IPv6 uses 128 bits, and so provides 2^{128} addresses(around 3.4×10^{38}).Of the tens of billions of devices that are expected to connect to the IoT over the next few years, many will be deployed in private networks that will use private address ranges and only communicate out to other devices or services on external networks by using gateways.
8. **Transmission Control Protocol(TCP) and User Datagram Protocol(UDP):** Both TCP and UDP sit at transport layer and use IP protocol. TCP is connection oriented and use handshake mechanism to set up the connection and guarantees message delivery. UDP is connectionless protocol and

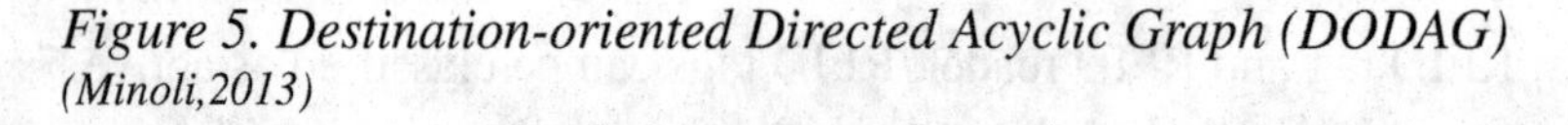
Figure 5. Destination-oriented Directed Acyclic Graph (DODAG)
(Minoli,2013)

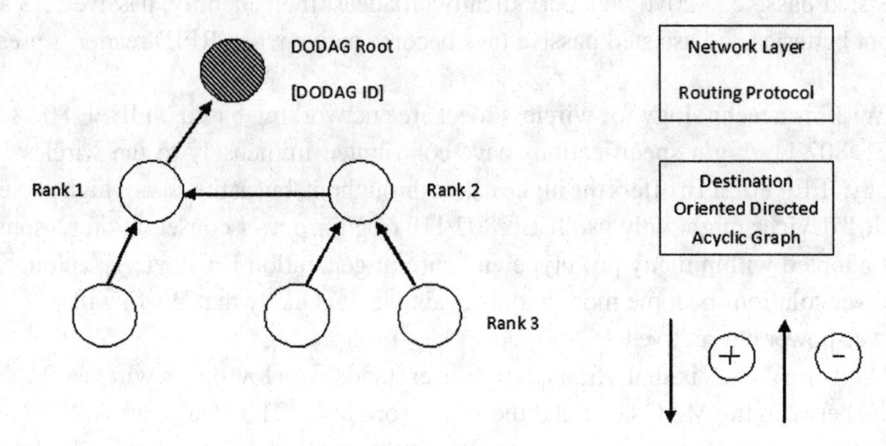

doesn't need to set up the connection. UDP is faster compared to TCP and results in less network traffic. IoT uses UDP as sensor type data could be sent efficiently and faster as compared to TCP.

9. **Message Queue Telemetry Transport(MQTT):** MQTT is a messaging protocol which utilizes publish/subscribe pattern to provide transition flexibility and simplicity of implementation. MQTT is built over TCP protocol. MQTT simply consists of three components: *subscriber, publisher*, and *broker*. As shown in Fig.6, an interested device would register as a subscriber for specific topics in order for it to be informed by the broker when publishers publish topics of interest. The publisher

Figure 6. MQTT publish/subscribe model

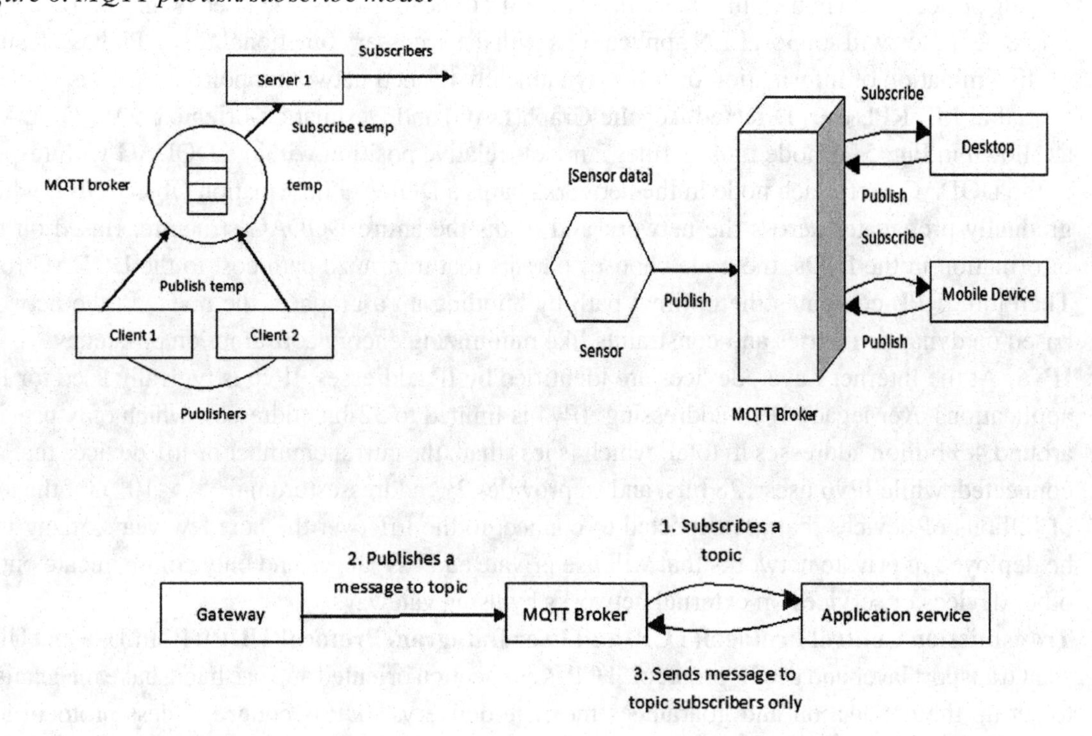

generates the interesting data. After that, the publisher transmits the information to the interested entities (subscribers) through the broker (Salman et al., 2017). Therefore, the MQTT protocol represents an ideal messaging protocol for the IoT and Machine-to-Machine communication and help in providing routing for small, cheap, low power and low memory devices in vulnerable and low bandwidth networks.

10. **Advanced Message Queuing Protocol (AMQP):** AMQP is an open standard application layer protocol that is designed to efficiently support various messaging applications and communication patterns. Middleware products written in different platforms and different languages are able to send the messages to one another supporting interoperability.
11. **Extensible Messaging and Presence Protocol(XMPP):** XMPP is an instant messaging (IM) standard that is used for multi-party chatting, voice and video calling and telepresence. XMPP allows users to communicate with each other by sending instant messages on the Internet no matter which operating system they are using. XMPP allows IM applications to achieve authentication, access control, privacy measurement, hop-by-hop and end-to-end encryption, and compatibility with other protocols.
12. **Data Distribution Service(DDS):** DDS is a standard that meets the features of IoT applications such as scalability, real time response, high performance and interoperable data exchanges using publish-subscribe architecture as shown in Fig.7. It's different from MQTT, where the broker component is absent and provides outstanding QoS, reliability and data urgency. DDS specifies two levels of interfaces, Data-centric publish-subscribe level(DCPS) and Data-Local Reconstruction Level(DLRL). The former is targeted towards efficient delivery of messages to the subscribers and later allows integration of DDS into the application layer.
13. **Constrained Application Protocol (CoAP):** The CoAP defines a web transfer protocol based on REpresentational State Transfer (REST) on top of HTTP functionalities. REST represents a simpler way to exchange data between clients and servers over HTTP(Ala Al-Fuqaha et al.,2015). REST can be seen as a cacheable connection protocol that relies on stateless client-server architecture. It is used within mobile and social network applications and it eliminates ambiguity by using HTTP get, post, put, and delete methods. REST enables clients and servers to expose and consume web

Figure 7. Data Distribution Service Architecture – Publish-Subscribe Model

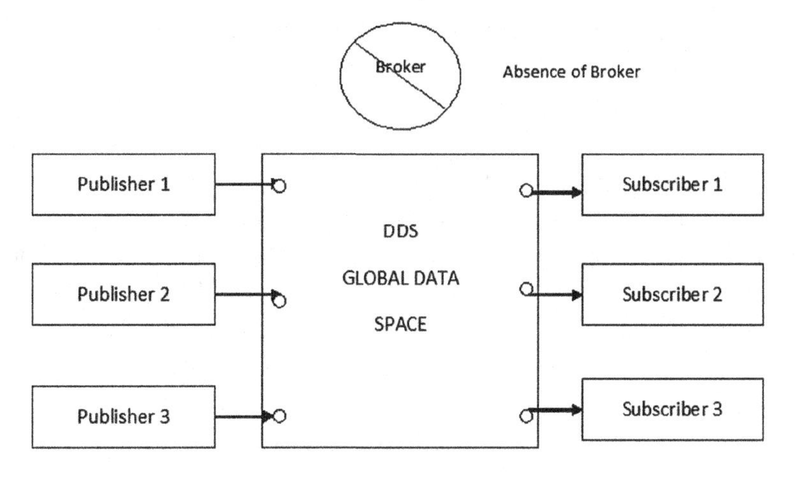

services like the Simple Object Access Protocol (SOAP) but in an easier way using Uniform Resource Identifiers (URIs) as nouns and HTTP get, post, put, and delete methods as verbs. REST does not require XML for message exchanges. Unlike REST, CoAP is bound to UDP (not TCP) by default which makes it more suitable for the IoT applications. Furthermore, CoAP modifies some HTTP functionalities to meet the IoT requirements such as low power consumption and operation in the presence of lossy and noisy links.

14. **Zigbee/IEEE 802.15.4:** As mentioned (Minoli,2013), Zigbee is a low-power, low data rate and close proximity wireless ad hoc network. It uses physical radio specified by IEEE 802.15.4 and utilizes globally available license free 2.4GHz frequency band to provide low data rate wireless applications. Zigbee offers low latency communication as compared to Bluetooth. Zigbee supports star, mesh and cluster-tree topologies. Zigbee mesh network connects sensors, actuators and controllers without being restricted by range.

INTEROPERABILITY AS A MAJOR CHALLENGE

There are various open issues in IoT related to security and privacy, real time processing of sensed data, heterogeneity and interoperability, performance, management at various levels of IoT architecture. Let us focus on interoperability as a major challenge which needs to handle large number of heterogeneous devices that belongs to various platforms. This issue should be considered by both application developers and IoT device manufacturers.

What Is Interoperability?

Interoperability is the key factor for IoT development. In an environment where enormous devices of heterogeneous nature operating in household appliances to wearables, from autonomous vehicles to the drones and others having capability to communicate with each other has a major challenge technically. Hence, Interoperability is defined as the ability of the systems or the components to communicate with each other irrespective of their vendor and technical specifications. Communication technologies and interoperability at various layers and interoperability of data generated by various heterogeneous devices is a major challenge to expand IoT in a major scale.

Need for Interoperability of Heterogeneous Devices

We all might have a question, what is the necessity or the need to solve the problem of interoperability. Consider an example, where we want an automated air conditioning system for home. Air conditioner installed in manufactured by company A and windows control system of the home are manufactured by company B. Both air conditioner and windows systems have their own communication language as they are manufactured by different companies. Both the systems are unable to communicate and coordinate with each other to ultimately meet the objective of automatic air conditioning system.

Government has taken initiative in ‘Smart City’ concept using IoT. Consider an example of traffic management in smart cities. If you are travelling in an autonomous vehicle and need to coordinate and communicate with the other vehicles which are moving on the road. If the technology does not allow them to communicate due to change in the brands and the types of vehicles running on the road, people's life is put at risk. Similarly, we can consider various applications of industries, e-Health sector, defence, automotive applications etc.

Another problem that arises due to standardization is, any strategy that is provided as a solution should be forwards compatible. i.e. devices and technologies which appears in near future should be added easily and flexible enough. The standards that are created should be trusted by every one based on the consensus.

Authors have done exhaustive literature survey on the issue of interoperability which arises due to communication between heterogeneous devices and researcher have shown (Veli-Matti Ojala, 2017) following is the solution given at various levels. The problem of interoperability has to be addressed at various levels in a specific way, which otherwise doesn't solve the purpose. Researchers have tried solving the interoperability issue between the heterogeneous devices at Network Layer, Messaging Protocol Layer, Application Layer and also at semantics level. There are various dimensions of interoperability. As discussed (Patel et.al,2016), there are different types of interoperability are technical interoperability, Syntactical interoperability, Semantic interoperability, Organisational interoperability. Technical interoperability deals with hardware and software components, Syntactical interoperability deals with data formats and Semantic interoperability deals with meaning of the content. Hence at various levels/ IoT layers the interoperability issue has to be handled accordingly. Following are the solutions and the proposed idea for solving interoperability issue are discussed in detail.

Network Layer Interoperability

Network layer has independent routing and switching mechanism. The protocols TCP/IP and UDP/IP are the common protocols used and interoperability between the devices manufactured by different vendors and with different specifications has to be attained. Network layer interoperability is a hardware problem. There is lack of interoperability between various protocols such as Wi-Fi, Bluetooth, Zigbee, Z-Wave etc. Each of these protocols functions in a different way and has its own strengths and weakness. Wi-Fi and Bluetooth are Wireless Local Area Network protocols where Bluetooth is used to exchange data over short distances and Wi-Fi provides high speed access to the internet.

Zigbee and Z-Wave are Wireless Personal Area Network standards which allows devices to have longer physical distance with lower data transmission rate. Various above mentioned technologies need to interoperate and the first layer where these wireless technologies could achieve interoperability is Internet Protocol.

IoT Gateway can provide the solution for the challenges occurring at the network layer. The IoT Gateway systems should have the features of *Data Forwarding, Protocol Conversion* and *Management and Control* as shown in the Fig.8.

Sensor node at the physical layer collects the sensor information and transfers information to the gateway. It also processes the data according to the commands that is sent by the gateway, hence sensor node should be deployed with data processing module. IoT Gateway at the middle level receives the sensor data and commands from the application. Based on the commands, it helps in configuration management, command mapping, protocol conversion with a defined format and uploads the data sensed by the sensors to the application platform.

Figure 8. Architecture of IoT Gateway System

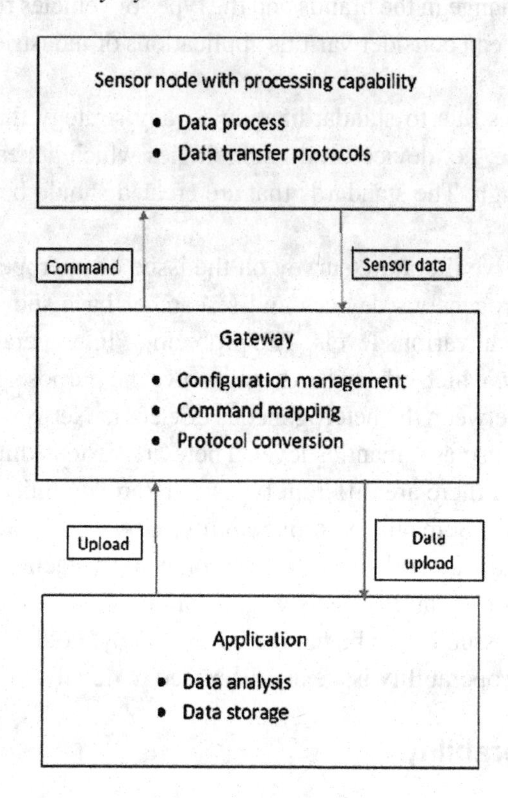

One of the important task is to know how the data should be structured, communicated and utilized by the IoT devices. Based on the issues at various levels of IoT communication researchers have considered data transfer protocols, data representation and semantics on different layers and look forward to provide solution for these challenges. Following sections discuss about messaging protocol interoperability, data format interoperability and semantic interoperability.

Messaging Protocol Interoperability

In IoT we have seen various messaging protocols standards such as CoAP, MQTT, AMQP, XMPP and others. Each and every messaging protocol will have its unique architecture and suits for a specific application. MQTT uses TCP and it favours to use in IoT application where data reliability is important. CoAP uses UDP and raises the issue of Network Address Translation(NAT) and can be solved using tunnelling with the added complexity. The aim is to bridge CoAP, MQTT and HTTP by defining simple REST API which converts the various data formats such as JSON,XML and others by offering security and authentication. As suggested in (thesis), mentions the solutions using proxy for protocol conversion and multi-protocol proxy. Fig.9, shows the Semantic Gateway as Service Architecture(SGS) and Fig, shows the simultaneous use of heterogeneous message protocols by multiple IoT systems. Multiprotocol proxy uses three components *Message Store, Topic Router, Message Broker*. *Message Store* is responsible in storing the entire information. Each device will have unique ID and topic registered on the topic router and provides information about all the connected IoT devices. *Message Broker* proxy different message protocols.

Figure 9. Concept of Multi-protocol Proxy
(Veli-Matti Ojala, 2017)

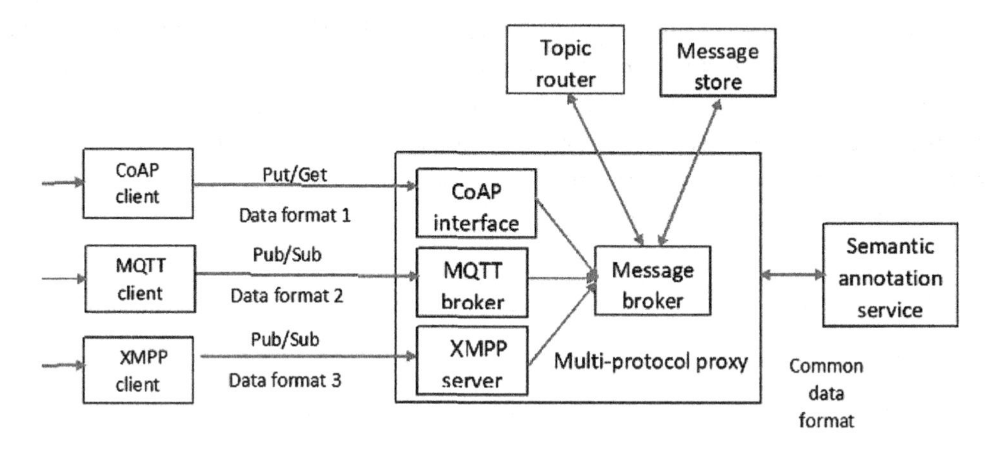

Interoperability at Application Layer and Proposed Idea of Implementation

Another approach to solve the problem of interoperability arising due to wide range of heterogeneous sensors which divided the entire scenario of constrained nodes to vertical silos being able to send data back and forth is by considering a general scenario on how data is sent to upper levels of the internet architecture i.e. application layer. System will have a set of power constrained battery operated devices that act as hosts for the various kinds of sensors. These sensors send their data using gateway as a bridge to the upper layers where the data is shared among the higher abstraction level of the applications.

Data generated by sensors of heterogeneous devices are quite invaluable from a programmer's point of view, hence we need to fine tune and refine the data that is being transmitted by exerting pressure on the gateway to do the necessary tidings.

Looking at the research papers related to data interoperability in IoT is the fact that, predominantly all the sensors are divided into two disjoint sets using protocols CoAP and MQTT. This leads to a potential solution of building a single consolidated protocol for data transfer between the Wireless Sensor Network and the Gateway. The implementation could be done by defining or programming web sockets which lies at the root of computer network development as the door for communication. There are two kinds of sockets TCP and UDP. Based on the type of sensor involved in communication, we need to decide which socket need to be opened and bind with the server socket. A server of our choice can be implemented that has enhanced capabilities to receive and forward Cross Origin Resource sharing. Whenever sensors decide to share the data, respective client socket of either TCP/UDP should be opened and then bind with the server and forward the details, then the corresponding server can forward the details to the requested resource. The sensor code can be implemented in C/C++, intermediate forwarding of the information to the server can be coded using scripting language of your choice, probably nodejs or python. Suitable web application as a client can be made to simulate the requests. Finally testing can be done in simulated environment using multiple requests from various clients to provide a realistic environment comparable to how it actually functions in the real world scenario.

IOT, MACHINE LEARNING AND DATA ANALYTICS

IoT generates big data with variety of modalities and varying data quality. Heterogeneity and interoperability plays an important role and in turn facilitates intelligent processing and analysis of the big data. Data science contributes in making IoT applications more intelligent. Data science includes the fields like data mining, machine learning and many other techniques in identifying the relevant patterns. Many industries have designated IoT as a significant domain and some of kicked off the pilot projects to map the potential of IoT into business applications.

Figure 10. IoT and Data Analytics

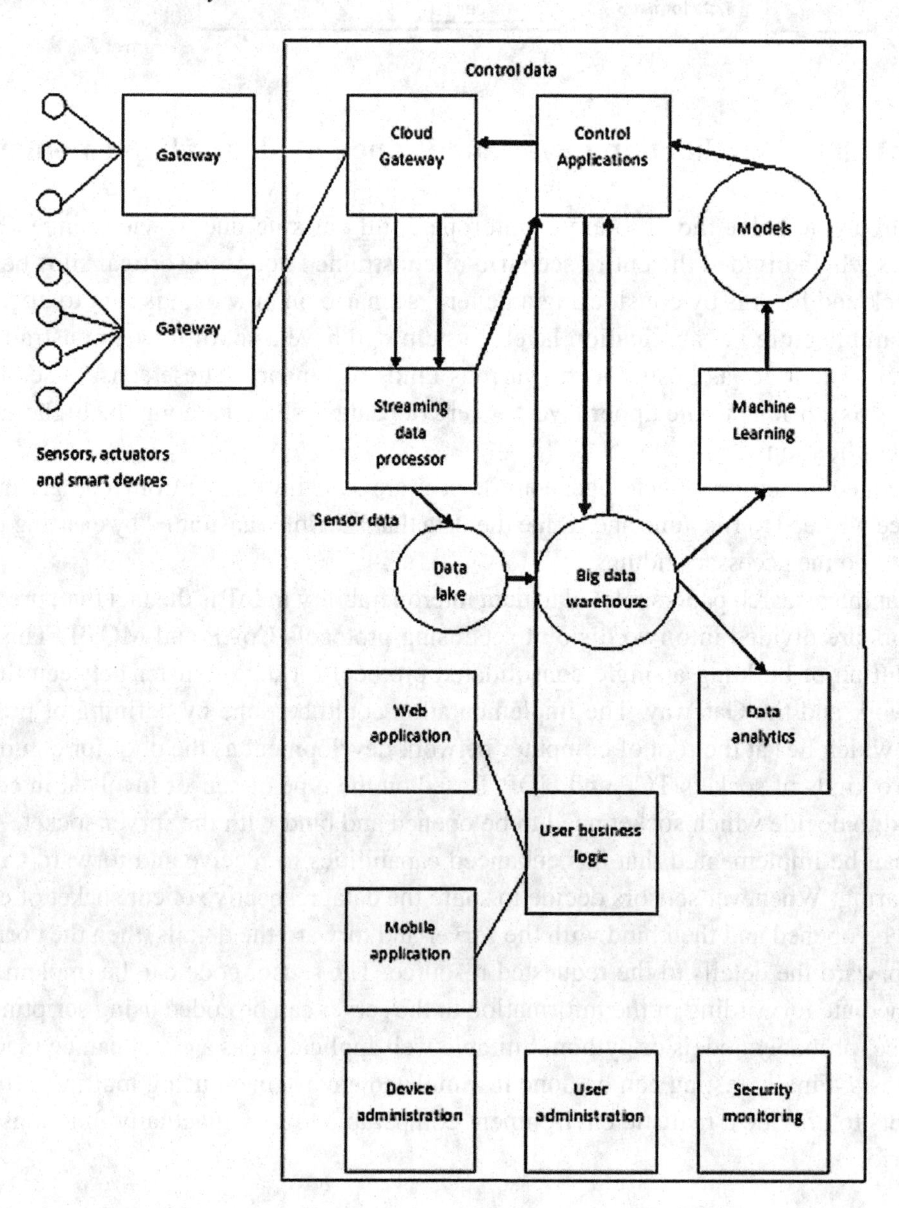

IoT devices generate huge volume of data as well as make use of the data. Data analytics plays a significant role in the growth and success of IoT applications. Data analytics helps in extracting relevant patterns from huge data set and analyses efficiently. It also helps the business units to gain insights on customer preferences and choices leading to improvement in the services and meet the customer demands and expectations.

There are various types of data analytics used along with IoT applications. They are as follows:

- Streaming analytics – Examples: Air fleet tracking, traffic analysis
- Spatial Analytics – Examples: Smart Parking, Sales at existing store location
- Time Series Analytics – Examples: Weather forecasting, health monitoring
- Prescriptive Analysis- Examples: Sales, finance, inventory and customers

Potential application areas of IoT covers various domains such as Smart Living, Smart Health, Smart Industry, Smart Cities, Smart Environment, Smart Agriculture, Smart Energy, Smart Environment and many other. One such case study of Smart Cities is discussed as follows with respect to the public transport system. The scenario of collecting the information, exchanging of the information, processing the collected information remains similar in domains enlisted above. Smart Living includes controlling of the remote appliances, home appliances, weather monitoring, safety monitoring, implementation of intrusion detection system etc. Smart Cities includes smart parking, waste management, public transportation system etc. Smart Environment includes air pollution monitoring, forest fire detection, water quality checking, protecting wildlife, alerting river floods etc. Smart Health covers surveillance of patients, fall detection by giving assistance to elderly and disabled, physical activity monitoring etc.

CASE STUDY: SMART PUBLIC TRANSPORT SYSTEM (SPTS)

Government of India has initiated a Smart Cities Mission which has the feature of applying smart solutions to infrastructure and services in area based development. The main objective of the mission is to make the governance citizen-friendly and cost effective which relies on online services to bring about accountability and transparency. Also, promotes a variety of transport options to reduce the urban heat effects in areas to promote eco-balance.

Let us consider the government as the customer who wishes to predict the demand for SPTS in the city. If the software is able to predict the demand with reasonable accuracy, customer can optimise the use of transport fleet by coordinating the man power and the number of vehicles. The software could be a web application and the mobile application.

Objective: Build a system that integrates various heterogeneous data sources, process using data analysis techniques and help customer in predicting for the transport demand and plan the arrangements. Data analysis could be carried out only if the following requirements are met. They are:

- System should support interoperability
- Data in XML, JSON and other formats has to be handled by the system.
- System needs to adjust for the changes in message protocol layer.
- If the data is not available, it should allow the system to function in normal way without effecting the functionality.

Table 1. Interoperability of data sources

	Network Layer	Message Protocol	Data Format and Semantics	Data Values
Transport Data Sheet	TCP/IP	SQL-query	SQL-result	Start time-end time, address, driver details, count of passengers
Weather Information	TCP/IP	HTTP	XML	Temperature, pressure, rain, wind
City and Vehicle Location	TCP/IP	HTTP/MQTT	JSON & GeoJSON	Location, Vehicle-GPS coordinates

Interoperability challenge exists due to various data sources with differences in message protocols, data formats and semantic representation. Table 1 shows the tabulation of various factors to be considered along with the data format and message protocol details.

- **Transport Data Sheet:** This is a data sheet which is a relational database, and all the actions of the transport is stored in the database. Example: If there is any public function or jatra in a city and the public movement towards these events is going to be dense. Once the event completes the same demand for the transportation will be required.
- **Weather Information:** Customer assumes that there is a correlation between the weather information and the demand for SPTS
- **City and Vehicle Location:** The information regarding the different routes, number of vehicles and the demand has to be analysed. There might be some routes where there is less demand for SPTS as compared to other routes and the vehicles can be diverted to the needy routes to help the public. GPS system helps in finding the current location of the vehicle in real time and helps in diverting the vehicles from low demand area to the higher demand regions.

GPS location data could be provided by using MQTT which runs over TCP/IP and there might arise a scenario where the data would modify its network layer to UDP/IP.

Interoperability challenges exists due to different messaging protocols from weather station and location of vehicles in HTTP and MQTT respectively. The transport data sheet has multiple relational database tables, into which entire information of transport actions will be logged in at real time. The parameters could be the driver information, the start time and end time of the journey by a particular vehicle, number of passengers travelled. This information could be obtained by using querying to the database by the custom written programs. The solution for handling these various data sources is through proxy or message stream. By using proxy, all requests could simply operate on HTTP. In few cases, we need to have dedicated proxy and other components can directly communicate using HTTP request. As shown in Fig.11, when receiving a query, the SQL-query to HTTP-proxy transforms it into suitable SQL-query and retrieves the data from the database. Similarly, vehicle information should be handled by MQTT to HTTP proxy.

The concept of proxy doesn't solve the problem of data format interoperability challenge. In message streaming, data sources will be connected to connecting software component. Each connector component is a separate software process. The main functionality of the connector component is to connect the data

Figure 11. Connecting to data sources using proxies
(Veli-Matti Ojala, 2017)

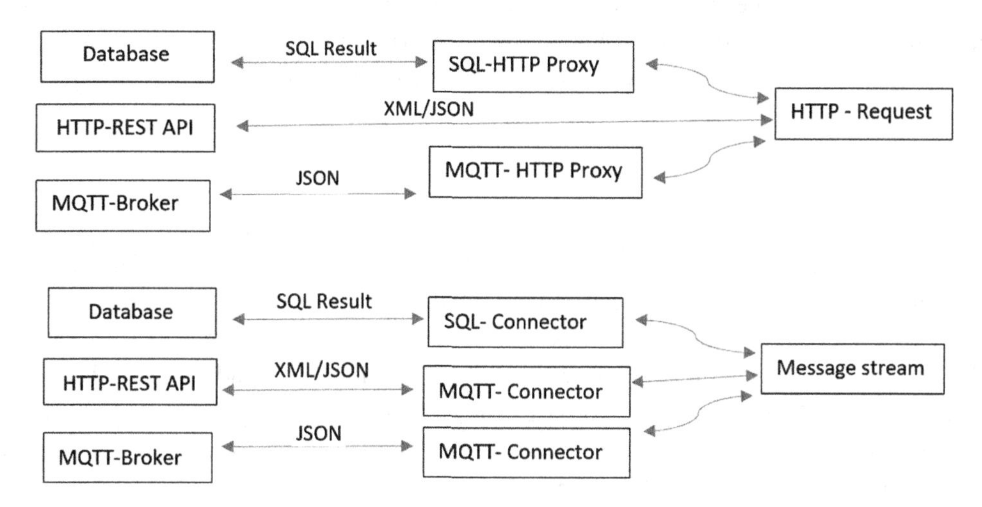

source using message protocol, parse the payload and publish a new message with the payload attached to the message stream component.

Data formats challenges also appear due to heterogeneous devices collecting information in different forms. Weather information in this case is considered to be received in XML and the vehicle information in JSON format. Transport data sheet were stored in relational database. The designed system is expected to cope with both the formats of XML and JSON and should be able to query the relational database with remote connection.

City and Vehicle information is a geospatial data representation and is represented in GeoJSON data format. GeoJSON is an extended version of JSON format which represents regions of space. Using these ontologies is very useful, but the challenge lies in unifying the semantic description from various data sources. The alternative to this challenge is presented by (Desai,2015) to use semantic gateway model which requires money and time.

Researchers have proposed many other ways to solve the problem of interoperability in IoT. There cannot be a unique solution for all kinds of wireless network topologies. IoT is implemented in diverse areas such as Body Area Network or e-Health, City Automation, Home Automation, Control Applications in industries and many others. The interoperability issue in the field of Body Area Network could be solved differently and efficiently as compared to City Automation applications. In some application especially in Wireless Sensor Networks and IoT where it involves the constrained resources, when the mobile devices enter into a particular network, the topology should be managed dynamically and communicate with each other.

FUTURE RESEARCH DIRECTIONS

Solutions to the problem of interoperability are already discussed in detail, but these solutions have to be standardised and performance has to be verified. The authors are intensively looking into how the two disjoint devices which uses CoAP and MQTT protocol with various data formats could be brought to the common format at the application layer. At network layer, gateway provides the solution. But is it possible to make use of devices at the network layer with similar capabilities behave as a gateway and forward the data to the intended device? Authors are exploring on various other possibilities how the overall interoperability issue could be overcome in a consolidated way.

CONCLUSION

Interoperable environment requires a well-established model that selects the right protocol by providing the message formats and the frameworks. IoT requires adaptive and dynamic models that enables diverse devices to interoperate and communicate. This is possible by the manufacturers of the devices by systematically and incrementally publishing the changes without affecting the advances in the development of applications. Horizontal linking of data through the standards is the key that enables heterogeneous devices from various industries to be on the common platform to improve the end-user experience. By implementing multi-vendor interoperability and added feature of authentication to each device helps in exploiting the full potential of IoT. Large enterprises embrace this technology, but are not adopting the technology to the mark due to its complexity of communication between heterogeneous devices.

REFERENCES

Al-Fuqaha, Guizani, Mohammadi, Aledhari, & Ayyash. (2015). Internet of Things: Architectures, Protocols, and Applications. IEEE Communication Survey and Tutorials, 17(4).

Desai, P., Sheth, A., & Anantharam, P. (2015). Semantic gateway as a service architecture for iot interoperability. *Mobile Services (MS), IEEE International Conference on*, 313–319.

Madakam, Ramaswamy, & Tripathi. (2015). Internet of Things(IoT): A Literature Review. *Journal of Computer and Communications, 3*, 164-173. Retrieved from http://www.ieccr.net/comsoc/ijcis/

Minoli. (2013). Layer 1/2 Connectivity: Wireless Technologies for the IoT. In *Building the Internet of Things with IPv6 and MIPv6 The Evolving World of M2M Communications*. Academic Press.

Ojala, V.-M. (2017). *Addressing the interoperability challenge of combining heterogeneous data sources in data-driven solution* (Master's Thesis). Academic Press.

Patel & Patel. (2016). Internet of Things – IOT: Definition, Characteristics, Architecture, Enabling Technologies, Application & Future Challenges. *IJESC, 6*(5).

Salman, T., & Jain, R. (2017). Advanced. *Computer Communications, 1*(1).

Chapter 2
Circular, Smart, and Connected Cities:
A Key for Enhancing Sustainability and Resilience of the Cities

Begum Sertyesilisik
Istanbul Technical University, Turkey

ABSTRACT

*Majority of the humanity is living in the cities. Cities have adverse environmental impact. Their environmental footprints need to be reduced. As the world's living conditions deteriorate, the survival of the humanity depends on the precautions taken. These precautions can include sustainable living styles, new technologies, and circular economy principles. Furthermore, climate change caused disasters can have adverse consequences as they can be deadly and as they can result in economic loss. The cities need to be resilient so that disasters adverse consequences can be reduced and the post-disaster phase rescue and recovery processes can be effectively carried out. Circular, smart and connected cities based on the new technologies such as big data, Information and Communication Technology (ICT), and Internet of Things (IoT) can contribute to the cities' sustainability and resilience performance. This chapter aims to investigate the roles of big data, IoT, ICT, as well as circular, connected and smart cities in enhancing sustainability and resilience of the cities. With this aim, based on the literature review, this chapter covers: need for, pillars of and aspects of smart, sustainable, circular, and resilient cities as well as ways for transforming the cities into smart, sustainable, circular, and resilient ones. This chapter can be beneficial to the researchers, academics, construction professionals, and policy makers.**Keywords: Disaster; Internet of Things; Information and Communication Technology; Smart Cities; Big Data; Resilience; Building Information Modelling; Circular Cities; Circular Economy*

INTRODUCTION

World has become a hub of interconnected cities. Quantities and sizes of the cities are being increased due to the humanity's trend in moving to urban land. This trend is expected to continue as more than half of the world's population is expected to be hosted in the cities (UN, 2014; Sertyesilisik & Sertyesilisik,

DOI: 10.4018/978-1-5225-9199-3.ch002

2015). The increasing population in the cities transforms the cities into resource exploitation hubs. Cities contribute to the causes of the climate change due their environmental footprint. Precautions have started to be taken to reduce the environmental footprint of the cities. These precautions include but they are not limited to: enhancing sustainability performance of the built environment through sustainable or green building assessment tools, sustainable or green material certificates. Most of these precautions could not be effective in preventing increase in the cities' environmental footprint. Furthermore, cities are vulnerable to climate change (e.g. risk of emergence of *biophysical hazards*) especially in case mainly only climate adaptation actions are taken (Moriarty & Honnery, 2015: 45). Circular economy principle based cities can contribute to the climate mitigation efforts and to the reduction in the cities' environmental footprint. Furthermore, circular economy based cities can contribute to the well-being of the citizens.

As cities cannot be isolated from the consequences of the climate change (e.g. disasters, extreme weather conditions), humanity has started to experience deadly disasters (e.g. tornado) and to observe how climate related disasters have caused disruption to the transportation in the cities and logistics to the cities as well as demolition of the built environment. Adverse consequences of these disasters emphasise the importance of adapting the cities to the climate change and of enhancing their resilience in addition to the climate mitigation efforts. Cities need to be smart and equipped with the infrastructure and technology needed to ensure their citizens' safety and well-being as well as to enhance the cities' resilience. Smartness and connectedness of the cities can contribute to their resilience, and to the effectiveness of the disaster management in all phases (in the pre- and post-disasters as well as during the disasters). They can reduce the adverse consequences of disasters and enable rescue and recovery processes to be effectively carried out.

Effective solutions and precautions are needed to reduce environmental footprint of the cities and to enhance resilience of the cities. Climate change avoidance actions (e.g. reducing environmental footprint), climate mitigation and adaptation actions are complementary actions for enhancing cities' circularity, sustainability, smartness, and resilience performances. Circular, smart and connected cities are based on the new technologies (e.g. big data, Internet of Things (IoT), Information and Communication Technology (ICT) which can contribute to the cities' sustainability and resilience performance. This chapter aims to investigate the role of big data, IoT, ICT, as well as circular, smart and connected cities in enhancing sustainability and resilience of the cities.

BACKGROUND

Smartness, circularity, sustainability, connectedness as well as disaster resilience of the cities are interrelated (Figure 1). Cities' performances in each of these fields influence each other's performances.

Smartness of the cities can enhance sustainability performance of the cities. Smart cities can be described as cities operating as well as solving their existing and potential problems with the help of smart technologies supported by big data, IoT, and ICT. Smart cities "...use digital data to deliver better public services and more effective uses of resources" (Alavi, Jiaob, Buttlar, & Lajnef, 2018). Ojo et al. (2014) have emphasized that smart cities can reduce their carbon emission and improve energy efficiency (Kumar, Singh, Gupta, & Madaan, 2018). Similarly, various researchers have emphasised contribution of the smartness of the cities to their sustainability performance (Ahvenniemi, Huovila, Aina, 2017: 50; Anthopoulos, 2017; Martin, Evans, & Karvonen, 2018: 269; Pinto Seppa, &Airaksinen, 2017). Furthermore, smart infrastructure of the cities can support the cities' liveability performance. Yigitcanlar &

Figure 1. Interrelation among the concepts of connectedness, smartness, circularity, sustainability, and resilience of the cities

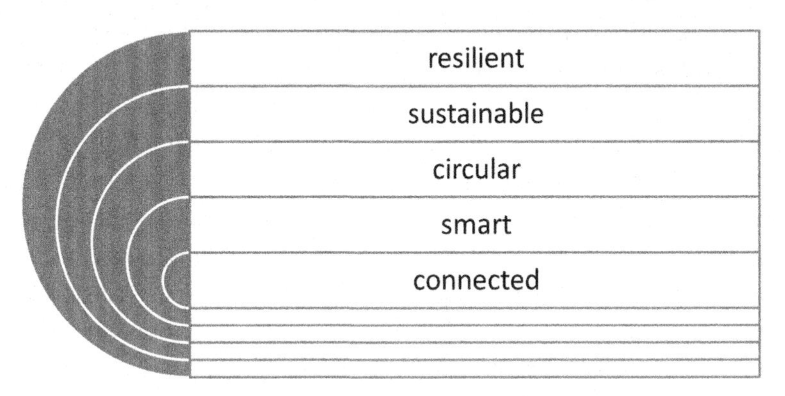

Kamruzzaman (2018: 49), on the other hand, have emphasized the importance of smart city agendas in achieving sustainability performance improvement through smartness of the cities. As smart cities' main pillars are smart governance, economy, citizens, mobility, environment, as well as smart living (Alavi, et al., 2018), smartness of the cities have influence on the citizens' wellbeing and welfare, sustainable development of the cities as well as sustainability performance of the production. Economist Intelligence Unit (2018)'s ranking is based on the following categories: stability; healthcare; culture and environment; education; infrastructure (The Economist Intelligence Unit, 2018). As this ranking is an indicator for the liveability of the cities, it provides an understanding of the welfare of their citizens. Mayors or Governors of the cities aiming to enhance their citizens' welfare can improve their cities performance on these five categories with the help of the smart, sustainable and resilient infrastructure and facilities. Furthermore, smart built environment enabling smart water and smart energy management can further contribute to the sustainability performance of the cities. Integration of smart energy management with the renewable energy usage can foster reduced dependence on fossil fuels. Renewable energy generation technologies' integration to the cities can contribute to the cities' sustainability performance as well as their resilience especially in case of disasters. New technologies emerge for increasing efficiency and effectiveness of the renewable energy generation. For example, there are various recent innovations in the field of the wind energy and wind turbines (Youtube websites). Similarly, in the field of solar energy new technologies and systems have been emerged and are being improved. Madrid's Solar tower energy as well as EnviroMission - Solar Tower are among these examples. Smart waste management can contribute to the effectiveness and sustainability of waste collection through directing the waste collector cars to the full bins only.

Circularity performance of the cities can enhance sustainability performance of the cities. Circular cities are cities which have minimized their waste generation due to their operation based on the circular economy principles. Circular economy is based on the eco-industrial development relying on eco-industrial development zones as well as on the cradle-to-cradle material cycle in the production processes (Sertyesilisik & Sertyesilisik, 2016). This cradle-to-cradle material cycle can be achieved through collaboration of the companies in sharing their resources in a way that output or by product materials of production processes can be used as input materials in other production processes. Smartness of the cities can contribute to the circularity performance of the cities.

Cities' performance for sustainability and disaster resilience need to be enhanced especially due to the environmental footprint of the cities as well as due to the increased magnitudes of the climate based disasters. Smartness of the cities can enhance resilience of the cities. Resilient cities are cities capable of resisting shocks (e.g. disasters) and able to restore and recover themselves in the post-shocks phase as well as cities capable of adopting themselves to the changing climatic conditions. Intensified disasters especially due to the climate change increases the need for disaster resilient cities. One effective way of enhancing the disaster resilience of the cities is transforming them into smarter ones. Smart technologies can enhance effectiveness of the relief and rescue efforts due to their help in identifying the state-of-the-art of the city, of the damaged areas and location of the people suffering. Smart technologies can make the city more agile in responding to the post-disaster relief and rescue efforts as well as in organizing the citizens enabling them to reach the first-aid and helping each other immediately.

Connectedness of the cities can influence and affect sustainability, circularity, smartness, and disaster resilience performances of the cities. It can enhance collaboration and information sharing among the cities contributing to the efficient use of resources, problem solving, immediate and effective response and relief activities.

PILLARS OF THE CIRCULAR, SMART, RESILIENT AND CONNECTED CITIES

Circularity, smartness and connectedness of the cities can contribute to the effective and efficient usage of their resources as well as to the resilience of the cities. As each city is unique, it has its own problems and solutions requiring city-based smart solutions to be executed. For this reason, accurate identification of the city's needs and problems can lead to effective solution of the problems through prioritization of the most important needs and problems complying with the Pareto principle. This can be achieved through IoT, ICT, big data, building information modeling (BIM) and geographic information systems (GIS) which are the main pillars and enablers of circular, smart, resilient and connected cities. These main pillars have been explained in the following paragraphs:

- **The Internet of Things:** IoT is an important enabler of smart, connected and circular cities. It enables city operations to be performed intelligently and effectively with relatively low human interaction (Silva, Khan, & Ha, 2018: 697). IoT can provide real-time data to recognize, locate, track, monitor and manage intelligently as well as enable the smart city to become interconnected, instrumented, and intelligent (Kim, Ramos, & Mohammed, 2017: 159). IoT transforms cities into connected intelligence spaces (connected devices, people and institutions) which can be described as cyber-physical spaces (Komninos, 2018). IoT, cyber-physical spaces, and citizens' participation are vital for the smart development (Antonelli & Cappiello, 2016; Komninos, 2018) IoT can provide smart solutions needed to solve the problems of the modern cities (Alavi, et al., 2018). Especially, its integrated usage with other technologies (e.g. cloud computing, robotics, micro-electromechanical systems, wireless communications, and radio-frequency identification) (Alavi, et al., 2018) can foster the effectiveness of the IoT and can play multiplier effect in its effectiveness in solving the problems of the city. For example, block chain technique enriched IoT can support automation of the business processes in a cost and time effective way enhancing users' experience (Kumar Sharma & Park, 2018: 650-651).

- **The Information and Communication Technology:** Smart cities are based on the ICT (Kramers et al., 2014; Osman, 2018: 1) and IoT (Elmaghraby & Losavio, 2014) based development (e.g. government functionality, city operations, etc.) (Kumar, et al., 2018). Kumar et al. (2018)'s research revealed that ICT and IoT are integrated especially in the fields of the government planning and policy design; citizens' participation; economic restricting; network and hardware; software components; sensor deployments; data centers and analytics; environment protection and safety; public safety and crime reduction; disaster alerts and control; tourism and entertainment; business activities; education facilities; cleanliness in city; health amenities; transport system; online civic amenities; utility supply. For example, ICT can support transportation (Battarra, Gargiulo, Tremiterra, & Zucar, 2018: 556) (e.g. parking, effective integration of different transportation types (Gohar, Muzammal, & Rahman, 2018: 114) or food system (Deakin, Diamantini & Borrelli, 2018). Furthermore, it can enhance sustainability performance of the smart cities (Aina, 2017; Bifulco et al. 2016; Kramers, et al., 2014; Papa, Galderisi, Majello, &Saretta, 2015). GeoICT, geoinformation embedded ICT (Navarra, 2013; Aina, 2017: 50), on the other hand, can contribute to the solution of the various problems included but not limited to the following: transportation problem, disaster management, renewable energy planning (Aina, 2017: 51). Furthermore, ICT can respond to the citizens' needs and it can contribute to the solution of the social issues (e.g. Aizuwakamatsu) in addition to it's contributions to the environmental, and economic issues (Trencher, 2018). Escolar, et al. (2018: 13) have emphasised the need for a smart city ranking index covering urban development criteria, and ICT usage criterion due to the importance of the ICT. They have suggested a new methodology based on these criteria.
- **The Big Data:** Big data, generated by the ICT, need to be analysed with the help of big data analytics or big data value chain so that data gathered can be used to solve the cities' problems (Osman, 2018: 1). As the amount of the big data being gathered is being increased due to the usage of the ICT, the big data analytics is becoming more and more important as well as challenging resulting in the emergence of new frameworks for big data analytics such as Smart City Data Analytics Panel (Osman, 2018) or the one for intelligent transportation systems (Gohar, Muzammal, & Rahman, 2018: 114). If used appropriately, big data can be useful in enhancing well-being of the people. Big data can enhance agility performance in the disaster and post-disaster phases. It can contribute to the sustainability performance of the cities. For example, smart lighting of the streets enables change in the illumination level based on the big data with respect to the quantity of the people walking in the street. Furthermore, big data can enable direct democracy through the citizens' direct participation in the decision making process and in the governance of their smart cities. Privacy related concerns have been raised in the big data usage, despite of these benefits of the effective usage of the big data.
- **BIM and GIS:** BIM and GIS as well as their integrated usage enable the built environment to be connected to and to become part of the smart cities. They contribute to the establishment of the smart cities at all levels starting from the building level. BIM based construction project management makes the construction project management processes smarter, integrating different disciplines to work in the same digital environment. In this way, it can contribute to the reduction in the variation orders and defective work. Based on the BIM level used, BIM does not only support construction project management but also the entire building's lifecycle including operation phase and facilities management (e.g. smart electricity/energy, smart lighting, smart water etc.). For this reason, BIM has started to become compulsory for bidding to the public entities tenders in various countries.

ASPECTS OF TRANSFORMING THE CITIES INTO MORE CIRCULAR, SMART, CONNECTED AND RESILIENT ONES

Transformation of the cities into circular, smart, sustainable and resilient cities has social, economic, technological, political and environmental aspects. If these dimensions are not well managed, they can act as challenges and obstacles in the transformation of the cities.

- **Social Aspects:** As the cities and their infrastructure are for the citizens / human beings and for their wellbeing and comfort, all infrastructures need to be user-friendly. Applications need to be usable by the citizens no matter what their education level, or age are. In this way, technology can enable all citizens to provide feedback on the problems their experience or observe in their cities as well as on their suggestions for and recommendations on the solutions of these problems and / or on the further improvements in and development of their cities. Big data gathered can be used for enhancing the citizens' wellbeing and comfort as well as for identifying and solving the problems of the city (e.g. resource efficiency, traffic, pollution, etc.). Furthermore, technology and big data can enable citizens' active and effective participation in the rescue and recovery phases in the post-disaster phase. Data privacy and ownership of data, however, are two main risks endangering humanity's freedom and privacy. "*Smart cities must ensure individual privacy and security … to ensure that its citizens will participate*" (Braun, Fung, Iqbal, & Shah, 2018: 499). For this reason, political and technological precautions need to be effectively taken not to endanger humanity's privacy.
- **Economic Aspects:** Smartness of the cities can enable data generation with respect to the by-products and outputs usable by other production processes as inputs. These data are vital for achieving circular cities and circular economy. Circular economy fosters synergy and efficiency in the tangible and intangible resources usage. At the company level, this synergy and efficiency can contribute to the companies' competitiveness through reduced production costs as well as to the sustainability performance of the industries. At the city level, circular cities can contribute to the citizens' welfare through effective usage of cities' resources and through potential jobs. Furthermore, they can contribute to their wellbeing through improved living conditions (e.g. sustainable environment). At the country level, countries having competitive companies can have strong economy fostering countries' sustainable development and contributing to the employment and welfare of their people. Furthermore, enhanced sustainability performance of the production processes can contribute to the wellbeing and health of the people. Circular economy can be achieved by relevant policies and technologies leading to the inventions and innovations in fostering circularity of the cities and of the production processes.
- **Technological Aspects:** Technology enables emergence of innovations needed for the smart infrastructure in the smart cities. Sometimes effective inventions cannot be transformed into innovations due to their lack of feasibility and their high initial investment costs. Furthermore, size of the city plays an important role in the economic development (Borsekova, Korónya, Vaňováb, Vitálišov, 2018: 17). Considering the economies of scale of the inventions, cities' size can influence feasibility of inventions due to the impact of the cities' size on the demand for particular invention. Furthermore, the bigger the cities are, the more money they have for allocating to the innovations (Borsekova, et al., 2018: 25). Inventions contributing to the social wellbeing, welfare of the citizens and to the circularity, smartness, sustainability and resilience of the cities need to be

supported so that they can be transformed into innovations. Feasibility of these type of inventions can be supported by relevant policies. There is interaction between technologies and transformation of the industries and human resources.

- **Political Aspects:** Policies have influence on the social, economic, technological, legal and environmental aspects of the circular, smart and connected cities. Policies need to be integrated with the citizens. Smart city and its ability to deal with the problems in a systematic and integrated way rely on governance enabled by stakeholder participation and involvement at the project and strategic levels (Fernandez-Aneza, Fernández-Güell, & Giffing, 2018: 4). For this reason, stakeholders' feedback and opinions are needed for accomplishment of the strategies of the smart cities (Fernandez-Aneza, et al., 2018: 4). In other words, support of politicians, administratives, citizens, and businessmen play vital role in the transformation of the cities into smart ones (Borsekova, et al., 2018: 17). Policies play important role in supporting and encouraging smart and green innovations which can enhance circularity of the production processes. Furthermore, policies play an important role in the establishment of the laws and regulations to secure data privacy enabling protection of the humanity's freedom and privacy. Policies need to support transformations needed in universities, and industries to achieve successful transformation of the cities. Policies need to be established with the systematic approach to analyze the impact and effectiveness of the policies in solving the problems. Global collaboration among politicians in solving the cities' problems can enable establishment of effective and smart policies. Global collaboration is needed to create synergy through sharing the experiences with respect to the lessons learned and best practices.
- **Legal Aspects:** Legal aspects complement and enable all other aspects (e.g. social, economic, technological, and environmental aspects) of the circular, smart and connected cities. They have impact on the social aspect as effective laws and regulations on data privacy security can protect humanity's freedom and privacy. They are related with the economic and environmental aspects as effective laws and regulations on environmental protection can reduce environmental footprint of the industries and of the cities. Furthermore, they have potential for fostering technologic developments and innovations as they can encourage the companies for research and development activities through subsidies, tax exemptions, etc.
- **Environmental Aspects:** Environmental regulations and restrictions can encourage and enforce the companies to reduce their waste as well as to operate based on the circular economy principles and in a more sustainable way. Their encouragement to enhance their sustainability performance can lead to the innovations to make their production processes more lean and smart. Furthermore, environmentally-friendly production processes can have social impact as it can contribute to the wellbeing of the people through enhanced sustainability performance of the living conditions as well as through innovation driven production having potential for new employment opportunities. Environmentally-friendly supply can be driven by the environmentally-friendly demand.

Figure 2. Ingredients (the social, economic, technological, political, legal, and environmental aspects) of the connected, smart, circular, sustainable, and resilient cities

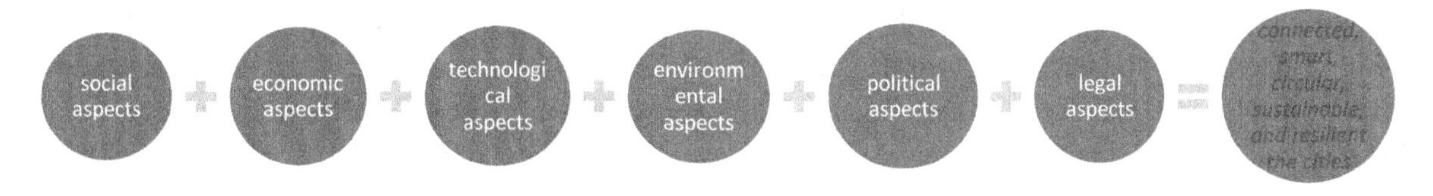

All these aspects influence each other. Their interactions need to be managed in an effective way to enhance creation of synergy. Their interactions can inhibit synergy and progress unless they are not well managed.

SOLUTIONS AND RECOMMENDATIONS

Transformation process of cities into more circular, smart and connected ones is based on the IoT, ICT, big data as well as BIM and GIS. It requires addressing social, economic, technological, political, legal and environmental aspects as well as their interaction. This process, however, is challenging. It requires construction industry and the universities' curriculums to be transformed in a way that it can address changing demand. Furthermore, cities' transformation process requires supporting strategic planning, policies and smart governance as well as smart city planning and management.

Transformation Needed in the Construction Industry

Construction industry needs to be based on and comply with the circularity and smartness principles. Characteristics of demand for construction is changed complying with the needs for circular, smart, sustainable and resilient cities and the built environment. Construction industry needs to be transformed into a more circular and smarter one so that it can comply with the change in the demand as well as enable the construction and transformation of the circular and smart cities. This transformation will result in more IT based constructions and materials as well as more IT based construction project management. Construction industry is one of the key enablers of the circular and smart cities due to its role in constructing the infrastructure and suprastructure of the cities starting from the material scale. IT based construction materials and buildings are important components of smart cities. Smart cities consist of smart built environment supported by smart materials. Smart and sustainable buildings consisting of smart systems and materials are connected to the smart neighborhood and to the smart cities.

Transformation in the construction industry can be accelerated by the change in the demand for smarter built environment, infrastructure and materials. The change in the demand can have influence on the innovation for achieving establishment of the cost effective smart built environment. Furthermore, change in the demand and in the innovation rate can influence the competition in the industry as well as investments for smarter production technologies and outputs. Companies' ability to survive in the highly competitive and ever changing construction industry depends on their agility to fit into the market of the today and of the future. Their ability to survive and their resilience to overcome intensive competition relies on the human resources' capability, and knowledge affected mainly by the higher educations' quality as well as on its contemporary curriculum and infrastructure.

Transformation Needed in the Curriculum of the Universities

Transformation needed in the construction industry can be achieved through transformation and adaptation of the universities' curriculum to equip the future professionals with the knowledge and skills needed for the establishment and construction of the circular, smart and resilient cities. Universities and higher education institutions play key role in the establishment of the smart cities' projects based on the knowledge-based ecosystems (Ardito, Ferraris, Petruzzelli, Bresciani, & Del Giudice, 2018). Ardito et al. (2018)'s study on the 20 smart city projects, revealed the universities' role as "... knowledge intermediaries, knowledge gatekeepers, knowledge providers, and knowledge evaluators." (Ardito, et al., 2018). Future construction industry professionals need to be capable of working in and for circular and smart built environment. They need to possess relevant knowledge and skills to enable their construction and refurbishment.

Need for Supporting Policies and Smart Governance

Policies and smart governance need to support transformation of the cities, construction industry as well as smart city planning and management. Smart governance dimension is one of the six main dimensions (e.g. smart governance, smart people, smart environment, smart economy, smart mobility and smart living) of smart cities (Giffinger et al., 2007; Kumar et al., 2018). Smart governance can directly influence and result in smart environment, economy, mobility and living. It can be enabled by the smart people. Smart governance enables the citizens to participate in the city development, laws and regulations as well as empowers them through this process. Empowered citizens participating in the smart city governance are important for smart and sustainable urban development which requires "*...environmental protection and social equity...*" (Martin, Evans, & Karvonen, 2018: 269).

Policies and politicians can encourage collaboration of the smart cities with each other in the form of networks. These networks (e.g. Spanish Network of Smart Cities) have potential for enhancing savings and efficiency in the governance as well as in solving current and future problems of the cities (Palomo-Navarro & Navío-Marco, 2017) as they can enable sharing experience and expertise in the relevant field.

Policies and laws can help establishment of the financial services needed for smart cities. Arora (2018: 57-58)'s research on India's 20 smart cities revealed the importance of establishing the financial services in the smart cities and in the transformation of the cities into smarter ones. Government needs to pay attention to the development of the financial system as it plays important role in required investments in various fields (e.g. education, health, etc.) (Arora, 2018: 57-58).

Policies, laws and regulations can encourage the cities' operations to be sustainable, smart and circular. Furthermore, they can encourage construction companies to become more innovative in enhancing their operations' and outputs' sustainability, smartness, resilience and circularity performances.

Construction industry's capacity and ability to construct smart cities based on circularity principle can be enhanced and supported by policies and smart governance. For example, infrastructure projects for enhancing smartness, circularity, resilience and sustainability of the cities can be encouraged via subsidies. Furthermore, human resources' capacity and expertise in the accomplishment of the construction projects can be achieved through innovative and contemporary education programmes at all levels as well as through relevant education policies.

Policies supporting smartness, circularity, sustainability and resilience of the cities can enhance success in solution of the cities' problems. Smart urban services are becoming effective solutions for the politicians and urban authorities to provide high quality services (Kumar et al., 2018).

FUTURE RESEARCH DIRECTIONS

Circular, smart, resilient and connected cities are the future cities having enhanced capacity in dealing with the challenges and severe problems (e.g. climate change) the cities are expected to face with in the near future. For this reason, future research is recommended to be carried out on the ways of:

- Enhancing and fostering collaboration among city majors, urban administrations, universities and construction industry at global level so that experiences gained can be shared and synergy can be created in enhancing welfare and wellbeing of the citizens and for enhancing performance of the cities with respect to their circularity, smartness, and resilience performances
- Monitoring effectiveness and challenges of the technologies applied for transforming cities into smarter as more circular and resilient ones
- Establishment of the laws and regulations on data privacy in the smart cities as well as enacting them globally
- Analysing future trends in technologies and innovations in the circular, smart, and resilient built environment
- Establishment of the systemic approach in the political and economic aspects in the urban development

CONCLUSION

This chapter aimed to investigate the role of big data, IoT, ICT, as well as circular, smart and connected cities in enhancing sustainability and resilience of the cities. This chapter emphasized need for and importance of circular, smart, sustainable, resilient and connected cities. This chapter put emphasis on the circular, smart and connected cities' role in enhancing resilience of the cities. The main pillars of these cities are: IoT, ICT, big data, BIM and GIS as they play important role in their establishment. Transformation of the cities into circular, smart, sustainable, connected and resilient cities has five main aspects (e.g. social, economic, technological, political and environmental aspects) which are interactive. The interaction of these aspects need to be well-managed to achieve synergy and not to cause entropy. The transformation of the cities requires addressing these main aspects, transformation in the construction industry and curriculum of the universities as well as supporting policies and smart governance. This chapter is expected to be beneficial to the researchers, academics, construction professionals, and policy makers.

REFERENCES

Ahvenniemi, H., Huovila, A., Pinto-Seppa, I., & Airaksinen, M. (2017). What are the differences between sustainable and smart cities? *Cities (London, England)*, *60*, 234–245. doi:10.1016/j.cities.2016.09.009

Aina, Y. A. (2017). Achieving smart sustainable cities with GeoICT support: The Saudi evolving smart cities. *Cities (London, England)*, *71*, 49–58. doi:10.1016/j.cities.2017.07.007

Alavi, A. H., Jiaob, P., Buttlar, W. G., & Lajnef, N. (2018). Internet of Things-Enabled Smart Cities: State-of-the-Art and Future Trends. *Measurement*, *129*, 589–606. doi:10.1016/j.measurement.2018.07.067

Anthopoulos, L. (2017). Smart utopia vs smart reality: Learning by experience from 10 smart city cases. *Cities (London, England)*, *63*, 128–148. doi:10.1016/j.cities.2016.10.005

Antonelli, G., & Cappiello, G. (Eds.). (2016). *Smart Development in Smart Communities*. Taylor & Francis. doi:10.4324/9781315641850

Ardito, L., Ferraris, A., Petruzzelli, A. M., Bresciani, S., & Del Giudice, M. (2018). The role of universities in the knowledge management of smart city projects. *Technological Forecasting and Social Change*. doi:10.1016/j.techfore.2018.07.030

Arora, R. U. (2018). Financial sector development and smart cities: The Indian case. *Sustainable Cities and Society*, *42*, 52–58. doi:10.1016/j.scs.2018.06.013

Battarra, R., Gargiulo, C., Tremiterra, M. R., & Zucar, F. (2018). Smart mobility in Italian metropolitan cities: A comparative analysis through indicators and actions. *Sustainable Cities and Society*, *41*, 556–567. doi:10.1016/j.scs.2018.06.006

Bifulco, F., Tregua, M., Amitrano, C. C., & D'Auria, A. (2016). ICT and sustainability in smart cities management. *International Journal of Public Sector Management*, *29*(2), 132–147. doi:10.1108/IJPSM-07-2015-0132

Borsekova, K., Korónya, S., Vaňováb, A., & Vitálišov, K. (2018). Functionality between the size and indicators of smart cities: A research challenge with policy implications. *Cities (London, England)*, *78*, 17–26. doi:10.1016/j.cities.2018.03.010

Braun, T., Fung, B. C. M., Iqbal, F., & Shah, B. (2018). Security and privacy challenges in smart cities. *Sustainable Cities and Society*, *39*, 499–507. doi:10.1016/j.scs.2018.02.039

Deakin, M., Diamantini, D., & Borrelli, N. (2018). The governance of a smart city food system: The 2015 Milan World Expo Mark. *City, Culture and Society*.

Elmaghraby, A. S., & Losavio, M. M. (2014). Cyber security challenges in smart cities: Safety, security and privacy. *Journal of Advanced Research*, *5*(4), 491–497. doi:10.1016/j.jare.2014.02.006 PMID:25685517

Escolar, S., Villanueva, F. J., Santofimia, M. J., Villa, D., del Toro, X., & Lopez Carlos, J. A multiple-attribute decision making-based approach for smart city rankings design. *Technological Forecasting and Social Change*. doi:10.1016/j.techfore.2018.07.024

Fernandez-Aneza, V., Fernández-Güell, J. M., & Giffing, R. (2018). Smart City implementation and discourses: An integrated conceptual model. The case of Vienna. *Cities (London, England)*, *78*, 4–16. doi:10.1016/j.cities.2017.12.004

Giffinger, R., Fertner, C., Kramar, H., Kalasek, R., Pichler-Milanović, N., & Meijers, E. (2007). *Smart Cities: Ranking of European Medium-Sized Cities. Centre of Regional Science (SRF)*. Vienna, Austria: Vienna University of Technology.

Gohar, M., Muzammal, M., & Rahman, A. U. (2018). SMART TSS: Defining transportation system behavior using big data analytics in smart cities. *Sustainable Cities and Society*, *41*, 114–119. doi:10.1016/j.scs.2018.05.008

Kim, T., Ramos, C., & Mohammed, S. (2017). Smart City and IoT. *Future Generation Computer Systems*, *76*, 159–162. doi:10.1016/j.future.2017.03.034

Komninos, N. (2018). Connected Intelligence in Smart Cities Shared, engagement and awareness spaces 4 innovation. URENIO Research, Aristotle University.

Kramers, A., Höjer, M., Lövehagen, N., & Wangel, J. (2014). Smart sustainable cities–exploring ICT solutions for reduced energy use in cities. *Environmental Modelling & Software*, *56*, 52–62. doi:10.1016/j.envsoft.2013.12.019

Kumar, H., Singh, M. K. S., Gupta, M. P., & Madaan, J. (2018). Moving towards smart cities: Solutions that lead to the Smart City Transformation Framework. *Technological Forecasting and Social Change*. doi:10.1016/j.techfore.2018.04.024

Kumar Sharma, P., & Park, J. H. (2018). Blockchain based hybrid network architecture for the smart city. *Future Generation Computer Systems*, *86*, 650–655. doi:10.1016/j.future.2018.04.060

Martin, C. J., Evans, J., & Karvonen, A. (2018). Smart and sustainable? Five tensions in the visions and practices of the smart-sustainable city in Europe and North America. *Technological Forecasting and Social Change*, *133*, 269–278. doi:10.1016/j.techfore.2018.01.005

Moriarty, P., & Honnery, D. (2015). Future cities. *Future*, *66*, 45–53. doi:10.1016/j.futures.2014.12.009

Navarra, D. D. (2013). Perspectives on the evaluation of Geo-ICT for sustainable urban governance: Implications for e-government policy. *URISA Journal*, *25*(1), 19–29.

Ojo, A., Curry, E., & Janowski, T. (2014). *Designing Next Generation Smart City Initiatives Harnessing Findings and Lessons From a Study of Ten Smart City Programs*. Academic Press.

Osman, A. M. S. (2018). A novel big data analytics framework for smart cities. *Future Generation Computer Systems*. doi:10.1016/j.future.2018.06.046

Palomo-Navarro, A., & Navío-Marco, J. (2017). Smart city networks' governance: The Spanish smart city network case study. *Telecommunications Policy*, *xxx*, 1–9.

Sertyesilisik, B., & Sertyesilisik, E. (2015). Sustainability Leaders for Sustainable Cities. In Leadership and Sustainability in the Built Environment. Spon Research, Routledge.

Sertyesilisik, B., & Sertyesilisik, E. (2016). Eco industrial Development: As a Way of Enhancing Sustainable Development. *Journal of Economic Development Environment and People*, *5*(1), 6–27. doi:10.26458/jedep.v5i1.133

Silva, B. N., Khan, M., & Ha, K. (2018). Towards sustainable smart cities: A review of trends, architectures, components, and open challenges in smart cities. *Sustainable Cities and Society*, *38*, 697–713. doi:10.1016/j.scs.2018.01.053

The Economist Intelligence Unit. (2018). The Global Liveability Index 2018 A free overview. *The Economist*.

Trencher, G. (2018). Towards the smart city 2.0: Empirical evidence of using smartness as a tool for tackling social challenges. *Technological Forecasting and Social Change*. doi:10.1016/j.techfore.2018.07.033

United Nations. (2014). *The future we want: Sustainable cities*. Retrieved from http://www.un.org/en/sustainablefuture/cities.shtml#facts

Yigitcanlar, T., & Kamruzzaman, M. (2018). Does smart city policy lead to sustainability of cities? *Land Use Policy*, *73*, 49–58. doi:10.1016/j.landusepol.2018.01.034

Youtube website. (n.d.a). *Anakata Wind Power Resources - Innovative Wind Turbine Technology*. Retrieved from https://www.youtube.com/watch?v=Rhh1zWM6SiQ

Youtube website. (n.d.b). *Baker Turbo-Vortex Wind Turbine Turbina*. Retrieved from https://www.youtube.com/watch?v=wTeiSHpbFt4

Youtube website. (n.d.c). *4 Most Popular Vertical Wind Turbines*. Retrieved from https://www.youtube.com/watch?v=a3n-VBpcqzM

Youtube website. (n.d.d). *Funnel wind turbine: radical new design harnesses 600% more electricity from wind – TomoNews*. Retrieved from https://www.youtube.com/watch?v=im8W4z4og-8

Youtube website. (n.d.e). *Future of Wind Energy - new Vertical axis Wind Turbine invention*. Retrieved from https://www.youtube.com/watch?v=Bmz_YZOWdV8

Youtube website. (n.d.f). *Floating wind turbine takes to the sky - BBC Click*. Retrieved from https://www.youtube.com/watch?v=RzCK9Ht0SWk

Youtube website. (n.d.g). *Heppolt Wind Turbine Progress Report - Wind Turbine New 2015*. Retrieved from https://www.youtube.com/watch?v=IaplRH7ldzQ

Youtube website. (n.d.h). *Wind Tulip in Jerusalem*. Retrieved from https://www.youtube.com/watch?v=28ok_7bSFHc

Youtube website. (n.d.i). *Introducing the Altaeros BAT: The Next Generation of Wind Power*. Retrieved from https://www.youtube.com/watch?v=kldA4nWANA8

Youtube website. (n.d.j). *New Wind Power System: Polish engineers develop more efficient wind turbine system*. Retrieved from https://www.youtube.com/watch?v=61Ekas-xbfU

Youtube website. (n.d.k). *New Wind Turbine FloDesign*. Retrieved from https://www.youtube.com/watch?v=WB5CawKfE2M

Youtube website. (n.d.l). *SeaTwirl puts a new spin on offshore wind turbines*. Retrieved from https://www.youtube.com/watch?v=Ccs3RP9LxIY

Youtube website. (n.d.m). *Sky Wolf Wind Turbine Animation 2014*. Retrieved from (https://www.youtube.com/watch?v=jGTO886FKMA)

Youtube website. (n.d.n). *Urban Power USA new 10 KW vertical wind turbine*. Retrieved from https://www.youtube.com/watch?v=oIAq3wvVOBA

Youtube website. (n.d.o). *Windjuicer a new wind power technology*. Retrieved from https://www.youtube.com/watch?v=NgOjbNy2HFk

Youtube website. (n.d.p). *WindTamer Turbines WindTamerTurbines.com*. Retrieved from https://www.youtube.com/watch?v=mpUPlHx_2gw

Youtube website. (n.d.q). *windtrap - a new wind turbine*. Retrieved from https://www.youtube.com/watch?v=ldGafXZdvlo

Youtube website. (n.d.r). *Wind turbines of the future*. Retrieved from https://www.youtube.com/watch?v=18ogee_Gj7k

KEY TERMS AND DEFINITIONS

Big Data: Sum of the digital data gathered.

Circular City: Cities that have minimized their waste generation due to their operation based on the circular economy principles.

Connected City: City digitally connected at all levels supporting efficient and effective operation.

Internet of Things: Network emerged through connected electronic devices.

Resilient City: Cities capable of resisting shocks (e.g., disasters) and able to restore and recover themselves in the post-shocks phase as well as cities capable of adopting themselves to the changing climatic conditions.

Smart City: Cities operating as well as solving their existing and potential problems with the help of smart technologies supported by big data, IoT, and ICT.

Smart City Governance: Interactive governance based on the contribution of the citizens to the government processes enabled by smartness of the cities.

Sustainable City: Environmentally friendly city that has lowered its environmental footprints.

Chapter 3
Smart Internet of Things (IoT) Applications

Rahul Verma
Mewar University, India

ABSTRACT

The internet of things (IoT) is the new buzzword in technological corridors with most technology companies announcing a smart device of sorts that runs on internet of things (IoT). Cities around the world are getting "smarter" every day through the implementation of internet of things (IoT) devices. Cities around the world are implementing individual concepts on their way to becoming smart. The services are automated and integrated end to end using internet of things (IoT) devices. The chapter presents an array of internet of things (IoT) applications. Also, cyber physical systems are becoming more vulnerable since the internet of things (IoT) attacks are common and threatening the security and privacy of such systems. The main aim of this chapter is to bring more research in the application aspects of smart internet of things (IoT).

INTRODUCTION

The Internet of Things (IoT) is the arrangement of home appliances, vehicles, physical devices, and other things implanted with connectivity, actuators, sensors, software, and electronics which empowers these things to collect, connect and exchange data (Brown, 2016; ITU, 2019; Hendricks, 2015), making openings for more coordinate integration of the physical world into computer - based frameworks, resulting in reduced human exertions, economic benefits, and efficiency improvements.

The number of Internet of Things (IoT) devices increased 31% year - over - year to 8.4 billion within the year 2017 (Köhn, 2018) and it is evaluated that there will be 30 billion gadgets by 2020 (Nordrum, 2016). The global market value of Internet of Things (IoT) is anticipated to reach $7.1 trillion by 2020 (Hsu & Lin, 2016).

Internet of Things (IoT) includes expanding web network beyond standard gadgets, such as tablets, smart phones, laptops, and desktops, to any extend of customarily stupid or non - internet - enabled physical gadgets and regular objects. Inserted with innovation, these gadgets can communicate and connect

DOI: 10.4018/978-1-5225-9199-3.ch003

over the web, and they can be remotely checked and controlled. With the entry of driverless vehicles, a department of Internet of Things (IoT), i.e. the Internet of Vehicle begins to pick up more consideration.

APPLICATIONS

The broad set of applications for Internet of Things (IoT) devices (Vongsingthong & Smanchat, 2014) is frequently partitioned into customer, commercial, industrial, and infrastructure spaces (Business Insider, 2015; Perera et al., 2015).

Consumer Applications

A developing portion of Internet of Things (IoT) gadgets are made for consumer use, including appliances with remote monitoring capabilities, connected health, wearable technology, home automation / smart home, and connected vehicles (Trak.in, 2016).

- **Smart Home:** Internet of Things (IoT) gadgets are a portion of the bigger concept of domestic mechanization, which can incorporate media and security systems, heating and air conditioning, lighting (Kang et al., 2017; Meola, 2016). Long term benefits might incorporate energy savings by automatically guaranteeing lights and electronics are turned off.
- **Elder Care:** One key application of smart home is to supply help for crippled and elderly people. These domestic frameworks utilize assistive innovation to suit an owner's particular disabilities (Demiris & Hensel, 2008). Voice control can help clients with mobility and sight restrictions whereas alert frameworks can be associated straightforwardly to Cochlear imparts worn by hearing impaired users (Aburukba et al., 2016). They can moreover be prepared with extra security highlights. These highlights can incorporate sensors that screen for medical crises such as falls or seizures (Mulvenna et al., 2017). Smart domestic innovation connected in this way can give clients with more opportunity and a better quality of life (Demiris & Hensel, 2008).

Commercial Applications

The term "Enterprise Internet of Things (EIoT)" alludes to gadgets utilized in corporate and business settings. By 2019, it is evaluated that EIoT will account for 9.1 billion devices (Business Insider, 2015).

- **Medical and Health Care:** The Internet of Medical Things (IoMT) (also called the Internet of Health Things (IoHT)) is an application of the Internet of Things (IoT) for health and medical related purposes, data analysis and collection for research, and monitoring. The futurologist's vision seems to be that before soon you may share your movement, heart rate, work out levels, and other essential information collected by your mobile gadget along with your specialist. "More and more care will be delivered outside clinics and hospitals". This implies mobile gadgets - from smart phones to monitoring gadgets - will become to be progressively vital as the number of patients cared for at domestic or in sheltered accommodation or other community centers increases. Internet of Things (IoT) gadgets can be utilized to empower remote health checking and emergency notification frameworks. These healths observing gadgets can extend from heart rate and blood pressure monitors to progressed gadgets able to checking specialized imparts, such as advanced hearing aids, Fit bit electronic wristbands, or pacemakers (Ersue et al., 2014). A few hospitals have started executing "smart beds" that can identify when they are possessed and when a patient is endeavoring to get up. It can moreover alter itself to guarantee suitable pressure and support is connected to the patient without the manual interaction of nurses (DaCosta et al., 2018). A 2015 Goldman Sachs report demonstrated that health care Internet of Things (IoT) gadgets "can save the United States more than $300 billion in yearly health care consumptions by decreasing cost and increasing revenue (Engage Mobile Blog, 2016; Roman & Conlee, 2015)." Later contributions indeed allude to Internet of Things (IoT) arrangements for medication as the Internet of Medical Things (IoMT).

Specialized sensors can moreover be prepared inside living spaces to screen the general well - being and health of senior citizens, whereas also guaranteeing that appropriate treatment is being managed and helping individuals recapture misplaced versatility by means of treatment as well (Istepanian et al., 2011). Other buyer gadgets to empower healthy living, such as wearable heart monitors or connected scales, are too a possibility with the Internet of Things (IoT) (Swan, 2012). End - to - end health observing Internet of Things (IoT) platforms are moreover accessible for antenatal and unremitting patients, helping one oversee wellbeing vitals and repeating medication necessities (IJSMI, 2018).

The Research & Development Corporation (DEKA), a company that makes prosthetic limbs, has made a battery - powered arm that uses myoelectricity, a gadget that converts muscle group sensations into motor control. The arm is nicknamed Luke Arm after Luke Skywalker (Star Wars).

- **Transportation:** The Internet of Things (IoT) can help within the integration of communications, control, and data preparing over different transportation frameworks. Application of the Internet of Things (IoT) expands to all angles of transportation frameworks (i.e. the user or driver, the infrastructure, and the vehicle (Mahmud et al., 2018)). Dynamic interaction between these components of a transport framework empowers intra and inter vehicular communication (Xie & Wang, 2017), safety and road assistance, vehicle control, logistic and fleet management, electronic toll collection systems, smart parking, and savvy traffic control (Ersue et al., 2014; Xie, 2016). In Logistics and Fleet Management for instance, the Internet of Things (IoT) platform can persistently screen the conditions and location of assets and cargo through wireless sensors and send particular alerts when administration exceptions happen (burglaries, harms, delays, etc.). In case combined with Machine Learning then it also helps in lessening traffic accidents by presenting laziness alarms to drivers and giving self driven cars too.
- **Building and Home Automation:** Internet of Things (IoT) gadgets can be utilized to control and monitor the electronic, electrical and mechanical frameworks utilized in different sorts of buildings (e.g., residential, institutions, industrial, or public and private) (Ersue et al., 2014) in building automation and home automation frameworks. In this regard, three fundamental areas are being covered in literature: (Haase et al., 2016)
  - The integration of the Internet with building energy administration frameworks in order to form energy efficient and Internet of Things (IoT) driven "smart buildings" (Haase et al., 2016).
  - The conceivable means of real - time checking for reducing energy utilization (Karlgren et al., 2008) and observing tenant behaviors (Haase et al., 2016).
  - The integration of smart gadgets within the built environment and how they might to know who to be utilized in future applications (Haase et al., 2016).

Industrial Applications

- **Manufacturing:** The Internet of Things (IoT) can realize the consistent integration of different manufacturing gadgets prepared with networking, actuation, communication, processing, identification, and sensing capabilities. Based on such a highly integrated smart cyber physical space, it opens the entryway to make whole new market and business openings for manufacturing (Yang et al., 2018). Network control and administration of manufacturing process control, asset and situation management, or manufacturing equipment bring the Internet of Things (IoT) inside the domain of industrial applications and smart manufacturing as well (Severi et al., 2014). The Internet of Things (IoT) intelligent frameworks empower real-time optimization of manufacturing production and supply chain networks, dynamic response to product demands, and rapid manufacturing of new products, by networking control systems, sensors, and machinery together (Ersue et al., 2014).

Digital control frameworks to automate service information systems, operator tools, and process controls to optimize plant security and safety are in the domain of the Internet of Things (IoT) (Gubbi et al., 2013). But it also expands itself to measurements to maximize reliability, statistical evaluation, and asset management via predictive maintenance (Tan & Wang, 2010). Smart industrial administration frameworks can too be coordinated with the Smart Network, subsequently empowering real - time energy optimization. Health and safety management, plant optimization, automated controls, measurements, and other functions are given by an expansive number of networked sensors (Ersue et al., 2014).

The term Industrial Internet of things (IIoT) is regularly experienced within the manufacturing businesses, alluding to the mechanical subset of the Internet of Things (IoT). IIoT in manufacturing might create so much business esteem that it'll inevitably lead to the 4th industrial transformation, so the so - called Industry 4.0. It is assessed that within the future, effective companies will be able to increase their income through Internet of Things (IoT) by making new business models and transform workforce, exploit analytics for innovation, and improve productivity (Daugherty et al., 2016). The potential of growth by actualizing IIoT will produce $12 trillion of worldwide GDP by 2030 (Daugherty et al., 2016).

While data acquisition and connectivity are basic for IIoT, they ought to not be the reason, or maybe the establishment and way to something greater. Among all the innovations, predictive maintenance is likely a moderately "easier win" since it is pertinent to existing management systems and assets. The objective of intelligent maintenance systems is to increase productivity and reduce unexpected downtime. And to realize that alone would produce around up to 30% over the total maintenance costs (Seebo Blog, 2018). Industrial big data analytics will play a crucial role in manufacturing asset predictive maintenance, in spite of the fact that it is not the only capability of industrial big data (Lee, 2015; Accenture, 2016). Cyber - Physical Systems (CPS) is the core technology of industrial big data and it'll be an interface between human and the cyber world. Cyber - Physical Systems (CPS) can be designed by following the 5C (Configuration, Cognition, Cyber, Conversion, Connection) design (Lee et al., 2015), and it'll change the collected information into noteworthy data, and eventually interfere with the physical assets to optimize processes. An Internet of Things (IoT) - enabled intelligent framework of such cases was proposed in 2001 and afterward illustrated in 2014 by the National Science Foundation Industry / University Collaborative Research Center for Intelligent Maintenance Systems (IMS) at the University of Cincinnati on a band saw machine in IMTS 2014 in Chicago (IMS Center, 2016; Lee, 2003; Lee, 2014). Band saw machines are not essentially costly, but the band saw belt costs are gigantic since they corrupt much quicker. In any case, without sensing and intelligent analytics, it can be only decided by encounter when the band saw belt will really break. The created prognostics framework will be able to recognize and screen the debasement of band saw belts even if the condition is changing, exhorting clients when is the most excellent time to supplant the belt. This will essentially progress client involvement and operator security and eventually spare on costs (Lee, 2014).

- **Agriculture:** There are various Internet of Things (IoT) applications in farming (Meola, 2016) such as collecting information on soil content, pest infestation, wind speed, humidity, rainfall, and temperature. This information can be utilized to robotize cultivating strategies, take informed decisions to reduce effort required to manage crops, minimize waste and risk, and improve quantity and quality. For instance, agriculturists can presently screen soil moisture and temperature from a remote place, and indeed apply Internet of Things (IoT) - acquired information to precision fertilization programs (Zhang, 2015).

In August 2018, Toyota Tsusho started a partnership with Microsoft to make fish farming instruments utilizing the Microsoft Azure application suite for Internet of Things (IoT) technologies related to water management. Created in portion by analysts from Kindai University, the water pump instruments utilize artificial intelligence to count the number of fish on a conveyor belt, analyze the number of fish, and derive the adequacy of water stream from the information the fish give. The specific computer programs utilized within the process fall beneath the Azure Machine Learning and the Azure Internet of Things (IoT) Hub stages (Quach, 2018).

Infrastructure Applications

Monitoring and controlling operations of feasible rural and urban infrastructures like railway tracks, bridges, on - and offshore - wind - farms is a key application of the Internet of Things (IoT) (Gubbi et al., 2013). The Internet of Things (IoT) framework can be utilized for checking any occasions or changes in basic conditions that can compromise security and increment risk. Internet of Things (IoT) can advantage the construction industry by increase in productivity, paperless workflow, better quality workday, time reduction and cost saving. It can assist in taking quicker decisions and save cash with Real - Time Data Analytics. It can also be utilized for planning repair and maintenance activities in a proficient way, by planning assignments between diverse service users and providers of these facilities (Ersue et al., 2014). Internet of Things (IoT) gadgets can also be used to control critical infrastructure like bridges to give access to ships. Utilization of Internet of Things (IoT) gadgets for observing and working infrastructure is likely to improve occurrence management and crisis response coordination, and quality of service, up - times and diminish costs of operation in all infrastructure related zones (Chui et al., 2014). Indeed areas such as waste management can advantage (Postscapes, 2014) from computerization and optimization that may be brought in by the Internet of Things (IoT) (Mukherjee, 2017).

- **Metropolitan Scale Deployments:** There are a few planned or continuous large - scale deployments of the Internet of Things (IoT), to empower better administration of cities and systems. For illustration, Songdo, South Korea, the first of its kind completely equipped and wired smart city is slowly being built, with around 70% of the business district completed as of June 2018. Much of the city is arranged to be wired and mechanized, with small or no human mediation (Poon, 2018).

Another application may be a currently experiencing venture in Santander, Spain. For this deployment, two approaches have been adopted. This city of 180,000 inhabitants has as of now seen 18,000 downloads of its city smart phone app. The app is connected to 10,000 sensors that empower services like digital city agenda, environmental monitoring, parking search, and more. City context data is utilized in this deployment so as to advantage vendors through a spark deals component based on city behavior that points at maximizing the effect of each notice (Rico, 2014).

Other illustrations of large - scale deployments underway incorporate the Sino - Singapore Guangzhou Knowledge City (Sino - Singapore Guangzhou Knowledge City, 2014); work on increasing transportation efficiency, reducing noise pollution, and improving air and water quality in San Jose, California (Intel Newsroom, 2014); and smart traffic administration in western Singapore (Coconuts Singapore, 2014). French company, Sigfox, commenced building an ultra - narrowband wireless information network within the San Francisco Bay Area in 2014, the primary business to realize such a deployment within the U.S. (Lipsky, 2015; Alleven, 2014) It in this way declared it would set up a total of 4000 base stations to cover a total of 30 cities within the U.S. by the end of 2016, making it the biggest Internet of Things (IoT) network coverage supplier within the nation thus far (Merritt, 2015; Fitchard, 2014).

Another case of a huge deployment is the one completed by New York Water Ways (NYWW) in New York City to associate all the city's vessels and be able to screen them live 24 / 7. The network was engineered and designed by Fluidmesh Networks, a Chicago - based company creating wireless systems for critical applications. The New York Water Ways (NYWW) network is right now giving coverage on the Hudson River, East River, and Upper New York Bay. With the wireless network in place, New York Water Ways (NYWW) is able to require control of its fleet and travelers in a way that was not previously possible. New applications can incorporate paperless ticketing, public Wi - Fi, digital signage, energy and fleet management, security and others (Securityinfowatch.com, 2012).

- **Energy Management:** Significant numbers of energy - consuming gadgets (e.g. televisions, bulbs, power outlets, switches, etc.) already coordinate Internet connectivity, which can permit them to communicate with utilities to adjust energy usage and power generation (Parello et al., 2014) and optimize energy utilization as a whole (Ersue et al., 2014). These gadgets permit for remote control by clients, or central management through a cloud - based interface, and empower capacities like scheduling (e.g., changing lighting conditions, controlling ovens, remotely powering on or off heating systems, etc.) (Ersue et al., 2014). The Smart grid is a utility - side Internet of Things (IoT) application; frameworks accumulate and act on energy and power - related data to improve the productivity of the generation and dispersion of electricity (Parello et al., 2014). Utilizing Advanced Metering Infrastructure (AMI) Internet - connected gadgets, electric utilities not only collect information from end - users, but also oversee distribution mechanization gadgets like transformers (Ersue et al., 2014).
- **Environmental Monitoring:** Environmental monitoring applications of the Internet of Things (IoT) ordinarily utilize sensors to help in environmental security (Davies, 2015) by checking water or air quality, soil or atmospheric conditions (Li et al., 2011), and can indeed incorporate regions like checking the developments of wildlife and their habitats (FIT French Project, 2014). Improvement of resource - constrained gadgets associated to the Internet also implies that other applications like tsunami or earthquake early - warning frameworks can moreover be utilized by emergency services to supply more compelling help. Internet of Things (IoT) gadgets in this ap-

plication ordinarily spans a huge geographic zone and can also be mobile (Ersue et al., 2014). It has been contended that the standardization Internet of Things (IoT) brings to wireless detecting will revolutionize this range (Hart & Martinez, 2015).

REFERENCES

Aburukba, R., Al-Ali, A. R., Kandil, N., & AbuDamis, D. (2016). Configurable ZigBee - based control system for people with multiple disabilities in smart homes. IEEE.

Accenture. (2016). *Industrial Internet Insights Report*. Author.

Alleven, M. (2014). Sigfox launches IoT network in 10 UK cities. *Fierce Wireless Tech*. Retrieved from https://www.fiercewireless.com/tech/sigfox-launches-iot-network-10-uk-cities

Brown, E. (2016). *21 Open Source Projects for IoT*. Retrieved from Linux.com

Business Insider. (2015). *The Enterprise Internet of Things Market*. Retrieved from https://www.businessinsider.in/The-Corporate-Internet-Of-Things-Will-Encompass-More-Devices-Than-The-Smartphone-And-Tablet-Markets-Combined/articleshow/45483725.cms

Chui, M., Loffler, M., & Roberts, R. (2014). The Internet of Things. *McKinsey Quarterly*. Retrieved from https://www.mckinsey.com/industries/high-tech/our-insights/the-internet-of-things

Coconuts Singapore. (2014). *Western Singapore becomes test-bed for smart city solutions*. Retrieved from https://coconuts.co/singapore/news/western-singapore-becomes-test-bed-smart-city-solutions

DaCosta, C. A., Pasluosta, C. F., Eskofier, B., DaSilva, D. B., & DaRosaRighi, R. (2018). Internet of Health Things: Toward intelligent vital signs monitoring in hospital wards. *Artificial Intelligence in Medicine, 89*, 61 - 69.

Daugherty, P., Negm, W., Banerjee, P., & Alter, A. (2016). *Driving Unconventional Growth through the Industrial Internet of Things*. Accenture.

Davies, N. (2015). *How the Internet of Things will enable 'smart buildings*. Extreme Tech.

Demiris, G., & Hensel, K. (2008). Technologies for an Aging Society: A Systematic Review of 'Smart Home' Applications. *IMIA Yearbook of Medical Informatics, 2008*, 33–40.

Engage Mobile Blog. (2016). *Goldman Sachs Report: How the Internet of Things Can Save the American Healthcare System $305 Billion Annually*. Engage Mobile Solutions, LLC. Retrieved https://www.engagemobile.com/goldman-sachs-report-how-the-internet-of-things-can-save-the-american-healthcare-system-305-billion-annually

Ersue, M., Romascanu, D., Schoenwaelder, J., & Sehgal, A. (2014). *Management of Networks with Constrained Devices: Use Cases*. IETF Internet Draft.

FIT French Project. (2014). *Use case: Sensitive wildlife monitoring*. Author.

Fitchard, K. (2014). Sigfox brings its internet of things network to San Francisco. *Gigaom.* Retrieved from https://gigaom.com/2014/05/20/sigfox-brings-its-internet-of-things-network-to-san-francisco

Gubbi, J., Buyya, R., Marusic, S., & Palaniswami, M. (2013). Internet of Things (IoT): A vision, architectural elements, and future directions. *Future Generation Computer Systems, 29*(7), 1645 - 1660.

Haase, J., Alahmad, M., Nishi, H., Ploennigs, J., & Tsang, K. F. (2016). The IOT mediated built environment: A brief survey. *IEEE 14th International Conference on Industrial Informatics (INDIN),* 1065 - 1068. 10.1109/INDIN.2016.7819322

Hart, J. K., & Martinez, K. (2015). Toward an environmental Internet of Things. *Earth & Space Science, 2*(5), 194 - 200.

Hendricks, D. (2015). *The Trouble with the Internet of Things. London Data store.* Greater London Authority. Retrieved https://data.london.gov.uk/blog/the-trouble-with-the-internet-of-things

Hsu, C. L., & Lin, J. C. C. (2016). An empirical examination of consumer adoption of Internet of Things services: Network externalities and concern for information privacy perspectives. *Computers in Human Behavior, 62*, 516–527. doi:10.1016/j.chb.2016.04.023

IJSMI. (2018). Overview of recent advances in Health care technology and its impact on health care delivery. *International Journal of Statistics and Medical Informatics, 7*, 1 - 6.

IMS Center, . (2016). *Center for Intelligent Maintenance Systems. Author.*

Intel Newsroom. (2014). *San Jose Implements Intel Technology for a Smarter City*. Retrieved from https://newsroom.intel.com/news-releases/san-jose-implements-intel-technology-for-a-smarter-city/#gs.6AIlWLiP

Istepanian, R., Hu, S., Philip, N., & Sungoor, A. (2011). The potential of Internet of m-health Things "m - IoT" for non - invasive glucose level sensing. *Annual International Conference of the IEEE Engineering in Medicine and Biology Society (EMBC)*, 5264 - 6.

ITU. (2019). *Internet of Things Global Standards Initiative*. Retrieved from https://www.itu.int/en/ITU-T/gsi/iot/Pages/default.aspx

Kang, W. M., Moon, S. Y., & Park, J. H. (2017). An enhanced security framework for home appliances in smart home. *Human-Centric Computing and Information Sciences, 7*(6).

Karlgren, J., Fahlén, L., Wallberg, A., Hansson, P., Ståhl, O., Söderberg, J., & Åkesson, K. P. (2008). Socially Intelligent Interfaces for Increased Energy Awareness in the Home. The Internet of Things. Lecture Notes in Computer Science, 4952, 263 - 275. doi:10.1007/978-3-540-78731-0_17

Köhn, R. (2018). *Corporations are joining forces against hackers*. Retrieved from https://www.faz.net/aktuell/wirtschaft/diginomics/grosse-internationale-allianz-gegen-cyber-attacken-15451953-p2.html?printPagedArticle=true#pageIndex_1

Lee, J. (2003). E - manufacturing - fundamental, tools and transformation. *Robotics and Computer - Integrated Manufacturing. Leadership of the Future in Manufacturing, 19*(6), 501 - 507.

Lee, J. (2014). Keynote Presentation: Recent Advances and Transformation Direction of PHM. *Road mapping Workshop on Measurement Science for Prognostics and Health Management of Smart Manufacturing Systems Agenda.*

Lee, J. (2015). *Industrial Big Data.* Mechanical Industry Press.

Lee, J., Bagheri, B., & Kao, H. A. (2015). A cyber - physical systems architecture for industry 4.0 - based manufacturing systems. *Manufacturing Letters, 3*, 18 - 23.

Li, S., Wang, H., Xu, T., & Zhou, G. (2011). Application Study on Internet of Things in Environment Protection Field. Lecture Notes in Electrical Engineering, 133, 99 - 106. doi:10.1007/978-3-642-25992-0_13

Lipsky, J. (2015). IoT Clash Over 900 MHz Options. *EETimes*. Retrieved from https://www.eetimes.com/document.asp?doc_id=1326599

Mahmud, K., Town, G. E., Morsalin, S., & Hossain, M. J. (2018). Integration of electric vehicles and management in the internet of energy. *Renewable and Sustainable Energy Reviews, 82*, 4179 - 4203.

Meola, A. (2016a). How IoT & smart home automation will change the way we live. *Business Insider*. Retrieved from https://www.businessinsider.com/internet-of-things-smart-home-automation-2016-8?IR=T

Meola, A. (2016b). Why IoT, big data & smart farming are the future of agriculture. *Business Insider.* Retrieved from https://www.businessinsider.com/internet-of-things-smart-agriculture-2016-10?IR=T

Merritt, R. (2015). 13 Views of IoT World. *EETimes*. Retrieved from https://www.eetimes.com/document.asp?doc_id=1326596

Mukherjee, B. (2017). The internet of things (IOT): Revolutionized the way we live. *Postscapes*. Retrieved from http://www.bizmak.xyz/internet-of-things-iot-revolutionized-the-way-we-live

Mulvenna, M., Hutton, A., Martin, S., Todd, S., Bond, R., & Moorhead, A. (2017). Views of Caregivers on the Ethics of Assistive Technology Used for Home Surveillance of People Living with Dementia. *Neuroethics, 10*(2), 255 - 266.

Nordrum, A. (2016). Popular Internet of Things Forecast of 50 Billion Devices by 2020 Is Outdated. *IEEE*. Retrieved from https://spectrum.ieee.org/tech-talk/telecom/internet/popular-internet-of-things-forecast-of-50-billion-devices-by-2020-is-outdated

Parello, J., Claise, B., Schoening, B., & Quittek, J. (2014). *Energy Management Framework*. IETF Internet Draft <draft-ietf-eman-framework-19>.

Perera, C., Liu, C. H., & Jayawardena, S. (2015). The Emerging Internet of Things Marketplace From an Industrial Perspective: A Survey. *IEEE Transactions on Emerging Topics in Computing, 3*(4), 585 - 598.

Poon, L. (2018). Sleepy in Songdo, Korea's Smartest City. City Lab. *Atlantic Monthly Group*. Retrieved from https://www.citylab.com/life/2018/06/sleepy-in-songdo-koreas-smartest-city/561374

Postscapes. (2014). *Smart Trash*. Retrieved from https://www.postscapes.com/smart-trash

Quach, K. (2018). *Google goes bilingual, Face book fleshes out translation and Tensor Flow is dope - And, Microsoft is assisting fish farmers in Japan*. Retrieved from https://www.theregister.co.uk/2018/09/01/ai_roundup_310818

Rico, J. (2014). *Going beyond monitoring and actuating in large scale smart cities*. NFC & Proximity Solutions - WIMA Monaco.

Roman, D. H., & Conlee, K. D. (2015). *The Digital Revolution Comes to US Healthcare*. Goldman Sachs.

Securityinfowatch.com. (2012). *STE Security Innovation Awards Honorable Mention: The End of the Disconnect*. Retrieved from https://www.securityinfowatch.com/video-surveillance/video-transmission-equipment/article/10840006/innovative-wireless-network-connects-new-york-waterways-ferries-for-safety-security-roi-and-more

Seebo Blog. (2018). *What's So Smart About Intelligent Maintenance Systems*. Retrieved from https://blog.seebo.com/whats-so-smart-about-intelligent-maintenance-systems

Severi, S., Abreu, G., Sottile, F., Pastrone, C., Spirito, M., & Berens, F. (2014). M2M Technologies: Enablers for a Pervasive Internet of Things. *The European Conference on Networks and Communications (EUCNC2014)*. 10.1109/EuCNC.2014.6882661

Sino - Singapore Guangzhou Knowledge City. (2014). *A vision for a city today, a city of vision tomorrow*. Retrieved from http://www.ssgkc.com/index.asp

Swan, M. (2012). Sensor Mania! The Internet of Things, Wearable Computing, Objective Metrics, and the Quantified Self 2.0. *Sensor and Actuator Networks, 1*(3), 217 - 253.

Tan, L., & Wang, N. (2010). Future Internet: The Internet of Things. *3rd International Conference on Advanced Computer Theory and Engineering (ICACTE)*, 5, 376 - 380.

Trak.in. (2016). *How IoT's are Changing the Fundamentals of "Retailing"*. Retrieved from https://trak.in/tags/business/2016/08/30/internet-of-things-iot-changing-fundamentals-of-retailing

Vongsingthong, S., & Smanchat, S. (2014). *Internet of Things: A review of applications & technologies. Suranaree Journal of Science and Technology*.

Xie, X. F. (2016). *Key Applications of the Smart IoT to Transform Transportation*. Retrieved from http://www.wiomax.com/what-can-the-smart-iot-transform-transportation-and-smart-cities

Xie, X. F., & Wang, Z. J. (2017). *Integrated in - vehicle decision support system for driving at signalized intersections: A prototype of smart IoT in transportation*. Transportation Research Board (TRB) *Annual Meeting*, Washington, DC.

Yang, C., Shen, W., & Wang, X. (2018). The Internet of Things in Manufacturing: Key Issues and Potential Applications. *IEEE Systems, Man, and Cybernetics Magazine, 4*(1), 6 - 15.

Zhang, Q. (2015). *Precision Agriculture Technology for Crop Farming*. CRC Press. doi:10.1201/b19336

Chapter 4
Smart IoT for Smart Cities Implementation:
Applications in Nutshell

V. V. Satyanarayana Tallapragada
Sree Vidyanikethan Engineering College, India

ABSTRACT

Internet of things (IoT) is the current area of research that allows heterogeneous devices to have a homogeneous connectivity based on the designed and desired application of the user. With the latest development in connectivity via smart phones, there is an exponential increase in users who access internet. However, various applications have already been designed based on the user's requirement. Therefore, this chapter intends to provide a detailed view on applications on IoT. Industrial applications help in monitoring the machinery so that production increases with minimum chaos if any error occurs. Safety helmet for mining based on IoT is used to measure the gas and temperature levels in the coal mines. Garbage management system is used for monitoring and clearing of dust bins. IoT-based domestic applications help users to have a better access over the equipment they use. As a business application, emotion analysis is performed to obtain the customers mood while shopping. Monitoring of crops from a remote location is another application which provides data on the health of the crop.

INTRODUCTION

With the development of technology, devices that connect to internet are increasing in an exponential manner. These devices in particular smart phones are having various built-in sensors that acquire data on a timely basis. Data acquired is processed within the system and results are provided to the user based on the apps that are being designed, developed, deployed based on the users need and intention. But, such data processing may cause overhead to the existing processor wherein the system may hang or may shutdown if all the apps continue to process the data at a time. Hence, there is a requirement for the development of new technique for processing of such data. In this juncture, sensor hub[Benini, L. (2013)] has been introduced in the recent past for the processing of the data acquired from various

DOI: 10.4018/978-1-5225-9199-3.ch004

Figure 1. Architecture of IoT

sensors. Sensor hub is a unit which combines data from different sensors and processes them such that the overhead on the main unit reduces. Hence, there is a rapid transformation from the devices which are manually controlled to the control using Internet based applications revealing the fact of the newly developed technology IoT. This has found its footprints in the various fields like education, travel, industries, health, business etc[Kishore Kumar Reddy N. G. and Rajeshwari K.(2017)]. To build applications which are designed for testaments have an architecture which creates a framework for the entire process. Figure 1 shows the basic architecture of the IoT.

Users communicate via the sensors which acquire the data required for the user. These sensors are installed in the devices which are designed for a particular purpose. The acquired data may be processed and will be stored on the cloud for further processing. The users further acquire the data that is stored on the cloud which is used for further analysis. As an example, if an automobile is newly designed by a company. The company installs it and will make it drove from one end to the other end of a place which may be of thousands of kilometers distance apart. While the vehicle is on move, the company will be assessing the performance of the designed machine by plotting the data which is acquired on the fly. Every instant the parameters will be acquired by the inbuilt sensors which is also a part of the design. As the vehicle completes a roundtrip the entire design data will be available to the company which makes the engineer to apply some corrections to the design enabling the users who buy the vehicle to have a less problems. The acquired data will show the stochastic nature of the vehicle at various places and under several conditions. Hence, the constraint need to be considered here is that the vehicle must travel in different road, climatic conditions and under different voltage levels of battery as well as fuel. Then, such data will be useful the company to have a worst case assessment, as the user's drive is also unpredictable.

Sensors may be broadly classified as industrial and domestic application sensors. Sensor is basically a device that can detect any change in the measurand whether it may be of electrical quantity viz., voltage, current etc., physical quantity viz., pressure, velocity, distance etc., thus providing an acknowledgment for the change in the measurand in terms of the same or may be of other type if so converted. Practically, industrial applications require to have better precision and accuracy over the domestic application sensors. Generally, domestic application sensors are designed in such a way that they are less reliable when compared to the industrial sensors which provide high accurate recordings/measurements. It is not always true that domestic application sensors are not reliable. With the inclusion of sensors onboard of every mobile the gap between these two have been reduced. Industrial sensors have high temperature capability so that their operating temperature range from -40 to 85 degrees, having 12 to 16-bit resolution, withstanding against high level of dusty environmental conditions. The latest industrial sensors are having contact less operation as one of the feature making them to have long life over the contact based operation, thus providing remote acquisition. In contrast to these domestic application sensors can be maintained at a less operating temperature and can also be maintained in non-hostile environment.

The data acquired from sensors is stored on the cloud which must be readily accessible to the user such that user can take appropriate action based on the system alert. Here arises a question “What is the need for cloud and What actually a cloud is?”. In general, cloud refers to the services that being hosted on to a server which is connected via internet. These services may differ based on the application and the user viz., storage or accessing of applications. Hence, cloud provides access to different applications wherein a remote user can also access as well as store the data. Computing using the cloud services is nothing but cloud computing, which provides access to various applications as well as storage enabling user’s system to have multiple capabilities which is similar to the usage of electricity without producing it and using it on a payment basis. Hence, most of cloud services are paid. A cloud service is an on-demand service that is provided to the user. There are various cloud service providers which are available online. The next sections of this chapter deals with the applications that are developed and deployed for domestic and industrial usage are clearly discussed.

SMART APPLICATIONS BASED ON IOT

1. IoT in Agriculture Sector

IoT has paved way to agriculture wherein the increasing population has to served with the increase in demand. Hence, there is a prime requirement to increase the production. As most of the population still depend on agriculture, there is a huge requirement for the development of new techniques to improve the productivity. Mainly, the crop yield depends on the weather conditions. The techniques that are developed and existing have constraint on these weather conditions. These weather conditions are of course predictable with the current day technology. But, if the growth of the crop as well as the condition of the crop is being viable then it will provide the farmer to take corrective actions against the upcoming threats due to the adverse weather conditions. In this module, an IoT application which is used in Agriculture is being discussed. Figure 2 shows the block diagram of a typical system which is used for telemetry of the crop[B. Dhanalaxmi and G. A. Naidu. (2017)].

The crop telemetry starts with acquiring the images from the crop viz., images of the leaf/crop. The image region is pre-processed for acquiring the region of interest. The region of interest may be homogeneous or non-homogeneous. Various segmentation techniques exist in literature which can be used for extraction of the region of interest. This region must be selected in such a way that it has to be processed by a pre-defined algorithm. The pre-processing stage also involves cropping the image to definite size so that any image which is acquired by a different camera can also be processed. The type of camera used for acquiring the image is of prominence here. The camera must be a high resolution camera so that each pixel will be used for a clear demarcation for the further segmentation process. In some of

Figure 2. Block diagram of typical system for crop telemetry

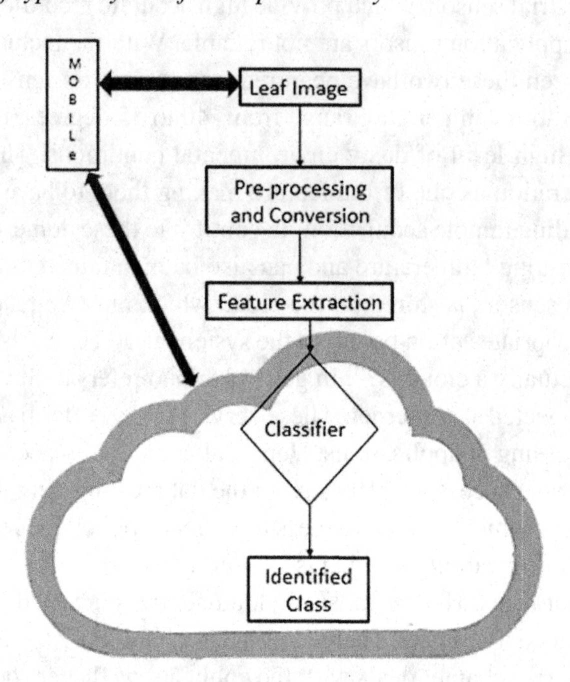

the applications, IR cameras are also used for acquiring the image. In this particular application, only a camera which acquires images in visible region are only used. The next step after pre-processing is extraction of the features from the acquired image. Features represent how a particular image is being described. For example, a leaf image can be described based on its color, which is the first and foremost classification feature which is normally used. Hence, a machine can also classify a basic leaf based on its color. Practically, all the leaf images may not be of green in color. Hence, in this application color models are used to extract the particular color components of the leaf images.

Figure 3 shows a leaf image. The leaf image is processed for various color model extraction which are further processed for feature extraction using wavelets or any other feature extraction application. After processing these images, different features are extracted using feature extraction techniques. In this particular application, wavelets are being used which decompose the image into various scales. These are further stored on the cloud for processing. Now such images are acquired on the fly and there may be as many images which can be trained on the fly such that as the number of images are increasing, number of test cases also increases, hence accuracy in recognition also increases. The number of features that are extracted must also be increased as the number of images increases which reduces the classification error. Hence, there is a need for selection of optimal features from the extracted features. Three types of classifiers are used in this particular application for classification of the trained images or leafs into diseased or non-diseased. Classifier is the one which recognizes the image based the features that are being extracted. There are basically two types of classifiers: Linear and Non-linear Classifiers. Kernel based classifiers are non linear classifiers which provide better accuracy as the number of test cases are increasing at a cost of computational complexity. Because of this reason it is intended to go for cloud based processing of the data reducing the computational complexity at the transmitting or receiving end. Figure 4 shows the performance of different classifiers in terms of recognition accuracy.

Figure 3. Brinjal Leaf Sample
(Courtesy: plantvillage.org)

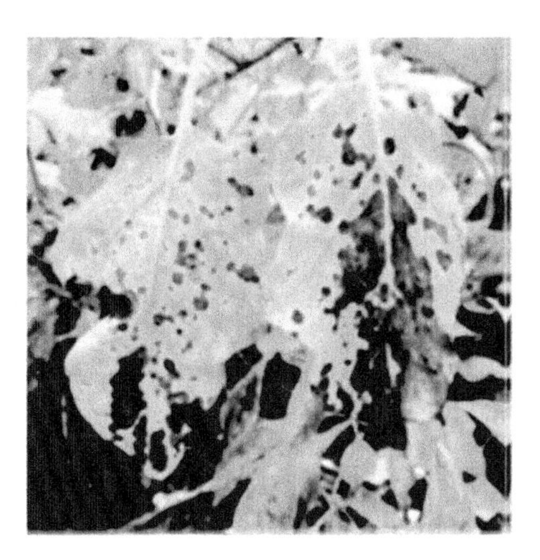

The number of test cases/classes are varied from 5 to 300 and the accuracy of the system is observed. It can be observed from the graph that Double Density dual tree complex Wavelet Transform (DDWT) with Euclidean distance based classifier outperformed the existing classifiers in terms of recognition accuracy[V. V. Satyanarayana Tallapragada, N. Ananda Rao and Satish Kanapala. (2017)]. Such a detection can also be extended to any crop based on the application. There are several types of fungal infections that can also occur in soyabean crop viz., downy mildew, frog eye and septoria leaf blight.

Figure 4. Performance of different classifiers with increase in number classes

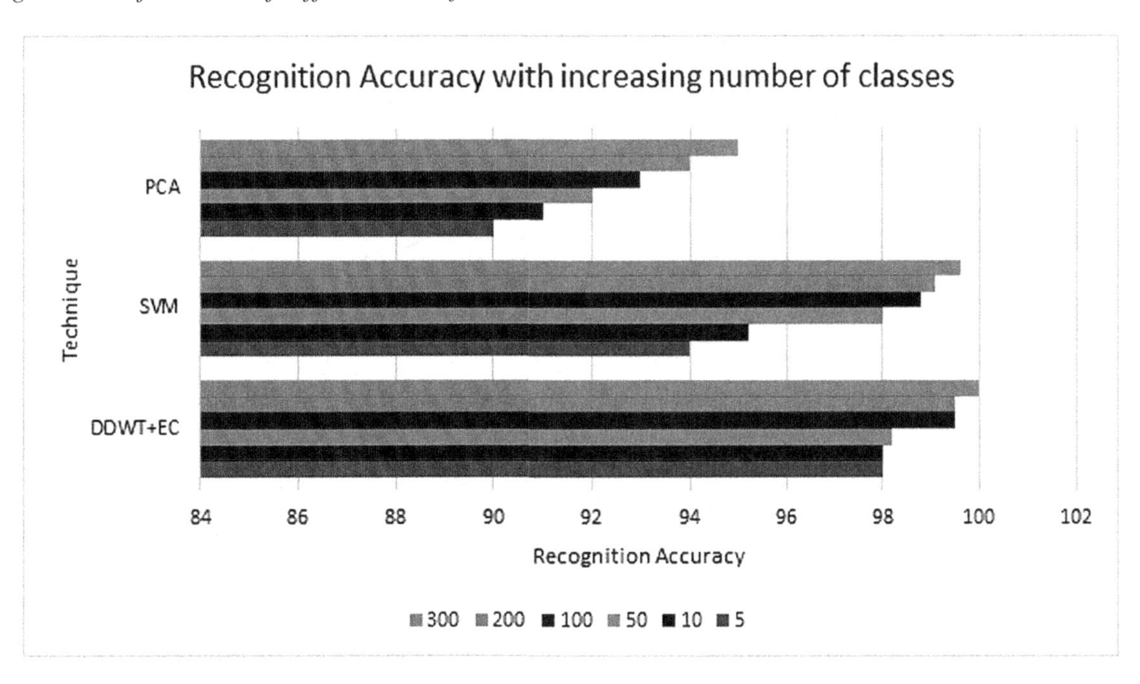

Figure 5. (a) Diseased leaf Image (c) Healthy Leaf (b) & (d) System response

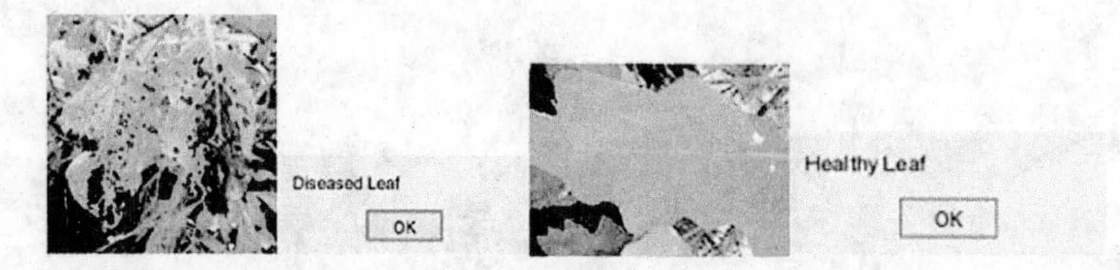

Using SVM classifier it is observed that approximately 90% classification accuracy is observed[Kaur, S., Pandey, S., & Goel, S. (2018)].

The system so designed processes the images and classifies them as diseased or non-diseased. Figure 5 shows the result displayed on the system.

2. IoT in Health Sector

IoT has made its way into the health sector enabling users to get required health monitoring parameters on the fly. As an example, a person can measure blood pressure by placing a watch on his wrist nowadays. The watch or the equipment consists of sensors which will be measure the systolic and diastolic values. The basic principle behind this measurement will be dealt later in this section. The basic architecture of mobile health using IoT is shown in the Figure 6 [S. H. Almotiri, M. A. Khan and M. A. Alghamdi (2016)].

The marked region of the figure 6 shows the mobile system or equipment which consists of inbuilt sensors that can measure various parameters like walking steps which is also called as pedometer. Pedometers are used by people who wants to reduce their Body Mass Index (BMI). It is also observed that the people who use these pedometers have reduced levels of blood pressure and BMI. The basic principle

Figure 6. Basic Architecture of Mobile Health Care System

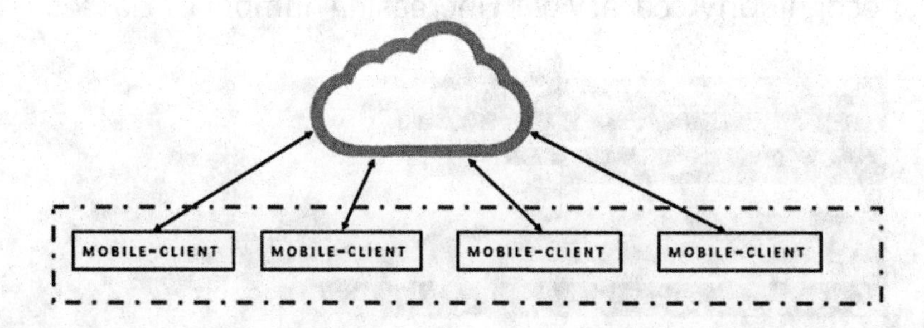

Figure 7. Accelerometer Internal structure

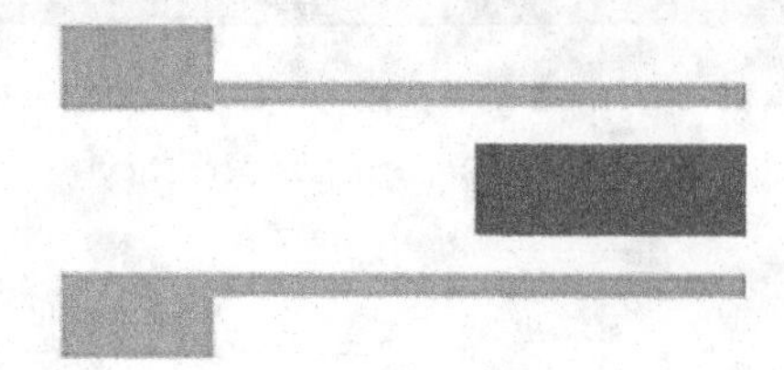

behind this pedometer is based on the Newton's II law of motion. There are two types of sensors used in pedometers. Mechanical and Electrical type of sensors. With the development of technology, electrical sensors have been widely used. These electrical sensors are accelerometers. The accelerometer work on capacitor principle. Figure 7 shows the basic construction of semiconductor based accelerometer.

The central beam which acts as a cantilever supported by the structure moves between the two electrodes – upper and lower electrode making the distance between the two electrodes to increase or decrease when a user moves his hand. The tilt in the body as well as hand is measured based on which the distance reduces or increases the capacitance value which is proportional to the distance travelled by the user. Every day the number steps is updated and saved on cloud for further analysis. User captures the daily as well as monthly report based on which corrective action will be taken to increase or decrease the number of daily walking steps. The number of steps that a user has taken till the start of the application will be displayed which can be set and rest as well. Mechanical sensors on the other hand cannot be used within small equipment like that of a watch which is a wearable device. Mechanical sensors contain a pendulum which oscillates and ticks every time when there is movement. Another type of mechanical sensor also uses a mass in conjunction with piezo electric crystal which produces a potential when there is movement of the user's hand. Such equipment cannot be miniaturized and also not rugged. Hence, it is obvious for the manufacturers to go with semiconductor based sensors which also provide long-life when compared to the mechanical based sensors.

As a proof of the application of pedometer which is intended for reduction in blood pressure, users will be trying to measure the blood pressure. Hence, as an additional application blood pressure measurement is also combined along with the same sensor which is included in the micro-chip. These sensors work on the principle of photoplethysmography. This technique is based on the light that senses the rate of flow of blood which is controlled by the pumping action of heart. Figure 8 shows the how light is focused to sense the heart beat from the human body which is installed in any equipment.

These lights actually on and off several times a minute. The light that is generated here is of green color which is absorbed by the blood. Red color is reflected by the blood hence looks red. Hence, the measurement of absorption between the heart beats is measured and displayed as a heart rate in pulses per minute. Similarly, in blood pressure is also measured, maximum and minimum values will be displayed on the application of the mobile. But it is observed that in most cases the actual measurement using sphygmomanometer will be ideal in contrast to the measurements done using such sensors. But it

Figure 8. PPG Sensor
[Peng, F., Wang, W., & Liu, H. (2014)]

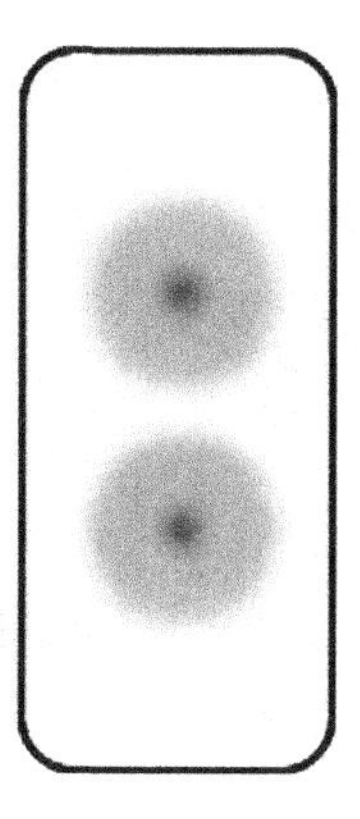

is quite necessary to upgrade and update with the latest development in technology. Hence, a 5% error is predicted in the measurement using these sensors.

Another application of IoT in health sector is the measurement of ECG signal[[Yasin, M., Tekeste, T., Saleh, H., Mohammad, B., Sinanoglu, O., & Ismail, M. (2017)]. ventricular arrhythmia in ECG signals is detected using an ASIC design. Such a design has provided 62.2% power consumption reduction and 16% area reduction. These designs are required for the latest IoT devices where the power requirements are of the major concern.

3. IoT in Coal Mines

The existing head protectors which are in implementation don't have any mechanism added to it to alarm excavators when a mining personnel is about to have an experience of unsafe location in terms of the gases that may get released in the near time and also when working with noisy equipment, being aware of one's surroundings is challenging. In this juncture, it is decided to provide better security to the personnel who are working in the mines. Improvement in the design of the existing mining helmet of the personnel is addressed in this project and the actual existing helmet is shown in Figure 9, such that more safety awareness is created between miners[R. S. Nutter (1983)]. In this module, a safety system for coal miners is specifically addressed.

A question arises that Why Coal Mining? Coal is the most important energy resource and remains the dominant fuel for power generation and many industrial applications. These are some of the most significant uses of coal. It is used for power generation, 58% of power in India is generated using coal. It is used in steel production, globally 74% of steel produced uses coal, it is also used in filters for water and air purification. Coal is used as energy source in cement production since large amounts of energy are required to produce cement.

The difficulties that are pone in coal mining are due to hazardous gases, mine wall failures and collisions of travelling vehicles in rare cases due to low visibility. The gases that are getting built up in the mining region are called as damps. There are different types of damps which need to be taken care and alert system need to be provided for the miner. The control room or the ground control must take care of these changes whenever a damp is released due to mining. Such alert system must also inform the miner prior in hand so that the miner gets psychologically strong. As damp is identified, the ground control must be capable of immediate action. This is possible only with appropriate detection and alert system.

Figure 9. Miners Helmet
[Courtesy: ark.gamepedia.com]

Figure 10. Voice Communication to the Miners from Helmet

A novel system is being developed for alerting miners taking into consideration of these damps and low visibility conditions as well. A RF based smart helmet is developed such that it creates a wireless sensor network, having an in-built voice module, speaker and a LCD screen which are helpful to alert the worker more clearly and accurately. The gases, temperature and humidity are detected through sensors and are transmitted wirelessly to the control room/ground station to take necessary appropriate corrective action.

It is obvious that a question arises, how power is provided to helmet? The solution to this problem is the usage of a normal 9V battery for testing purpose is used in this case. A Lithium Ion battery can also be used in the place of a 9V batter which can be a rechargeable one. Another problem comes into picture that how to know the battery level every time. Is it possible to monitor the health condition of the entire system which includes sensors, battery etc.? Yes, it is possible using IoT, wherein the data is transmitted to the control room every time. The battery level concerned with a particular helmet is monitored and will be alerted.

The existing head protector which are in implementation do not have any mechanism added to it for alarming the excavators when a mining personnel is about to have an experience of unsafe location in terms of the gases that may get released in the near time and also when working with noisy equipment, being aware of one's surroundings is challenging. The information cannot be a wired communication as the miners will be travelling hundreds of kilometers into mines where in chaotic situations will be occurring in between may cause damage to the wired connection. Hence, wireless connection is the only way out. Figure 10 clearly depicts the solution for the proposed method.

This system is designed in such a way that the miner can notify himself about the environmental conditions near him. This is done by the help of the speaker which will be continuously alerting the miner about the gas release or any other hazardous condition. Figure 11 shows the complete hardware built system with the helmet.

The miners helmet shown in figure 11 consists of transmitter that transmits the parameters that are acquired from the sensors. Figure 12 shows the block diagram on the transmitter side which consists of Arduino Board which is equipped with the different sensors. The sensors that are connected to the Arduino Board are gas, temperature, humidity, light sensors. The gas sensors sense the hazardous gases that are releasing out of the mine when the miner is in work. The temperature sensor every instant senses the temperature. The humidity sensor senses the humidity in the air surrounding the miner. It is already notified to the reader that rather than having different sensors, it is now feasible to use a sensor hub in the place of the different sensors. Such a change may lead to less circuit complexity and less processing

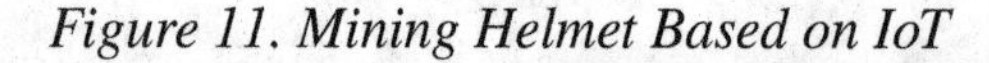

Figure 11. Mining Helmet Based on IoT

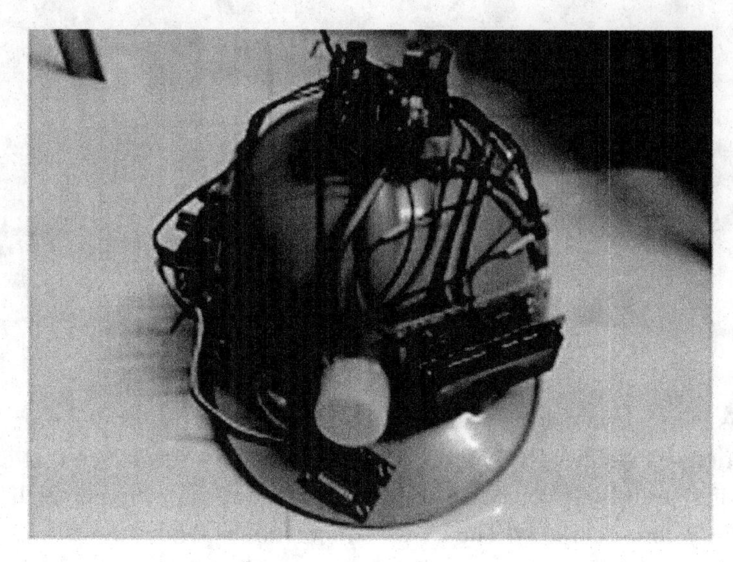

Figure 12. Block Diagram Representing Transmitter Side on Helmet

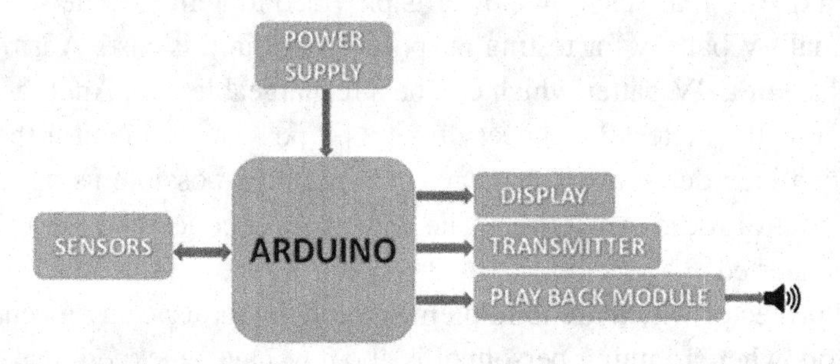

overhead to the processor. Care must be taken in selection of the components as well as the modules as the entire system need to be mounted on the helmet of the miner which is again an overhead to the weight that the miner is carrying.

Figure 13 shows the receiver side block diagram which consists of only two components. One is the receiver which is a RF receiver. This will be receiving the telemetry data sensed by the sensors which are onboard of the miners' helmet. Figure 14 shows the completely installed sensors and equipment for sensing. It is the cut-way view wherein all the components are visible. This will be inside a light weight chamber, so that all these components will be invisible. As the data is received it will be logged on to the cloud for further analysis. As the data will be acquired every instant, it is also suggested to use big data analytics for further analysis. Figure 15 shows the helmet when the system is under on condition.

Figure 13. Block diagram of receiver

Figure 14. Completely installed sensors

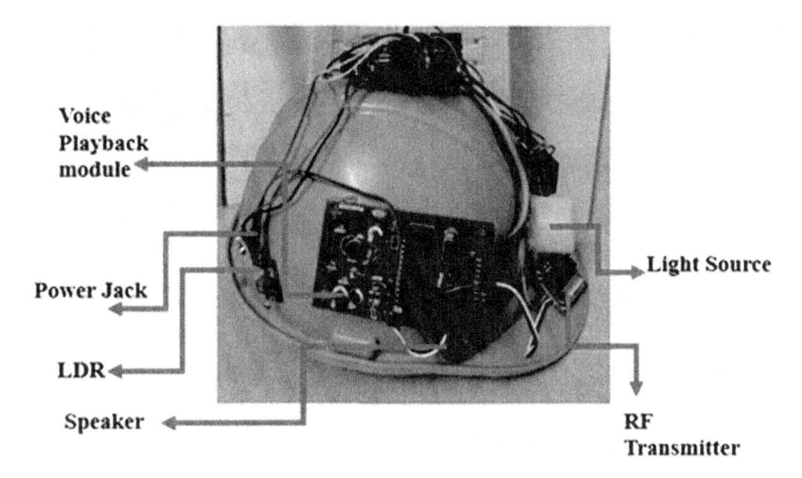

Figure 15. Helmet under ON condition

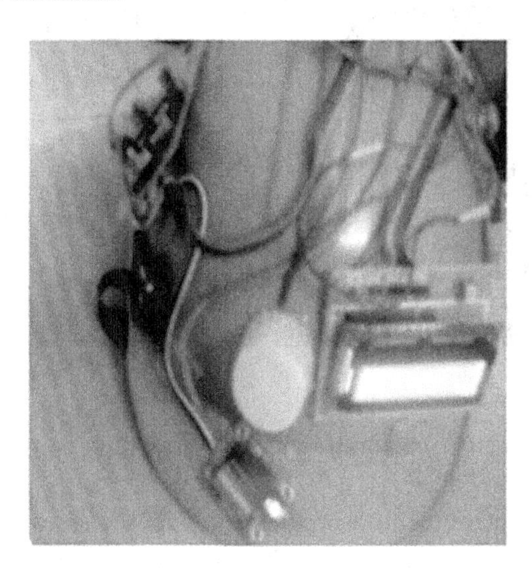

Figure 16 shows the result of the miner helmet which shows the data is logging continuously on the computer screen. If the data crosses a pre-determined threshold, then the system alerts the ground control office for necessary action.

4. IoT Based Garbage Monitoring System

In order to provide services to the public, government adopts various methods which involve technology based improvements. Such an improvement that help in improving the living standard is the garbage monitoring system using IoT[Parkash, Prabu V.(2016)]. The current system involves dumping of garbage by the public themselves in most of the cities by segregating it. But there is no provision in places where there is bin for dumping of the garbage to notify the authorities when it is full. Hence, this application helps in alerting the authorities by sending an alert notification stating that the bin is full. The level

Figure 16. Monitored Data Alert on Serial Monitor Wirelessly

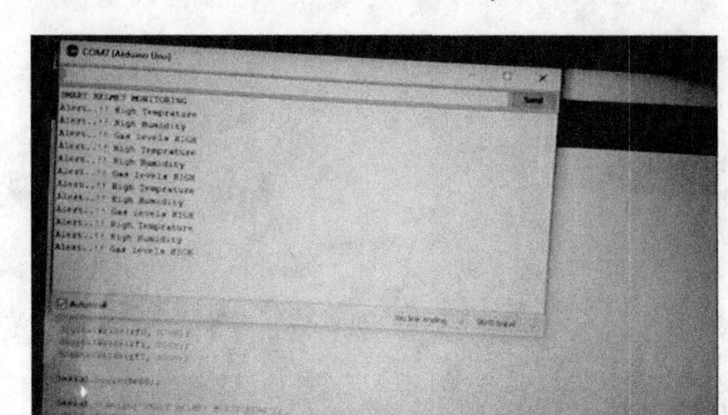

Figure 17. Trash Dumping into the Bins

of garbage can be detected using two methods. One of the method is by the use of ultrasonic sensors. Another one is by the use of load cell.

Let us consider the application using ultrasonic sensors. These sense the amount of garbage collected in the bin which is placed in the public area. When the bin is filled up to the threshold level or maximum, a message will be updated on the webpage.

Four trash cans are considered here which are separated by 24 cms apart. There is no certain rule that these must be of the same distance apart, randomly, it was considered. Initially, when the system is switched on all the bins, numbered as B1 – B4 will be displayed as 0. When the cans are slowly being filled up by garbage, the same will be displayed on the LCD displayed which is further transmitted via wifi to the cloud intimating the authorities to take appropriate action. The results are shown in the figures 17 and 18.

Another part of the design is by the use of load cell. Load cell produces an electrical signal which is proportional to the force that is applied on it. This force in this particular application will be due to the filled in garbage of the bin. Hence, as the force increases, a display that is calibrated according to the force that is applied shows how much full the bin is and will be intimated via internet. Such a load cell can also be easily interfaced with that of an Arduino board for a particular design.

Figure 18. LCD Display Indicating the Trash Levels

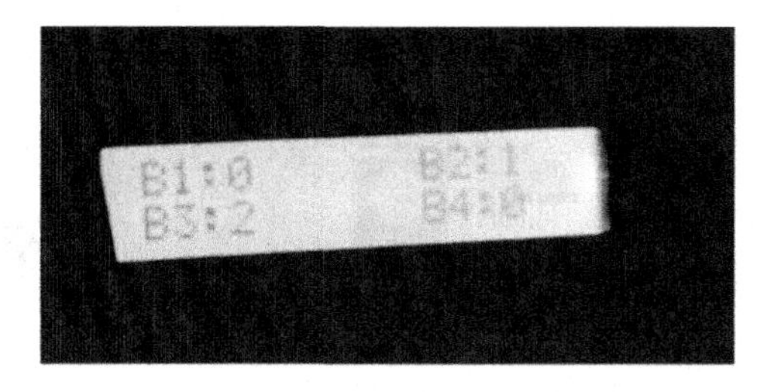

Weight of the basket also plays a major role in waste management using IoT. Based on the weight change it is observed that the bin is filled or not. Hence, weight sensors are used for such measurement. Chemical, temperature and humidity sensors are also used for solid waste management.[Anagnostopoulos, T., Zaslavsky, A., Kolomvatsos, K., Medvedev, A., Amirian, P., Morley, J., & Hadjieftymiades, S. (2017)]. The results that are stored in the cloud will be monitored daily by the municipal authorities which provide necessary information to take corrective actions.

5. IoT in Business Applications

Business Intelligence is one of the major aspect nowadays in any of the well-developed commerce in which retail is one of the leading industries that has drawn the attention in the recent research for big data analytics. Introducing e-commerce has led to the improvement in sales besides reducing the operational cost and also improving the customer relationship with better services. So, it is now a part of the retail business to understand the customer behavior and their sentiments as well for improvement in sales. In this juncture, a novel technique is proposed EMOMETRIC[V. V. Satyanarayana Tallapragada, N. Ananda Rao and Satish Kanapala (2017)], which is an intelligent trolley that tracks the emotions of the customer. Such a technique provides the behavioral insight only when combined with IoT. Thus, IoT based customer emotion analysis is performed that runs on Apache Spark Cluster. Real time emotion and face tracking is performed. It is observed that the accuracy of the system is limited to 95%. A novel solution for the analysis of the customer emotion by the integration of the IoT based technique is shown in Figure 19 and 20. Figure 19 shows the overall architecture and Figure 20 shows the technical aspects of the system. The face is recognized using a camera that captures the frames on real time and on fly. These frames are sent to the local server using MJPG. Such streams are captures from all the trolleys which are enabled with this feature. As the system is proposed for analysis of human emotion in retail business, the trolleys that are used by the users are fitted with a camera that captures the emotions. The local server runs a face detection algorithm and captures the face boundary and maps it to that particular trolley id, such that, each record contains trolley id and emotion data. Now, the analysis is done based on the emotion and the purchase of the customer. As the number of trolleys that are being used will be large, the data that is captured in every frame will also be large. Hence, big data system is used to process the data. For this reason, Apache Spark system is used which outperforms the Hadoop cluster in processing.

In this mode, hourly data aggregation is performed and regression analysis is done. This regression correlates the weight of the trolley with the customer emotions and provides details on how the customer is thinking of a particular item while shopping. The emotion tracking is done by fixing a facial model on to the captures customer face. This algorithm fits the tracked face with different light weight classifiers which means that each point that is marked on the face will be having a weight. The system is designed

Figure 19. Overall Architecture

in such a way that the customer emotions can be captured in low light region also. Table 1 shows the customer emotions in varying light conditions at different instances of a day. It can be observed that this system is independent of light intensity and pose variation. The system can correctly track the customers pose every time and under real time conditions. The table 1 also shows the plot of the data based on the mood of the customer. This system is combined with the trolley and analysis is performed.

Figure 21 shows the relat time tracking session of a customer emotion. The camera which is posited on the trolley captures the image first, then processes it in such a way that it will recognize the face. The facial points are marked, processed for emotional analysis. The last row of the table above shows the combined emotions of a person.

Figure 20. Technical Overview of the proposed system

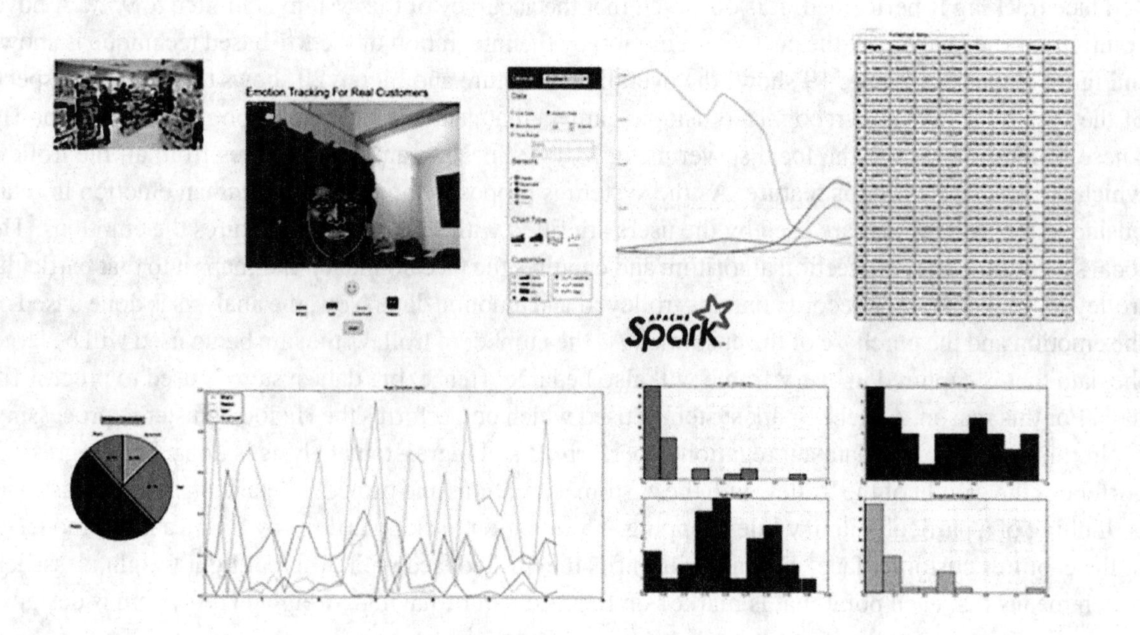

Table 1. Mood variation plot with variable illumination conditions

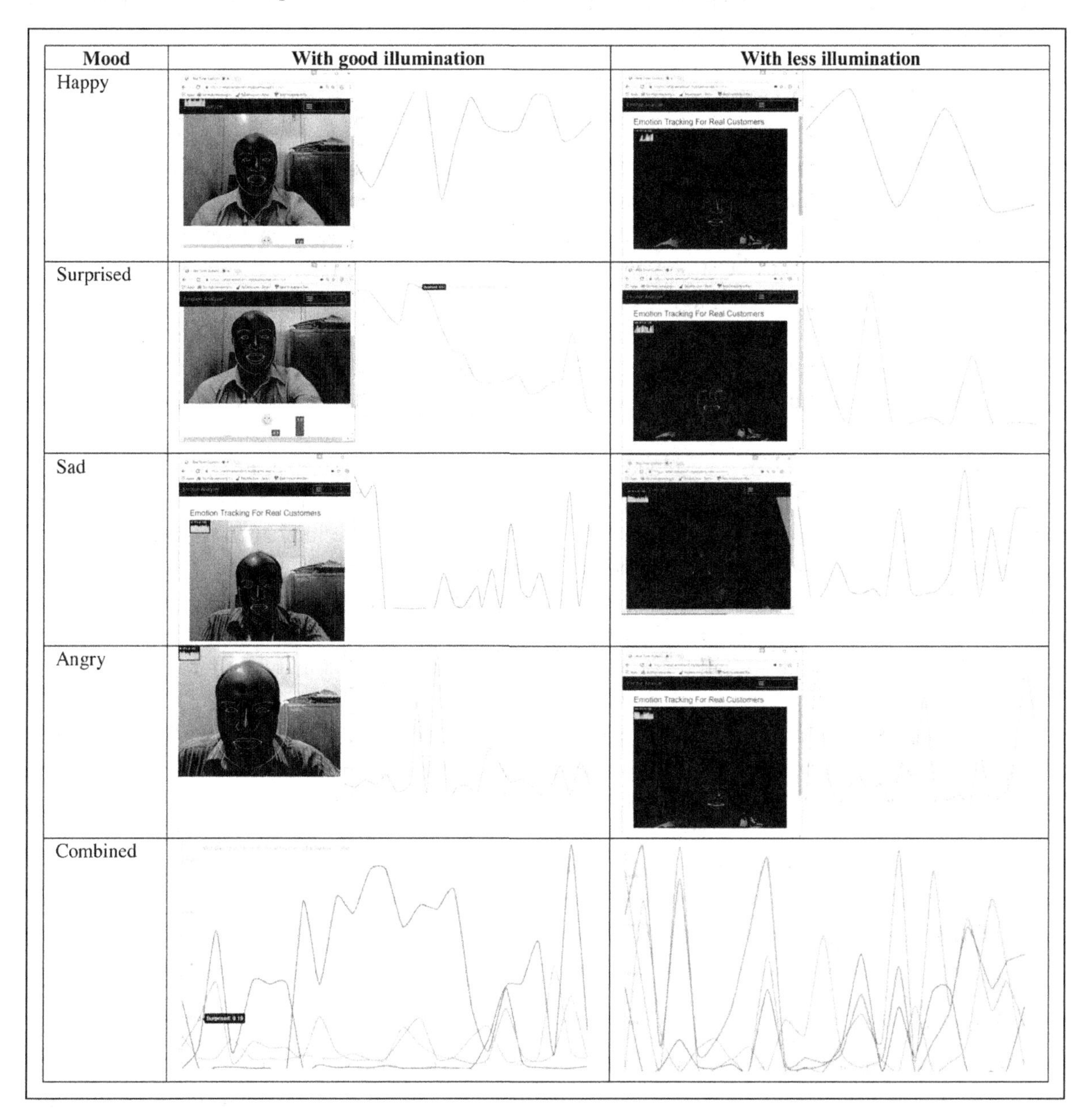

Mood	With good illumination	With less illumination
Happy		
Surprised		
Sad		
Angry		
Combined		

Figure 21. Sequence of Real time tracking session of customer emotion

This technique proves to be the best for real time business analysis wherein based on customer emotions the mart can have more goods which also improves the sales.

6. IoT in Military Applications

In the current scenario of the usage of latest technology in the war has become complex, multidimensional making it highly dynamic and also disruptive. Such an environment creates wide adversaries with unexpected enemies. Hence, the use of IoT has paved new and an intelligent way to evade the enemies. The system so developed must be strong enough to identify the enemy radar as well. The system will be having less time to react as well to respond in the war field. This is one of the major challenge that makes the system to be developed in such a way that its switching time must be fast. The war personnel will be having shorter timeframes to assess the situation and make an appropriate decision.

The use of IoT in military applications can be divided into two parts. One with the vehicle or plane that they use in war, another one with the soldier. First, the system that is implanted with the solider will be discussed clearly. The system is designed so that it will monitor the health of the solider. A pulseoximeter is used to monitor the muscle activity based on the sensors. It will also be monitoring the heart rate based on the sensor located on the chest. All these sensors will be implanted on the bulletproof jacket that they wear. It is a well-known fact that, in a troop, there will be a doctor, a communication engineer and remaining will be soldiers. The communications person takes care of the communication equipment and the doctor takes care of the health of the troop personnel on the spot under emergency. Hence, it is proposed to transmit all the troop health condition via cloud and monitor remotely [F. T. Johnsen et al.,(2018)].

The use of sensors can be made available in collaborative and non-collaborative mode. The former mode connects the mobile to the other units in the field by the use of RFID. This helps in connecting the other equipment in the vicinity of the primary equipment to get connected. Now, this is done automatically to reduce the time lapse in scanning. In such a collaborative mode, which is generally a friendly mode of operation which manages our friends or our operations such as command, transportation of materials as well as the supply of the warfare, maintenance of the equipment.

Another mode of operation is Network transmission mode which is in particular used for communication between the physical layer which captures the information to the application layer. Network transmission mode is used mainly for M2M communication. Such communication is mainly used in weapon handling and control[L. Yushi, J. Fei, Y. Hui(2012].

Another important question arises here by the usage of these sensors in such a sensitive war field is that "Whether the way of communication is secure"? Hence, there is need for the development of appropriate secure way of communication as the enemy every time tends to hack the system for acquiring information before in hand or else they can modify our systems parameters which may lead to catastrophe.

The security is provided in IoT based on the Object Level Protetion which was developed in support of the NATO NNEC which provides protection for the single objects rather than considering a group of objects and systems. Metadata which is linked to the objects related to data is used for protection mechnanisms that examine the requirements for a particular object. It is observed that the grant of cryptographic access provides better security rather than the usage of confidentiality over the network. Such a cryptographic security based access provides control over the information and the data object and will be protected from end to end[Wrona, K (2015)].

7. IoT Applications in Smart City

Smart city is the term that is widely used in the recent days which gives a view of cameras, sensors that acquire the data on fly providing a healthy and well secured environment to the public in that area. Smart cities make use of the modern communication tools like that of ICT which make them even more smart. Governance is required for the proper maintenance and operation of the smart cities. Many queries are build up on the smart cities such as the data ownership which acquires data from various sensors which may be public data. When a problem persists what is merit of that and the treatment that is being provided? Who prescribes the rules on these data collection and security based issues? How to identify the exact stakeholders? What are the technologies to be used and how they can be assessed? The humane smart city takes the citizen as the prime element keeping his data in mind providing security for such data acquired from various sensors. Citizens are involved in decision making and are processed by using a pre-defined algorithms based on pre-defined criteria [Almeida, V. A. F., Doneda, D., & Moreira Da Costa, E.(2018)].

Smart cities are proposed in a way to improve the standards of the people. Smart city includes smart surveillance, smart electricity, water and distribution, smart building, smart healthcare, smart services, smart infrastructures, smart transportation [R. Morello, S. C. Mukhopadhyay, Z. Liu, D. Slomovitz, and S. R. Samantaray (2017)]. Smart surveillance provides better security which enhances the security interms of monitoring the traffic and providing better security without any latency. As an example, every part of the city will be monitored via cameras so that any theft will be recorded in the camera so that the police will be alerting the nearby control room or monitoring station to have hold on the situation to make it under control without having any problem of law and order. Smart electricity uses smart grids which makes the distribution of power to the smart city by monitoring the loads at every instance making changes in the load imbalance and also provides continuous power to such a smart city. Smart billing or prepaid billing is one of the example of a smart electricity. Smart water is another part of the smart city wherein the billing is done based on the usage of the water by a consumer. A meter for such measurement can be placed at the distribution point into a home line. Billing is done based on commercial and noncommercial use. The buildings must be constructed in such a way that they are fit for such smart city applications. Hence, smart buildings which are built with inbuilt sensors that acquire data, process it and also store in the cloud for further processing and access. Now, the data that is acquired is a vast data, hence, big data analytics is required for the processing of such data. Smart transportation is another part of the smart city which provides better transportation for a better living. Such transportation is based on the GPS data wherein the such data provides users where the required transport is, making them to have a knowledge on how much time to wait for the transport to reach their destination. Further, recently one of smart cities have been provided by electric vehicles to the municipal authorities to monitor the working personnel actions via GPS. Working personnel will be reaching each and every home daily for collecting the waste and hence, the vehicle will be everyday covers a particular longitude and latitude point which will be monitored by the higher authorities. Further, if the working personnel is not moving in that area, alert will be provided to the higher authorities to take corrective action making the entire system smart. Smart infrastructure is another area which holds all these individual parts of the smart city marking and making them into a smart infrastructure. Hence, it can be concluded that smart cities provide better living based on the information that is acquired from various sensors that are placed at various places in the city based on the application.

CONCLUSION

Internet of Things (IoT) is the current research area which is of prominence. IoT enables to connect heterogeneous devices enabling them to have a homogeneous type of connectivity based on the application that is desired by the user. With the latest development of connectivity in smart phones via Long Term Evolution (LTE) and Large Scale Convergence (LSC) there is an exponential increase interms of users who access internet, resulting in globally connected Ethernet. It is expected that almost one to two trillion devices will get interconnected by the end of 2025. Such a connectivity will improve the life style of humans providing them better solutions to their requirements. In this juncture, IoT requires a better framework which can connect these devices which are heterogeneous in nature. Further, the applications that are being designed for these devices must also be capable enough interms of processing, storage and other better capabilities. Applications that are designed based on IoT can be broadly classified as Industrial and Domestic. Industrial applications are these applications which help industry personnel in analyzing the industrial day to day activity such that the production is increased. The machinery used in the industry need to be observed daily on a routine basis which require some maintenance. Hence, it is required to monitor them and check them for wear and tear resulting in a sequential maintenance, thus, improving the budget allocated for maintenance. It is to be known that in some areas of the industry that human intervention is not practically possible. In such areas, there is a requirement for machinery to be designed in such a way that it itself will monitor other equipment that are interconnected. Such systems are developed and deployed in some of the industries which reveals the telemetry data. Such telemetry data consists of health of the system, giving out various parameters that are predetermined. A smart helmet which provides the environmental conditions in a coal mine is discussed. Gas, temperature and humidity levels are being continusouly monitored and transmitted to the ground control station for further necessary action if the monitored levels crosses a pre-determined threshold.

Monitoring and acquiring data from fields is also a prime industrial application related to farming industry which increases the productivity of farms resulting in more crop production. Such a monitoring system also acquires information from the fields where farming is of prominence and compares with the known and predetermined parameters, providing the farmer information required for better yield. An image processing system which can acquire images of the crop at various stages of the crop growth are processed and can be compared with that of the templates that are being stored in the database. Such a system provides results on the growth of the crop before in hand, sharing farmer the details about the crop, making him aware of crop condition. This system can also be designed in such a way that it can also provide suggestions based on the data/parameters from the images acquired.

Waste Management is another industrial application which provides details on how to process the waste that is being produced day-to-day. IoT system that is being developed in this connect must be capable enough to identify and segregate industrial waste. These systems must be intelligent enough to process different types of waste based on the inputs that are being provided by correlating with the existing template. A smart waste management system is developed for the purpose of alerting the authorities to clear the filled in dustbins. Such a monitoring system every instant scans the level of the bin and transmits the levels in a monitoring system so that the respective authorities can clear the bins. Till date, it is being done manually and now new technology need to be implemented.

Recent boom in the sales in particular in retail sector has given place the for the development of applications based on IoT for better understanding of customer preferences and their behaviour. Such a system improves the business. In order to address this problem of behavioural analysis an intelligent trolley that incorporates low cost, yet efficient facial emotion tracking combined with IoT and Big Data is provided for providing a meaningful customer insight. The system also acquired the images of the customer under low illumination and the algorithm is so developed that it can process the low resolution and less illuminated images. Results show better emotion analysis.

The applications that are discussed are not limited and the entire design depends on the user and their intention and usage. The better the sensor selection, more precisely the parameters can be acquired. The usage of sensor hub is also discussed which reduces the processing overhead on the system.

Smart city is the currently widespread word wherein lot of funds are being provided for the development of such cities. Smart cities improve the living standards of the people. Such a provision can only be done by the use of IoT which ultimately provides better security with better living conditions. As a security application smart military applications can be also be designed paving way to obtain telemetry data of the soldier by acquiring the parameters from the soldiers body at every instant remotely.

REFERENCES

Almeida, V. A. F., Doneda, D., & Moreira Da Costa, E. (2018). Humane smart cities: The need for governance. *IEEE Internet Computing*, *22*(2), 91–95. doi:10.1109/MIC.2018.022021671

Almotiri, S. H., Khan, M. A., & Alghamdi, M. A. (2016). Mobile Health (m-Health) System in the Context of IoT. *2016 IEEE 4th International Conference on Future Internet of Things and Cloud Workshops (FiCloudW),* 39-42.

Anagnostopoulos, T., Zaslavsky, A., Kolomvatsos, K., Medvedev, A., Amirian, P., Morley, J., & Hadjieftymiades, S. (2017). Challenges and Opportunities of Waste Management in IoT-Enabled Smart Cities: A Survey. *IEEE Transactions on Sustainable Computing*, *2*(3), 275–289. doi:10.1109/TSUSC.2017.2691049

Benini, L. (2013). Designing next-generation smart sensor hubs for the Internet-of-Things. *5th IEEE International Workshop on Advances in Sensors and Interfaces IWASI.* 10.1109/IWASI.2013.6576075

Dhanalaxmi, B., & Naidu, G. A. (2017). A survey on design and analysis of robust IoT architecture. *2017 International Conference on Innovative Mechanisms for Industry Applications (ICIMIA)*, 375-378. 10.1109/ICIMIA.2017.7975639

Johnsen, F. T., & … . (2018). Application of IoT in military operations in a smartcity. *2018 IEEE International Conference on Military Communications and Information Systems (ICMCIS)*, 1-8.

Kaur, S., Pandey, S., & Goel, S. (2018). Semi-automatic leaf disease detection and classification system for soybean culture. *IET Image Processing*, *12*(6), 1038–1048. doi:10.1049/iet-ipr.2017.0822

Kishore Kumar Reddy, N. G., & Rajeshwari, K. (2017). Interactive clothes based on IOT using NFC and Mobile Application. *2017 IEEE 7th Annual Computing and Communication Workshop and Conference (CCWC),* 1-4.

Morello, R., Mukhopadhyay, S. C., Liu, Z., Slomovitz, D., & Samantaray, S. R. (2017). Advances on Sensing Technologies for Smart Cities and Power Grids: A Review. *IEEE Sensors Journal*, 99.

Nutter, R. S. (1983). Hazard Evaluation Methodology for Computer-Controlled Mine Monitoring/Control Systems. *IEEE Transactions on Industry Applications*, *IA-19*(3), 445–449. doi:10.1109/TIA.1983.4504222

Parkash, P. V. (2016). IoT Based Waste Management for Smart City. *International Journal of Innovative Research in Computer and Communication Engineering*, *4*(2), 1267–1274.

Peng, F., Wang, W., & Liu, H. (2014). Development of a reflective PPG signal sensor. *2014 7th International Conference on Biomedical Engineering and Informatics,* 612-616.

Tallapragada, V. V. S., Rao, N. A., & Kanapala, S. (2017). Leaf Disease Detection Using Combined Feature of Texture, Colour and Wavelet Transform. *International Journal of Control Theory and Applications*, *10*(21), 159–167.

Tallapragada, V. V. S., Rao, N. A., & Kanapala, S. (2017). EMOMETRIC: An IOT Integrated Big Data Analytic System for Real Time Retail Customer's Emotion Tracking and Analysis. *International Journal of Computational Intelligence Research*, *13*(5), 673–695.

Wrona, K. (2015). Securing the Internet of Things a military perspective. *Proceedings of the IEEE 2nd World Forum on Internet of Things (WF-IoT)*, 502–507. 10.1109/WF-IoT.2015.7389105

Yasin, M., Tekeste, T., Saleh, H., Mohammad, B., Sinanoglu, O., & Ismail, M. (2017). Ultra-Low Power, Secure IoT Platform for Predicting Cardiovascular Diseases. *IEEE Transactions on Circuits and Systems. I, Regular Papers*, *64*(9), 2624–2637. doi:10.1109/TCSI.2017.2694968

Yushi, L., Fei, J., & Hui, Y. (2012). Study on application modes of military internet of things (miot). *IEEE International Conference on Computer Science and Automation Engineering (CSAE)*, 630–634. 10.1109/CSAE.2012.6273031

Chapter 5
Deployment of Narrowband Internet of Things

Sudhir K. Routray
https://orcid.org/0000-0002-2240-9945
Addis Ababa Science and Technology University, Ethiopia

ABSTRACT

Internet of things (IoT) is an integral part of modern digital ecosystem. It is available in different forms. Narrowband IoT (NBIoT) is one of the special forms of the IoTs available for deployment. It is popular due to its low power wide area (LPWA) characteristics. For new initiatives such as smart grids and smart cities, a large number of sensors will be deployed and the demand for power is expected to be high for such IoT deployments. NBIoT has the potential to reduce the power and bandwidth required for large IoT projects. In this chapter, different practical aspects of NBIoT deployment have been addressed. The LPWA features of NBIoT can be realized effectively if and only if its deployment is done properly. Due to its large demand, it has been standardized in a very short span of time. However, the 5G deployment of NBIoT will have some new provisions.

INTRODUCTION

Internet of things (IoT) is a pervasive application of the Internet connected through several sensors and actuators. Initially, it was started as a value added service over the cellular communication networks. Over the time, it has evolved as a new dimension of the cellular networks. Now, it can be applied for several applications which are very attractive in the new initiatives such as smart cities and smart grids. The modern day cellular networks are in a very suitable position to provide the IoT deployment over their infrastructure. However, it can be deployed both in the cellular and non-cellular frameworks. Cellular networks provide cost effective deployment as their infrastructure is already mature. Non-cellular deployment needs a new infrastructure for the IoT deployment.

Right now, there are several types of IoT networks. One of them is narrowband IoT (NBIoT). This is a new type of radio access technology which needs a very few resources for its deployment when compared with other forms of IoTs. Therefore, it is considered as a low power wide area (LPWA) technol-

DOI: 10.4018/978-1-5225-9199-3.ch005

ogy. That means it needs a very low power levels for its operations while covering a large area. LPWA features are essential for large scale deployment of IoTs for long term sustainability. In the modern digital ecosystem, wide spread deployment of sensors and actuators for all the essential facilities of the cities would be very expensive and energy consuming. Therefore, energy and other resource saving technologies are essential for modern projects such as smart cities, smart retail, smart transport systems, smart grids and smart policing.

The term "narrowband" is due to the use of 200 kHz bands for NBIoT. In Long Term Evolution (LTE), bands are comparatively much wider than 200 kHz. The time when NBIoT was standardized by the Third Generation Partnership Project (3GPP) in 2016, the typical 4G LTE channels used to be of around 10 MHz. Therefore, 200 kHz bands are much narrower than the LTE bands. Of course, when compared with the GSM and UMTS bands, the 200 kHz may not be that narrow. However, the name (i.e., NBIoT) was given based on the comparisons with the 4G LTE bands. In 5G, the bands will be even wider and the 5G associated NBIoT will be really very narrowband application over it.

The main objective of this chapter is to provide the deployment techniques of NBIoT. We provide a detailed study and analysis of NBIoT features, principles, potentials and some typical applications. We also present the bandwidth and other essential requirements of NBIoT for practical deployment. We show the typical models for NBIoT and their utilities for practical deployment. We also considered the energy efficiency and other related aspects which are important considerations for its long term sustainability.

The reminder of the chapter is organized in five sections. In Section 2, the literature review of NBIoT deployment has been presented. In Section 3, the bandwidth and band selection related issues of NBIoT deployment have been presented. It is shown that in three different ways it can be deployed effectively, and even hybrid deployments are also available. In section 4, the methods and techniques for the deployment of sensors, actuators and servers of NBIoT have been shown. The roles of edge computing facilities have been presented in this section. In Section 5, the future research directions of NBIoT and its deployment techniques have been presented. Finally, in Section 6, the chapter has been concluded with the main points.

LITERATURE REVIEW

Due to its importance for the modern digital ecosystem, a lot of research is going on different types of IoTs. Every year new outcomes of research change the IoT landscape to a large extent. Overall, the IoT research is quite mature now. Almost all aspects of IoTs are being investigated for better performances. In comparison to the other forms of IoTs, NBIoT is relatively new. NBIoT was proposed as an alternative solution for massive machine type communications (Li et al., 2018). Due to its suitability for large coverage using a small amount of power, it became popular for several applications in the cellular communication framework. In Ratasuk et al. (2016a), the basic ideas behind NBIoT are presented. Its deployment issues too have been addressed in this paper. The deployment of NBIoT can be done over three different types of bandwidth allocation: standalone deployment using dedicated carriers for NBIoT; in-band deployment using the LTE bands when they are not used for LTE communication; and guard band deployment using the band gaps provided between the LTE bands. Depending on the situation, two or more hybrid of the above methods can also be used. In the standalone mode, NBIoT normally uses one GSM or LTE equivalent channel of 200 kHz known as a physical resource block (PRB). In the two other modes, it normally uses a GSM or LTE band of 180 kHz. NBIoT normally uses low cost systems and can provide enhanced coverage. The coverage can be 20 dB more above the existing GPRS coverage (Ratasuk et al.,

2016a). A properly deployed NBIoT can use the batteries for the sensors for 10 years while serving a large coverage area. Similarly, latency too can be improved with proper deployment.

NBIoT was first standardized in 3GPP Release 13 (Ratasuk et al., 2016b). As NBIoT is an LTE technology, it can be used over the LTE and its legacy systems. However, when compared with other forms of IoTs, it will be much simpler and uses fewer resources (i.e., power and bandwidth). Even the infrastructure for NBIoT is very much cost effective when compared with other forms of IoTs. With proper optimized deployment, NBIoT reduces the overheads significantly and increases the coverage. It uses low complexity components. Furthermore, its performances can be improved through low latency initiatives. In Ratasuk et al. (2016b), the cross layer design features of NBIoT have been shown to achieve the optimal performances. Mostafa et al. (2017), addressed the connectivity maximization issues of NBIoT under non-orthogonal multiple access (NOMA). NBIoT was developed as an alternative to the wireless sensor networks for massive machine type communications under the cellular framework. However, there are several challenges for high density connectivity in NBIoT. The authors tried non-orthogonal schemes of multiple accesses to increase the connectivity. In their paper, they formulated joint allocation of sub-carriers and power for maximizing the connectivity. With this scheme, they also achieved good quality of service and found to be very effective in connecting a large number of NBIoT devices.

Petrov et al. (2018) use the NBIoT networks for crowd-sensing applications. In their paper, they used several IoT sensors and collected their outcome to analyze the vehicular movements along the roads. The vehicles had their sensors mounted over the top and their signals were received by the NBIoT sensors deployed around the roads ubiquitously. They found that the connectivity of vehicles using NBIoT is very effective, and it produces very accurate results of the vehicular movements. Thus they suggest that NBIoT can be used for crowd-sensing applications such as vehicular monitoring to provide assistance to the road traffic. Wang et al., (2017) have studied the air interface of NBIoT for practical applications. According to them, NBIoT provides better deployment flexibility, very long battery lives, low power wide area coverage, high device density and low device complexities. These are the features, normally not found in case of other forms of IoTs. This technology has great potentials for ubiquitous deployment across a large geographical area. It can also be deployed much beyond the cellular infrastructure. Thus it is one of the main technologies for connected living. Zayas et al. (2017), studied the coexistence of NBIoT with the contemporary cellular technologies such as GSM, UMTS and LTE. It was found that NBIoT is compatible with all these cellular technologies. Thus, it can be deployed alongside 2G, 3G and 4G communication networks. It was also found that NBIoT is better than the machine type communications (MTC) in terms of the massive deployments, low complexity in connectivity and much reduced data rates. They showed that deployment of NBIoT is simpler when compared with other similar technologies.

Schlienz and Raddino (2016) analyzed the machine type traffic of NBIoT. They emphasized the analysis from the similarities points of views of LTE and NBIoT. It is found that the NBIoT traffic is very similar to the LTE traffic despite the differences in the data rates. They showed that NBIoT has the ability to operate in difficult situations. Thus it has the potential to provide services in the rural and remote areas where other options are not available. Routray and Sharmila (2017), presents the green features of NBIoT. In large scale IoT deployment, a huge number of sensors are required. These sensors need a significant amount of energy to operate for years. The authors have shown that NBIoT is the practical solution for these scenarios. NBIoT sensors can operate for more than 10 years with normal batteries for sensors. Similarly, both the transmission and reception of information in NBIoT uses much lower energy than other IoTs. Furthermore, the bandwidth and data rates of NBIoT are exemplary. Only 200 kHz is enough for NBIoT devices. Overall, NBIOT can be considered as a green technology due to its energy efficient features.

Ramnath et al. (2017) analyzed the tracking accuracies using IoT networks. Tracking is an obvious application for cellular networks due to their near ubiquitous presence. IoTs are deployed over the cellular infrastructure and use a large number of sensors. These sensors can be used for tracking of objects which can provide better accuracies. Normally, the sensors used by the IoTs are much higher in number than the sensors of cellular networks. Thus the accuracies of tracking can be improved significantly using IoT networks. NBIoT is even better for these applications as it supports massive deployments of sensors and it is an LPWA technology. Sharma et al. (2017) studied the applications of NBIoT and other forms of IoTs for location based services. IoTs can provide better accuracy of locations when compared with the outcomes of other available alternatives such as global positioning system (GSM) and cellular network based localization. The main reason behind this improvement in the accuracy is due to the presence of large number of sensors across the coverage area. The location based services can be added to the IoT networks as value added services. Anand and Routray (2017), presents the utilities and applications of IoT for healthcare. In the modern healthcare there are several applications where sensors are used. In this framework, IoT can be very useful in supporting several needs of healthcare. There are many instances in healthcare provisioning where continuous monitoring of patients is essential. This can be done through NBIoT sensors. The information obtained from monitoring should also be sent to the healthcare providers. This can be done through NBIoTs. For a large scale deployment of healthcare, NBIoT is preferred over the other forms. It is cost effective and uses low power which suitable for human uses. It can be used for both rural and urban areas with low cost of deployment. However, it has some challenges such as latency. In critical healthcare monitoring, the latency has to very low. At this moment, the latency in NBIoT is not that low. So, it is expected to be better in the future.

Routray et al. (2017) considered the applications of quantum cryptography for the security related aspects of IoT. Security is a big concern for IoT these days. A robust security is required to provide ubiquitous IoT coverage. Quantum cryptography is a new security system which has the capability to provide perfect security. For ubiquitous applications such as IoT it seems to be a right choice. However, the cost for deployment may be expensive when compared with other alternatives. In the coming years, it will be used in several critical applications. Ayoub et al. (2018) analyzed several IoTs for widespread applications. In case of pan-country deployment, the costs become too high. In order to reduce the costs, economical versions of the IoTs are advisable. There are several forms of IoTs which are very much energy and bandwidth efficient. Thus in the long term, their costs too become lower in comparison to the other forms of the IoTs. LPWA technologies are the choice for a large scale deployment of IoTs across a large geography. There are several LPWA technologies available for large scale deployment. Some of the examples are: NBIoT, LoRa, LTE Cat-M, DASH7, EC-GSM-IoT, SigFox, Wi-SUN. Some of these technologies use the licensed bands such as the NBIoT. Several other technologies use unlicensed bands for their operation. The authors analyzed both the types of IoTs for practical deployment. They compared their technical specifications and standards. They also provided the requirements of mobile IoTs and their research challenges.

Li et al. (2018) analyzed several LPWA technologies available for the practical deployment. The LPWA technologies have several attractive features such as low power consumption, wider coverage, and a large capacity of carrying information. The authors observed that the LPWA technologies find a boom in the IoT related applications. The LPWA technologies are popular in all the applications including the power grids and smart grids. In mission critical applications too these technologies can be deployed due to their high reliability, good quality of service (QoS), and safety related features. These technologies can be deployed using both licensed and unlicensed bands. However, the licensed bands are preferred due to their complete standardization. NBIoT is the currently available as a licensed LPWA technology which provides both energy and bandwidth efficiency. In addition to this, its QoS is as good as the QoS of the cellular networks. Therefore, NBIoT is considered as the smart choice for the control and management of smart grids. The authors analyzed both the qualitative and quantitative aspects of NBIoT and found it to be better than other available alternatives. NBIoT can be deployed in smart grids in both the urban and rural environments. Mekki et al.(2018) analyzed the large scale deployment of IoT devices around 2020. According to their estimate, more that 50 billion devices will be interconnected using several radio technologies. Out of them, the LPWA technologies will be leading the connectivity. Out of all available LPWAs, the authors analyzed the top three they considered suitable for the large scale deployment. These three are: LoRa, Sigfox and NBIoT. In terms of battery lives, costs, and capacity LoRa and Sigfox can be considered for practical deployment. However, when it comes to the QoS and latency related aspects NBIoT outperforms other alternatives. Therefore, depending on the application scenarios one of these LPWAs can be chosen for practical deployment.

Kumar et al. (2018) studied the practical deployment of an IoT based solution to monitor the air pollution. Their aim is to choose an IoT which is very much ubiquitous so that the air pollution related parameters can be measured over a large coverage area. Due to the rapid industrialization, the air pollution these days have gone up very fast. It is essential to monitor the level of air pollution to control the pollutions at their sources themselves. It is possible only through some technologies which are widespread. In order to use the IoT for this, the authors divided the geography in to small sectors in the shape of a grid. In every grid, they deployed several sensors to take the reading of the pollution level. Depending on the priorities, each grid is assigned with more or less number of sensors. Effectiveness of this arrangement was checked through simulation and found that air pollution can be monitored through IoTs. Kiss et al. (2018) discuss about the future of IoT. They suggested that in the future the IoT architectures are going to be dynamic and self-organizing instead of the static ones which are seen initially. The main aim of the future IoT networks is to harness the outcomes optimally. That can be done using devices of many different capabilities and collaboration between different parts which would communicate in a very flexible manner. In order to provide all these facilities, the networks have to be very agile and resource efficient. The authors also proposed an offloading framework for wide variety of tasks using the IoT devices. They also suggested for a new computing paradigm for these versatile applications over the IoT networks. According to the authors, edge computing would ease these complexities and can provide the IoT networks to work faster. So, in the fifth generation networks, edge computing facilities should be provided at the border or edge nodes to carry out the offloading efficiently. Several clustering techniques as well as the artificial intelligence (AI) can be used in these cases. Finally, the authors proposed some technical solutions to these versatile IoT networks.

Venticinque and Amato (2018) studied the growth trends of IoT networks over the existing ICT infrastructure. They find that the existing infrastructure is not sufficient and efficient for the forecasted IoT growth. Dense connectivity in the networks reduces their reliability of service provisioning. The QoS also goes down with densification. The cloud connectivity for IoT solutions affects the QoS directly. Fog related solutions have been proposed for IoT, and it suggested the presence of an intermediate node to ease these connectivity related complexities. The authors suggested new methodologies for optimal deployment of IoTs in a complex and dense scenario. These methodologies they provided through an extended IoT platform of CoSSMic project, which is a pan-European project of effective IoT deployment. They provided the new functionalities for the future IoTs which can be executed efficiently. Hoglund et al. (2018) provided the relevant standards for IoT deployment which are framed by several standardization bodies across the world. The authors mainly focused on the 3GPP standards for IoT deployment. In LTE Release 13, there were several provisions for the deployment of the low power wide area network technologies such as NBIoT and massive machine type communications. In LTE Release 14, these standards were further enhanced for better compatibility with the existing devices. These standards collectively known as LTE-M became very popular. In Release 14, new voice services were added to LTE-M and the data rates too were enhanced to meet the 5G demands.

Fattah (2018) describes the 3GPP specifications for the new technologies such as IoTs which can be compatible with the 5G networks. The author starts with the basic design techniques of IoT and then shows their protocol stacks. The detailed functions of each layer and protocol in the IoTs are explained in this book. The function of the cellular networks is pivotal in the success of the IoTs. The LPWA technologies such as NBIoT too have been addressed in this book. The wide range of services and functions NBIoT can carry out have been explained with proper justification. Finally, the author shows the reasons and methods to make all the IoTs compatible with the 5G networks. Cao and Li (2018) analyzed the traffic from the new technologies such as IoT. They characterized the traffic from different sources. Then they compared the characteristics of the traffic from different sources such as the Internet and the IoT. The traffic from NBIoT too has been analyzed in this study. They generated the traffic for NBIoT using mesh devices and networks having different energy levels. In their experimentation, they also tried the moving IoT nodes using cars. The network of cars using NBIoT was deployed and their characteristics were measured under different conditions. The Internet of Vehicles (IoV) was tested in the NBIoT framework. Feltrin et al. (2019) studied several features of NBIoT in different setups. They started with the analysis of common features of NBIoT such as energy efficiency, larger coverage area, and several other key functionalities. They deployed the proposed model of NBIoT in the laboratory and tested their performances. Both the uplink and downlink functions for the remote sensors and actuators were carried on and they found the performances to be satisfactory. Then they also tested the real applications and noted down the performance metrics. These tests were also done for unicast and multicast conditions. Overall the performances were found to be quite good when compared with other cellular alternatives.

Bello et al. (2018) studied the key features of NBIoT. They started with the common features such as energy efficiency, reduced complexities, and large coverage. Then they analyzed the provisions available to provide those features. For instance, to enable energy efficiency, NBIoT has power saving modes (PSMs) and extended discontinuous reception (eDRX) modes which work in tandem to save energy when there is no meaningful task. In order to test these performances, the authors developed a Markov model based on four different probable outcomes. They also tested the optimized performances of NBIoT under a long enough duration. They found that the available provisions such as PSM and eDRX perform according to the standards set for NBIoT in 3GPP. Popli et al. (2018) studied the energy efficient NBIoT and analyzed their architecture, technological features and overall advantages over other similar alternatives such as LoRa and SigFox. They analyzed the virtues and demerits of NBIoT as an LPWA technology. They found that NBIoT has several provisions to save energy in its operations. The energy efficiency of NBIoT was investigated for the large scale deployment. They analyzed the future success of NBIoT based on the current specifications and future developments. They identified the main bottlenecks of NBIoT on its way towards success. They proposed two techniques for energy efficiency of NBIoT which would make it green for large scale applications.

NBIoT DEPLOYMENT BANDS

In 3GPP Release 13 and 14, NBIoT has been specified with its technological details. The bands for deployment too have been proposed in these standards. According to 3GPP Release 13, NBIoT can be deployed in three different bands in the following ways. They are: standalone deployment, guard band deployment, and in-band deployment. We have shown these three different types of band deployment in Figure 1. The success of NBIoT is very much dependent on the choice of its spectrum. In case of the standalone and guard band deployment, there is not much restriction over the selection of NBIoT bands for an emerging demand. However, the in-band deployment is a complicated process as NBIoT and LTE/ UMTS/ GSM bands compete for the common spectrum. When the demand for LTE services is low (mainly during the late night and early morning), there is no big problem as the NBIoT services can use those common bands. However, during the peak hours when LTE services are using all their bands, NBIoT services cannot be provided. This failure can have very serious consequences on the NBIoT services. Therefore, in such cases an alternative should be there to handle the bottlenecks.

Standalone Deployment

In this method, a new and unused band is allocated for the use of NBIoT operations (see Figure 1 (a) for standalone deployment). The cellular bands for GSM are located around 900 MHz and 1800 MHz. So, right now, the bands between 700 MHz and 800 MHz can be allocated for the NBIoT applications. This band is suitable for several radio applications using sensors and actuators. Several countries in Europe such as Sweden and Czech Republic use this band for NBIoT. Deployment in this band is also fast as it is not widely used for other applications at the moment. The current 3GPP plans for this deployment is based on the both uplink and downlink segregation.

Guard-Band Deployment

In this method, the NBIoT services use the guard bands available between the LTE/ UMTS/ GSM communication bands (see Figure 1 (b) for guard-band deployment). Normally, guard bands are provided in the LTE to avoid interference between the adjacent communication bands. These bands used to be idle and the operators do not like this (because they pay for the licenses of these bands and in return, it does not produce any revenue). Now, NBIoT can use these narrow bands which are win-win situations for both the operators and the NBIoT users. Guard band deployment does not interfere with the existing services.

In-Band Deployment

In this method, NBIoT is deployed over the LTE/ UMTS/ GSM channels. So, the spectrum becomes common for both the LTE as well as the NBIoT applications (see Figure 1 (c) for in-band deployment). In this case, priority is given to the LTE services. When LTE services are not provided over a certain band then it can be used for NBIoT applications. However, if a demand comes for that specific band then NBIoT applications have to migrate to another channel which is free at that moment. This spectrum jumps can be provided through frequency hopping algorithms. In the in-band deployment, normally one PRB of NBIoT is a fraction of the single LTE frequency slot (i.e., one LTE frequency slot is an integral multiple of an NBIoT PRB). At this moment, this is not a big problem as the available spectrum can be managed through frequency hopping. However, in the future when the LTE services become crowded, NBIoT services will be directly affected. Thus in-band deployment may not be a suitable option during the peak hours.

Hybrid Deployments

Hybrid deployments of NBIoT are also possible. For instance, NBIoT can be deployed over both in-band and guard bands together. When the in-band allocations are saturated or LTE services do not leave any in-band spectrum holes then guard bands can be used for NBIoT. So, based on the above three scenarios, the following hybrid combinations are possible:

1. Standalone and guard band deployment;
2. Standalone and in-band deployment;
3. Guard band and in-band deployment;
4. Standalone, in-band and guard band deployment.

In the hybrid deployment cases, the sensors, actuators, servers and the other components should have compatibility over all the three bands. For hybrid deployment, the spectrum and other resource management get complex. However, performance wise, the hybrid methods are better than the individual deployment techniques. They provide better flexibility over spectrum selection and service blocking due to lack of spectrum may be completely removed.

The above cases are mainly for the NBIoT deployment over the cellular networks. However, as we have mentioned before, it is also possible to deploy NBIoT in a non-cellular framework. In such cases, both licensed and unlicensed bands may be used. Non-cellular deployment is more expensive than the cellular deployment as the complete NBIoT infrastructure is custom built for it. In such cases, the network will have better flexibility. Service blocking is not at all a problem in the custom built NBIoT network with its own spectrum.

DEPLOYMENT OF SENSORS AND SERVERS

Practical deployment of NBIoT components can be followed from its deployment architecture. In fact, architecture based deployment is very much flexible for the protocol design and implementation. In this section, we first show the architecture of NBIoT and then show its practical deployment techniques. In case of Internet, we have normally two models: one is the open systems interconnections (OSI) model which provides the understanding of its features; and the other is the Transmission Control Protocol / Internet Protocol (TCP/IP), which is used for practical design. In case of NBIoT too, there exist both the OSI model and the practical protocol based model. OSI model provides the basic functions of NBIoT and the protocol based model is used for its practical design. We show both these models (i.e., OSI and the protocol based model) in the following subsections.

Figure 1. Different deployment options for NBIoT in the cellular framework

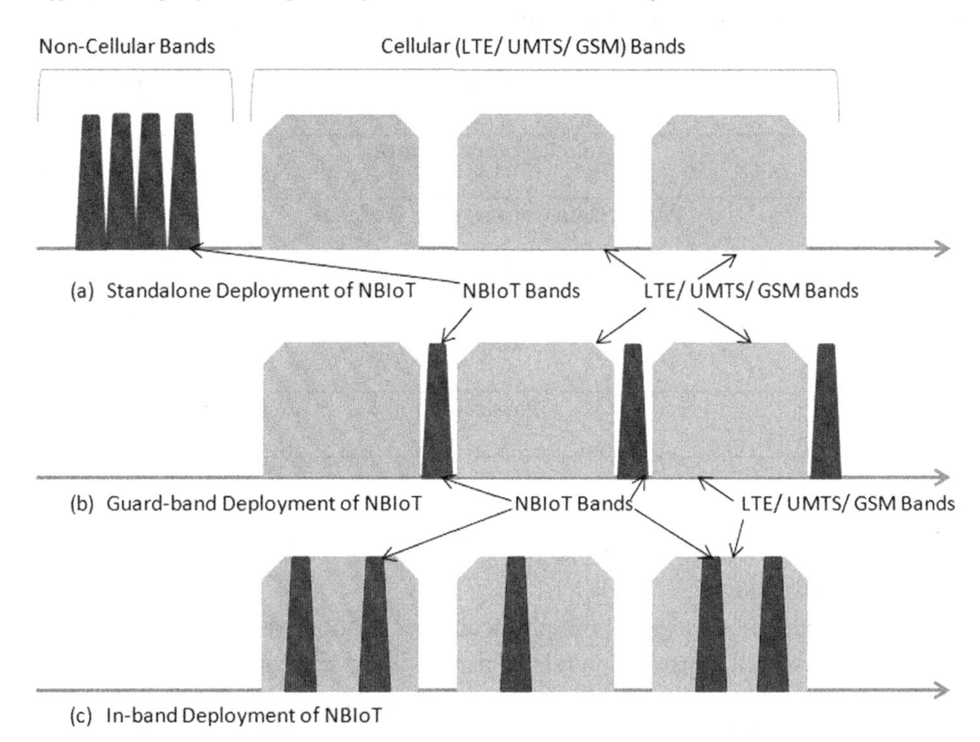

The OSI Model of NBIoT

The OSI model for NBIoT has six different layers (Routray & Sharmila, 2017). The bottom one is the physical layer which is the radio interface for communicating between the NBIoT end devices with the NBIoT server. Just above the physical layer is the medium access control (MAC) layer. It has almost the similar functions as the MAC layers of other systems. Above the MAC layer, radio link control (RLC) layer is present. This layer takes care of the radio link establishments and deletion of existing links between different devices/ parts of NBIoT network. Above the RLC layer, packet data convergence protocol (PDCP) layer is present. PDCP layer takes care of the transport aspects of the packets from one node to the other in the NBIoT. Wireless packet communications are managed through the protocols of PDCP such as the user datagram protocol (UDP). Just above the PDCP layer there is a radio resource control (RRC) layer. The main functions of RRC are to control the resources of NBIoT such as power and the allocation of bandwidths for different applications. The top-most layer of NBIoT OSI is the non-access stratum (NAS) layer. This layer is responsible for the communication between the user equipment (UE) and the NBIoT server. NAS layer also takes care of security and privacy related aspects of the NBIoT such as encryption and decryption. The OSI layers of the NBIoT architecture are shown in Figure 2.

The Protocol Based Implementable Model

Just like the TCP/IP model of the Internet which is practical for network design, NBIoT has also protocol based implementable models. Based on this framework, we have also six different layers of this model.

Figure 2. The six layers of NBIoT architecture. In the respective layers, the associated protocols are present which are essential for the overall function on NBIoT.

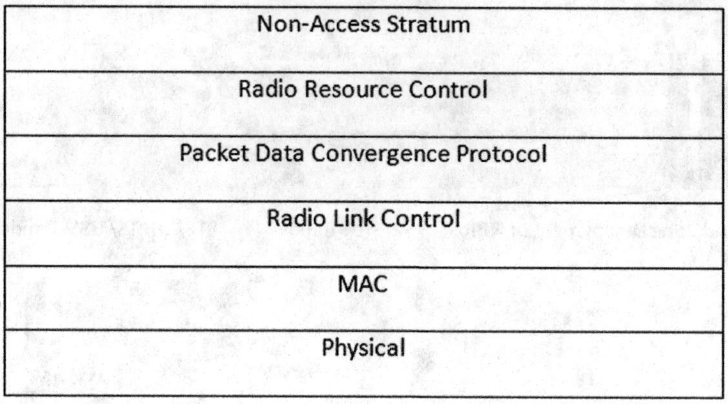

Non-Access Stratum
Radio Resource Control
Packet Data Convergence Protocol
Radio Link Control
MAC
Physical

The bottom layer is the physical layer just like the TCP/IP model. This protocol based implementable model has been shown in Figure 3.

Just like the OSI model, in this model the physical layer is at the very bottom. It takes care of the physical layer connectivity and transmission related aspects. Above the physical layer is the 6LoWPAN (IPv6 over Low power Wireless Personal Area Network) layer which is the low power version of the IPv6. It takes care of the all IP related functionalities of NBIoT. User datagram protocol (UDP) takes care of the wireless transport aspects of the packets from the end devices to the servers. The security related aspects of the packets are taken care of by the datagram transport layer security (DTLS). In fact, DTLS is a stream oriented communication protocol which takes care of the security aspects of the datagrams. Above the DTLS, we have the constrained application protocol (CoAP). It is specially designed for low power constrained networks. CoAP is very similar to Hypertext Transfer Protocol (HTTP), but it is much lighter than HTTP and uses only UDP. The end devices are at the top of this hierarchy. This protocol based model is in a direct form and ready for practical implementation.

Figure 3. The practical protocol based implementable layers of NBIoT.

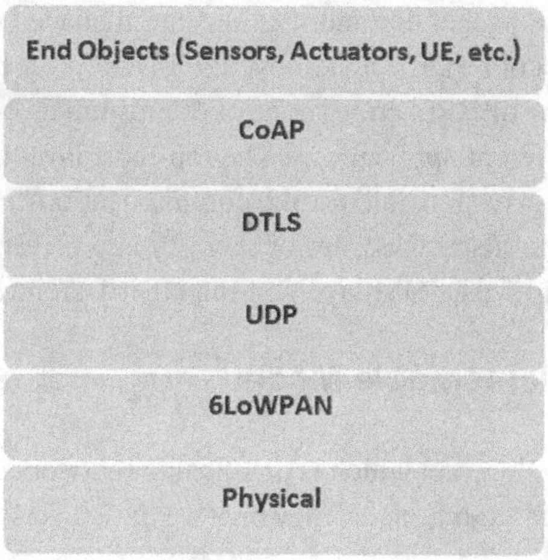

End Objects (Sensors, Actuators, UE, etc.)
CoAP
DTLS
UDP
6LoWPAN
Physical

Practical Deployment of Sensors and Servers

NBIoT is an LPWA technology. Therefore, it is expected to be deployed for a large coverage area comprising of a large number of sensors and actuators. Under such complex situation, it is advisable to reduce the whole task of deployment in to several smaller ones. In fact it is the common principle of all large scale deployment problems. In case of NBIoT the whole architecture can be divided into three different layers. This would divide the whole architecture into three different parts:

1. The client side architecture (Deployment of sensors and actuators) also known as the client layer
2. The server side architecture (Deployment of the servers) also known as the gateway layer
3. The connectivity between the clients and the servers (Deployment of the physical connections) also known as the platform layer

This layered structure is quite essential to provide a systematic design for the NBIoT networks to facilitate different services. The client side architectures normally have the sensors and actuators to collect the data or to carry out some specific pre-defined functions. This is the most complex part when the deployment is concerned. For large projects such as smart city or smart grid, millions of sensors and actuators have to be deployed. The client side architecture is well supported by the connectivity through the platform layer. The connections can be either wired or wireless. However, for NBIoT (a radio access technology), wireless connections are the preferred choice. Connectivity can be provided through a special infrastructure such as cellular communication systems or may also be standalone infrastructure for just NBIoT. The server side architecture deals with the main control systems and data acquisition systems. It consists of clouds and data centers in case of large NBIoT. In order to facilitate better control and data management, edge computing facilities have to be provided at the appropriate border locations. A typical NBIoT architecture has the following parts:

1. End elements (typically sensors and actuators);
2. Connections through some physical media normally which connects with the Internet or the NBIoT data centers;
3. IT and computing facilities at the edge to facilitate edge computing;
4. Data center and other storage facilities such as a cloud.

Each of the parts can be deployed independently or in collaboration with the other parts. However, for the deployment of any part, complete specification of the whole architecture is required. In Figure 4, we show the typical deployment scenario for the above four parts.

Figure 4. The four main parts of NBIoT in its practical component deployment

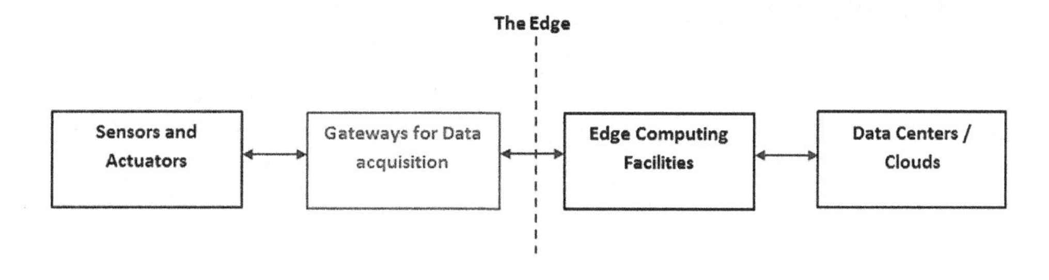

Deployment of Energy Management Systems

Energy management is an essential part of any large IoT network. In NBIoT, effective energy consumption is even more critical as it is an LPWA technology. The sensors are used to collect the required data in NBIoT. However, the sensors are not required to record all the information that comes in their ranges. Rather only the specific required data has to be collected and sent to the nest processing unit. This is a very tricky situation. Using appropriate logical settings, the required data can be selected only when it is available. Thus the sensors/ actuators can be kept in the sleep/ idle mode during the time when the meaningful data is not available. Similarly for parts too can be kept in inactive mode when there is no function to carry out. The sleep mode in deployed through PSM which is normally set with the help of a timer. Each PSM follows a specific time duration known as PSM cycle. In each cycle, the PSM sends a few low power sensing pulses. If it detects any activities through those sensing pulses the sensor is awaken. Otherwise, the PSM cycles are repeated and sensor remains in the sleep mode. The power spent by a sensor during the sleep mode is much lower than power spent during its awake/ active mode.

eDRX is another mechanism to reduce the power consumption during the inactive period. In fact, it is an LTE feature which is also used for NBIoT. It can be used alone or in collaboration with PSM. Use of both the mechanisms can save some extra power than any one of them. Lime PSM, eDRX also follows a eDRX cycle. That means in every cycle, the devices remain active only for a very short time. During that active interval if they detect any signal then the device is switched on to the active/ awake mode. Otherwise, the devise is allowed to remain in the inactive/ sleep mode. The difference between the PSM and eDRX is that PSM uses longer cycles and multiple sensing pulses. However, eDRX uses a shorter cycle and only one or two sensing pulses. Depending on the devices and the functions the eDRX cycles can be of different time durations.

Deployment of Other Supporting Infrastructure

NBIoT deployment normally deals with a large coverage area with a large number of sensors, actuators and connections. A lot of data is collected, analyzed, and based on the analysis; decisions are taken for different activities through the actuators. Such a huge amount of data cannot be stored in traditional storage systems. Rather, large data centers and clouds are required to store the collected data effectively. So, cloud based storage systems are very common for NBIoT networks. The processing systems too need a lot of computing power. Special data analysis interfaces are required to do these complex tasks. These days, cloud based storage is preferred for NBIoT. The cloud deployment for NBIoT is very much similar to the typical like other cases. Control mechanisms, edge computing infrastructure, coverage management, cell selection/ reselection, and data rate control mechanisms are deployed either in the central office or in a distributed fashion. In fact, this is very much dependent on the type of application for which NBIoT is deployed.

Application Specific Deployment

In the above sections, the deployment issues addressed are the typical ones. In practice, the deployments are very much application specific. Depending on the applications several performance related parameters are decided. For instance, in smart city projects the sensors are deployed around the city for various data measurement and monitoring purposes. However, when it comes to the smart grids, the deployment is very much within the grid coverage area. The grid parameters are measured and monitored in this case. Therefore, the application specific deployments are very much depending on the applications and the output oriented.

Industry Specific Deployment

In the above subsection, we have mentioned the application specific deployments. Similar approaches are also needed for industry specific deployments of NBIoT. For instance, the smart cities need several types of sensors and actuators deployments distributed across the cities through appropriate edge computing and server facilities. However, for the industrial control and monitoring for manufacturing a very different approach is required. Industrial needs are very much real time and the delays have to be the minimum possible. Both the latency and reliability requirements of NBIoT are critical in the industrial environment. In the industrial environment a multi-vendor scenario is very much common as a single vendor normally does not provide all the advanced specifications. Best possible solutions are acquired through multiple vendors. In order to keep the performances at their optimum interoperability development testing (IODT) is carried out for these multi-vendor services. IODT is essential for every industrial deployment to perform optimally and efficiently. In addition to it, several other conformance tests are carried on in the industrial deployment of NBIoT. For better reliability and disaster resilience, industrial deployments are provided with backups. If the first choice of monitoring fails, the backup alternatives are switched on through some feedback mechanism.

FUTURE RESEARCH DIRECTIONS

NBIoT is a highly demanded technology. Its features are very much attractive as a green technology. It provides a lot of hope for energy and bandwidth efficient deployment of large scale IoT projects. Its coverage span is very impressive and the power consumption is very attractive. Its future looks very bright due to these eye-catching features. In the future, this technology is expected to get much better. Some of its future research directions are presented in this section.

Further Enhancement of Efficiency

NBIoT is an energy efficient technology which reduces the energy consumption by increasing the sensitivities of the receivers and sensors. However, there are still further scopes for the improvement of energy efficiency. For the remote connections, wireless transmission may not be very much energy efficient. In such cases, a wired medium may be used. Special antennas can be designed for NBIoT applications with better energy efficiency. Similarly, energy per bit can be reduced using appropriate waveforms. Energy harvesting technologies can also be incorporated in the sensors and actuators. Right now, NBIoT mainly uses BPSK as the carrier modulation scheme. Only at a few instances QPSK is used. Spectral efficiency of NBIoT can also be improved using the new techniques. It would further narrow the required bands for communication. However, it is noteworthy that the use of advanced techniques and waveforms may increase the complexities of the transceivers. Finally, there will be a trade-off between efficiency and the costs.

Conjugation With Artificial Intelligence and Machine Learning

Artificial Intelligence (AI) is a powerful tool for several network related applications. In the present day networks, the flexibility is very limited as they are able to carry out only a few numbers of preset functions. Using AI these networks can provide several new functions. In addition to that several network parameters can be manipulated remotely. Similarly, machine learning can be used for better performances in the autonomous systems. These are the learning techniques to train the machines and devices to perform optimally. AI and machine learning have the ability to increase and improve the tasks over the IoT networks. In the coming years, both AI and machine learning are going to play significant roles in NBIoT.

Expansion in to Unchartered Territories

NBIoT has several applications in the recent times. New applications are getting added every year. It is expected that in the future several new applications will be there for NBIoT. It may find its place in very new areas such as underwater communication, sensing and monitoring. In the defense and sports sector too it is expected to find some place. Here we have mentioned the foreseeable applications of NBIoT. However, there will be several new applications which are completely out of imagination of the recent developments. NBIoT is a better option over other alternatives due to its high efficiency and large range of coverage.

CONCLUSION

In this chapter, we presented the main features and characteristics of NBIoT. Now, NBIoT is a very popular technology and ready for deployment over the cellular networks. It provides several advantages over the other forms of IoTs such as energy efficiency, bandwidth efficiency, fast deployment, low cost of components, and longevity of the network components. We analyzed several aspects of practical deployment of NBIoT. Bandwidth for NBIoT deployment is decided according to the availability of the bands. If LTE bands are very much occupied during the peak hours then guard band or standalone deployments are preferred. However, the hybrid deployment over several bands is better than just one type of band deployment. Sensor, actuator and server deployments should be followed using the layered structures. Energy saving mechanisms are essential for long term sustainability of NBIoT. Though we show the general trends of deployment to be more or less uniform, in practice, almost all the deployments are very much application specific. Industrial deployments are normally done through multiple tests and backups. Depending on the applications, the data demands, and other specific requirements, the deployments are customized.

REFERENCES

Anand, S., & Routray, S. K. (2017, March). Issues and challenges in healthcare narrowband IoT. In *Inventive Communication and Computational Technologies (ICICCT), 2017 International Conference on* (pp. 486-489). IEEE. 10.1109/ICICCT.2017.7975247

Ayoub, W., Samhat, A. E., Nouvel, F., Mroue, M., & Prévotet, J. C. (2018). Internet of Mobile Things: Overview of LoRaWAN, DASH7, and NB-IoT in LPWANs standards and Supported Mobility. *IEEE Communications Surveys & Tutorials.*

Bello, H., Jian, X., Wei, Y., & Chen, M. (2018). Energy-Delay Evaluation and Optimization for NB-IoT PSM with Periodic Uplink Reporting. *IEEE Access: Practical Innovations, Open Solutions.*

Cao, X., & Li, Y. (2018, January). Data Collection and Network Architecture Analysis in Internet of Vehicles Based on NB-IoT. In *2018 International Conference on Intelligent Transportation, Big Data & Smart City (ICITBS)* (pp. 157-160). IEEE. 10.1109/ICITBS.2018.00048

Fattah, H. (2018). *5G LTE Narrowband Internet of Things (NB-IoT).* CRC Press. doi:10.1201/9780429455056

Feltrin, L., Tsoukaneri, G., Condoluci, M., Buratti, C., Mahmoodi, T., Dohler, M., & Verdone, R. (2018). *NarrowBand-IoT: A Survey on Downlink and Uplink Perspectives.* IEEE Wireless Communication Network, ToAppear.

Hoglund, A., Bergman, J., Lin, X., Liberg, O., Ratilainen, A., Razaghi, H. S., ... Yavuz, E. A. (2018). Overview of 3GPP Release 14 Further Enhanced MTC. *IEEE Communications Standards Magazine, 2*(2), 84–89. doi:10.1109/MCOMSTD.2018.1700050

Kiss, P., Reale, A., Ferrari, C. J., & Istenes, Z. (2018, January). Deployment of IoT applications on 5G edge. In *Future IoT Technologies (Future IoT), 2018 IEEE International Conference on* (pp. 1-9). IEEE.

Kumar, M., Sabale, K., Mini, S., & Panigrahi, T. (2018, January). Priority based deployment of IoT devices. In *Information Networking (ICOIN), 2018 International Conference on* (pp. 760-764). IEEE. 10.1109/ICOIN.2018.8343220

Li, Y., Cheng, X., Cao, Y., Wang, D., & Yang, L. (2018). Smart choice for the smart grid: Narrowband Internet of Things (NB-IoT). *IEEE Internet of Things Journal, 5*(3), 1505–1515. doi:10.1109/JIOT.2017.2781251

Mekki, K., Bajic, E., Chaxel, F., & Meyer, F. (2018). *A comparative study of LPWAN technologies for large-scale IoT deployment.* ICT Express.

Mostafa, A. E., Zhou, Y., & Wong, V. W. (2017, May). Connectivity maximization for narrowband IoT systems with NOMA. In *Communications (ICC), 2017 IEEE International Conference on* (pp. 1-6). IEEE. 10.1109/ICC.2017.7996362

Petrov, V., Samuylov, A., Begishev, V., Moltchanov, D., Andreev, S., Samouylov, K., & Koucheryavy, Y. (2018). Vehicle-based relay assistance for opportunistic crowdsensing over narrowband IoT (NB-IoT). *IEEE Internet of Things Journal, 5*(5), 3710-3723.

Popli, S., Jha, R. K., & Jain, S. (2018). A Survey on Energy Efficient Narrowband Internet of things (NBIoT): Architecture, Application and Challenges. *IEEE Access: Practical Innovations, Open Solutions.*

Ramnath, S., Javali, A., Narang, B., Mishra, P., & Routray, S. K. (2017, May). IoT based localization and tracking. In *IoT and Application (ICIOT), 2017 International Conference on* (pp. 1-4). IEEE. 10.1109/ICIOTA.2017.8073629

Ratasuk, R., Mangalvedhe, N., Zhang, Y., Robert, M., & Koskinen, J. P. (2016, October). Overview of narrowband IoT in LTE Rel-13. In *Standards for Communications and Networking (CSCN), 2016 IEEE Conference on* (pp. 1-7). IEEE. 10.1109/CSCN.2016.7785170

Ratasuk, R., Vejlgaard, B., Mangalvedhe, N., & Ghosh, A. (2016, April). NB-IoT system for M2M communication. In *Wireless Communications and Networking Conference (WCNC)*, 2016 *IEEE* (pp. 1-5). IEEE.

Routray, S. K., Jha, M. K., Sharma, L., Nyamangoudar, R., Javali, A., & Sarkar, S. (2017, May). Quantum cryptography for IoT: A Perspective. In *IoT and Application (ICIOT), 2017 International Conference on* (pp. 1-4). IEEE. 10.1109/ICIOTA.2017.8073638

Routray, S. K., & Sharmila, K. P. (2017, February). Green initiatives in IoT. In *Advances in Electrical, Electronics, Information, Communication and Bio-Informatics (AEEICB), 2017 Third International Conference on* (pp. 454-457). IEEE. 10.1109/AEEICB.2017.7972353

Schlienz, J., & Raddino, D. (2016). *Narrowband internet of things whitepaper.* White Paper, Rohde&Schwarz.

Sharma, L., Javali, A., Nyamangoudar, R., Priya, R., Mishra, P., & Routray, S. K. (2017, July). An update on location based services: Current state and future prospects. In *Computing Methodologies and Communication (ICCMC), 2017 International Conference on* (pp. 220-224). IEEE.

Venticinque, S., & Amato, A. (2018). A methodology for deployment of IoT application in fog. *Journal of Ambient Intelligence and Humanized Computing*, 1–22.

Wang, Y. P. E., Lin, X., Adhikary, A., Grovlen, A., Sui, Y., Blankenship, Y., ... Razaghi, H. S. (2017). A primer on 3GPP narrowband Internet of Things. *IEEE Communications Magazine*, *55*(3), 117–123. doi:10.1109/MCOM.2017.1600510CM

Zayas, A. D., & Merino, P. (2017, May). The 3GPP NB-IoT system architecture for the Internet of Things. In *Communications Workshops (ICC Workshops), 2017 IEEE International Conference on* (pp. 277-282). IEEE.

ADDITIONAL READING

Al-Fuqaha, A., Guizani, M., Mohammadi, M., Aledhari, M., & Ayyash, M. (2015). Internet of things: A survey on enabling technologies, protocols, and applications. *IEEE Communications Surveys and Tutorials*, *17*(4), 2347–2376. doi:10.1109/COMST.2015.2444095

Chen, S., Xu, H., Liu, D., Hu, B., & Wang, H. (2014). A vision of IoT: Applications, challenges, and opportunities with china perspective. *IEEE Internet of Things journal, 1*(4), 349-359.

Da Xu, L., He, W., & Li, S. (2014). Internet of things in industries: A survey. *IEEE Transactions on Industrial Informatics*, *10*(4), 2233–2243. doi:10.1109/TII.2014.2300753

Islam, S. R., Kwak, D., Kabir, M. H., Hossain, M., & Kwak, K. S. (2015). The internet of things for health care: A comprehensive survey. *IEEE Access: Practical Innovations, Open Solutions*, *3*, 678–708. doi:10.1109/ACCESS.2015.2437951

Xu, J., Yao, J., Wang, L., Ming, Z., Wu, K., & Chen, L. (2018). Narrowband internet of things: Evolutions, technologies, and open issues. *IEEE Internet of Things Journal*, *5*(3), 1449–1462. doi:10.1109/JIOT.2017.2783374

Zanella, A., Bui, N., Castellani, A., Vangelista, L., & Zorzi, M. (2014). Internet of things for smart cities. *IEEE Internet of Things journal, 1*(1), 22-32.

KEY TERMS AND DEFINITIONS

Bandwidth Efficiency: It can be defined as the effective and useful use of every Hertz of frequency available for communication. In radio communication, bandwidth efficiency is essential as the available spectrum for communication is very limited. Bandwidth efficiency can be increased through spectral efficiency (by increasing the number of bits per hertz).

Deployment of Narrowband IoT: It is the planning, design, testing, and on-site practical implementation of the NBIoT servers, sensors, actuators and other supporting parts of NBIoT to make it fully functional for practical applications. A full-scale deployment may take a long time (a few years) for a large project like the smart cities and smart grids.

Energy Efficiency: Energy efficiency can be defined as the effective use of energy to carry out any task. The lower is the consumption, the better is the system. In communication engineering, energy efficiency is very important as the system needs a large amount of energy to provide services to the people.

Internet of Things: Internet of things is a new technological paradigm in which several objects and devices are connected to the Internet using sensors and transceivers. The connectivity can be wireless or wired. However, now the popularity of wireless connectivity is dominant over the wired connections. Similarly, there are several possible connection technologies available. In the recent years radio technologies are more popular than any other alternatives. However, several other technologies are available in the market.

Low Power Wide Area Networks: Low power wide area networks are the ones which can cover a large area for communication using a very small amount of power. These technologies are very popular for the large-scale deployments such as smart grids and smart cities.

Narrowband IoT: This is a recent version of IoT. As the name suggests this version uses narrow bands for its operation. It is a popular LPWA technology due to its large coverage, low energy consumption, and overall simplicity. It is compatible with almost all kinds of cellular communications.

Chapter 6
Smart Cities Project:
Some Lessons for Indian Cities

Mahima Nanda
Guru Nanak Dev University, India

Gurpreet Randhawa
Guru Nanak Dev University, India

ABSTRACT

The smart cities mission of the Government of India has opened up new pathways for urban redevelopment and transformation. But given the limited resources available with a developing country, a more pragmatic approach would be to first learn from the best international experiences and approaches and then implement those in Indian context. With this view, the chapter examines some of the best practices related with different aspects of a smart city and suggests their relevancy for the development of smart cities in India. The study found that by focusing on the five core areas (i.e., urban mobility and public transport, safety and security of citizens, health and education, water management, and robust IT connectivity and social networking) the concerned authorities in India can successfully achieve their goal of urban redevelopment and transformation with scarce resources. Limitations and scope for future research are discussed in the end.

INTRODUCTION

As per the Census of 2011, around 31 percent of Indian population resides in urban areas and they contribute 63 percent to India's GDP. But with increasing urbanization, this number is expected to rise by 2030 with 40 percent people living in urban areas while contributing 75 percent to India's GDP (Ministry of Urban Development, 2015). This increased influx in the cities certainly requires proper planning of all aspects including development of economic, social, physical and other infrastructure. Well planned cities would not only enhance the quality of life of the residents but would also attract tourists and investments. Thus, by aiming at Smart Cities, the Indian Government has taken a right step for uplifting the face of the country by developing citizen friendly and sustainable cities in India.

DOI: 10.4018/978-1-5225-9199-3.ch006

A Smart City refers to a city that "uses information and communications technology (ICT) to enhance its livability, workability and sustainability" (Berst et al., 2013, p. 2). In simpler terms, this job consists of three steps: "collecting, communicating and crunching" (Padode et al., 2016, p. 12). The first step involves collecting information through sensors and other devices. In the next step, data is communicated using wireless and wired networks. Last step involves crunching or analysing the collected data in order to understand the current as well as future situation. For this, a new type of intelligent infrastructure is required - "an innovative and open platform based on smart sensor networks that can help forward-looking cities more predictably integrate a complex suite of services cost-effectively, at pace and at scale" (Sensors for Smart Cities, 2015, p. 1).

Ideally, a Smart City aims at developing a comprehensive eco-system which requires development of institutional, physical, social and economic infrastructure (Ministry of Urban Development, 2015). It's quite a big challenge for cities with limited resources to conceptualize and adopt practices and technologies to have a right-mix of infrastructure which would enable them to transform into a Smart City. As per the United Nations ESCWA Report (2015, p. 29), there are five "High Priority Pillars" of a Smart City development. The report emphasises that if focus is put on these dimensions or pillars, then the transformation planning and strategic development becomes more feasible and successful. Similarly, the Ministry of Urban Development in India has also given ten core infrastructure elements for a Smart City. These elements in alignment with the "High Priority Pillars" are shown in the table 1 below:

Data in table 1 shows that there are a number of core infrastructural elements essential in the development of a Smart City, e.g., public transport, sanitation, affordable housing, good governance, sustainable environment, etc. The present paper focuses on one area under each broad priority area category. For instance, under transport area the paper examined some of the best ways to channelize urban mobility and public transport. In case of public safety and security, the paper examined best techniques of taking care about safety of citizens. Among public services, health and education are chosen because these form the core elements for building up social infrastructure of the country (World Economic Forum, 2016). Out of various utilities, water management is taken up in this paper as there are around 76 million people in India who do not have access to safe water (Burgess, 2016). Moreover, provision of safe drinking water and proper disposal of waste water is regarded as a precondition for healthy and disease free citizens of a nation. In the last, under IT connectivity and social networking the paper focuses on significance of IT infrastructure and digitalization.

Since the announcement of Smart Cities Mission in India in 2014, various studies have been conducted which have vividly described the essentially required components of Smart Cities and how they can be

Table 1. High priority pillars and core smart city infrastructure elements for India

S. No.	High Priority Pillars	Core Infrastructure Elements
1.	Transport	Urban Mobility and Public Transport
2.	Public Safety and Security	Safety and Security of Citizens
3.	Public Services	Health and Education, Sanitation, Affordable Housing & Good Governance
4.	Utilities	Water Supply, Electricity Supply & Sustainable Environment
5.	Social networking	Robust IT Connectivity and Digitalization

Source: Adapted from United Nations ESCWA (2015) & Ministry of Urban Development (2015).

built in India. However, keeping in view the limited resources available with a developing country like ours, a more pragmatic approach would be to first learn from the best international experiences and approaches, and then apply those in Indian context. Limited studies have tried to focus on this aspect. Thus the present study aims to bridge this crucial gap in research. Specifically, the main objective of this study is to examine the best practices related with some core areas of a Smart City and to provide suggestions regarding their applicability for the development of Smart Cities in India.

BACKGROUND

The existing literature highlights the multifaceted nature of 'Smart Cities'. From multiple definitions to dimensions, existing studies have discussed Smart Cities and its components in their own different ways. The essential components vary not only from country to country but also from city to city, even within the same country because of the different requirements of every city. With the aim to select and discuss the important Smart City components essential for Indian cities, table 2 shows the recent studies conducted on Smart Cities.

For the purpose of this study, data was collected from secondary sources by reviewing the literature related to Smart Cities. This includes reports of United Nations ESCWA, Inter-American Development Bank (IDB), Ministry of Urban Development, NASSCOM & Accenture, etc. In addition, various case studies on usage of modern ICTs used to build Smart Cities around the world were also reviewed.

CORE SMART CITY INFRASTRUCTURE ELEMENTS

The successful international approaches and experiences related with the major thrust areas of a Smart City are considered as a guiding light for the Indian Smart Cities in near future. In the following section, first of all, best practices from across the world are presented under different aspects and then their applicability in Indian context is discussed.

Urban Mobility and Public Transport

The success of a modern city is significantly based upon its mobility effectiveness i.e. the efficiency with which goods and people move throughout the city (Wipro Earthian, 2015). Growing urbanisation, modern facilities and various job opportunities have made cities the hub of economic growth. The World Bank has estimated that India will have 600 million people living in cities by 2031 (Urban Transport, 2014). However, only around 20 Indian cities with over 500,000 population have any sort of systematic transport systems for public. Further, India's fatality and accident rates are extremely high and in fact, one of the highest in the world, mostly affecting the vulnerable and the poor who do not own any means of transportation (Padode et al., 2016). Furthermore, according to global current traffic index, five Indian cities are in the list of top 20 cities that have worst traffic conditions in the world with Mumbai topping the list (Traffic Index Rate, 2017).

Specifically, the main challenges to Indian transportation system include road congestion, inadequate public transportation systems, low and inefficient use of technology in transportation systems, increasing level of air pollution due to automobiles, etc. In this regard, innovative and technology driven mobility

Table 2. Studies related with smart cities

Author (Year)	Objective	Findings
Mahizhnan (1999)	To discuss Smart Cities with reference to the transformation of Singapore towards an Intelligent Island.	A massive infrastructure development programme by the government involving IT education, IT infrastructure, and IT economy is responsible for transforming Singapore into a Smart City. Improving the quality of life of the people should be the main focus of all the Smart Cities.
Nam & Pardo (2011)	To explore how a particular city can be transformed into a smart one, by drawing on recent practices of smart cities.	Strategic principles aligning to the three main dimensions (technology, people, and institutions) of smart city are integration of infrastructures and technology-mediated services, social learning for strengthening human infrastructure, and governance for institutional improvement and citizen engagement.
Chourabi et al. (2012)	To propose a framework to understand the notion of smart cities.	Eight critical factors of smart city initiatives include: economy, management and organization, governance, technology, policy context, people and communities, natural environment, and built infrastructure. These factors form the basis of an integrative framework that can be used to examine how local governments are envisioning smart city initiatives.
Cocchia (2014)	To study the literature about Smart City and Digital City and to investigate the birth and the development of these two concepts along with finding their shared features and the differences between the two.	Regarding the definition of both Smart City and Digital City, it is observed that a shared and acknowledged definition of the two is largely missing. Regarding the development of the two concepts, Digital City had a quite stable development while Smart City had a slow development phase till 2010.
Mundhe et al. (2014)	To provide Smart Water solutions for Aurangabad City, Maharashtra, India.	The insufficient water condition of the city can be improved by using Geospatial technologies for water management and working with the KML (Keyhole Markup Language) and Google API (Application programming interfaces) for locating the available ground water resources and analysing their properties for water usage.
Neirotti et al. (2014)	To provide a comprehensive understanding of the concept of Smart City through the study of suitable application domains, namely: natural resources and energy, transport and mobility, buildings, living government, and economy and people.	The evolution patterns of a Smart City mainly depend on its local contextual factors. In particular, economic development and structural urban variables are likely to influence a city's digital path, the geographical location to affect the strategy, and density of population, with its associated congestion problems, is an important component to determine the routes for the Smart City implementation.
Albino et al. (2015)	To discuss the definitions, dimensions, performance measures and few initiatives of Smart Cities.	The meaning of a Smart City is multi-faceted. A universal fixed system of smart cities is very difficult to define with the variety of characteristics of cities worldwide. Also, the different definitions articulated by particular cities calling themselves "smart cities" lack universality.

solutions can help to provide our cities with efficient and smart transport systems (Mehra and Verma, 2016). Smart and intelligent cities, all around the world, have adopted efficient systems and techniques for smart transportation. Analysing their strategies and methods can help the Indian cities in their endeavour to enhance their transport systems.

On-Going Plans in India

Indian government had planned to launch National Common Mobility Card which would act like "a single e-purse" and would enable the user to reach his destination by travelling in all required public transport

Table 3. Smart Mobility initiatives from famous cities

<table>
<tr><th>City (Country): Smart Initiative</th><th>Impact / Outcomes</th><th>Source(s)</th></tr>
<tr><td colspan="3">1. Traffic Management</td></tr>
<tr><td>a) London (England):
Transport for London (TfL): An innovative open platform that makes real-time information available that is used for Congestion Charging (done using number plate recognition), intelligent road network systems, etc.
Pedestrian SCOOT (Split Cycle Offset Optimization Technique), a TfL initiative, uses advanced cameras and smart sensor technologies to keep an eye on the pedestrian traffic at busy roads and junctions.</td><td>• Their website provides integrated data that has helped travellers in effectively planning their journey.
• Data feed has helped in saving millions of pounds.
• Pedestrian SCOOT has enhanced road safety for both cyclists and pedestrians.</td><td>Microsoft Services (2011); TfL (2014); United Nations ESCWA (2015)</td></tr>
<tr><td>b) Pittsburgh (USA)
Adaptive Traffic Signals are deployed in city's busy areas. It aims to optimize operations without having to widen the roads, re-route traffic, etc.</td><td>• 40% reduction in wait time for vehicles.
• 26% decreased travel time.
• 21% lesser vehicle emissions.</td><td>Berst et al. (2013)</td></tr>
<tr><td colspan="3">2. Smart Parking</td></tr>
<tr><td>a) San Francisco (USA):
SFPark: For managing parking. Sensors are used to get real time information that helps to detect nearest accessible parking slot. Motorists get information via text messages so that they find a parking slot easily, saving fuel and time.</td><td>It has led to reduction in parking rates, ease in finding a parking space, improvement in parking availability, reduction in greenhouse gas emissions, etc.</td><td>Sensors for Smart Cities (2015); SF Park Pilot Evaluation (n.d.)</td></tr>
<tr><td>b) Boulevard Victor Hugo, Nice (France):
Over 200 sensors and detecting instruments are used to get information on parking, traffic, street-lighting as well as environmental quality as experienced in real time.</td><td>'Smart Parking services' has resulted in 30 percent decrease in traffic congestion, reduction in air pollution, along with a growth in revenues from parking.</td><td>Berst et al. (2013)</td></tr>
<tr><td colspan="3">3. Last Mile Connectivity</td></tr>
<tr><td>a) Bogota (Colombia):
(i)TransMilenio: It is a bus transportation system that runs on dedicated lanes in main city roads and utilizes more than 400 km of bike lanes in the city.
(ii) Moovit: It is the name given to a website and a mobile application that combines TransMilenio with the integrated routes allowing city-people to decide their daily route.</td><td>20 percent residents, attracted by speed and low costs, have shifted to public transportation systems.</td><td>Bouskela et al. (2016)</td></tr>
</table>

systems by paying a single fare (National Common Mobility Card, 2016). This system is expected to provide an ecosystem in the city for digital payment of government fees & taxes, utility bills, transit charges, parking and other services charges. In this context, various Indian smart cities have initiated common cards like Bhubaneshwar's Common Payment Card System (CPCS), Ahmedabad's Janmitra Card, smart cards in Raipur, Chandigarh, etc. These are prepaid debit cards that facilitate everyday transactions, such as payment of fees, taxes, utility bills, transit charges, parking and other services charges. These are expected to reduce cash transactions and promote digital economy (Chatterji, 2018).

Indian traffic is very different from the western traffic due to its peculiar characteristics of being non-lane based and chaotic. Thus, Intelligent Transport Systems (ITS), used for efficient traffic management in developed countries, cannot be used as such in India. ITS techniques have to undergo innovation and adaptation to complement the distinct traffic characteristics of Indian roads (Sen and Raman, 2012). Intelligent Traffic Management Systems (ITMS) have been successfully implemented in cities like Pune, Jaipur, Kolkata and Ahmedabad, etc. to decongest the city roads and are soon expected to be installed in other cities like Vijayawada, etc. (Traffic Cops, 2018).

As far as traffic in a highly populated country like India is considered, an effective way to manage it would be to have a properly running public transport system, like BRT (Bus Rapid Transit) system. Although it was launched successfully in some cities like Pune, Ahmedabad, etc. but these are hardly few examples of a country having more than 4000 cities and towns. Thus, efforts should be made to expand this concept to other highly populated and polluted cities of India, at the earliest. There has also been cases where the infrastructure for BRT is in ready form but lack of political will acts as a hurdle in its proper working, for instance, the recent case of Amritsar city where its infrastructure was laid throughout the city before February, 2017 elections in the state but later on this dream project became a white elephant for district administration.

Way-Forward

In terms of ICTs, smart traffic lights (that regulate and manage traffic depending upon real-time traffic information), video-analytics based surveillance (that compensate for the traffic personnel's low capacity to handle traffic at peak hours and can help in getting real-time simulations and traffic projections), smart tolls (that facilitate in reducing the queues and can effectively collect user charges), GPS-based vehicle tracking (help in identifying the violators of traffic-rules), online fines and dues payment (provides transparency in punishing the traffic-rules violators thereby decreasing corruption instances) etc. can be used for managing the traffic (NASSCOM & Accenture, 2015). In terms of physical infrastructure for intelligent transit, smart parking and single fare card systems should be adopted. The former involves cameras, parking sensors, etc. to provide smart management of parking spaces while the latter comprises of a single card for fare payment on different participating public transport systems.

Safety and Security of Citizens

From the point of view of an average citizen, safety of public is one of the most noticeable and perhaps a key responsibility of the cities (Berst et al., 2013). It is also considered as a prerequisite to have a socially and economically attractive environment that attracts investments for the growth and development of a city (NASSCOM & Accenture, 2015). With surging crimes and extra pressure on cities due to fast modernization, an integrated framework of public safety and a secure environment for businesses and citizens is vital. That's why modern day disaster prevention personnel, fire and police departments, surveillance and monitoring agencies, NGOs and others use ICTs to enhance security of the citizens (Padode et al., 2016). In this context, the key practices across nations which can help India to build safe and secure cities are mentioned in table 4.

On-Going Plans in India

The Integrated Control Command Centres (ICCCs) aims to enhance safety, security and provide better public services in the cities. Such centres assist in city surveillance which ensures citizen safety by acting as a deterrent to criminal activities and provide environment sensors which help in disaster management as well as in controlling environmental pollution. Such centres have already become operational in various cities, e.g., Ahmedabad, Vadodara, Surat, Pune, Nagpur, Visakhapatnam, Bhopal and Kakinada, etc.

Table 4. Smart safety and security initiatives from famous cities

City (Country): Smart Initiative	Impact / Outcomes	Source(s)
a) Thailand: Department of Special Investigation (DSI) had large data sets of more than a million records that were collected from multiple sources in structured and unstructured formats, like videos, images, documents, etc. This voluminous data made data-mining and extraction difficult and gave broad and unclear results. In response to all this, DSI implemented a Big Data Solution. It gave investigating officers self-service business intelligence tools and data-management capabilities.	DSI is able to improve its accuracy and has shortened the criminal case investigation time from 2 years to just 15 days. It also plans to implement and manage its own private cloud to manage its confidential data.	Padode et al. (2016)
b) Miami (USA): Advanced analytics are being used by Miami-Dade Police Department to close the toughest robbery cases. Advanced models are used to analyse tough robbery cases against the department's historical crime data.	With this, the robbery unit is able to uncover insights that are vital to solve the cases. The detectives use this tool as a second-chance on what they considered to be dead-end cases.	IBM (2013b)

Way-Forward

With respect to India, the top 53 cities with population of more than one million, also known as mega cities, have an average crime rate of 345.9 against national average of 215.5 (National Crime Records Bureau, 2014). Thus, public safety, particularly of women in India, using smart tools is the need of hour. This requires coordination of different agencies to supervise and act in concerned public areas while respecting citizens' rights. Using CCTV surveillance (security cameras with video surveillance data being saved and monitored at control centres), mobile applications (that allows citizens to interact with authorities to report on city's safety problems, request services or other concerns), analytics (that report crime distribution by location, frequency etc. to facilitate planning and to help project crime patterns by identifying areas with recurring issues), sensors and panic buttons in public places (that trigger alert to policemen in case of an emergency), etc. can help to build an effective service system with smaller teams (Bouskela et al., 2016; NASSCOM & Accenture, 2015). One of India's projects named Police Crime and Criminal Tracking Network & Systems (CCTNS) that was launched in 2009, is already a good move in this direction. It aims at having an integrated and comprehensive system that would enhance policing functions by adopting e-Governance practices (National Crime Records Bureau, n.d.). However, till February, 2017 only 77 percent of data migration to online platform has been done and citizen portal services have still not been launched in all the states (National Crime Records Bureau, 2017). Thus, this issue need to be addressed by the concerned authorities at the earliest in order to make Indian cities safer and secure.

Health and Education

Health and education have been recognized as two strong pillars of sustainable development of a nation. However, India's innovation-lacking education and poor health outcomes are the most important developmental challenges which can be addressed using ICTs (Padode et al., 2016). New age ICTs aims at strengthening and transforming the delivery of vital healthcare, educational and other human services and Smart Cities paved the way to adopt these ICTs to improve the life of its citizens.

Table 5. Smart health and education initiatives from famous cities

City (Country): Smart Initiative	Impact / Outcomes	Source(s)
1. Smart Healthcare		
Estonia: 100 percent population has a digital identity through an ID card that provides electronic medical record for all citizens. This record is accessible to patients, doctors, hospitals, clinics and even pharmacies to monitor each citizen's health.	**i)** It has helped doctors to access the patient's X-ray images or blood test results directly from their offices. **ii)** It provides vital information about the patient like blood group, allergies, medical history, etc. in emergency situations. **iii)** It helps patients get medicines from pharmacies using e-prescription system.	Bouskela et al. (2016)
2. Smart Education		
Tainan (Taiwan): **City Education Center:** This is responsible for all the technology needs of public schools in the city. They have also migrated to a centralized IT infrastructure based on a private cloud model.	(i) The city officials expected to annually save US $ 344,000 in support and hardware costs. **(ii)** This facilitates the students to have an increased access to educational resources and teachers can improve classroom materials using this new technology.	Berst et al. (2013)

Smart healthcare aims to provide effective assistance in real time. Interconnected healthcare systems that use ICTs to improve emergency and various other health related services are its key components. These mainly include electronic patient records that facilitate sharing of information across hospitals, clinics, and pharmacies; telemedicine that helps to extend the reach of medical services etc. (Bélissent, 2010). Smart education is also equally essential tool as it helps in not only improving the physical performance of the cities but also prepare students for life in a highly complex, and highly technological future-world (Smart Education, n.d.). In this regard, table 5 shows some of the prominent practices from other nations in the areas of health and education.

On-Going Plans in India

ICCCs also contribute to smart health and smart education initiatives. For instance, smart city Rajkot is working in the arena of smart healthcare and already has smart schools that promotes e-education and also has an e-library. On the other hand, Ahmedabad is set to launch smart card based immunisation tracking system for children till the age of 15 years. This immunisation tracking system aims to facilitate all the paediatricians in the city to register dose of vaccination given to a child with a mobile app. The QR code enabled immunisation cards can be accessed by any doctor using the mobile phone for decoding the QR code. All details of immunisation will be available on his phone. Even the parents will be getting the reminders for any pending immunisation until their child turns 15 years.

Way-Forward

In India, the major challenges in the way of these essential services include low access to healthcare services, high out-of-pocket expenditure, increase in non-communicable diseases like heart diseases, diabetes, etc., low public investment and education supply and demand gap (Randhawa & Sidhu, 2011;

Padode et al., 2016). The situation can be dramatically improved by using information insights, coupled with clinical collaboration, thereby leading to improvement in patient safety and outcomes, quality of care, etc. which also ensures cost-effective healthcare. Analytics can help arrange population from high risk to low risk and accordingly doctors and health workers can be deployed. Other solutions include smart health card, e-health records, medical simulation, online databases with real-time information on availability, etc. for making healthcare services in urban India more effective and efficient. For improving the education sector in India, ICTs can be used in multiple ways like smart classrooms, CCTV surveillance in schools, GPS tracking in school buses, online recruitment of teachers, online centralized student admission system, etc. (NASSCOM & Accenture, 2015).

Water Management

Often referred as 'Elixir of Life', water is indispensable for humans, animals and plants. Worldwide, various international and national organizations have emphasized on 'Right to Water and Sanitation' and have recognised that clean drinking water and sanitation are fundamental to the attainment of all human rights. However, global urbanization has posed a wide array of water-related challenges – principal among them is supply of clean drinking water and disposal of waste water. In this context, the Indian government has listed core infrastructural components for effective water management of a Smart City which include "adequate water supply" and "sanitation, including solid waste management" (Ministry of Urban Development, India, 2015, p. 5). Thus, it is imperative to properly manage the inflow and outflow of water in order to transform cities into Smart Cities. Smart and intelligent cities from other nations have adopted various systems and techniques for efficient water management. Few of them are mentioned in Table 6.

On-Going Plans in India

Water supply in India is characterized by several deficiencies, like poor availability, inadequate treatment, etc. However, 24x7 water now exists in some Indian cities like Hubli Dharwad, Belgaum, and Gulbarga and it has depicted plethora of benefits, for example, introduction of water meters and payment of tariffs based on water use reduces citizens' private expenses for securing water, generates revenues for municipal water service provider, and also promotes water conservation. Also, another important component of smart cities: i.e., water metering (though mechanical) now exists in cities like Udaipur, Pimpri Chinchwad, Bangalore etc. Even ICCCs are also being used for waste water management practices in cities like Vadodara, Rajkot, etc. in order to build upon water conserving infrastructure (Smart City, 2017).

Way-Forward

By taking clues from some of the best practices mentioned in table 6, all the Indian smart cities can make effort for 24X7 water supply system along with waste water recycling and storm water drainage system. Proper water infrastructure comprising of water treatment plant, pumping station, clear water and recycled water reservoirs, and transmission network should be made operational in all cities to make them 'Smart'. In urban areas, the implementation of smart water management systems can make significant improvements in water distribution, helping to decrease losses due to non-revenue water, and helping to enhance waste-water and storm water management (Berst et al., 2013). As per Indian Ministry of Urban

Table 6. Smart water management initiatives from famous cities

City (Country): Smart Initiative	Impact / Outcomes	Source(s)
1. Conserving Water		
a) Barcelona (Spain): Sensor technology is applied in the irrigation system which offers live data on temperature, humidity, sunlight, atmospheric pressure and wind velocity. Real time data is transmitted and a system of electro valves is remotely controlled for water delivery across the city.	This initiative has helped gardeners to decide what the plants require so that they can adjust to avoid overwatering. The data can be checked on an online map by interested citizens, sitting anywhere in the world. This initiative has helped the city to attain a 25 percent increase in water conservation.	Adler (2016); Laursen (2014)
b) Tel Aviv (Israel): Remotely controlled irrigation system is used to conserve the water that is being used for irrigating the public gardens. The system works in real time and supervises the sprinklers in all gardens.	This irrigation method helps in opening and shutting the sprinklers automatically. It also monitors the quantity of water allotted to each section of the garden, and receives detailed reports that help to track water consumption.	Tel Aviv Smart City (n.d.)
2. Water Storage and Purification		
a) Singapore: **NEWater and Desalination:** The former is a wastewater recycling initiative that involves transforming water from toilets/drains in multiple stages and the latter involves desalinating sea water to use it for drinking purposes. To avoid wastage and reduce water leaks in the distribution system, sensors are used which alert the central server via Wi-Fi network in case of any leakage.	In Singapore, around 25 percent of the every-day consumption of water comes from the sea and 40 percent from the NEWater initiative. Monitoring through sensors assures the residents that treated water reaching them is safe for drinking purpose.	Bouskela et al (2016); Singapore Water Story (n.d.)
b) The Netherlands: **Digital Delta:** It is an innovative cloud-based system that harnesses insights from Big Data to transform flood control and management of the entire Dutch water system. This system helps to integrate and analyse water data from a wide range of data sources, that includes precipitation measurements, water level and water quality monitors, radar data, model predictions as well as current and historic maintenance data from pumping stations, dams, etc.	The initiative provides water experts with a real-time intelligent dashboard which combines processes and visualizes data from multiple organizations. This new management system addresses concerns ranging from the quality of drinking water, to the increasing frequency and impact of extreme weather-related events, to the risk of floods and droughts. By modelling weather events, they are able to determine the best course of action.	IBM (2013a)

Development (2015, p. 6), the water management in Smart Cities can be done through "Smart meters & management, leakage identification, preventive maintenance and water quality monitoring". Smart metering and recycling of waste water can make a considerable contribution to efficiently manage water in cities. For water conservation, sensors could be used. Presently, pipe leakages in India contribute up to 50 percent water wastage in major Indian cities (Sensors for Smart Cities, 2015). With sensors fitted on each pipe, leakages can be easily detected and corrected before any heavy loss. Also, the irrigation network in public parks can automatically be turned off whenever rain is observed to save water.

Robust IT Connectivity and Social Networking

Social networking as another strong pillar of a Smart City involves interaction and communication among people and residents in digital arenas. For this to effectively take place, building social capital in the form of smart people is essential. A Smart City would be short-lived and will fail if focus is just put on

Table 7. Smart IT connectivity and social networking initiatives from famous cities

City (Country) Smart Initiative	Impact / Outcomes	Source(s)
1. Access to Internet Services		
a) Maribor (Slovenia): Free wireless internet access to students, citizens and tourists.	This has led to round the year availability of important services, e-services for disadvantaged people, enhanced interaction among local communities, increased city visibility, tourism development, etc.	Maribor (n.d.)
b) Chihuahua (Mexico): Chihuahua Digital City program: Free Wi-Fi internet access to residents at various public places.	It has led to increased internet access and has encouraged citizens to use public spaces for various things including study, business, communication, social activities promotion, citizen participation, assistance to residents through various organisations that support pensioners, women, etc.	Bouskela et al. (2016)
2. Well Connected People		
a) Scottsdale (Arizona): **Speak Up Scottsdale:** It is a website that functions as a discussion forum and provides citizens a platform to give new ideas and provide feedback. It also has the capability to launch surveys and polls.	Officials can take notice of things that are considered important by several residents. A discussion thread led to enhanced safety for pedestrians after a shopping centre was newly expanded.	Padode et al. (2016)
b) Ningbo (China): **iCityBoss app:** The app facilitates connection between citizens and city and ensures participatory citizenship. It collects and aggregates data from multiple sources: government, business and institutions and act as a centre point for interaction for the various Smart City services.	Citizens can effectively interact with local authorities to communicate their problems and provide suggestions. It has also led to effective traffic management, resource sharing between different municipal departments, etc.	Bouskela et al. (2016)

its planning, designing, building infrastructure, technologies, etc. while its citizens remain unaware with a lack of commitment towards these efforts. People, in an inclusive society, should essentially acquire e-skills so that they can utilize, manipulate and personalize the available information that facilitates them in decision making (United Nations ESCWA, 2015). Table 7 highlights some of the best examples of smart social-networking techniques adopted by cities of modern world.

On-Going Plans in India

In India, free access to public Wi-Fi is offered at limited places. Budding smart cities like Ahmedabad, Pune, Nagpur, Bhopal, Jaipur, etc. have started providing free city wifi from their ICCCs. They are also providing various citizen services via a single platform which in turn assists in smart governance of cities.

Way-Forward

In India, there are several issues like poor information accessibility, increased digital divide, lack of responsible usage of internet, etc. Various Government and private firms aim to deliver many internet based services especially to urban residents, but they have been unable to do so. Thus, in order to strengthen this pillar of Smart City development, public Wi-Fi at no or affordable price could be provided to citizens that could be accessed anywhere and from any device. The access to various city services can be significantly improved by giving residents multi-channel (mobile, web and face-to-face) service facilities (NASSCOM & Accenture, 2015). This would help to improve their interaction with the state authorities. Also, ICTs should be used for training and capacity building of municipal authorities so that they can provide smart citizen services.

LIMITATIONS AND FUTURE RESEARCH DIRECTIONS

The study is limited to only five aspects which are considered vital for transforming a city into a Smart City. However, there are also other essential infrastructural elements required in a Smart City (e.g., affordable housing, 24X7 electricity supply, solar/wind energy management, rain water harvesting, good governance, proper solid waste management, sustainable environment, etc.) which are not considered in this study due to time constraint and these can be taken up in future research.

The study highlighted only few best examples for each of the five areas considered most important for Smart City development. Other best practices from developed and technologically advanced nations could be studied in future research.

CONCLUSION

Smart City projects are regarded as a key strategy of governments around the world to enhance and improve the quality of life of billions of people who are living in cities (Cocchia, 2014). Similarly, the Indian Smart Cities mission is an essential urban renewal and transformational program of the Indian government that focuses on retrofitting and redevelopment of Indian cities (Ministry of Urban Development, 2015). The usage of the word 'Smart' describes the essentiality of innovative and transformational technologies in cities that would revolutionize all forms of city governance. For this, the need of the hour is to use appropriate ICTs in transportation, public safety and security, public services, utilities, and social networking of residents that would enable cities to use technologies, data and information to improve services and infrastructure for its citizens. This comprehensive development will have multiple benefits like better quality of life, employment creation (IT professionals would be required for programming, data analytics, network and system integration, high-end consulting, etc.) and enhanced incomes for all (as Smart Cities would surely attract investors which will boost industry and that will raise demand for workforce), leading to inclusive and sustainable growth.

Further, other government initiatives like AMRUT i.e. Atal Mission for Rejuvenation and Urban Transformation (that aims to develop urban infrastructure), HRIDAY i.e. Heritage City Development and Augmentation Yojana (that aims to holistically improve important heritage cities), Swachh Bharat Mission (for cleaning roads, streets and other infrastructure), which are co-existing, need to synergise with Smart City mission to develop and transform Indian cities more effectively. The Economic Survey of 2016-17 emphasized cities as being dynamos for moving from "Competitive Federalism to Competitive Sub-Federalism" which would facilitate faster growth and development of the nation (Government of India, 2017, p. 300). The competitive Smart Cities mission, which aims to select and develop 100 cities in India, is thus a right move. The development of smart systems in various cities of our developing nation will have a reciprocating effect on its social as well as economic growth trajectory.

ACKNOWLEDGMENT

This research received no specific grant from any funding agency in the public, commercial, or not-for-profit sectors.

REFERENCES

Adler, L. (2016). *How smart city Barcelona brought the internet of things to life.* Retrieved July 3, 2017, from http://datasmart.ash.harvard.edu/news/article/how-smart-city-barcelona-brought-the-internet-of-things-to-life-789

Albino, V., Berardi, U., & Dangelico, R. M. (2015). Smart cities: Definitions, dimensions, performance, and initiatives. *Journal of Urban Technology*, *22*(1), 3–21. doi:10.1080/10630732.2014.942092

Bélissent, J. (2010). *Getting clever about smart cities: New opportunities require new business models.* Retrieved July 10, 2017, from http://193.40.244.77/iot/wp-content/uploads/2014/02/getting_clever_about_smart_cities_new_opportunities.pdf

Berst, J., Enbysk, L., & Williams, C. (2013). Smart cities readiness guide: The planning manual for building tomorrow's cities today. *Smart Cities Council.* Retrieved from http://www.corviale.com/wp-content/uploads/2013/12/guida-per-le-smart-city.pdf

Bouskela, M., Casseb, M., Bassi, S., Luca, C. D., & Facchina, M. (2016). The road toward smart cities: Migrating from traditional city management to the smart city. *Inter-American Development Bank (IDB).* Retrieved June 5, 2017, from https://publications.iadb.org/bitstream/handle/11319/7743/The-Road-towards-Smart-Cities-Migrating-from-Traditional-City-Management-to-the-Smart-City.pdf?sequence=3

Burgess, T. (2016). Water: At what cost? The state of the world's water 2016. *WaterAid.* Retrieved July 4, 2017, from https://www.wateraid.org/uk/~/media/Publications/Water--At-What-Cost--The-State-of-the-Worlds-Water-2016.pdf?la=en-GB

Chatterji, T. (2018). Digital urbanism in a transitional economy– A review of India's municipal e-governance policy. *Journal of Asian Public Policy*, *11*(3), 334–349. doi:10.1080/17516234.2017.1332458

Chourabi, H., Nam, T., Walker, S., Gil-Garcia, J. R., Mellouli, S., Nahon, K., ... Scholl, H. J. (2012). Understanding smart cities: An integrative framework. In *Proceedings of 45th Hawaii International Conference on System Sciences*. Maui, HI, USA: IEEE.

Cocchia, A. (2014). Smart and digital city: A systematic literature review. In R. P. Dameri & C. R. Sabroux (Eds.), *Smart city: How to create public and economic value with high technology in urban space* (pp. 13–43). Springer International Publishing.

Government of India. (2017). *Economic Survey 2016-17.* Retrieved June 6, 2017, from http://indiabudget.nic.in/es2016-17/echapter.pdf

IBM. (2013a). *IBM harnesses power of big data to improve Dutch flood control and water management systems.* Retrieved June 29, 2017, from https://www-03.ibm.com/press/us/en/pressrelease/41385.wss

IBM. (2013b). *Miami-Dade police department: New patterns offer breakthroughs for cold cases.* Retrieved June 21, 2017, from http://smartcitiescouncil.com/system/tdf/public_resources/Miami_Dade%20 police.pdf?file=1&type=node&id=200

Laursen, L. (2014). *Barcelona's smart city ecosystem.* Retrieved June 5, 2017, from https://www.technologyreview.com/s/532511/barcelonas-smart-city-ecosystem/

Mahizhnan, A. (1999). Smart cities: The Singapore case. *Cities (London, England), 16*(1), 13–18. doi:10.1016/S0264-2751(98)00050-X

Maribor. (n.d.). *Free wireless internet access in the municipality of Maribor.* Retrieved May 26, 2017, from http://www.smartcitymaribor.si/en/Projects/Smart_Living_and_Urban_Planning/Free_wireless_internet_in_the_Municipality_of_Maribor/

Mehra, S., & Verma, S. (2016). Smart transportation - Transforming Indian cities. *Grant Thornton India.* Retrieved May 24, 2017, from http://www.grantthornton.in/globalassets/1.-member-firms/india/assets/pdfs/smart-transportation-report.pdf

Microsoft Services. (2011). *London transport manages 2.3 million website hits a day with new data feed.* Retrieved May 26, 2017, from http://smartcitiescouncil.com/system/tdf/public_resources/London%20 Transport%20and%20its%20very%20busy%20website.pdf?file=1&type=node&id=542

Ministry of Urban Development. (2015). *Smart cities: Mission statement & guidelines.* Retrieved May 10, 2017, from http://164.100.161.224/upload/uploadfiles/files/SmartCityGuidelines(1).pdf

Mundhe, M., Pandagale, P., & Pathan, A. K. (2014). Smart water for Aurangabad city. *International Journal of Advanced Research in Computer Science and Software Engineering, 4*(10), 649–652.

Nam, T., & Pardo, T. A. (2011). Conceptualizing smart city with dimensions of technology, people, and institutions. In *Proceedings of the 12th Annual International Digital Government Research Conference: Digital Government Innovation in Challenging Times.* New York, NY: ACM. Retrieved April 6, 2018, from https://dl.acm.org/citation.cfm?id=2037602

NASSCOM & Accenture. (2015). *Integrated ICT and geospatial technologies framework for 100 smart cities mission.* Retrieved May 28, 2017, from http://agiindia.com/pdf/Integrated-ICT-Geospatial-Technologies-2015%20(Nasscom-Accenture).pdf

National Common Mobility Card. (2016, November 23). *Press Information Bureau.* Retrieved May 24, 2017, from http://pib.nic.in/newsite/PrintRelease.aspx?relid=154158

National Crime Records Bureau. (2014). *Crime in India 2013 Statistics.* Ministry of Home Affairs. Retrieved June 27, 2017, from http://ncrb.nic.in/StatPublications/CII/CII2013/Statistics-2013.pdf

National Crime Records Bureau. (2017). *CCTNS Pragati Dashboard (31.05.2017).* Retrieved July 4, 2017, from http://www.ncrb.gov.in/BureauDivisions/CCTNS/CCTNS_Dashboard/PRGATI%20dashboard%2028.02.2017%20ver%209.0%20for%20MHA.pdf

National Crime Records Bureau. (n.d.). *Crime and Criminal Tracking Network & Systems (CCTNS).* Retrieved July 4, 2017, from http://www.ncrb.gov.in/BureauDivisions/CCTNS/cctns.htm

Neirotti, P., De Marco, A., Cagliano, A. C., Mangano, G., & Scorrano, F. (2014). Current trends in smart city initiatives: Some stylised facts. *Cities (London, England)*, *38*, 25–36. doi:10.1016/j.cities.2013.12.010

Padode, P., Padode, F., Mali, D., Gawde, S., & Yadav, S. (2016). *Smart cities India readiness guide: The planning manual for building tomorrow's cities.* Retrieved June 7, 2017, from http://india.smartcitiescouncil.com/system/files/india/premium_resources/Smart_Cities_Council_India_Readiness_Guide_v2016-02.pdf?file=1&type=node&id=3330

Randhawa, G., & Sidhu, A. S. (2011). Status of public health care services during the process of liberalization: A study of Punjab. *Pravara Management Review*, *10*(2), 10–15.

Sen, R., & Raman, B. (2012). *Intelligent transport systems for Indian cities.* Paper presented at 6th USENIX/ACM Workshop on Networked Systems for Developing Regions.

Sensors for Smart Cities. (2015). *ESRI India*, 9. Retrieved May 24, 2017, from http://www.esri.in/~/media/esri-india/files/pdfs/news/arcindianews/Vol9/sensors-for-smart-cities.pdf?la=en

SF Park Pilot Evaluation. (n.d.). Retrieved July 5, 2017, from http://sfpark.org/about-the-project/pilot-evaluation/

Singapore Water Story. (n.d.). Retrieved July 4, 2017, from https://www.pub.gov.sg/watersupply/singaporewaterstory

Smart Cities Mission Integrated Command and Control Centre. (2017). In *National workshop for project management consultants of smart cities.* Retrieved http://smartcities.gov.in/upload/uploadfiles/files/ICCC.pdf

Smart Education. (n.d.). Retrieved June 24, 2017, from http://www.ibigroup.com/new-smart-cities-landing-page/education-smart-cities/

Tel Aviv Smart City. (n.d.). Retrieved May 27, 2017, from https://www.tel-aviv.gov.il/en/WorkAndStudy/Documents/Tel-Aviv%20Smart%20City%20(pdf%20booklet).pdf

Traffic cops mull over intelligent system to decongest city. (2018, September 10). *The Times of India.* Retrieved February 27, 2019 from https://timesofindia.indiatimes.com/city/visakhapatnam/traffic-cops-mull-over-intelligent-system-to-decongest-city-roads/articleshow/65746292.cms

Traffic Index Rate. (2017). Retrieved March 21, 2017, from https://www.numbeo.com/traffic/rankings_current.jsp

Transport for London. (2014). *Tfl to launch world-leading trials of intelligent pedestrian technology to make crossing the road easier and safer*. Retrieved May 24, 2017, from https://tfl.gov.uk/info-for/media/press-releases/2014/march/tfl-to-launch-worldleading-trials-of-intelligent-pedestrian-technology-to-make-crossing-the-road-easier-and-safer

United Nations ESCWA. (2015). *Smart cities: Regional perspectives.* Retrieved April 5, 2017, from https://worldgovernmentsummit.org/api/publications/document/d1d75ec4-e97c-6578-b2f8-ff0000a7ddb6

Urban Transport. (2014). Retrieved March 24, 2017, from http://www.worldbank.org/en/country/india/brief/urban-transport

Wipro Limited. (2015). *Mobility in Urban India*. Retrieved May 24, 2017, from http://www.wipro.org/earthian/pdf/mobility-in-urban-India.pdf

World Economic Forum. (2016). *Inspiring future cities & urban services.* Retrieved March 21, 2017, from http://www3.weforum.org/docs/WEF_Urban-Services.pdf

KEY TERMS AND DEFINITIONS

Information and Communication Technology (ICT): ICT refers to technologies that capture, transmit and display data and information electronically and includes all devices, applications, and networking elements that allow organizations as well as people to connect in the digital world.

Public Services: Services like health care, education, sanitation, waste disposal, etc. organized by a government or an official body in order to benefit all the people of society.

Smart Cities Mission: A mission launched by Government of India in 2015 which aims to convert 100 urban areas into Smart cities through extensive renewal and retrofitting programs.

Smart City: An urban area using high-end information and communication technologies to efficiently manage resources leading to enhanced government services and citizen welfare.

Smart Education: A technology driven learning system that enhances the capability of the educators while enabling the learners to learn more efficiently, effectively, comfortably, and flexibly.

Smart Mobility: A technology driven transport system that aims to transport people and goods around the globe further and faster.

Social Networking: Social networking is the use of internet-based platform for the purpose of connecting to people you may or may not know for personal or professional purpose.

Water Management: Water management includes various activities like proper planning, efficient distribution, and optimal use of water resources so that it can meet current and future needs.

Chapter 7
Municipal Solid Waste Management:
Case Study on Smart City Tirunelveli

Raghavi K.
Anna University Chennai – Regional Office Tirunelveli, India

Anie Gincy V. G.
Anna University Chennai – Regional Office Tirunelveli, India

Rajesh Banu J.
Anna University Chennai – Regional Office Tirunelveli, India

Dinesh Kumar M.
Anna University Chennai – Regional Office Tirunelveli, India

ABSTRACT

Smart city technology evolved with the developments in wireless sensor networks (WSN) and the internet of things (IoT). IoT-based waste management is an advanced waste management system offered in smart cities. The practice of monitoring, transporting, and processing of solid waste are included in the waste management. Litter bins play an indispensable role in the waste collection process at the primary level. The process of monitoring litter bins would become difficult for the ones placed at out of reach areas and remotely located sites. Smart litter bin (SLB) is generally embedded with different types of sensors where used for sensing the garbage levels and locating the bins location. Radio frequency identification (RFID), sensors, global positioning systems (GPS), general packet radio service (GPRS) are the components in smart waste management system and are discussed in this chapter. These components were used to monitor the collection, transportation, processing, and dumping. This chapter also focuses on the perception of IoT architecture to upgrade waste management in smart cities.

DOI: 10.4018/978-1-5225-9199-3.ch007

INTRODUCTION

The rapid population growth and economic development of developing countries over the last several decades has led to an unprecedented increase in solid waste production which in turn has led to rapid deterioration of environmental quality (Wang et al., 2018). The quantity and quality of the solid waste generated depends upon the livelihood of people, urbanization, commercial activities and socio-economic status. The generated solid waste consists of wet waste as well as dry waste. The wet waste is biodegradable and can be employed in the production of biogas and biofuels whereas the dry waste can be recycled or landfilled. The major problem encompassed by the city municipalities includes the collecting garbage, transporting and disposing it (Bharadwaj et al., 2016). Smart city by definition, has the following characteristic of smartness in administration, living, environment, transportation and economy (Perera et al., 2014). Smart city concept incorporates the information and communication technology (ICT), and various physical strategies connected to the network, called Internet of things (IoT). The Internet of Things is a combination of different hardware & software technology. The Internet of Things provides solutions based on the incorporation of information technology, which refers to hardware and software used to store, retrieve, and process information and communications technology which includes electronic systems used for communication between individuals or groups. This concept could optimize the efficiency of city operations and services. Smart city technology allows city executives to interact directly with public, to monitor the activities in city and to manage assets and resources efficiently. IoT plays a substantial role in reforming the livelihood of people with new technologies. IoT is widely used in various applications such as home automation, health care, environmental monitoring, manufacturing, energy management, agriculture and transportation. IoT reduces the challenges in waste management such as monitoring the garbage bins, transporting waste and processing. Waste management in smart cities mainly aimed on monitoring of waste, effective collection and separation with the benefit of IoT based solutions. This chapter deals with the waste management and the integration of IoT in waste management turns into smart waste management (SW) system in smart cities.

MUNICIPAL SOLID WASTE MANAGEMENT

Municipal solid waste may include wet waste such as (food waste, leaves, vegetables, meat residues), recyclable waste recyclable waste (paper, plastic bottles, metal, cloth), and non-recyclable waste (plastic bags, glass, metal). Prevention, end-of-pipe strategies and restoration practices are the three main existing waste management practices. Generally, the prevention practices involves waste minimization, awareness, and legislations. End-of-pipe focuses the collection of waste at sources, recycling, waste separation, reuse, recycling and energy from waste. Environmental restoration strategies aims to restore the damaged environment. In this practices, prevention practices deal the highest efficiency with the lowest cost, while environmental restoration is the most expensive practice with the lowest efficiency. According to Chen, 2010, municipal waste mainly contains kitchen waste. Normally, solid waste disposal methods include selling, recycling, landfill, burning and dumping in open areas without any treatment. Recycling includes treating biosolids by composting residues of vegetables, anaerobic fermentation (organic substance) to obtain bio-fertilizers, biogas and other resources. But landfill and burning is expensive and disposed to generating secondary pollution such as air and soil pollution. Dumping solid waste in open areas without treatment is common developing countries, may causes pollution and affect the surrounding environment. Municipal solid waste management (MSWM) refers to the effective management of generation, handling, collection, storage, transportation, processing and disposal of solid waste using the best principle and practices of economics, public health, conservation, engineering, aesthetics and other environmental conditions. The stages of solid waste management are segregation and storage at source, collection from source, transportation from source to storage points, disposal of different categories of solid waste and monitoring mechanism. Segregation of solid waste can be classified into biodegradable waste, non-biodegradable waste, recyclable waste, inert and hazardous waste. Improper management of municipal solid waste causes hazards to the environment, increase the accumulation of waste, spreading health issues and contaminate underground water resources through leaching. Converting the waste into useful products through chemical or biological method is an attracted way of management before disposing it (Banu et al., 2007). Several researchers have carried out the various treatment process efficiently to treat the solid waste and convert into energy like biogas (Kavitha et al., 2017; Shanthi et al., 2018). Yeom et al. 2018 reported about the various process in municipal waste management.

MSWM problems are increasing rapidly in Asian countries. As per current scenario in India, various household activities, institutional and commercial activities generates 1,27,486 tonnes per day of municipal solid waste (CPCB 2012, Nandan et al., 2017). The Ministry of Environment, Forest and climate change of India (MoEF) has released a notification on 8th April, 2016 which is commonly called as the Solid Waste Management Rules, 2016. The objective of these rules was to develop an effective infrastructure for the collection, segregation, storage, transportation and disposal of solid waste.

INTERNET OF THINGS

Internet of things (IoT) can be defined networking of physical substances with the use of embedded sensors. IoT is another level of innovation by internet in which the objects make themselves identifiable and communicate the data about themselves. They can access data that has been accumulated by other things, or they can be components of complex services. Internet of things (IoT) is a global infrastructure for the information society, empowering advanced services by interconnecting physical and virtual things based on existing and evolving information and communication technologies. These sensors collect information about the substances, their surroundings and communicate the data to other stations, linked by wired or wireless networks (Singh et al., 2016). IoT allows to connect people with each and everything, at any time, at anywhere. The concept of IoT was first put forth by Kevin Ashton, a member of Radio Frequency Identification (RFID) Development community in 1999. Because of the rapid development in the embedded communication, cloud computing and data analytics, IoT has become a more relevant practical technology recent days. Based on stipulated protocols, these objects are connected to each other anytime anywhere, thereby sharing information in order to achieve smart reorganizations, positioning, tracing and process control (Patel et al., 2016). The fundamental characteristics of IoT includes interconnectivity, things related services such as privacy protection and semantic consistency, heterogeneity, dynamic changes, enormous scale, safety and connectivity. IoT could be incorporated into three categories viz., people to people, people to machine and machine to machine interacting through internet. The IoT application is possible due the technologies such as wireless sensor networks, sensor networks, 2G/3G/4G, GSM, GPRS, RFID, WI-FI, GPS, microcontroller, and microprocessor. Internet of Things can be grouped into three categories: (1) technologies that assist "things" to obtain related data, (2) technologies that assist "things" to process related data, and (3) technologies to develop security and privacy. The current popularity under IoT is machine to machine (MTM) communication which simplifies the human work. This technique has already been implemented in some of the home automation. For example one can switch off an air conditioner (AC) in his home from his workplace. In this case two machines are under communication and the operator is the human with less effort. Other Machine to machine (MTM) applications are smart energy, smart traffic, smart payment, smart health care and smart environment which together constitute the smart city. The arrival of IoT provides new applications to world.

IOT BASED SMART WASTE MANAGEMENT

In smart city concept, IoT enables efficacious management of solid waste generated and improves the lifestyle of the people living around (Bharadwaj et al., 2016). IoT based waste management is mainly depend on the development of data gaining and sensor-based technologies, development of communication technologies and data transmission infrastructure, testing the skills of IoT systems in field experiments and truck routing and scheduling for waste collection operations. Hong et al. 2014 reported that IoT executes sensing, stimulating, gathering, storing, and processing the information by linking virtual devices to the Internet. The author also explained about an experiment about the IoT-based smart garbage system to improve the effective waste management. The results from the experiment shows not only the reduction of waste and also saves energy in waste management The current system is not capable of managing the waste management in urban area as it causes health and environment issues. The main advantages of IoT is the way of communication techonology and the device used for it with in one fnctional system by storing, transferring the data in smart and secure way. Fujdiak et al. 2016 demonstrated about a genetic algorithms as an effective tool for the waste collection. In this experiment the results show that genetic algorithms helpful in tracking the waste collection vehicle more efficiently. Figure 1 displays the IoT based waste management system.

The Ministry of Electronics and Information Technology, India is taking a forward step in smart city by integrating the IoT in waste management. Providing regular interval disposal of waste can control the waste accumulation in one place. This can be solved by using IoT based waste management. This technology provides the municipalities about the waste dumping information in bins at regular intervals, whether it is full or not (Manasi et al., 2016). Segregating and collecting the waste from bins are the main drawbacks in waste management. IoT based waste management helps to get the data about the bins and to direct the collection vehicles to the routes using mobile application that ensure proper waste collection takes place. This can help the vehicle to avoid unnecessary time traveling by visiting the not filled bins. The above advantages can considerably help to reduce the time period and traffic in transportation. This scenarios will make improve the idea of environmental friendly smart cities in developing countries like India.

The following are some of the merits of IoT based waste management system:

- It easily identifies the filled bins and send information to vehicles, so that can save the time. This will reduces the operation and maintenance costs.
- By two communication, it shows the filled bins location and reduces the unwanted traffic and travelling distances during the waste collection process.
- It will show the traffic on the roads and also alternate way to avoid the traffic to collection vehicles.

Figure 1. Internet of Things based waste management system

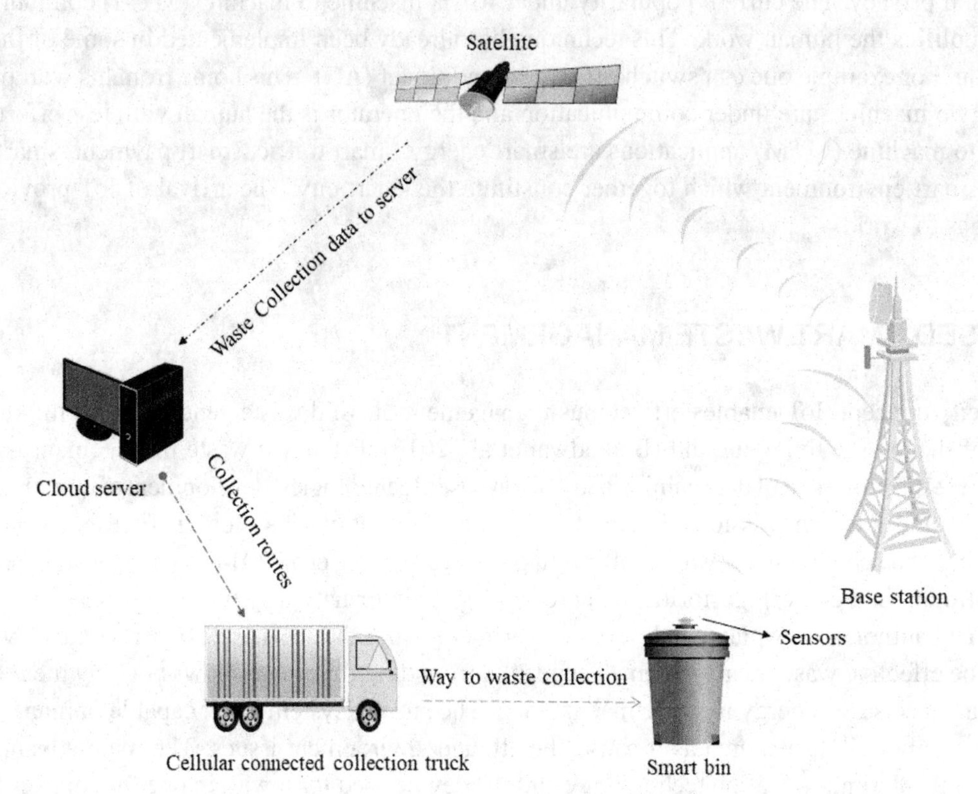

- Sensors present in the bins will informed about the waste conditions such as weight, odour and moisture.
- By applying the smart waste management, then the city becomes smart city.
- It helps the administration to gain extra revenue.
- Web cameras shows the conditions about the waste bins and its surroundings which will helps to improve the clean environment.

ARCHITECTURE OF IOT BASED MSWM

The primary objective of the proposed distributed architecture is to take advantage of the deployed infrastructure of things and the cloud computing resources. It is used to minimize the computing costs and improve the overall performance. To share the application's workload between the server-side and the rest of things with computing capabilities is the main idea of this system. The devices used for this communication are smartphones, wearables, tablets, smart sensors, and other embedded devices. This workload-sharing among the things enables a horizontal scaling to reduce costs, rather than resort to remote servers. Thus, in accordance with our proposal, these kinds of devices perform more processing tasks than the server-side layer. In addition, cloud computing is available to use only as a last resort if needed. In the case of the asynchronous synchronization needs between cloud server computing and the different devices, our system develops a push notification-based approach (Mora et al., 2018). This system gives the results for harmful environmental conditions. Anagnostopoulos et al. 2015 revealed that the proposed system assumes that the city is divided in sectors which carry a certain number of homes, public buildings and places. Each sector has LCTs (Low Capacity Trucks) and HCTs (High Capacity Trucks). Waste is transported to several dumps outside of the city. Truck navigation is attained through an Android app. With using GPS the position of truck was tracked which is incorporated with truck. Each sector contains a certain number of bins. Each and Every bin has a RFID tag, a capacity sensor and an actuator. The RFID tag is used to identify the bin which is filled. Capacity sensors are used to estimate the volume of waste stored. Actuators lock the lid of the bin when the bin become full in order not to allow more waste storage till the bin gets empty. When the bin is full of waste the system is informed; through a wireless sensor network, and initiates a dynamic route in order to empty the bin. Bins are cleared by LCTs. Vasagade et al. 2017 discussed that the proposed system consists of ARM (Advanced RISC Machine) is the master control. The system also consists of IR sensor which working is based on the optical principle of reflected wave. The sensor is placed in inside and outside of the dustbin. It aims to clean the outside of the bin also. If the garbage is present outside of the bin it is detected using the sensor which is placed outside the bin. In this system the bin is designed as a half circular shape. Some systems used rectangular shape litter bins such system use Arduino UNO micro controller which is used to read the data from sensor (Kumar et al., 2016). The IoT architecture is made up of several layers such as sensor layer, gateways and networks and management service layer. Figure 2 illustrates the system architecture of smart waste management system

Figure 2. System architecture of smart waste management system

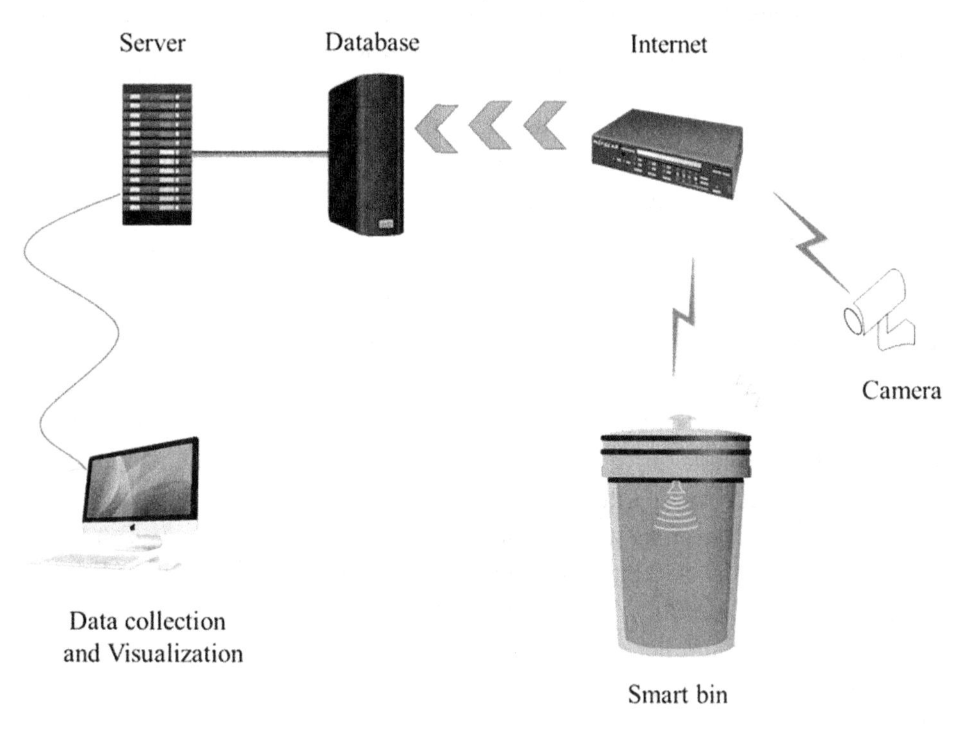

Smart Device/Sensor Layer

Sensor layer consists of an IoT based model with sensors measuring the waste volume in bin or containers, with the ability of transmitting data to the internet through a wireless link. This data is used to enhance the management and schemes of waste collection logistics (Gutierrez et al., 2015). This layer consists of smart devices integrated with sensors. Sensors have the capability of measuring temperature, pressure, humidity, flow etc. Each garbage bin is fitted with a gas sensor to detect harmful gases, IR sensors at the middle and top of the bin to measure the level of waste, weight sensor at the bottom of the bin to detect the weight of waste and a microcontroller to collect all the sensor data and a module for sending the data to a gateway device. For smart city, temperature and humidity sensors can be used to collect temperature and humidity details of that area.

Gateways and Networks

The data produced by the tiny sensors in each garbage bin is collected by the gateway devices. High performance wireless or wired network systems are used as transport medium. This type of networks supports the machine to machine networks and transfer the data. These networks can be in the arrangement of a private, public or hybrid models and are fabricated to support the communication requirements for potential waste management.

Management Service Layer

The management layer consists of device and application management as well as identity and access management. The device and application management tracks the operational devices and applications. If any problems arise, the concerned authorities will be notified via messages and relevant alerts will be generated. The identity and access management ensures that only authorized users and applications have access to each layer. If any unauthorized user tries to access the system, an alert will be generated and the access will be denied. IOT brings connection and interaction of objects and systems together by providing information in the form of data such as temperature of waste, current location of bins or collection trucks and traffic data.

SMART BIN

Smart Litter Bin or smart bin is one of the advanced technology used in smart waste management, with the help of IoT. Smart litter bin helps to maintain the surroundings clean and also used to reduce the pollution. These bins are annexation with micro controller based system having infra-red (IR) sensor and radio frequency (RF) module along with intermediate system broadcasting live condition of litter on web browser. Figure 3 shows the working of smart bins. The sensors are placed in the litter bins situated at public places (Shyam et al., 2018). When the litter comes to the level of sensor then that signal will be display to controller. The controller will give notification to the driver of litter collection truck as to that litter bin is become full and requires urgent attentiveness. This set-up maintains a dry waste and wet waste separately for that we are using a moisture sensor if that sensor noticed then the lid will open for a weighed waste otherwise lid will open for dry waste. Some systems used gas sensor for detecting presence of harmful gases (Bharadwaj et al., 2016). This will come to the aid of minimize the overflow of litter bin and thus keeping the surroundings clean. In this system the litter bins which are placed in every part of the city. These bins are made with a sensor which is helping to track the level and weight of litter. The unique ID will be given for each and every dustbin in city (Dilip et al., 2018). It is very easy to find which garbage bin becomes full. When the level and weight of the bin get to the threshold limit, the device will transfer the reading along with the unique ID provided. In order to stay away from the putrescible smell around the bin safe chemical sprinkler is used that will sprinkle the chemical as soon as the smell sensor detect the decaying smell. Once the bins are become full then the user will not be able to obtain the bins. In such set of conditions the bin put on view to the direction of the nearby bins on LCD display also creates the voice message if the user location the waste on the floor. The condition of the bin was concerned authorities from their place with the help of internet and an instant action will be taken to replace spilling over the bins with free bins. Some systems used ultrasonic sensor and gas sensor which is used to detect the hazardous gas. This system uses the cloud and mobile app (Misra et al., 2017). Rupa et al. 2018 discussed that the system manages the garbage collection through Municipal Corporation incorporated with an IoT based embedded device which is attached to dustbin in each area. This device continuously updates the condition of dustbins in each area. This device detects the level of dustbin using ultrasonic sensor. If the dustbin becomes full it will update its status of getting full on the website designed for garbage management along with date and time and will go to waiting state and remain in this state till dustbin gets free. A timer is also set simultaneously in this state for a fixed duration within which dustbin must be cleaned by the Municipal Corporation. If this timer gets become invalid and dustbin is not cleaned by their employees on given time then device will dispatch a message to the higher authority that dustbin not cleaned on time and again set the timer for the same duration and remain in waiting state. Once the dustbin is cleaned by the employees the device will comes out of waiting state and will update its status of getting cleaned on the website along with date and time. Thus a record is maintained regarding dustbin status for each area in the website in tabular form using IOT technology along with embedded system which will efficiently manage the garbage collection by the Municipal Corporation and will resolve the major environmental issue of inefficient garbage collection leads to a clean and healthy environment.

Figure 3. Smart bin system

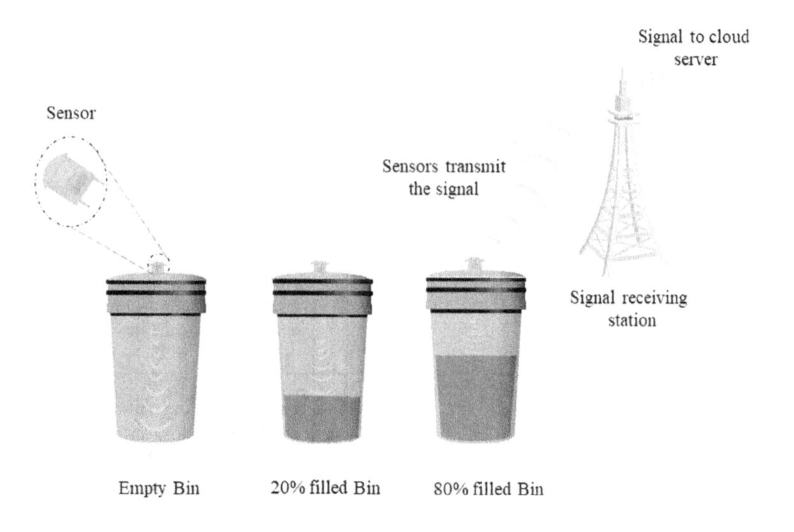

COMPONENTS OF IOT BASED SWM

The components of IoT based solid waste management are i) Radio Frequency Identification (RFID) (ii) Sensors (iii) Web cameras (iv) Global Positioning System (GPS) (v) General Packet Radio System (GPRS). Table. 1 shows features and components of IoT waste management system.

Radio Frequency Identification (RFID)

Radio Frequency Identification (RFID) refers to the technology whereby digital data encoded in RFID tags are captured by a reader via radio waves. It comprises of a reader and one or more tags. RFID is similar to barcoding in that data from a tag or label. This data can be read outside the line of sight whereas barcodes must be scanned to read out. The RFID tag is a tiny device that consists of microchip and antenna. Two types of tags are generally used. Active tag contains inbuilt battery and do not demands power from the reader. They have longer distance range than passive tags. Passive tags do not have internal battery and depends upon the RFID reader for its operating power and have a lower range limited up to few metres (Kumar et al., 2016). When the tag is placed close to the reader, the radio frequency transmitted by the reader is collected by the RFID tag and transmit to antenna, then it activates the transponder (Islam et al., 2012). The received radio frequency is converted into electric power enough for the tag to transmit the data to the reader which in turn transmits the tag ID to the external device by serial communication. This RFID technology is used as an alternative for the bar code technology and used in various sort of applications including automotive, payments, laundry, library, retail supply chain management, ticketing and in industry (Hannan et al., 2011).

RFID belongs to a technology called as AIDC (Automatic Identification and Data Capture). AIDC automatically identifies objects, collects data and convey the data to the system. From the system the data are directly read out with human involvement. It is a powerful weapon for waste collection, disposal and management, delivering unique and compelling benefits to governments and private authorities handling waste collection. This technology increases the efficiency of waste management operations and it also coincides with the disposal policies and practices. In RFID system, the waste collecting bags

Table 1. Features and components of Internet of Things based waste management

Categories	Components	Features
Physical arrangements	• Waste Bins and containers with sensors • Waste collection Trucks with RFID tag • System based depots and dumping station	• Waste storage • Waste transportation • Collection of wastes storage
IoT based technologies	• RFIDs • Sensors • Actuators • Cameras • GPS • GPRS	• Sensing the volume, moisture, temperature and weight of waste • Locating the bin and station • Monitoring the bins
Software used	• DSS • GIS • Scheduling • Routing	• Data about the collection of waste • Scheduling for waste collection • Directing the vehicles to the location

Figure 4. Radio Frequency Identification system in smart waste management

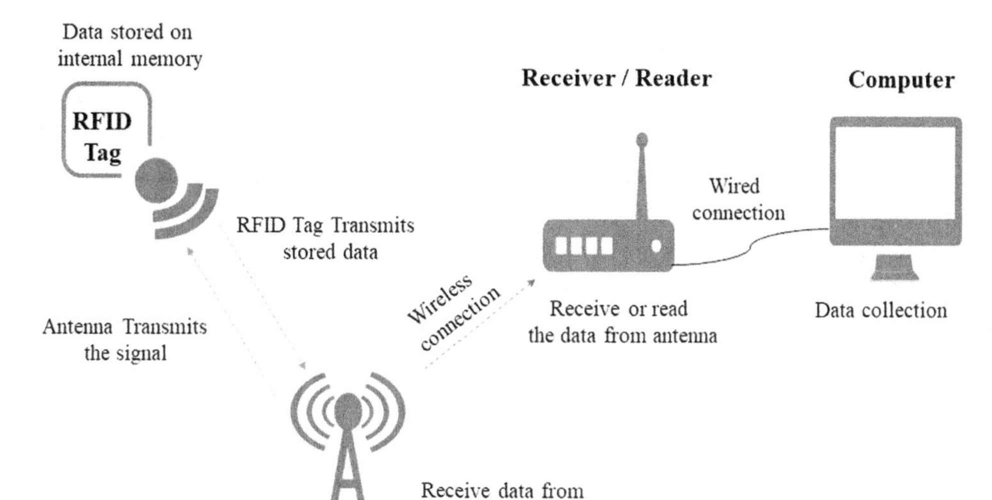

are closed with Waste Identification Stickers (WIS). When these bags are get full, then thrown into the regular garbage containers. The RFID tag is attached to the garbage container (Likotiko et al., 2017). A reader, antenna and scale are fixed in a garbage truck. The readers and antenna communicates with the RFID tags. The RFID transmits a unique code. The reader on the trucks receives this code once the bin is full and the lid gets closed. Through this RFID process, time, type of container, weight and load and customer and location information are collected. These information are then transferred to the central waste management system. The wastes are transported to the recycling centres. In recycling centres, wastes are collected and empty bags are returned to the waste management centres for the reuse. The turn bags are conveyed into the container. The wastes are mixed and crushed and finally enters into compactors. Figure 4 illustrates the working process of RFID system in smart waste management.

Sensors

Sensors plays a vital role in IoT based smart waste management system. Sensors enable the physical quantities and transforms it into digital signals, which are managed wirelessly by a network (Ma, 2014). Through sensor, the waste management people obtain the information about the filling level of smart bins (Zavare et al., 2017). The litter bin is made with the sensor which is used to track the position of the litter (Folianto et al., 2015). This system maintains the dry waste and wet waste by the use of moisture sensor. Sensors placed on the bins and by using a microcontroller that sends data to a control station. In IoT based waste management, sensors are used to measure a set of physical quantities such as capacity, weight, temperature, humidity, chemical reactions, and pressure. Likewise, Actuators are used in smart waste management where it can automatically lock the bins when it identifies the bin is full. Various types of sensors are used in solid waste management such as Ultrasonic (US) sensor, Infrared (IR) sensor, weight sensor and moisture sensor.

US Sensor

The bin in each area is connected to a sensor which identifies the level of waste collected in the bin. The working of this ultrasonic sensor is based upon the frequency of sound. The sound wave is generated from the sensor and it hits the waste and reflects back. The sensor calculates the time taken by the sound wave to travel and reflect back and also measures the exact distance at which the waste is present. This sensor prevents the waste from overflowing and also sends an alert to the authority concerned for collecting the waste. The range of US sensor varies from 2cm - 400cm with contactless collection (Kumar et al., 2016). This sensor unit consists of transmitters, receivers and circuits. The sensor is set at a predefined level. For example, when garbage is of size 20cm, the level will be set at 15cm. When the level in the garbage is less than 15cm, the bin is empty and when the garbage level is higher than 15cm, the bin is full. The alert messages are received in the control room. The main disadvantages of this type of sensor is high cost.

IR Sensor

Infrared (IR) sensor which works through the emission and detection of infrared radiation by an instrument. It intellects the characteristics about its surroundings. IR sensors uses the wavelengths of 700 nanometers to 1 millimeters in light spectrum. IR sensor measures the actual level of garbage present in the dustbin. The sensor is placed at three different levels to show the level of garbage filled in it (Navghane et al., 2016). Singh et al. 2016 reported that in nearby range of coverage IR sensors shows better results than US sensors and also IR sensors are cheap as compared to the other one. The authors also explained the working of IR sensors, the device emits the light which reflected on the object that are in closeness of the device. Then the light reflected from the surface enters into the light sensor. This causes a peak in the intensity. The sensor can return the intensity value of the reflected light from the close objects. This helps to determine feature of the object. Normally, IR sensors and its emitters are positioned on the bin or container top where the transmitter faces inside the bin. IR rays are continuously transmitted by the emitter. Once the bin gets filled up to the engaged sensor level, the rays are get reflected back. At a particular interval of time, the bin status are notified by system.

Weight Sensor

Weight sensor which works on the principle of piezo-resistivity is used to detect the amount of garbage in dustbin. Deka et al. 2018 reported that when the waste level increases it shows the increasing weight of the waste and if it over- flows the bin after certain period of time, the sensors collect the details of the volume and weight of the bin, which is displayed through the LCD panel interfaced with the micro-controller and it send the data to the system.

Moisture Sensor

Waste can be physically characteristics as wet waste and dry waste. These differentiations can be achieved by fitting a moisture sensor that indicates the type of waste. The process involved under the moisture sensor is, once the waste is collected in the bin, the moisture sensor detects the type of waste whether wet or dry and separates them according to their moisture content (Surapaneni et al., 2018). Thus the processing of waste becomes easier when the collected waste is separated. The liquid waste can be easily recycled whereas the solid waste is difficult to crush and recycle. This process usually takes place at the waste water treatment plant. But through the introduction of IoT in SWM, the process is made simple at the beginning level so that the tedious works are neglected at the treatment plant. Dry and hard wastes and other plastic wastes can be recycled separately so that the time consumed in the treatment plant is comparatively reduced.

Cameras

Cameras are used in monitoring the bin and to get images for further identification. Anagnostopoulos et al. 2017 reported that the cameras can be act as a special sensors and it can be used to measure the volume of bins. In bin monitoring low cost cameras are used and to get the clear images about the bins it is mounted on the top of the truck along with RFID reader (Hannan et al., 2011). It is connected with the GPS and GPRS. The author also explained that the camera can cover the bin nearby area of 3 m^2. Two types of images can be employed in analysing the waste estimation viz. grey and binary images. Grey image represents an image as a matrix of bright and dark pixels. The images are uploaded to the designated server from the truck via GPRS connectivity.

Global Positioning System (GPS)

GPS, a satellite-based navigation system which is operated through the satellites. It may use to get the locations and its information about the places on the earth. In this system, the satellites periodically emit radio signal to receivers to get the information. Currently GPS is widely being used in vehicles for providing emergency road side assistance, determining the vehicle's position and helping drivers to keep track of their position. Advanced GPS systems help to show the alternate routes or create automated routes to designated places (Arebey et al., 2010). In SWM, GPS is effectively used to identify the position of the truck and bin locations. GPS is used to improve, monitor the efficiency, and utilize the resources and to bring out cost benefit and services by implementing GPS and vehicle tracking in solid waste collection system. Using GPS the bin location can be classified according to the basis of population, density and waste generation. It optimizes the trip distance and the time to ensure maximum utilization of resources. Carbon emission is reduced as the waste is collected time to time. The primary data that is needed to bring out the GPS and vehicle tracking in SWM are Google APP for raster generation, Road network for garbage location and reports generated from RFID helps to calculate the amount of garbage that can be collected. The software used are dashboard module, live real time vehicle tracking module, alerts and warning module. GPS is necessary for tracking the location of bins and nearby recycling stations during dynamic routing (Anagnostopoulos et al., 2017). It is used to identify the alternate routes when heavy traffic goes on.

General Packet Radio System (GPRS)

GPRS can be related with the internet and is developed from the existing GSM system. Every user can adopt number of wireless channels at the same time through GPRS. GPRS is a packet overlay on the top of the existing digital circuit switched to a voice based network. It allows data packets to be conveyed across the mobile network using packet switching. Using packet data access, the garbage bins are segregated into groups and time slots are fixed so that the GPRS transmits the data to the collector regarding the amount of separated waste to the control room. The working principle of GPRS and GPS are similar where GPS tracks the data during mobility and GPRS tracks the data of statutory waste bins. Using the data gathered, the vehicles are scheduled at times to collect the garbage (Visali et al., 2017).

INTEGRATION OF IoT COMPONENTS AND SOFTWARES INTO WASTE MANAGEMENT

The development of IoT technologies have been incorporated into the waste management from collection, transportation and disposal.

IoT Application in Waste Generation

A unique RFID tag was fitted to every bin to identify the waste generation details. The RFID tag contains the data about the waste producers such as name, address, weight of waste produced, temperature etc. Such information are read through the help of RFID readers which is installed in the collection vehicle. This data will sent to the stations along with the time when the vehicle collects the waste. Solid waste management of the city is the authority to issue the tag and managing the data server regularly. Moreover this has the advantages to know about the amount of waste collected from a place or a bin which helps in the waste segregation.

IoT Application in Waste Collection and Transportation

IoT plays a vital role in waste collection and transporting it from the source. The components such as RFID reader, camera, GPS unit, a GPRS unit and sensors. The RFID reader will help to collect the data in the RFID tag about the bin. Camera in the vehicle helps to find out the actions take place in collection and transportation process. If any problems like over dripping of waste from the vehicle can be easily identified through the camera. Weight of the waste collected by vehicle are identified by weight sensors and send back the details to server automatically. The units of GPS and GPRS are used to record the location of vehicle, their routes and operational statuses. Then transmit this information to the control centre, where it is properly controlled. This will helps to identify whether collection vehicles are following designated routes, however it also provides data about the vehicle allocation and route adjustment in case of emergencies such as breakdowns of vehicle, non-routine collection tasks besides route optimization in the event of traffic incidents.

IoT Application in Waste Disposal

After returning from the waste collection, the vehicle is identified by tag reader and checking the weight is done by weighing bridge. The weight information and data about the vehicle is sent to the server. This can be done to identify the amount of waste collected in a day from a location. Various sensors are used to identify the conditions of waste.

CASE STUDY ABOUT THE SOLID WASTE MANAGEMENT IN TIRUNELVELI

Tirunelveli district is unique in nature because of the presence of crawling clouds of the Western Ghats, Courtallam waterfalls and the Pothigai hills. It is a major city in the South Indian state of Tamilnadu. The city covers an area of 189.9 km^2 and had a population of 4,73,637 as of 2011 census. Tirunelveli city is known to be the Twin city of Tamilnadu and has highly provoked several business giants across the state. Palayamkottai has been called as Oxford of South India since most of the educational institutions, government departments and administrative units are established in these areas. Many small scale and large scale industrial sectors were located Tirunelveli district. The city which is the fifth largest municipal corporation in the state after Chennai, Coimbatore, Madurai and Trichy. It has been divided into four zones namely Tirunelveli, Thachanallur, Palayamkottai and Melapalayam. Tirunelveli Municipal Corporation is responsible for the management of solid waste generated in the city. Tirunelveli Municipal Corporation consists of 55 wards and the city sanitation is effectively operated by 17 sanitary units each one composing of two to four wards. Every sanitary unit is headed by a sanitary inspector who were assisted by a sanitary supervisor. For every units, separated equipment and vehicles has been provided for the solid waste management. Without proper solid waste management in the corporation, the natural scenic beauty of the city has been deteriorated with the drastic increase in solid waste generation. There are different types of solid waste such as municipal solid waste, hazardous waste, industrial waste, agricultural waste and biomedical waste. Thamirabarani a perennial river, located in Tirunelveli district is highly polluted with untreated solid waste dumping which in turn leads to the spread of water borne diseases. The enormous increase of solid waste in recent days is mainly due to increasing urbanization, improved livelihood of people and consequent urban growth with more and more new colonies and extension areas. This is significantly expected to affect the quality of environment as well as health of the people. Table.2 displays the detail about the zone wise population and number of bins used.

In total waste generated, 65% wastes were collected and 34% of the waste were not collected. Normally, 340 grams per capita waste generated and approximately 161 tonnes per day in Tirunelveli Corporation. Sanitary workers has been used to collect the solid waste generation using pushcarts or tricycles with coloured bins for the collection of garbage from each house of the street or door-to-door collection. Metallurgical industries, petrochemical industries and solid waste yards are identified as the main sources of hazardous wastes in Tirunelveli. Except the plastic waste, which is collected separately once a week, other mixed litter consisting of organic materials, paper, low-value plastics, domestic hazardous and inert waste is being dumped in the Ramyanpatti Compost Yard, landfill on the outskirts that has been converted into a waste management processing site. For every day, over 150 tonnes of garbage

Table 2. Population and number of dustbins in each Zones of Tirunelveli Corporation

S.No	Zones	Population* (in lakhs)	Households (in numbers)	Dumper Bins (in numbers)
1	Thachanallur	1.06	28513	31
2	Tirunelveli	1.04	33627	27
3	Melapalayam	1.44	33602	52
4	Palayamkottai	1.19	33205	45

*- 2011 census

was dumped and this created the surroundings very uneasy, especially when fire broke out there. During monsoon, the problem was severe as the nauseating smell and swarms of houseflies invading the houses from the garbage dump. Tirunelveli Corporation had compacted the waste on 32.50 acres of the total extent of 180 acres. The corporation moved the entire quantity of 2.20 lakh cubic metres of waste to just six acres by compacting it.

Steps Taken by Tirunelveli Corporation

Tirunelveli Corporation found that it was difficult to clean all areas regularly. Based on the discussion in various forums, corporation started concept of promoting residential associations to take up sanitation in their colonies by engaging private sanitary workers. In Tirunelveli around 1.7 lakh household took up a challenge to eradicate the solid waste issues. Based on Solid Waste Management Rules 2016, the corporation made a step that it was mandatory to isolate biodegradable and non-biodegradable wastes separately before the disposal or handover the waste to the authorised waste pickers. A fine of amount Rs. 10 were collected if unsegregated waste were handover to the waste pickers. The municipal corporation took several steps such as campaign, loud speaker announcement to reach the public. Beyond that the corporation took a forward step or mission called "Litter free Tirunelveli" (LIFT) to control the waste accumulation. Under this mission, the biodegradable waste is collected every day and non-biodegradable waste like plastic is collected every Wednesday. In LIFT mission, every week 9-10 tonnes of plastic waste are collected. While it is commendable that people from around 178,000 households hand over segregated plastic waste to some 300 waste collectors, there is no separation or safe destruction of glass, metal or other dry waste. the collectors, on the field from 6 am every day can sell the plastic waste directly to the recognised scrap dealers for Rs. 2-3 per kg.

Disposal of Waste

Degradable Waste

The degradable waste is being disposed at the garbage composting yard. This waste is converted into manure through the Open Windrow Method. At Present a small amount of the collected waste is utilized only for manure generation. (About 2 - 4 tonnes of garbage collected from market area) Windrows are formed at the new platform and the manure is obtained through the aerobic process which is sieved through the sieving machine and is sold to private agency. The residue (Refuse) is classified to be disposed off a Land fill which is to be constructed as per the standards.

Non-Degradable Waste

Non degradable wastes are dumped separately in the composting yard. Self Help Groups are given loan for purchasing the Shredder Machine to be put into use to shred the non degradable garbage like Plastic, Polythene, Termocole etc. shredded product is planned to be utilized for laying plastic roads and to sell to other Local Bodies. Table.3 displays the detail about the waste produced in a day.

This all the steps helped the city corporation to achieved the recognition of first Indian city to achieve 100% segregation of waste at source. With technical help from National Institute of Technology, Tiruchirapalli, compacted clay liner was provided over the compressed waste and covered by vegetative soil layer for about 450 mm thickness was provided on the HDPE liner. Grass was planted on the vegetative

Table 3. Solid waste collection and treatment method in Tirunelveli Municipal Corporation

S.No	Zones	MSW Generated/ Day (tonnes)	MSW Collected / Day (tonnes)	Treatment Method
1	Thachanallur	41	39	Composting
2	Tirunelveli	40	38	
3	Melapalayam	54	51	
4	Palayamkottai	45	42	

soil to make each heap look like a grass mound. This innovative step makes the dumping wad into an eco park, gets attraction by the public. Effectiveness of the final cover on mounds is being monitored periodically to prevent it from eroding and ensure healthiness of leachate collection system. Moreover, groundwater quality in this region is being periodically checked. By the effective step of corporation, the waste dumping yard has been transformed into a greenish park. Tirunelveli has undertaken different waste separation initiatives and has been praised by the Swachh Bharat Abhiyan and the US Green Building Council. Table.4 shows the solid waste management steps taken by Tirunelveli Corporation.

The city corporation bagged national award for the waste management from Swachh Bharat Mission (SBM), an initiative movement by Government of India and "Hon'ble Chief Minister's Award" for the best corporation in 2016 by Government of Tamil Nadu, India. The Government of Tamil Nadu, India has announced a project about waste to energy for Tirunelveli Municipal Corporation at an estimated cost of Rs.55 crore in 2014. As an initiative of converting waste to energy, a five million tonnes bio-methanation plant is under operation and a 2000 kWh plant is in progress.

Future Plans About the Smart Waste Management in Tirunelveli Smart City

The Government of India selected 11 cities from Tamil Nadu to be developed as smart cities. Tirunelveli has been selected as one of the smart city among them. The future of civil habitation is in the hands of smart cities. Anagnostopoulos et al. 2015 reported that smart cities aims to use information and communication technologies and other means to improve quality of life, efficiency of urban operations, services and competitiveness. The technology will ensures that it meets the needs of the present and future generations with respect to economic, social and environmental aspects. The fundamental component which is relevant to environmental pollution is the smart environment. Tirunelveli Smart City Limited has been formed by the state government in Tirunelveli Municipal Corporation. At a cost of 986.18 crore,

Table 4. Solid waste management steps by Tirunelveli Municipal Corporation

Name of the Corporation	MSW Collection (tonnes)	MSW Segregation (tonnes)	MSW Storage Detail	Transportation Detail	Processing of MSW	Disposal of MSW
Tirunelveli	Door-to-door collection	1 M.T. of plastic waste is segregated daily and driven to shredding for laying of road	4.5 cu.m containers of 220, and 1.1 cu.m bins of 410 are used.	Transportation done through 7 mini lorries, 3 tipper lorries, 3 non tipper lorries, 2 autos and 5 dumper placers.	Open Windrow Formation	Composting

30 works have been sanctioned to improve the city infrastructures under the smart city mission. Various solid waste management projects at a projected cost of 12.36 crore have been sanctioned under SBM. For collecting solid waste, purchase of primary and secondary vehicle has been proposed in SWM projects. On an average, 100 M.T. of degradable waste and 2.5 MT of Non degradable wastes are eliminated from the City daily to the compost yard. Under SWM Action plan, for door to door Collection, 250 Nos. of Tricycle with 1000 Bins were newly purchased at an calculate cost of Rs.28.45 Lakhs and is now set into use. To process the organic waste, 37 Micro Compost Centers have been sanctioned at a projected cost of 6.6 crore. This case study proves that Tirunelveli Municipal Corporation takes necessary step to handle the waste management in effective way. By implementing the smart city mission, Tirunelveli city may improve its waste management into smart.

FUTURE PERSPECTIVES

The major problems in waste management is public participation and lack of responsibility towards waste in the community. For sustainable waste management, need to promote community awareness and change the attitude of people towards waste management. Sustainable and economically feasible waste management must ensure maximum resource extraction from waste, safe disposal of waste by the development of engineered landfill and waste-to-energy facilities. India faces challenges related to waste policy, waste technology selection and the availability of appropriately trained people in the waste management sector. There has been technological advancement for processing, treatment and disposal of solid waste. Energy-from-waste is a crucial element of SWM because it reduces the volume of waste from disposal also helps in converting the waste into renewable energy and organic manure. The Indian Government decided to make 100 cities as smart cities. The Government have necessary step to redraw its long term vision in solid waste management and rework their strategies as per changing lifestyles. They should reinvent garbage management in cities so that the processing of waste may improve instead of landfill it. The future technological developments for IoT should be made in order to allow for physical devices to operate in changing environment and to be connected all time everywhere. The evolution of smart city into zero-waste sustainable cities requires four linked primary strategies - waste prevention, upstream waste separation, on-time waste collection, and proper value recovery of collected waste. The aim is to predict the design and development of an IoT-enabled waste management structure for smart and sustainable cities on connecting waste management practices to the whole product life-cycle. Constrained application protocol, Message queue telemetry transport, Extensible messaging presence protocol, Advanced massage queuing protocol and Data distribution service are some of the application oriented advancement in IoT waste management.

The following are some of the demerits of IoT based waste management system:

- Sensors used in waste management system are very high cost.
- Sensor used in smart bins have limited memory size.
- RFID tags used in waste management are affected by metal objects.
- Some of the wireless technologies like WIFI, Zigbee have shorter range.
- Smart waste management requires skilled labours to operate the machines.
- As the area of waste collection increases, the installation of smart bins will also increase and it requires high cost.

Future work will progress these framework by taking a nearer look at the effects of other factors such as regulation, policy, product design strategies, and technology on waste management. Future development in IoT will focus on the limitations mentioned above to provide smart and better waste management in smart cities. Future waste collection and management infrastructures should connected with existing systems. These connection will facilitate the integration of waste management practices with other activities within smart communities.

CONCLUSION

Waste collection has become dynamic due to the proliferation of sensors and other components of IoT in recent days. The IoT sector is just in the first steps of development and has a lot more for potential development. Solid waste management should be mechanically treated in mere future where manual and physical cleaning should be avoided. IoT provides a good scope for SWM using GPS and GPRS. The internet of things is a computing concept that describes the idea of everyday physical objects being connected to the internet and being able to identify themselves to other devices. The term is closely identified with RFID as the method of communication, although it also may include other sensor technologies, wireless technologies or QR codes. In this chapter, solid waste management is enhanced with integration of IoT technologies such as RFID, sensors, GPS, GIS and cameras were discussed. IoT can be used in waste collection, bin and vehicle tracking and efficient waste management. IoT will improve the city into smart and provide better waste management.

REFERENCES

Anagnostopoulos, T., Zaslavsky, A., Kolomvatsos, K., Medvedev, A., Amirian, P., Morley, J., & Hadjieftymiades, S. (2017). Challenges and opportunities of waste management in IoT-enabled smart cities: A survey. *IEEE Transactions on Sustainable Computing*, *2*(3), 275–289. doi:10.1109/TSUSC.2017.2691049

Anagnostopoulos, T., Zaslavsky, A., & Medvedev, A. (2015). Robust waste collection exploiting cost efficiency of IoT potentiality in smart cities. In *Recent Advances in Internet of Things (RIoT), 2015 International Conference on* (pp. 1-6). IEEE.

Arebey, M., Hannan, M. A., Basri, H., Begum, R. A., & Abdullah, H. (2010, June). Solid waste monitoring system integration based on RFID, GPS and camera. In *Intelligent and Advanced Systems (ICIAS), 2010 International Conference on* (pp. 1-5). IEEE.

Banu, J. R., Raj, E., Kaliappan, S., Beck, D., & Yeom, I. T. (2007). Solid state biomethanation of fruit wastes. *Journal of Environmental Biology*, *28*(4), 741–745. PMID:18405106

Bharadwaj, A. S., Rego, R., & Chowdhury, A. (2016, December). IoT based solid waste management system: A conceptual approach with an architectural solution as a smart city application. In *India Conference (INDICON), 2016 IEEE Annual* (pp. 1-6). IEEE. 10.1109/INDICON.2016.7839147

Chen, C.C. (2010). A performance evaluation of MSW management practice in Taiwan. *Resour. Conserv. Recycl.*, *54*(12).

Deka, K., & Goswami, K. (2018). IoT-Based Monitoring and Smart Planning of Urban Solid Waste Management. In *Advances in Communication, Devices and Networking* (pp. 895–905). Singapore: Springer. doi:10.1007/978-981-10-7901-6_96

Dilip, K. P., Dnyandeo, J. M., Changdev, J. P., & Lavhate, S. S. (2018). IoT based solid waste management for the smart city. *International Journal of Advance Research, Ideas And Innovations in Technology*, *4*(2).

Folianto, F., Low, Y. S., & Yeow, W. L. (2015). Smartbin: Smart Waste Management System. *IEEE Tenth International Conference on Intelligent Sensors, Sensor Networks and Information Processing*.

Fujdiak, R., Masek, P., Mlynek, P., Misurec, J., & Olshannikova, E. (2016, July). Using genetic algorithm for advanced municipal waste collection in smart city. In *Communication Systems, Networks and Digital Signal Processing (CSNDSP), 2016 10th International Symposium on* (pp. 1-6). IEEE. 10.1109/CSNDSP.2016.7574016

Gutierrez, J. M., Jensen, M., Henius, M., & Riaz, T. (2015). Smart waste collection system based on location intelligence. *Procedia Computer Science*, *61*, 120–127. doi:10.1016/j.procs.2015.09.170

Hannan, M. A., Arebey, M., Begum, R. A., & Basri, H. (2011). Radio Frequency Identification (RFID) and communication technologies for solid waste bin and truck monitoring system. *Waste Management (New York, N.Y.)*, *31*(12), 2406–2413. doi:10.1016/j.wasman.2011.07.022 PMID:21871788

Hong, I., Park, S., Lee, B., Lee, J., Jeong, D., & Park, S. (2014). IoT-based smart garbage system for efficient food waste management. *The Scientific World Journal*. PMID:25258730

Islam, M. S., Arebey, M., Hannan, M. A., & Basri, H. (2012). Overview for solid waste bin monitoring and collection system. In *Innovation Management and Technology Research (ICIMTR), 2012 International Conference on* (pp. 258-262). IEEE. 10.1109/ICIMTR.2012.6236399

Kavitha, S., Banu, J. R., Priya, A. A., Uan, D. K., & Yeom, I. T. (2017). Liquefaction of food waste and its impacts on anaerobic biodegradability, energy ratio and economic feasibility. *Applied Energy*, *208*, 228–238. doi:10.1016/j.apenergy.2017.10.049

Kumar, N. S., Vuayalakshmi, B., Prarthana, R. J., & Shankar, A. (2016, November). IOT based smart garbage alert system using Arduino UNO. In Region 10 Conference (TENCON), 2016 IEEE (pp. 1028-1034). IEEE.

Likotiko, E. D., Nyambo, D., & Mwangoka, J. (2017). *Multi-agent based IoT smart waste monitoring and collection architecture*. Academic Press.

Ma, J. (2014). Internet-of-Things: Technology evolution and challenges. *IEEE MTT-S International Microwave Symposium*, 1-4.

Manasi, H. K., & Smithkumar, B. S. (2016). A Novel approach to Garbage Management Using Internet of Things for smart cities. *International Journal of Current Trends in Engineering & Research*, *2*(5), 348–353.

Misra, D., Das, G., Chakrabortty, T., & Das, D. (2018). An IoT-based waste management system monitored by cloud. *Journal of Material Cycles and Waste Management*, 1–9.

Mora, H., Signes-Pont, M. T., Gil, D., & Johnsson, M. (2018). Collaborative Working Architecture for IoT-Based Applications. *Sensors (Basel)*, *18*(6), 1676. doi:10.339018061676 PMID:29882868

Nandan, A., Yadav, B. P., Baksi, S., & Bose, D. (2017). Recent Scenario of Solid Waste Management in India. *World Scientific News*, (66), 56-74.

Navghane, S. S., Killedar, M. S., & Rohokale, D. V. (2016). IoT based smart garbage and waste collection bin. *Int. J. Adv. Res. Electron. Commun. Eng*, *5*(5), 1576–1578.

Nirde, K., Mulay, P. S., & Chaskar, U. M. (2017). IoT based Solid Waste Management System for Smart city. *International Conference on Intelligent Computing and Control Systems*.

Patel, K. K., & Patel, S. M. (2016). Internet of things-IOT: definition, characteristics, architecture, enabling technologies, application & future challenges. *International Journal of Engineering Science and Computing*, *6*(5).

Perera, C., Zaslavsky, A., Christen, P., & Georgakopoulos, D. (2014). Sensing as a service model for smart cities supported by internet of things. *Transactions on Emerging Telecommunications Technologies*, *25*(1), 81–93. doi:10.1002/ett.2704

Rupa, Ms., Rajni Kumari, Ms., Nisha Bhagchandani, Ms., & Ashish Madhur, Mr. (2018, May). Smart Garbage Management System Using Internet of Things (IoT) For Urban Areas. *IOSR Journal of Engineering*, *08*(5), 78–84.

Shanthi, M., Banu, J. R., & Sivashanmugam, P. (2018). Effect of surfactant assisted sonic pretreatment on liquefaction of fruits and vegetable residue: Characterization, acidogenesis, biomethane yield and energy ratio. *Bioresource Technology*, *264*, 35–41. doi:10.1016/j.biortech.2018.05.054 PMID:29783129

Shyam, G. K., Manvi, S. S., & Bharti, P. (2017, February). Smart waste management using Internet-of-Things (IoT). In *Computing and Communications Technologies (ICCCT), 2017 2nd International Conference on* (pp. 199-203). IEEE.

Singh, A., Aggarwal, P., & Arora, R. (2016, September). IoT based waste collection system using infrared sensors. In *Reliability, Infocom Technologies and Optimization (Trends and Future Directions), 2016 5th International Conference on* (pp. 505-509). IEEE. 10.1109/ICRITO.2016.7785008

Solid Waste Management Rules. (2016). *Ministry of Environment*. Forest and Climate Change Government of India.

Surapaneni, P., Maguluri, L. P., & Symala, M. (2018). Solid Waste Management in Smart Cities using IoT. *International Journal of Pure and Applied Mathematics*, *118*(7), 635–640.

Vaisali, G., Sai Bhargavi, K., Kumar, S., & Satyanarayana, S. (2017, December). Smart solid waste management system by IOT. *International Journal of Mechanical Engineering and Technology*, *8*(12), 841–846.

Vasagade, T. S., Tamboli, S. S., & Shinde, A. D. (2017). Dynamic solid waste collection and management system based on sensors, elevator and GSM. In *Inventive Communication and Computational Technologies, 2017 International Conference on* (pp. 263-267). IEEE. 10.1109/ICICCT.2017.7975200

Wang, F., Cheng, Z., Reisner, A., & Liu, Y. (2018). Compliance with household solid waste management in rural villages in developing countries. *Journal of Cleaner Production*, *202*, 293–298. doi:10.1016/j.jclepro.2018.08.135

Yeom, I. T., Sharmila, V. G., Banu, J. R., Kannah, R. Y., & Sivashanmugham, P. (2018). *Municipal waste management. In municipal and industrial waste: source, management practices and future challenges*. Nova Science Publisher.

Zavare, S., Parashare, R., Patil, S., Rathod, P., & Babanne, V. (2017). Smart City Waste Management System Using GSM. *Int. J. Comput. Sci. Trends Technol.*, *5*, 74–78.

Chapter 8
Smart Grid:
A Study on Communication Technologies, Security, and Privacy

P. J. Beslin Pajila
Francis Xavier Engineering College, India

E. Golden Julie
Anna University, India

ABSTRACT

Smart grid is a new emerging trend as it is more efficient because of its computation process and energy utilization. It provides many benefits due to the communication between the service company and users. Smart grid includes smart metering, which is used to collect the data from the appliances, devices, sensors, etc. and transfer the data to the network, and the data is analyzed for energy consumption. The transmission and distribution of energy; the smart grid architecture; communication technologies like LAN, HAN, and NAN; and security and privacy challenges are discussed in this chapter.

INTRODUCTION

Smart City is created by the integration of new technologies. Smart city concept is a new article because it is formed by combining innovative technologies. Such as in recent years government has been taking lot of initiatives to make the cities into smart cities, with the support of information and internet, to attain linkage of intelligence, self sensing and adaptation etc. Smart cities are preferred because of its advantages like environment protection, security for public and new life style it offers to the inhabitants. People around the world have started sensing the importance of smart city. Humans are connected with the smart objects like smart phones, gadgets etc. Moreover all the objects in this world are connected to one another each other for monitoring, control and automation. The life style of people has improved because of the new innovations and technology. In smart cities, the heterogeneous elements are combined together to form an IoT. Microcontrollers, transceiver and protocols that stack for communication are equipped together in IoT. It will manages and optimizes the services like transport, electricity, structural

DOI: 10.4018/978-1-5225-9199-3.ch008

Figure 1. IoT Based Architecture for Smart Grid

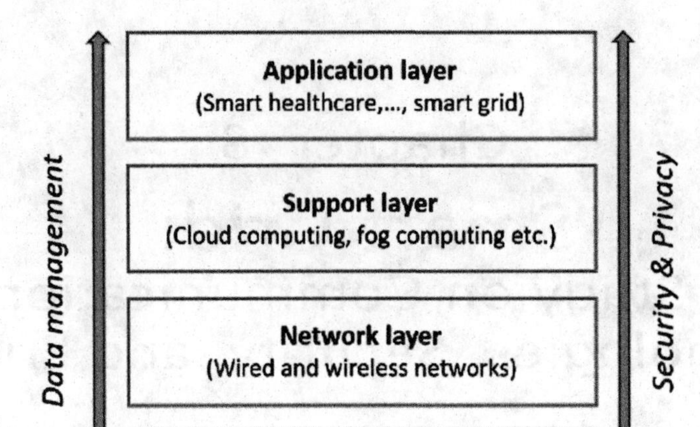

health monitoring, waste management, noise monitoring, traffic congestion, energy consumption, smart parking, smart lightening etc.. The above mentioned management makes the city into smart city. The smart Grid will overcome all the problems that occur in the power Grid like insecurity, damage in the transmission power line, physical and cyber-attack etc. (Anastasia Mavridou & Mauricio Papa, 2012). An IoT based architecture in figure 1, which has four layers. They are Network layer, Application layer, Support layer and Perception layer. The bottom layer is the perception layer; it has sensors and WSN etc. The main purpose of this layer is to collect data from the end devices (things in the real world) and transmit it to the upper layer (network layer). The root layer in IoT architecture is network layer, which is connected with the servers, devices and smart things, and it transfers the data from the perception layer to the higher layer. It builds upon mainly on internet, Communication Networks and WSN which is called as basic networks. The next higher layer is Support layer, it functions along with higher layer called application layer. It has the computing techniques like cloud computing, edge computing fog computing etc. The top most layers are the application layer, which serves the end users based on their demand (Lei Cui, Gang Xie, Youyang Qu, Longxiang Gao & Yunyun Yang, 2018).

Smart grid is one of the applications of smart city that makes a way to change the city smart. When a traditional power grid is replaced by smart grids, it reduces energy utilization, and unwanted costs. Smart meters are used by the consumers to share about their utilization of the energy to the providers. Since, multiple smart meters are linked with each other and it is computerized, it gets vulnerable to several attacks. The large scale data generated by the grid is stored in the cloud. By using Anomaly detection techniques the anomalies can be detected from the data that is stored in the cloud. It is also supports forensic investigation. This technique is applied to all the IoT components to detect the compromised devices (Zubair A. Baig et al., 2017).

Electricity is the necessary thing in daily life. Without electricity water services and other basic amenities will be a problem. Electricity helps to maintain the quality of life and productivity. The fundamental grid operations are not changed since 1930, the general infrastructures are same but some of the technologies have changed because of new inventions. The significant issue is fusing the renewable energy such as wind, sun and water into grid (Dr. Sanjay Goel, Dr. Stephen F. Bush & Dr. David Bakken, 2013). Grid uses communication technologies to emerge the grid with new functions and it expands toughness to many anomalies. Electricity generation sources, transmission system and distribution system are the three main components that are used for the creation of a grid. Many problems raised in power grid tend to the innovation of new smart grid. It is a smart two way communication and it has multiple sources and one end user/customer. An end user receives the benefit from the smart grid from more than one source (Anastasia Mavridou & Mauricio Papa, 2012).

The Smart meters and IoT devices are located insecurely in the customer's home. The adversary will use these devices to steal the private data, because the grid maintains a communication between the intelligent components and the cloud. The household information such as number of individual's livings in the home can be extracted by the adversary by using the consumption patterns. The ownership details and the customer data stored in the cloud are also at risk of privacy. Denial of Service attack (DoS) will affect smart grid by delaying the legitimate messages and also create a jamming or traffic (Zubair A. Baig et al., 2017).

Information is shared using smart meter in smart grid and moreover, the smart meters interconnected will make the smart grid vulnerable to attacks. Two main challenges in smart grid because of the data stored in cloud are 1) Pattern will take off the data of the individuals living in a house and appliances used in a house 2) Theft of personal data like accessibility data that are already stored in cloud. Smart grid uses protocols like IPV4, IPV6, TCIP and 6LoWPAN, but still smart grids are vulnerable to attack (Zubair A. Baig et al., 2017).

GOALS

Electricity is the most severe framework where other infrastructures like transportation waters supply and communications trust in. It is very essential for the economy, stability and citizens' comfort. Maintaining energy will scale down the need for investment in new generation and it will mitigate the electricity cost. Large percentage of electricity is produced from the inexpensive power that was formed from the greenhouse gas emissions, coal. Power grids are to be protected from potential malicious attacks. The critical requirements for the nations are detecting and destroying power grid cyber attacks. The blackouts occur in the country causes serious economic losses and destruction of public confidence in the economy. The quick detection and correctness of the anomalies are terminated due to the lack of visibility of the grid. The need for additional power plants and transmission capacity is decreased due to the composition of the alternative power sources and the improved efficiency. The smart grid is used to manage the usage of energy by the home user and maintain/reduce the cost of usage. Even though the smart grid has many benefits technically and beneficially, it has many security and privacy issue. Because of the information from the home devices is collected by the smart meters and it is connected to the vast network, the smart meter act as the usage –reporting device, because information is collected by the smart meters from the home appliances. It is connected with the home appliances to control those (Patrick McDaniel & Sean W. Smith, 2009; Vincent J. Forte, Jr., 2010).

The quality of the power is very important, because fluctuations in power damage the equipment. Some appliances will be shut off during peak power demand. The usage of power ie., the detailed information about the electricity usage and pricing information's can be tracked by the customers using

Figure 2. Standard Smart Grid Architecture
(Zubair A. Baig, 2017)

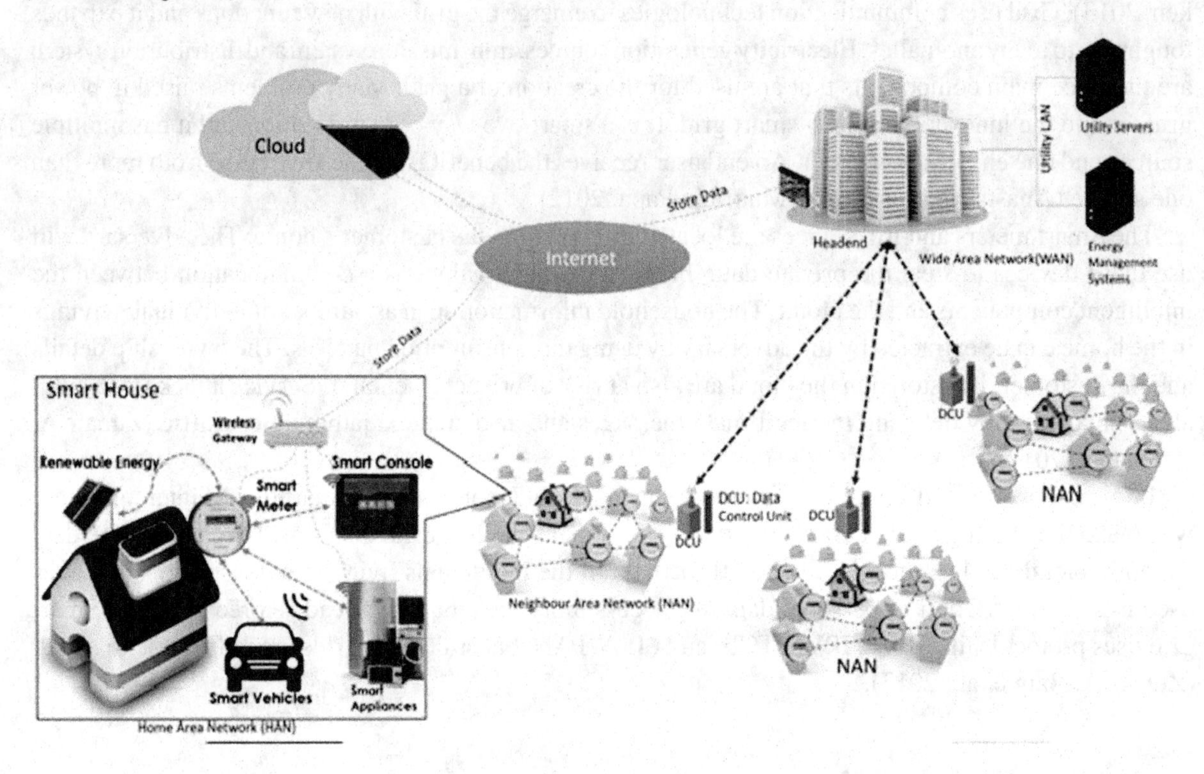

smart appliances. Such transparency information is useful for the customers to manage their current bill. The stability of the grid is improved by matching the supply and demand. Low latency and high communication infrastructure are used for the accomplishment of stability. The control of the grid will be information driven by adding information and communication infrastructure over grid and it become more efficient and responsive. The anomalies and monitor the grid operations can be tracked in the smart grid since the visibility is improved. Sensor is placed in the smart grid to improve the visibility. So, sensor is considered as the key feature for the clarity of the smart grid. Moreover data collection also considered as the key feature of the smart grid clarity. The grid is divided into smaller portions and it is called as micro grids. The smaller portion is isolated from each other. Each sub grids are self-sufficient; survive in case of large-scale failures. The micro grids will survive even when the large power plants failed. The standard smart grid architecture in figure 2 represents the communication of HAN, LAN and NAN using real time applications.

As observed in figure 3, the generation region contributes data with regional system, control centre and power system. If there is any generation collapse, actions will be arrested immediately by the regional system operator and power system. Regional system operator will monitor the transmission domain by exchanging data with the control center. In the near future transmission domain will be implemented with Wide Area Situation Awareness (WASA) System in association with EMS (Advanced Energy Management System). Energy Services Interface (ESI) is available in customer domain and through the AMI or internet it will communicate with other regions (ParikhPP et al., 2010).

Figure 3. Smart Grid Frameworks
(Parikh PP, Kanabar MG & Sidhu TS, 2010)

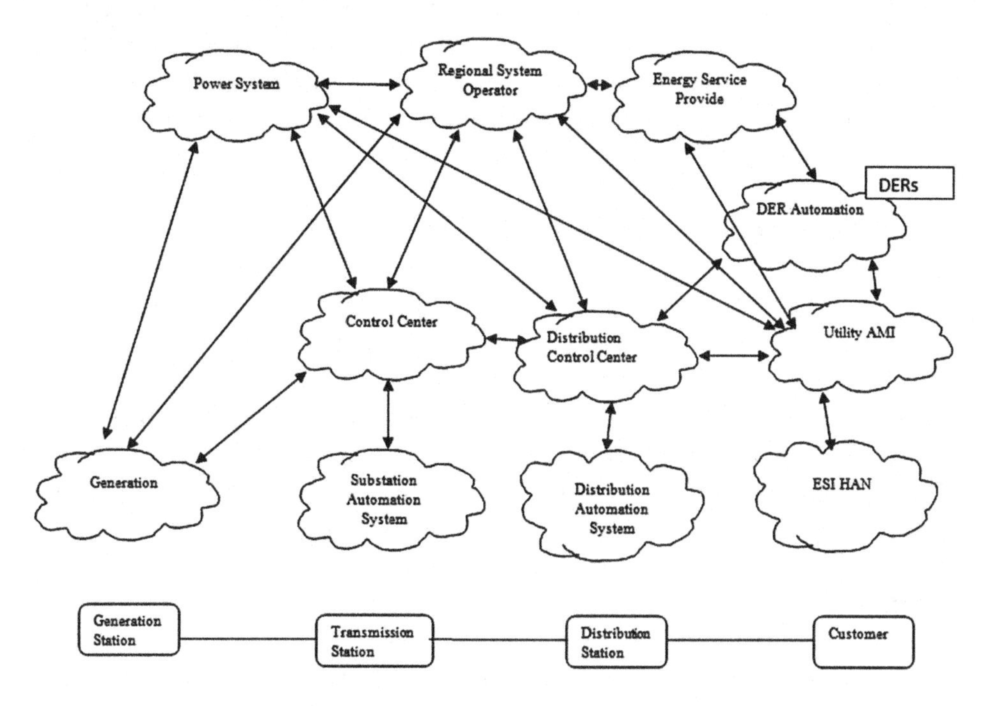

COMMUNICATION INFRASTRUCTURE

Huge number of sensors and actuators are available in smart grid; they are installed in transformers, generators, equipment, substation, power plants and home users. The data transfer between devices and data centers and the control of all the grid components are handled by sensors and actuators. The smart grid must be upgraded to handle huge volume of data. It is upgraded for secure connection (Dmitry Baimel, Saad Tapuchi & Nina Baimel, 2016). The communication infrastructure depends on three main types of networks: They are Home Area Network (HAN), Neighborhood Area Network (NAN) and Wide Area Network (WAN).

Home Area Network

HAN is for small area (within 10 meters), especially for a house or a small office. It has very minor data rate transmission than others, ie, 100 bits/second. Through wireless or wired modem, internet connection is shared between multiple users. Distribution of resources or information between computers, mobile and other devices through internet is enabled by HAN. Energy consuming smart phone devices and smart meters are connected to HAN. Data from the smart home devices are gathered and forwarded to smart meters via HAN. Home energy is managed efficiently by HAN by using the communication technologies like ZigBee or Ethernet (Dmitry Baimel et al.,2016). Topologies exists in HAN are Bus Topology, Tree Topology and Mesh Topology (Tsuyoshi Masuo, 2011). In a commonplace usage, a HAN comprises a broadband. Web association is common between different clients through a wired or remote modem. It empowers the correspondence and distribution of assets between PCs, versatile and different gadgets over a system association. In keen framework execution, all shrewd home devices that expend vitality and savvy meters can be linked or inter linked with HAN. The device information is obtained and broadcasted through HAN to the keen meters. HAN permits increasingly proficient home vitality the executives. HAN can be executed by ZigBee or Ethernet advancements. In the future, HAN will function ahead than the smart grid and will be a guide for the advanced utilization. Figure 4 represents the transmission infrastructure based on HAN, LAN and NAN.

Neighborhood Area Network (NAN)

NAN covers large area than HAN (upto 100 meters) usually for few urban buildings.NAN is also used to connect one or more HANs so that energy from the smart home is transmitted to NAN. This data is stored in Local Data Centers. The transmission data rate of NAN is upto Kbps. The communication technologies used by NAN are PLC, Wi-Fi and Cellular (Dmitry Baimel et al., 2016). The Topologies of NAN are Tree Topology and Mesh Topology. It is narrated by using the functionalities like ESI function, Relay Function and Collector Function. One or More ESI is connected with the collector and the collectors are also connected with each other (Tsuyoshi Masuo, 2011).

Wide Area Network (WAN)

WAN is for vast areas that covers up to ten kilometers. It connects several NAN and LDCs. The transmission data rate of WAN is in Gbps. The communication technologies used by WAN are Ethernet networks, Wi-MAX, 3G/LTE and microwave transmission (Dmitry Baimel et al., 2016). Expected usages

Figure 4. Smart Grid Communication Infrastructure
(Dmitry Baime et al.,2016)

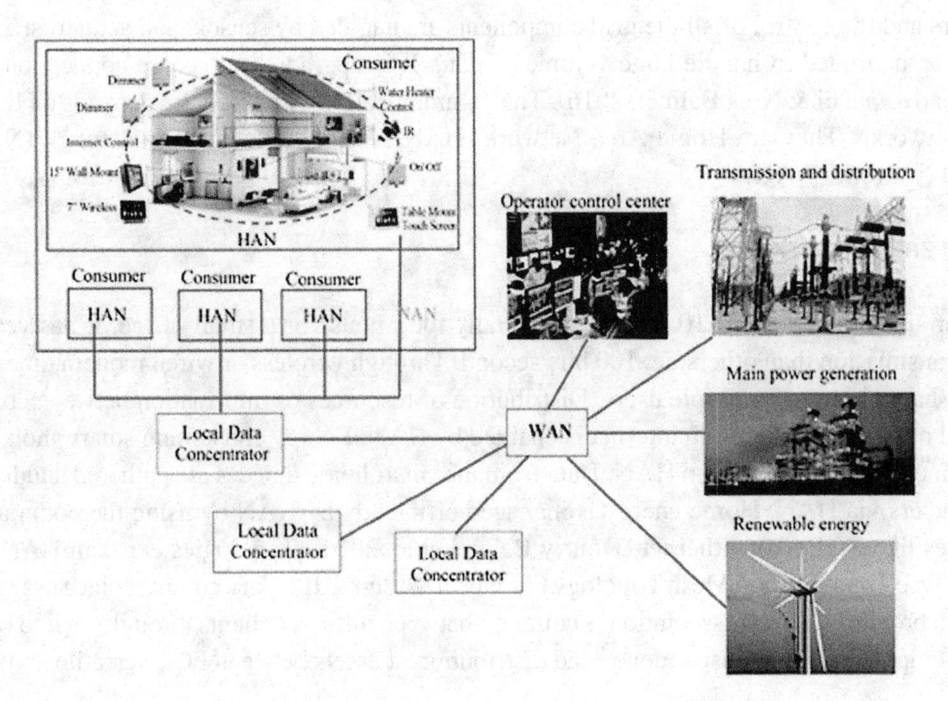

for Smart Grid WAN frameworks make them defenseless against security dangers in the type of physical or remote assaults on equipment, programming/firmware, trading off of certifications, and setup and system assaults (e.g., denial of-service). The framework should bolster confirmation and approval of the exchanged data. For the secure access by the smart grid gadgets or smart grid gateway, different levels of useful authentication must be made by the Smart Grid WAN. The intercommunication framework of smart grid is also called the components of smart grid. They are: a WAN, a FAN and a HAN.

WAN, FAN and HAN connects geographically distant sites, intelligent electronic devices (IEDs) and the customer networks simultaneously. HAN also connects the networks within the customer domain. Each component has independent characteristic and requirements (Dr.Sanjay Goe et al., 2013). According to the authors Kwang-Cheng Chen et al, the main components of smart grid are smart meters, distribution and transmission control centers, transmission and distribution lines, electric power generators, controllers, electric power substations and collector nodes. Two way communications between each of the above components can be done using Zigbee and mesh wireless (Anastasia Mavridou & Mauricio Papa, 2012).

COMMUNICATION TECHNOLOGIES

ZigBee

ZigBee uses IEEE 802.15 standard. Applications that has very low data rate, long battery life, low cost and secure networking uses ZigBee as a communication technology. ZigBee is applicable for wireless transfer of data at relatively little rates. It connects around 60,000 gadgets to the network. ZigBee applications are simpler and less expensive than other technologies. 128bit symmetric encryption keys are used by ZigBee for security purpose. ZigBee Smart Energy is an application that integrates smart meter with other devices .The smart meters collects information from the integrated devices and controls them (Dmitry Baimel et al., 2016).

Zig:Bee is a wireless transmission innovation. It has free wireless recurrence system and open worldwide standard and coincides with various technologies at specific groups. It is a basic, less power, inexpensive, low battery, less information rate (20-250 Kbps) computers functioning with 2.4 GHz, 868MHz, and 928 MHz under IEEE802.15.4 standard inside limits of 10 meters. It is acknowledged as a absolute technology for conversation between smart meter and apparatuses in home area network because of the simplicity of establishment, and in economic compound commitment wireless connection to their customers. In home area network, it is utilized among home machines easily, for example, smart lightning, vitality checking, home computerization, and electronic meter scanning, and so on. In Smart Grid, HANs Zig-Bee might be utilized in different home machines. For example, washer dryers, air conditioning, water radiator, PHEV and so on. Zig-Bee is appropriate for cyclic or discontinuous information as in Wireless Personal Area Networks (WPANs).Smart meters is not applicable for Bluetooth and Wi-Fi (Mukta Jukaria, Prof. B. K. Singh & Prof. Anil Kumar, 2017).

WLAN

WLAN links two or more devices and connects it through wider internet. The users are connected to the network and have the ability to move within a network. WLAN is easy to install so it is mostly used in home and also for commercial purpose. Due to its limitless establishment around the world, WAN can be easily combined into the smart grid. This gives clients the capacity to move everywhere inside a nearby coverage area and it is associated with the system. Generally present day WLANs depend on IEEE 802.11 standards, marketed under the Wi-Fi brand name. WLANs get turned out to be well known in the home because of simplicity of establishment, and in financial structure offering wireless access to their clients. The upsides of WLAN are ease, tremendous arrangement around the globe, fitting, play gadgets etc. The major impediment of WLAN is high portable for clashing with different gadgets that convey on similar frequencies (Dmitry Baimel et al., 2016; R. Bayindiretal, 2016).

Cellular Networks

Cellular networks have well-established infrastructure and are largely established in most countries. The data rate for cellular network is up to 100 Mbps. The components and gadgets in the smart grid used to communicate with each other using cellular networks. GSM, GPRS, 2G, 3G, 4G and WiMAX are the existing technology for cellular networks (Dmitry Baimel et al., 2016). Cell systems have profound steered foundation that permit high information rate up to 100 Mbps what's more, utilized for correspondence among various parts and gadgets in savvy network. GPRS, GSM, 2G, 3G, 4G furthermore, Wi-Max are different current cell advances, out of which the Wi-Max innovation is the most fitting and of enthusiasm for smart matrix usage. The critical characteristics of the cellular networks are slarge region of organization, huge amount of information exchange, and accessible security algorithms in the cellular conversation. The significant disadvantage is that cell systems are regular with different clients and are not completely committed to the Smart Grid interchanges (Mukta Jukaria et al., 2017). Cellular systems are generally conveyed in many nations and have settled foundation. In addition, they permit high information rate correspondences up to 100Mbps. Along these lines, the cellular systems is utilized for correspondence between distinctive components and gadgets in smart matrix. There are a few existing advances for cellular correspondence such as GSM, GPRS, 2G, 3G, 4G and WiMAX. The WiMAX innovation is the ultimate impressive for smart grid usage. It is taking a shot at 2.5 and 3.5 frequencies, with information conversion rate of 70Mbs and inclusion up to 50km. A WiMAX chips is integrated inside the smart meters. The benefits of the networks are as of now existing foundation with wide zone of sending, high rates of information exchange, accessible security algorithms that are as of now executed in the cell correspondence. The real drawback is that cell systems are shared with different clients and are not completely devoted to the keen network correspondences. This can be severe issue if there should arise an occurrence of crisis condition of the grid.

Power Line Communications

Data transfer between devices through electrical power lines can be handled by PLC.A modulated carrier signal is added to the power cables in the PLC for the implementation purpose. PLC uses different frequencies for different power line.PLC connects Local Data Concentrator and smart meters in Neighborhood Area Network (Dmitry Baimel et al., 2016). Since PLC allows the connectivity to any device connected to the power grid, it plays an essential part in the Smart Grid. Power line communication can be installed in transmission, distribution and customers which are the physical Smart Grid domains. PLC is installed over the High Voltage (HV), medium voltage (MV), and low voltage (LV) lines and it

Table 1. Smart Grid Communication Technologies

Communication Technologies	Advantages	Disadvantages
ZigBee	• The price is very less. • It is smaller in size. • Bandwidth is very limited.	• Less battery lifetime • Memory size too small. • Very small data rate • Processing capacity is also very less
WLAN	• Cost is very low. • broad establishment around the world,	• The conflicts with other devices are high while communicating with the same frequencies.
Cellular Networks	• Data transfer are with high rate • Security algorithms are available.	• Cellular Networks will cause series problem in case of emergency case
Power Line Communications	• Establishment cost is low	• less frequency of communication

supports two way communications. PLC is used in FAN and HAN which are the Smart Grid Communications infrastructure (Dr. Sanjay Goel et al., 2013). PLC module is connected with the appliances, the utilized electricity information from the appliances send the control signaling (Kwang-Cheng Chen, Ping-Cheng Yeh, Hung-Yun Hsieh & Shi-Chung Chang 2010).

The functions of smart meters are Power Line Communication, Wireless Communication and a Processing Unit. Wireless Communication module performs similar operation like PLC, it gathers information about control signaling and send it for power operation. The processing unit is for distributed computing (Kwang-Cheng Chen et al., 2010). The communication and networking are handled by two types. They are

1. Information's are gathered from the sensors, appliances etc. and the collected information is send to the smart meters, and it predicts urgent need for electricity. Smart meters send the data to the smart grid for the control operation of the appliances.
2. Data from smart meter to smart grid infrastructure, proper power generation and storage is guarded and adapted in real time in the best possible way.

Messages will be passed between gadgets and smart meters that are encouraged by wireless communication and power line communication. Errors (noise, interference, fading, distortion) will occur in gadgets and smart meters to overcome this reliable data link control protocol is used to make the protocol more reliable. The mechanism that is used to enhance the secure communication between gadgets and smart meters is called FEC-Based error correction. The FEC based error correction used because the traffic between gadgets and smart meters are abnormal in nature. Load Management allows the customer to adapt the electricity utilization of their appliances in real time. It balances the load (power) even in a high or low voltage. The loads (power) information of different areas are collected by the power grid and it make decision whether to diminish the current load or not and send request to the particular area. For this a power grid has extra "intelligence", for making decision. An advanced smart metering needed in the smart grid to do all those works (Kwang-Cheng Chen et al., 2010). Various drawbacks and benefits of Smart Grid communication technologies are shown in the above Table 1.

POWER GRID OPERATION

We need to learn the operation of power grid to understand the communication of smart grid. Multiplication of the speed of rotation of the machine produces electricity at very low voltage. The amount of steam let into the turbine is used to determine the power output. By rotating the winding of the generator, we can find the voltage of the power. Fixed winding is for finding the output. The lower voltage is stepped up into a higher voltage and it is connected to a substation (switching station). Switching devices has circuit breakers, which are heavy-duty switches that open the circuit when there is a fault. The switching devices provide redundancy, in which they are interconnected with each other. Even some components gets failed, some substations can operate properly. Grid consists of interconnected high voltage power lines joined at other switching stations. Redundancy is very essential so multiple lines interconnect substations. Grid usually connects to bulk supply stations to step down the voltage by transformers. Mostly in cities bulk supply stations are used to bring reliable power into it. In high-voltage side the bulk supply stations are well-interconnected and vice versa for low side. Since the low side has few interconnections, it forms sub transmission system (Dr. Sanjay Goel et al., 2013).

Medium voltage (MV)(between 1 kV and 35 kV) distribute power to the large substations. The MV normally stepped down the power to the end user at low voltage (LV).High Voltage (HV) is above 100 kV and if the voltage is above 345kV it is called extra-high voltage (EHV). If the voltage exceeds above 1000kV, then it called ultra-high voltage (UHV).

SMART GRID ARCHITECTURE

Smart metering is an important aspect that makes a smart house. The architecture given below is the architecture diagram for Energy Smart House (A. M. Carreiro, C. H. Antunes, & H. M. Jorge 2012). This ESH will utilize all the energy that is produced and there won't be any loss in the energy. The electricity bill will also get reduced because energy that is produced by the entire home appliances are utilized by it and the stored extra energies will be utilized for future. The usage of electricity is measured and controlled by sensors and actuators that are available in smart house. The data available in the sensors are extracted by a device called Energy Box Device (EBD). That extracted data is processed and communication is made with the resource utility, where EBD will act as a gateway (A. M. Carreiro et al., 2012).

Situational Awareness Architecture has SCADA gateway, network sensors, database and a command centre located in the control center. The network sensors will arrest the network traffic in promiscuous mode. Network sensor acquires information about the service of device, state, unit configuration etc. The collected data is analyzed and it is forwarded to the SCADA gateway. The malicious traffic is identified by attack signature. The information gathered by the sensors is translated into a canonical form and it is moved into the database. There is a coupling between command centre and SCADA gateway. Rules and fault conditions of the system are loaded in the critical stable table. The system may or may not enter into a critical state that was verified the command center (Anastasia Mavridou & Mauricio Papa, 2012).

The Simplified Smart Grid Domain Model Diagram given below shows the coupling between different domains that is denoted by the circle as shown in the figure 5.

Reference Point 1: It represents coupling between Grid domain and Communication Network. Data's and control signals are exchanged between gadgets available in Grid domain and the service provider domain.

Reference Point 2: It represents coupling between smart metering domain and communication network. Metering Data is transferred from the smart metering domain to the customer domain. The Communication between the customers in the customer domain through the operators and service providers in the service provider domain.

Reference Point 3: It represents coupling between customer domain and communication Network. The communication between the gadgets in the service provider domain and the gadgets in the customer domain was enabled.

Reference Point 4: It represents coupling between service provider domain and communication network. The interaction between the services and applications in the service provider domain with other domains.

Reference Point 5: It represents coupling between smart metering domain and customer domain (Tsuyoshi Masuo, 2011).

Figure 5. Simplified Smart Grid Domain
(Tsuyoshi Masuo, 2011)

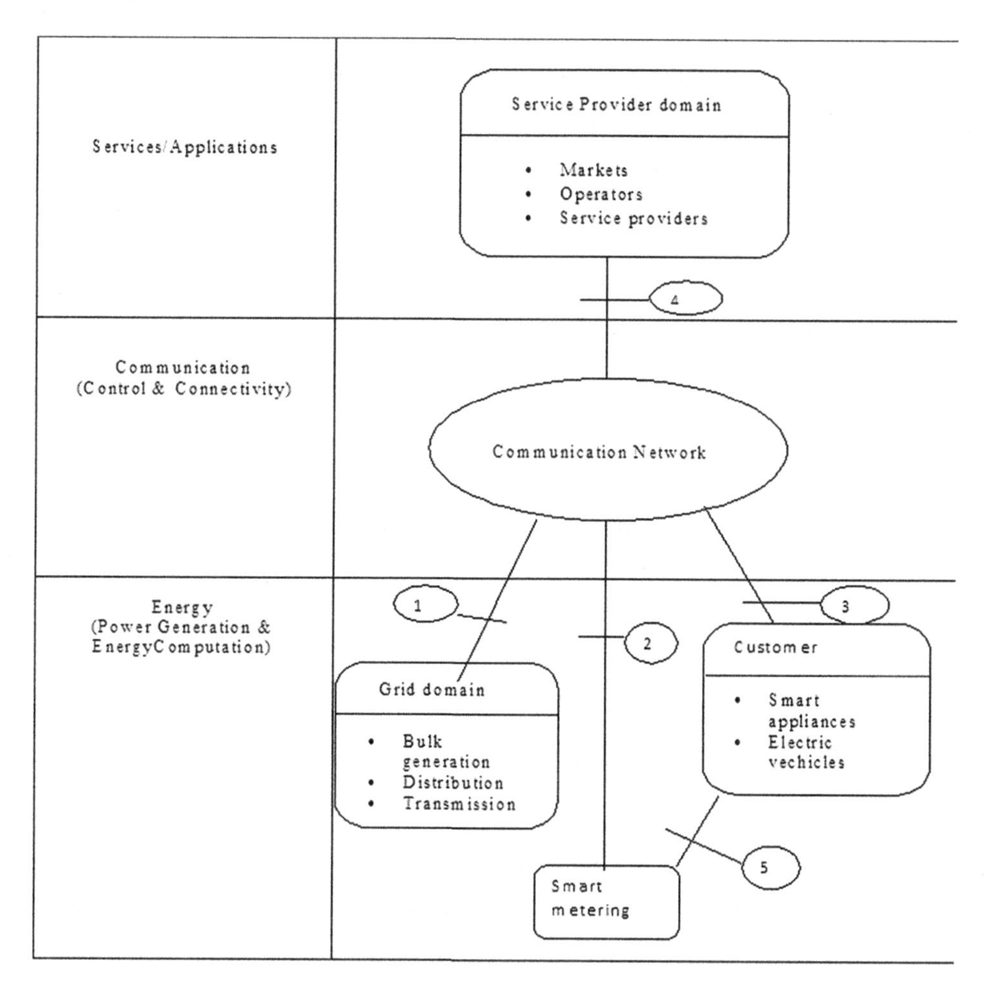

Figure 6. Simplified Reference for Smart Grid
(Tsuyoshi Masuo, 2011)

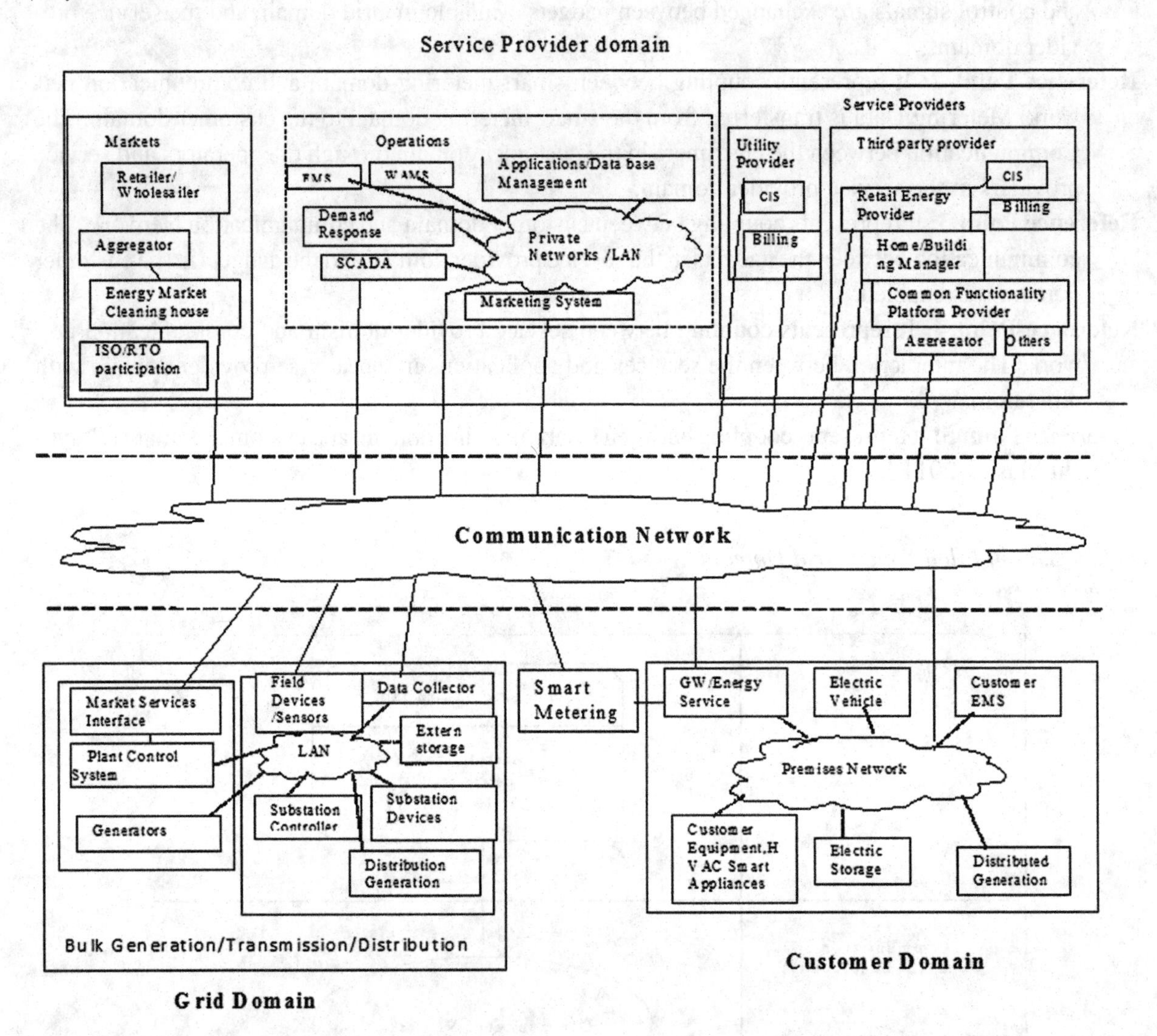

The simplified reference model shown in the figure 6 is the modified architecture of smart grid domain model. It impersonates the logical perspective of the smart grid; the communication network makes a bridge between the logical devices that exists in the smart grid. There is a reference design outline demonstrating applied information stream between areas (Tsuyoshi Masuo, 2011).

SMART GRID SECURITY AND PRIVACY

Since, the smart meters connect all the home appliances to control, to manage and to collect the information from it. Further, the users are connected above a huge network of smart meters; they are vulnerable to network-borne attacks. Some of the attacks that intrude and affect the smart meters are distributed denial-of-service attacks, meter bots, smart meter rootkits, meter-based viruses. When misinformation injected into the control system, it harms the electrical infrastructure. Since the customer's information is available in the vast area, the privacy of them is also in risk. The information in the smart meters is distributed across the side-channel that contains the behavior and habits of customer. Existing rule for privacy policy in U.S is not in a proper format, there is no clear idea how their policy will be helpful in privacy and security. The Government established a national pedestrian for customer privacy protections; such rule is identical to a HIPAA (Health Insurance Portability and Accountability Act) for smart grid. It identifies how the information is collected, to whom it is opened and the result of information misuse (Patrick McDaniel & Sean W. Smith, 2009). The major threats that occur during the smart grid installation is cyber-attacks, physical attacks or natural disasters. That tends to the infrastructure failure, energy theft, customer privacy breach etc. The threats are inspected by authorization; authentication and privacy based on the security levels. Some of the planned crimes are rioting, hacking, terrorism, vandals, cybercrime, disruption of services, energy theft etc. The purposeful causes are the operation dictum due to fake data insertion, data leakage etc. The protected smart grid must be authenticated, confidential; maintain great level of honesty, efficiency, maintain huge level of authentication, robustness, accuracy, resilience, self-remedial, huge level of accessibility, controllability, observability etc. US Department of Energy (DOE), Electric Power Research Institute (EPRI), Institute of Electrical and Electronics Engineers (IEEE), National Institute of Standard and Technology (NIST),), North American Electric Reliability Corporation (NERC), Federal Energy Regulatory Com-mission (FERC) etc are some the standard bodies that play an great role in the installation of Smart Grid System (Abdulrahaman Okino Otuoze, Mohd Wazir Mustafa & Raja Masood Larik, 2018). Some of the threats that occur due to the unprotected physical infrastructure are floods, fire outbreaks, earthquakes, tsunamis, explosions, dangerous radiation leakage, population, landslides, dust erosion etc. Some of the techniques that are used to enchance the security ability of the Smart Grid by researchers are, Petri Net (PN), which analysis the system in detail and identify the failures. Another system that was developed by DeSantis in Rome Italy, detects the faults in the grid operation. The evolutionary learning and clustering concept was hired for handling the fault event. Communication is the main feature in Smart Grid installations; data management and the ability in communications are increased. Internet of Things (IoT) are exploited and installed for smart grids in connection to smart cities, home automation network, buildings etc. It is also useful for checking the energy consumption, control and improves the safety, flexibility. Many Information and Communication Technologies (ICT) are used in the guarantee framework that was proposed by (Abdulrahaman et al., 2018). The ICT contains the ZigBee, Internet, wireless mesh networks, 3G, 4G, WiMax, Bluetooth, WiFi, distributed intelligence techniques, virtual power plants, smart metering, micro-grids, demand response technology etc. Distributed and Renewable energy is an important ICT tool that was used to distribute the electricity grid, the operations are more flexible and it is strong enough to the various security threats.

Based on threat sources the smart grid can be classified into two types. They are technical source of threats and non-technical source of threats. Technical source of threats classified into three types, they are infrastructural security, technical operational security and system data management security. Non-Technical source of threats are classified into two types they are environmental security and Government administrative policies and implementation. The structure for identifying the smart grids threats are shown in the Table 2.

Table 2. Structure for finding the smart grid threats

Sl.No	Security Challenges That Check in Fact of Following Terms	
1	Security level	Authentication, Authorization and Privacy
2	Sources of threats	Technical and non-Technical
3	Cause of threats	Human factors and non-human factors
4	Unit Failure/Breakdown	Generator units, Transmission Lines, Distribution Lines and Substation units
5	Factors leading to attacks	Natural and non-Natural
6	Intent cause of attacks	Deliberate and non-Deliberate
7	Impact level	High, Medium and Low
8	Any other noble factor	-

(Abdulrahaman et al., 2018).

Technical Source of Threats

If the threats that are identified occurred based on the technical aspects then it is called as technical source of threats. The important features in technical source of threats are Infrastructural security, Technical operational security and systems' data management security.

Infrastructural Security

Since Smart Grid infrastructure is formed by the correlation between users, substation, utilities, transformers, transmission etc. Moreover fiber optic, wireless, power Line carrier etc also used for making the Smart Grid. So Security is the most important thing we need to address because the infrastructure is vulnerable to attacks. Some of the attacks are cyber security rupture, collapse etc. AMI is available at the central part of smart grid operation and it is more vulnerable. Energy is a necessary feature for everyone in this universe, so the need for energy is an ever-growing worldwide.

Advanced Metering Infrastructure Security (AMI)

AMI is the fundamental part of Smart Grid; it includes smart meters, data management systems, and communication networks. Information's are exchange between customers and utilities. It also contains sensors, computer to store the data, devices and meters. So it is vulnerable to attacks like cyber-attacks. The Smart Meter is the primary target component for attacks to steal electricity.

Smart Meters and Energy Theft

The Smart Meters replaces the normal digital and analogue meters. That gives information to the customers and the company about the usage of electricity in the customer's home. The other advanced functions of Smart Meters are power quality monitoring, energy theft detection, remote command operations etc.

Attacks are making use of the Smart Meters for power theft. Smart Meters will deal with the data of both customer and service. So the Smart Meters need to be installed in the secure environment for secure communication.

Cyber-Attacks

Cyber-attacks are the common attack that occurs in smart grids. Mostly the attacks will mislead the services in fault decisions about the capability and usage. The authentication, confidentiality and privacy of data must be done to make the data protected from the unauthorized modifications. The cyber-attack reported in US power system network, also in Georgia, the United Kingdom etc. are some of the examples for cyber-attacks.

Technical Operational Security

This security binds the infrastructural deployment and operational policy, control initiations depend on reliability, system status and flexibility of operations, regular checkup, regular maintenance etc. Defects also occur in the protected devices, so construct the system with self-healing.

Systems' Data Management Security

In this security concept the data's are recorded, stored, and monitored. Rules and regulations are followed for handling data policy, privacy attachments by performing individuals, customer's satisfaction.

Non-Technical Source of Smart Grids Threats

The threats that countered against the installation or operations are called as Non-technical source of smart grids hazard. It has natural risk like floods, earthquake, bush burning, falling of trees etc.

Environmental Security

This type of security issues occur due to the natural or artificial dynamite such as tremors, earthquakes, floods, landslides, burning of bushes, falling of trees etc. These types of hazards are handled by restoring the services and data in case of any failures in the system.

Government Regulatory Policies and Implementation

The Smart Grids faces many challenges due to the enhancement of new technologies and services. So the government engages and provides administrative policies to favor continuous market procedure and private sector change.

A framework is proposed by (Abdulrahaman et al., 2018), which identify the threat and clear it. Its main objective is to discover the threat then at the mentioned input, there is warn from the AMI data. Threat feature extraction, Data processing and classification, Identification of threats, threat clearance and grouping of threats were done in the central processing. The threat clearance and feedback maintains the output data and do the data analysis.

CONCLUSION

The smart grid communication technologies, architecture and privacy of the smart grid were discussed in the above sections. The operations of the power grid and the functionality of the smart grid are also discussed. The Smart Grid is very useful in the future because it collects all the energy from the home appliance and the devices and those energies are stored and collected grid through the smart meters. It consumes the energy; it prevents the wastage of energy. It replaces the normal power grid. Sensors and Actuators used by the smart meters for collecting, storing and communication purpose. Smart meters are vital subject in smart grid to gather ongoing information and execute the guidelines from control focus through sensors, for example, current, voltage to get easy to understand organize. To diminish the specialized trouble happening in channel and to prevent the illegal utilization of energy, besides smart meters, it is important to take a look at the social and social conduct of clients. Traditional structures of society will likewise diminish losses. Smart metering framework can peruse every required information from the network. Distinct smart metering framework can likewise go about as a sensor. Utilization of smart metering in intelligent grid gives huge application opportunities. The application field of smart metering can be broadened for example, by utilizing smart meters in the lattice as sensors to distinguish issues and varieties.

REFERENCES

Baimel, D., Tapuchi, S., & Baimel, N. (2016). Smart grid communication technologies- overview, research challenges and opportunities. *International Symposium on Power Electronics, Electrical Drives, Automation and Motion (SPEEDAM)*. 10.1109/SPEEDAM.2016.7526014

Baimel, D., Tapuchi, S., & Baimel, N. (2016). *Smart Grid Communication Technologies*. Scientific Research Publishing. doi:10.4236/jpee.2016.48001

Bayindiretal, R. (2016). Renewable and Sustainable Energy Reviews.*Smart grid technologies and applications. Renewable & Sustainable Energy Reviews*, *66*, pp499–pp516. doi:10.1016/j.rser.2016.08.002

Carreiro, Antunes, & Jorge. (2012). Energy Smart House Architecture for a Smart Grid. *IEEE International Symposium on Sustainable Systems and Technology (ISSST)*.

Chen, Yeh, Hsieh, & Chang. (2010). Communication Infrastructure of Smart Grid. *Proceedings* of *the 4th International Symposium* on *Communications, Control* and *Signal Processing*.

Cui, Xie, Qu, Gao, & Yang. (2018). Security and Privacy in Smart Cities: Challenges and Opportunities. *IEEE Access*.

Forte, V. J. Jr. (2010). *Smart Grid at National Grid*. IEEE. doi:10.1109/ISGT.2010.5434729

Goel, Bush, & Bakken. (2013). *IEEE Vision for Smart Grid Communications: 2030 and Beyond*. IEEE Communication Society.

Jukaria, Singh, & Kumar. (2017). *A Comprehensive Review on Smart Meter Communication Systems in Smart Grid for Indian Scenario*. Academic Press.

Masuo, T. (2011). *Deliverable on Smart Grid Architecture*. International Telecommunication Union.

Mavridou & Papa. (2012). *A Situational Awareness Architecture for the Smart Grid*. Institute for Computer Sciences, Social Informatics and Telecommunications Engineering.

McDaniel, P., & Smith, S. W. (2009). *Security and Privacy Challenges in the Smart Grid*. IEEE Computer and Reliability Societies. doi:10.1109/MSP.2009.76

Otuoze, Mustafa, & Larik. (2018). Smart grids security challenges: Classification by sources of threats. *Journal of Electrical Systems and Information Technology*, 468–483.

Parikh, Kanabar, & Sidhu. (2010). Opportunities and challenges of wireless communication technologies for smart grid applications. *Proceedings of power energy soc. gen. meet. IEEE*, 1–7.

Zubair, Szewczyk, Valli, Rabadia, Hannay, Chernyshev, … Peacock. (2017). Future challenges for smart cities: Cyber-security and digital forensics. Security Research Institute & School of Science, Edith Cowan University.

KEY TERMS AND DEFINITIONS

3G/LTE: LTE represents long term evolution. It's a term utilized for the specific sort of 4G that conveys the quickest versatile Internet experience. You'll normally observe it called 4G LTE. Utilizing a 4G cell phone on Verizon's 4G LTE systems implies you can download documents from the Internet up to multiple times quicker than with 3G.

LDC: LDC stands for load duration curve. It is utilized in electric power generation to represent the connection between creating maximum requirement and maximum utilization. The LDC curve demonstrates the capacity use prerequisites for every increase of load.

Microgrids: A microgrid is a little scale control lattice that can work freely or cooperatively with other little power matrices. The goal of micro grid is reliable, reduction of cost and carbon emission.

Power Line Carrier: Power line carrier portrays the whole procedure of communication utilizing high-voltage power lines as the methods for transmission. Power lines give a dependable connection as a result of their abnormally rough development, and furthermore offer opportunity from limitations forced by common-bearer guidelines.

SCADA: SCADA is an abbreviation for supervisory control and data acquisition. SCADA frameworks are utilized to screen and control a plant or hardware in ventures, for example, media communications, water and waste control, vitality. A normal SCADA framework involves Controllers, I/O signal equipment, programming, etc.

Smart Meters: A smart meter is an electronic device that stores the utilization of electric energy and transfers the data to the power vendor for billing and monitoring. It also stores the usage of the power periodically.

Wi-Max: WiMAX is a group of remote broadband correspondence norms dependent on the IEEE 802.16 arrangement of measures, which give various physical layer (PHY) and media access control (MAC) alternatives.

Chapter 9
Block-Chain-Based Security and Privacy in Smart City IoT:
Distributed Transactions

Thangaraj Muthuraman
Madurai Kamaraj University, India

Punitha Ponmalar Pichiah
Meenakshi Government Arts College, India

Anuradha S.
Mobius Knowledge Services Pvt. Ltd., India

ABSTRACT

The current technology has given arms, hands, and wings to the smart objects-internet of things, which create the centralized data collection and analysis nightmare. Even with the distributed big data-enabled computing, the relevant data filtering for the localized decisions take a long time. To make the IOT data communication smoother and make the devices talk to each other in a coherent way the device data transactions are made to communicate through the block chain, and the applications on the localized destination can take the decisions or complete transaction without the centralized hub communication. This chapter focuses on adding vendor-specific IOT devices to the public or private block chain and the emerging challenges and the possible solutions to make the devices talk to each other and have the decision enablement through the distributed transactions through the block chain technology.

INTRODUCTION

In recent years, there has been large numbers of people moving towards urban living. As forecasted by 2030 more than 60% of the population will reside in an urban environment. The challenges related to increased population in the urban will throw in to the development of the Smart City. The Smart City concept incorporates several multifaceted systems of infrastructure, human activities, technology, public and political structures, financial system and provides collaborations between citizens and government.

DOI: 10.4018/978-1-5225-9199-3.ch009

Smart City provides a smart way to manage components such as transport, health, energy, homes and buildings and the environment. Smart technology allows the officials to observe the city happenings and to interrelate with community and city infrastructure. Smart cities become smarter with the recent advancements in the digital equipment. The data generated by this equipment are primarily by wireless sensor networks. Wireless sensor networks have been deployed in many consumer applications and industrial such as smart home, health monitoring, smart people, smart economy, smart governance, smart parking, smart roads, smart mobility, smart environment, and smart living.

BACKGROUND

Wireless Sensor nodes deployed with different Smart City applications generate large amounts of heterogeneous data. However, it's difficult to connect the WSN and Internet because it lacks of uniform standardization in communication protocols. In the usage of entrenched devices, wireless communication and obtainable internet infrastructure the Internet of things (IoT) connects PC's and other surrounding electronic devices easily. Internet of Things is composed of hundreds of millions of objects that can be identified, sensed and connected based on standardized and interoperable communication protocols. Internet of things allow things and people to be interconnected anytime, anywhere, anything and anyone, ideally using any path/network and any service. These devices transmit the data to the internet through specific protocols to achieve monitoring, management and location tracking. With the support of internet connection, and cloud computing technologies, the IOT system can intelligently process the objects' state, and control for decision-making, separately without human's intervention.

IOT Gateway

The devices adhere to different protocols makes the security mechanism of device to multi chain communication complex. In 2016, the expense for the IOT devices was 120 billion, it is expected to grow by 253 billion by 2021. It indicates the complexity and maintenance cost of the IOT devices in the forthcoming industrial scenarios.

IOT Gateway plays a very important role in interconnecting multiple smart devices together to form network and share resources and information among the equipment with different network protocols. Smart city is interconnected and instrumented with the support of IOT. Performance and reliability of the smart city are improved with broadly scattered, low storage capability and processing capability of IoT devices. So IoT provides more efficient operations based on different aspects, such as energy saving policies, economic considerations, reliability levels, etc. IoT Architecture is shown in figure 1.

IoT devices exchange vast amounts of critical data as well as sensitive information, and hence are appealing targets of various cyber-attacks. These devices spend their available energy in executing application functionality, and supporting security and privacy.

Security Issues in IoT

IoT devices are resource-constrained. Therefore, using the conventional security mechanisms in the smart devices are not easy. The major security constraints of IoT devices are Limitations based on hardware,

Figure 1. IoT Architecture

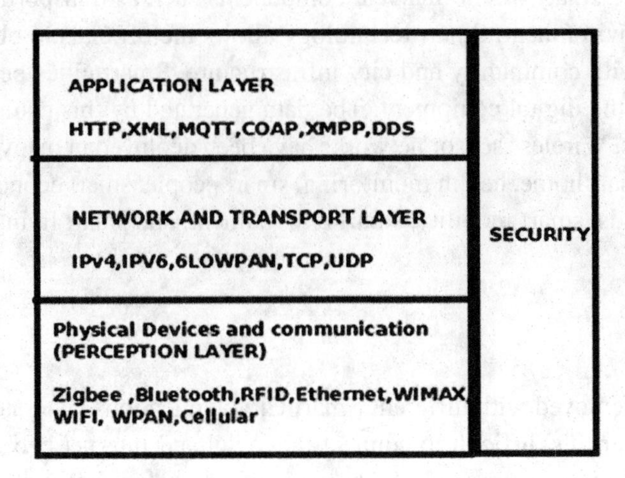

Table 1. Security issues in IoT

	Perception Layer	Network Layer	Application Layer
Purpose	The layer is used to acquire the data through sensors and actuators from the environment.	This layer is used in data routing and transmission of data to the IoT hubs and devices over the internet.	The application layer provides the authenticity, integrity, and confidentiality of the data.
Issues	Attacks from the external network such as replay attack, node capture attack, timing attack and deny of service also bring new security problems. In the other hand sensor data still need the protection for integrity and authenticity.	The network layer of IoT is also inclined to DoS attacks, traffic analysis, eavesdropping and passive monitoring. The network layer is highly susceptible to Man-in-the-middle attack.	Malicious Code Injection, Denial-of-Service (DoS) Attack, Spear-Phishing Attack, Sniffing Attack. Integration of applications are very difficult to ensure data privacy and identity authentication among different applications, The huge amounts of connected devices that share data will cause large overhead, another important issue in IoT are interaction between users, revealing the data, managing the applications
Security Mechanism	Public key encryption algorithm, hash algorithms, intrusion detection and risk assessment to provide security protection.	This can be achieved by p2p encryption, data integrity and intrusion detection.	Security concerns are Authentication, key agreement, firewalls, intrusion detection and risk assessment.

software and network. IoT operates on three layers namely perception, network and application layers. Each layer of IoT has intrinsic security issues shown in Table 1.

In terms of energy consumption and processing overhead traditional security methods are expensive for IoT and many of the security frameworks are highly controlled. They are not well-suited for IoT due to its huge expansion, many-to-one nature of the traffic, and single point of failure. Existing security methods often provide noisy data or incomplete data, which may potentially thwart some IoT applications from proposing personalized services and also very expensive for IoT devices.

Trustworthiness of IoT data is also important issue. IoT database are in centralized architectures. Untrusted entities can alter information available in the database according to their concerns, so the information available in the database is not reliable. This conveys about the need to verify that the information has never been altered. Trustworthiness in IoT data can be provided by distributed service. Because it is trusted by all its participants that guarantees that the data remains absolute. Security concerns like privacy, confidentiality, authentication, access control, end-to-end security, trust management, global policies and standards are addressed with this distributed service security and privacy protection.

Block chain is a data structure that creates a distributed, immutable, transparent, secure and auditable ledger. The block chain can be consulted openly and fully, allowing access to all transactions and can be verified and collated by any entity at any time. It uses public key cryptography to exchange transaction among parties. The transactions are time stamped and stored cryptographically on a distributed ledger called blocks, the linked blocks form a block chain. To change or remove blocks of data that are recorded on the block chain ledger is extremely difficult. Figure 2 shows the structure of a block chain.

RELATED WORK

Rathore et al. (2016) discussed a combined IoT-based system for smart city development and urban planning using Big Data analytics. For urban planning or city future development. Talari, et al. (2017) discussed review on the concept of the smart city besides their different applications, benefits, and advantages. In addition, most of the possible IoT technologies are introduced, and their capabilities to merge into and apply to the different parts of smart cities. Meanwhile, some practical experiences all across the world and the key barriers to its implementation are thoroughly expressed. Alavi et al. (2018) discussed comprehensive literature review of key features and applications of the IoT paradigm to support sustainable development of smart cities. Saravanan & Srinivasan (2018) listed the different kind of IoT applications used in different domains such as healthcare, smart cities, transport, e-governance and entertainment. These applications use the cloud services for faster IoT transactions. Cloud IoT technologies plays vital role in the implementation of smart cities by tagging every device in the cloud (Saravanan, et al., 2019)

Figure 2. Structure of Block chain

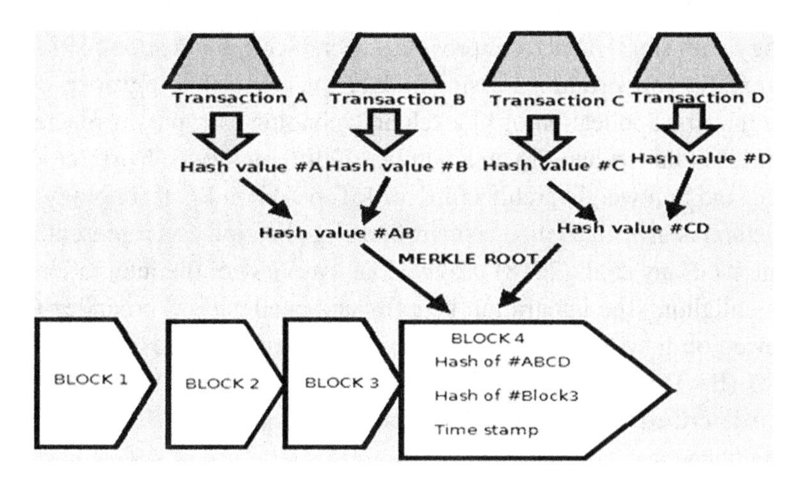

An emphasis is placed on concomitance of the IoT solutions with other enabling technologies such as cloud computing, robotics, micro-electromechanical systems (MEMS), wireless communications, and radio-frequency identification (RFID). Byun, et al. (2016) in the proceeding paper outlined IoT based smart city business (service) model. Gauer et al. (2016) stated in their paper that the IoT and wireless sensor network contribute a large amount of data. In order to take advantage of the increasing amounts of data, there is a need for new methods and techniques for an effective data management for the IoT. Meanwhile, Park & Rue (2015) introduced in their paper, Analysis on Smart City Service Technology with IoT", the technologically advanced countries" IoT based smart city service models.

Examples of this are the smart grid, parking management system, smart home, smart farms using foreign cases, and basic introduction. Zhanlin, et al. (2014) presented the generic concepts of using cloud-based intelligent car parking services in smart cities as an important application of the Internet of Things paradigm. They showed a high-level view of the smart parking system middleware and demonstrated the provision of car parking services. Nathalie, et al. (2012) contributed the design of a pervasive infrastructure where new generation services interact with the surrounding environment, thus creating new opportunities for contextualization and geo-awareness. Ammar et al. (2018) discussed the security of the main IoT frameworks, and the essentials of developing third-party smart apps, the compatible hardware, and the security features. Comparing security architectures shows that the same standards used for securing communications. Dorri, et al. (2017) outlined the various core components and functions of the smart home tier and proposed BC-based smart home framework is secure by thoroughly analysing its security with respect to the fundamental security goals of confidentiality, integrity, and availability. Atzori, et al. (2010) addressed the Internet of Things as the main enabling factor of this promising paradigm and it integrates several technologies and communications solutions. Bellavista et al. (2013), proposed a solution to integrate and opportunistically exploit MANET overlays, impromptu, and collaboratively formed over WSNs, to boost urban data harvesting in IoT.

Overlays are used to dynamically differentiate and fasten the delivery of urgent sensed data over low-latency MANET paths by integrating with latest emergent standards/specifications for WSN data collection. Schaffers et al. (2013) explored "smart cities" as environments of open and user-driven innovation for experimenting and validating future Internet-enabled services. Mischa, et al. (2014) designed to support the Smart City vision, which aims at exploiting the most advanced communication technologies to support added-value services for the administration of the city and for the citizens. Mulligan & Olsson (2013) discussed the architectural evolution required to ensure that the rollout and deployment of smart city. Biswas, & Muthukumarasamy (2016) proposed a security framework that integrates the blockchain technology with smart devices to provide a secure communication platform in a smart city. Jaag, & Bach (2017) explored opportunities arising from blockchain technology for postal operators (POs) and discussed about the first application of blockchain technology, crypto currencies like Bitcoin. but POs may be able to exploit this technology in a number of different ways. Dorri, et al. (2016) proposed a new secure, private, and lightweight architecture for IoT, based on BC technology.

The described method is investigated on a smart home application as a representative case study for broader IoT applications. Hany et al. (2018) provided an overview of the integration of the blockchain with the IoT with highlighting the integration benefits and challenges. Fernández-Carames & Fraga-Lamas (2018) reviewed on how to adapt blockchain to the specific needs of IoT in order to develop Blockchain-based IoT (BIoT) applications. After describing the basics of blockchain, the most relevant BIoT applications are described with the objective of emphasizing how blockchain can impact traditional cloud-centered IoT applications.

CHALLENGES IN BLOCK-CHAIN AND IoT INTEGRATION

All the interactions between devices go through blockchain, enabling an undisputable record of interactions. This approach ensures that all the chosen interactions are noticeable as their details can be enquired in the blockchain, and additionally it increases the autonomy of IoT devices. Nevertheless, recording all the interactions in blockchain would involve an increase in bandwidth and data, which is one of the well-known challenges in blockchain. Incorporating blockchain in IoT is challenging. Some challenges are,

- **Storage Capacity and Scalability:** IoT devices can generate gigabytes (GBs) of data in real time; this limitation represents a great barrier to its integration with blockchain. Current blockchain implementations can only process a few transactions per second, so this could be a potential bottleneck for the IoT.
- **Processing Power and Time:** Required to perform encryption for all the devices involved in a blockchain-based system. IoT ecosystems are very diverse.
- **Security:** IoT applications have security problems at different levels, due to the lack of performance and high heterogeneity of devices. The IoT architecture is affected by many factors such as the environment, participants, vandalism and the failure of the devices. IoT devices should be tested before their integration with blockchain and they should be located and encapsulated in the right place to avoid physical damage.
- **Anonymity and Data Privacy:** Blockchain is the ideal solution to address identity management in IoT, where anonymity needs to be guaranteed.
- **Data Integrity:** Ensure data access at the same time as they avoid overloading blockchain with the huge amount of data generated by the IoT.
- **Lack of Skills:** Few people understand how blockchain technology really works and when you add IoT to the mix that number will shrink drastically.
- **Legal and Compliance Issues:** This challenge alone will scare off many businesses from using blockchain technology.

PROPOSED WORK

Smart city has the ability to monitor waste management, car parking, and traffic signalling and structural health monitoring. Sensors are placed to collect information and perform communication. Communications between devices are known as transactions. Transactions are classified into four categories such as Store, access, add and remove. Store transaction is generated by devices to store data. An access transaction is generated to access the local storage. A add transaction is generated when adding a new device to the smart city and a device is removed via a remove transaction. All the transactions use a shared key to secure the communication. Hashing issued to detect any change in transactions content during transmissions. Store and access transactions to or from the local storage of smart city. Add and remove transactions are stored in Blockchain (BC) explained in figure 3.

To perform monitoring in smart city, is further categorized into areas. Block chains are created for each area. Starting from the genesis transaction, each device's add and remove transactions are chained together as a distributed ledger in the BC. Each block in the BC contains block header. The block header has the hash of the previous block and authorizing policy.

Figure 3. Proposed Architectural Frame work

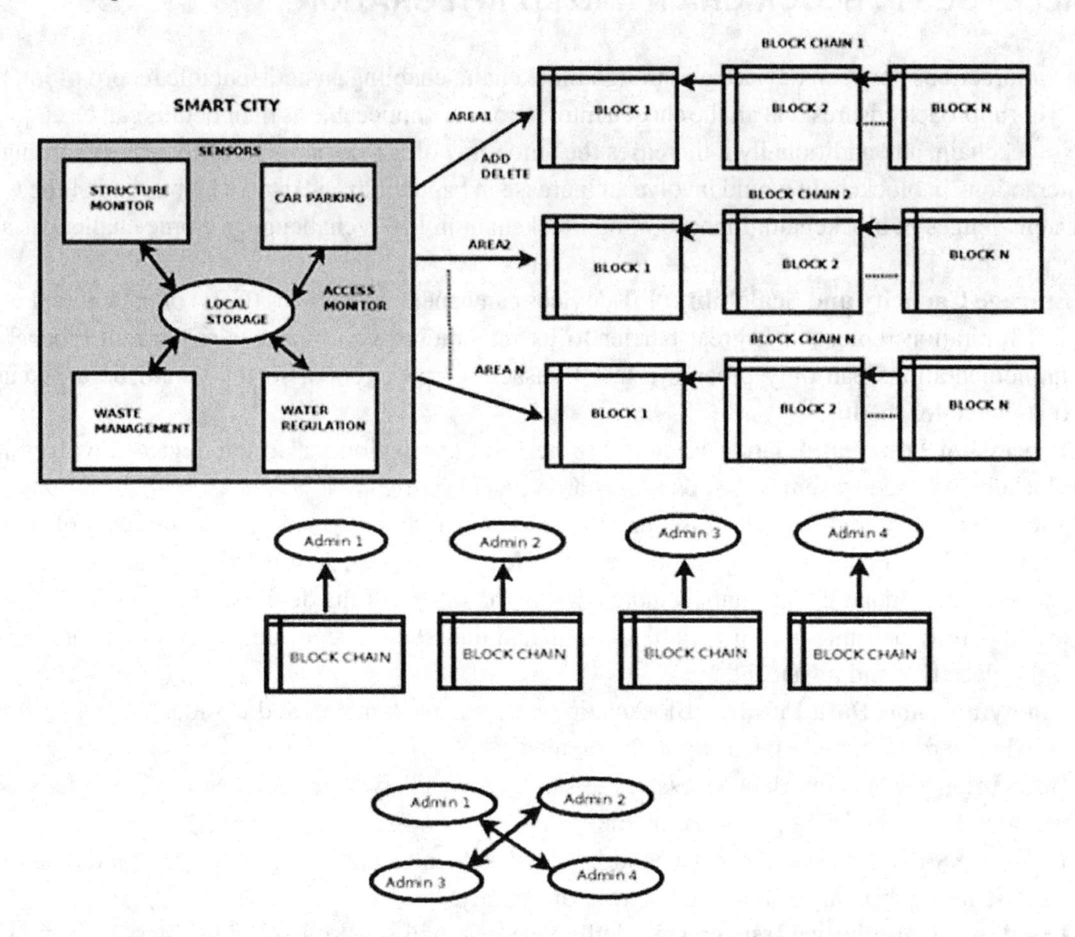

Distributed Database

In the smart city scenario of greater number of sensors, the cost and communication are still more complex. The data leaks of the centralized network of devices and the interoperability is the matter of question always. This paper discusses on the transition of the security mechanism from 'security through obscurity' to 'security through publicity' and the centralized communication of IOT devices to M2M communication mode of making the devices to talk. The main focus areas could be the transition of centralized computing to decentralized computing, centralized storage to distributed ledgers, centralized messaging to decentralized whispers.

The block chain keeps the pool of transactions, every entry into the transaction is verified and make them into blocks. This transaction fingerprint is maintained by the hash function. From the device reading, using the device specific messaging system, the data records are to be formed and chunked into blocks. From the transactional buffer, a set of transactions are considered for the block chunking and the blocks are formed with the unique hash functions. The chunked blocks are getting transmitted to the public or private blockchain through the block chain client created.

Every chunk is verified against the validation and verification set. The signature of the sender, the hash value of the block is verified and validated in the target block chain. If it is the financial transaction

Figure 4. Block chain Transactional preambles and transactions

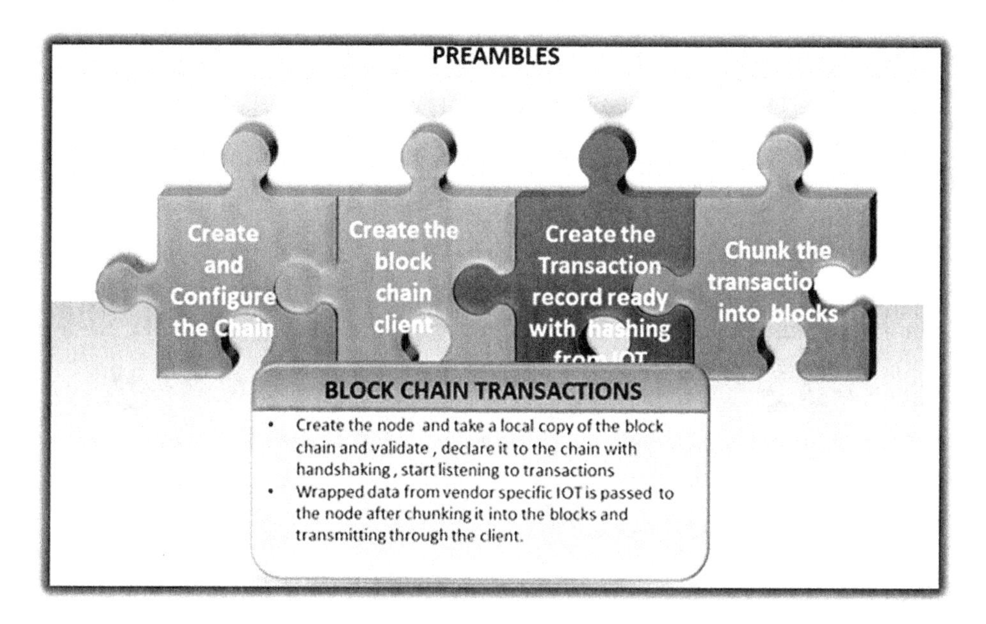

the keys purchased are to be used. Genesis blocks are the arbitrary blocks and are being used to initiate the block chain transmission and the other blocks of transactions are connected either with parent -child link or with the linked list kind of reference. Any node in the network, take the copy of the block chain which it intends to join and verify the following aspects, Check hash, check block, check chain for the validity and correctness. Then it announces its presence in the network and start listening the transaction.

Decentralized Messaging (Whisper)

Current IOT infrastructure is based on centralized architecture of either client-server or cloud which makes the IOT network very expensive. Even we are talking for decades about M2M, there is no single platform that connects all vendor devices and set the single communication platform. Peer to peer communication of devices, literally make devices talk to each other will reduce the cost of the centralized data processing servers. Distributed digital ledgers are created for the transactional data between the devices which authorized and validated by the registered nodes in the block chain network. The regulatory group that works on streamlining the IOT data is chain of things. We created a block chain node and created a private block chain with initially 8 nodes and added a random factor to that to expand. Formed the block chain client daemon to connect to the block chain nodes. Created initially 5 devices to register and transmit the transactional traffic data in blocks to the block chain, later added a random factor to mimic the devices. The node address of the daemon we copied into the 3D visualization block chain traceable framework and traced the communication as depicted in the following figures.

```
>>> Creating the private chains
[root@bigdataman multichain-1.0.6]# multichain-util create chain1
MultiChain 1.0.6 Utilities (latest protocol 10011)
Blockchain parameter set was successfully generated.
```

```
You can edit it in /root/.multichain/chain1/params.dat before running multi-
chaind for the first time.
>>> Setting the Parameters of the Chain created
 [root@bigdataman multichain-1.0.6]# cat ~/.multichain/chain1/params.dat
# ==== MultiChain configuration file ====
# Created by multichain-util
# Protocol version: 10011
# The following parameters can only be edited if this file is a prototype of
another configuration file.
# Please run "multichain-util clone chain1 <new-network-name>" to generate new
network.
>>> Basic chain parameters
# Chain protocol: multichain (permissions, native assets) or bitcoin
chain-protocol = multichain
 # Chain description, embedded in genesis block coinbase, max 90 chars.
chain-description = MultiChain chain1
 # Root stream name, blank means no root stream.
root-stream-name = root
 # Allow anyone to publish in root stream
root-stream-open = true
 # Content of the 'testnet' field of API responses, for compatibility.
chain-is-testnet = false
# Target time between blocks (transaction confirmation delay), seconds. (2 -
86400)
target-block-time = 15
# Maximum block size in bytes. (1000 - 1000000000)
maximum-block-size = 8388608
>>> Global permissions
# Anyone can connect, i.e. a publicly readable blockchain.
anyone-can-connect = false
# Anyone can send, i.e. transaction signing not restricted by address.
anyone-can-send = false
# Anyone can receive, i.e. transaction outputs not restricted by address.
anyone-can-receive = false
# Anyone can receive empty output, i.e. without permission grants, asset
transfers and zero native currency.
anyone-can-receive-empty = true
 # Anyone can create new streams.
anyone-can-create = false
 # Anyone can issue new native assets.
anyone-can-issue = false
# Anyone can mine blocks (confirm transactions).
anyone-can-mine = false
 # Anyone can grant or revoke connect, send and receive permissions.
```

```
anyone-can-activate = false
# Anyone can grant or revoke all permissions.
anyone-can-admin = false
# Require special metadata output with cached scriptPubKey for input, to sup-
port advanced miner checks.
support-miner-precheck = true
 # Allow arbitrary (without clear destination) scripts.
allow-arbitrary-outputs = false
 # Allow pay-to-scripthash (P2SH) scripts, often used for multisig. Ignored if
allow-arbitrary-outputs=true.
 # Allow bare multisignature scripts, rarely used but still supported. Ignored
if allow-arbitrary-outputs=true.
allow-p2sh-outputs = true
allow-multisig-outputs = true
>>>Consensus requirements
# Length of initial setup phase in blocks, in which mining-diversity,
# admin-consensus-* and mining-requires-peers are not applied. (1 - 31536000)
# Miners must wait <mining-diversity>*<active miners> between blocks. (0 - 1)
setup-first-blocks = 60
mining-diversity = 0.3
 # <admin-consensus-upgrade>*<active admins> needed to upgrade the chain. (0
- 1)
admin-consensus-upgrade = 0.5
 # <admin-consensus-admin>*<active admins> needed to change admin perms. (0
- 1)
admin-consensus-admin = 0.5
# <admin-consensus-activate>*<active admins> to change activate perms. (0 - 1)
admin-consensus-activate = 0.5
 # <admin-consensus-mine>*<active admins> to change mining permissions. (0
- 1)
admin-consensus-mine = 0.5
 # <admin-consensus-create>*<active admins> to change create permissions. (0
- 1)
admin-consensus-create = 0.0
 # <admin-consensus-issue>*<active admins> to change issue permissions. (0
- 1)
admin-consensus-issue = 0.0
>>> Defaults for node runtime parameters
lock-admin-mine-rounds = 10
# Ignore forks that reverse changes in admin or mine permissions after this
many mining rounds have passed. Integer only. (0 - 10000)
mining-requires-peers = true
# Nodes only mine blocks if connected to other nodes (ignored if only one
permitted miner).
mine-empty-rounds = 10
```

```
# Mine this many rounds of empty blocks before pausing to wait for new trans-
actions. If negative, continue indefinitely (ignored if target-adjust-freq>0).
Non-integer allowed. (-1 - 1000)
mining-turnover = 0.5
# Prefer pure round robin between a subset of active miners to minimize forks
(0.0) or random equal participation for all permitted miners (1.0). (0 - 1)
# Native blockchain currency (likely not required)
initial-block-reward = 0
# Initial block mining reward in raw native currency units. (0 -
1000000000000000000)
first-block-reward = -1
# Different mining reward for first block only, ignored if negative. (-1 -
1000000000000000000)
reward-halving-interval = 52560000
 # Interval for halving of mining rewards, in blocks. (60 - 1000000000)
reward-spendable-delay = 1
 # Delay before mining reward can be spent, in blocks. (1 - 100000)
minimum-per-output = 0
 # Minimum native currency per output (anti-dust), in raw units.
 # If set to -1, this is calculated from minimum-relay-fee. (-1 - 1000000000)
maximum-per-output = 100000000000000
 # Maximum native currency per output, in raw units. (0 - 1000000000000000000)
minimum-relay-fee = 0
 # Minimum transaction fee, per 1000 bytes, in raw units of native currency.
(0 - 1000000000)
native-currency-multiple = 100000000    # Number of raw units of native cur-
rency per display unit. (0 - 1000000000)
>>>Advanced mining parameters
skip-pow-check = false
# Skip checking whether block hashes demonstrate proof of work.
pow-minimum-bits = 8
# Initial and minimum proof of work difficulty, in leading zero bits. (1 - 32)
target-adjust-freq = -1
# Interval between proof of work difficulty adjustments, in seconds, if nega-
tive - never adjusted. (-1 - 4294967295)
allow-min-difficulty-blocks = false
 # Allow lower difficulty blocks if none after 2*<target-block-time>.
>>>Standard transaction definitions
only-accept-std-txs = true
# Only accept and relay transactions which qualify as 'standard'.
max-std-tx-size = 4194304
# Maximum size of standard transactions, in bytes. (1024 - 100000000)
max-std-op-returns-count = 32
# Maximum number of OP_RETURN metadata outputs in standard transactions. (0 -
1024)
```

```
max-std-op-return-size = 2097152
# Maximum size of OP_RETURN metadata in standard transactions, in bytes. (0 -
67108864)
max-std-op-drops-count = 5
# Maximum number of OP_DROPs per output in standard transactions. (0 - 100)
max-std-element-size = 8192
 # Maximum size of data elements in standard transactions, in bytes. (128 -
32768)
>>> CHAIN UTIL PROPERTIES
default-network-port = 9571
 # Default TCP/IP port for peer-to-peer connection with other nodes.
default-rpc-port = 9570
 # Default TCP/IP port for incoming JSON-RPC API requests.
chain-name = chain1
 # Chain name, used as first argument for multichaind and multichain-cli.
protocol-version = 10011
 # Protocol version at the moment of blockchain genesis.
network-message-start = f7fbe4f0
# Magic value sent as the first 4 bytes of every peer-to-peer message.
address-pubkeyhash-version = 0042fb51
 # Version bytes used for pay-to-pubkeyhash addresses.
address-scripthash-version = 05b07713
# Version bytes used for pay-to-scripthash addresses.
private-key-version = 8077c3e7
  # Version bytes used for exporting private keys.
address-checksum-value = 7ef3706e
# Bytes used for XOR in address checksum calculation.
>>> Genesis Block Parameters
genesis-pubkey = [null]
# Genesis block coinbase output public key.
genesis-version = [null]
 # Genesis block version.
genesis-timestamp = [null]
  # Genesis block timestamp.
genesis-nbits = [null]
   # Genesis block difficulty (nBits).
genesis-nonce = [null]
   # Genesis block nonce.
genesis-pubkey-hash = [null]
  # Genesis block coinbase output public key hash.
genesis-hash = [null]
  # Genesis block hash.
chain-params-hash = [null]
   # Hash of blockchain parameters, to prevent accidental changes.
[root@bigdataman multichain-1.0.6]# multichaind chain1 -daemon
```

```
MultiChain 1.0.6 Daemon (latest protocol 10011)
Starting up node...
Looking for genesis block...
Genesis block found
Other nodes can connect to this node using:
multichaind chain1@192.168.1.52:9571
This host has multiple IP addresses, so from some networks:
multichaind chain1@192.168.122.1:9571
Listening for API requests on port 9570 (local only - see rpcallowip setting)
Node ready.
```

Figure 5. Block chain 3D view of IOT Transactions

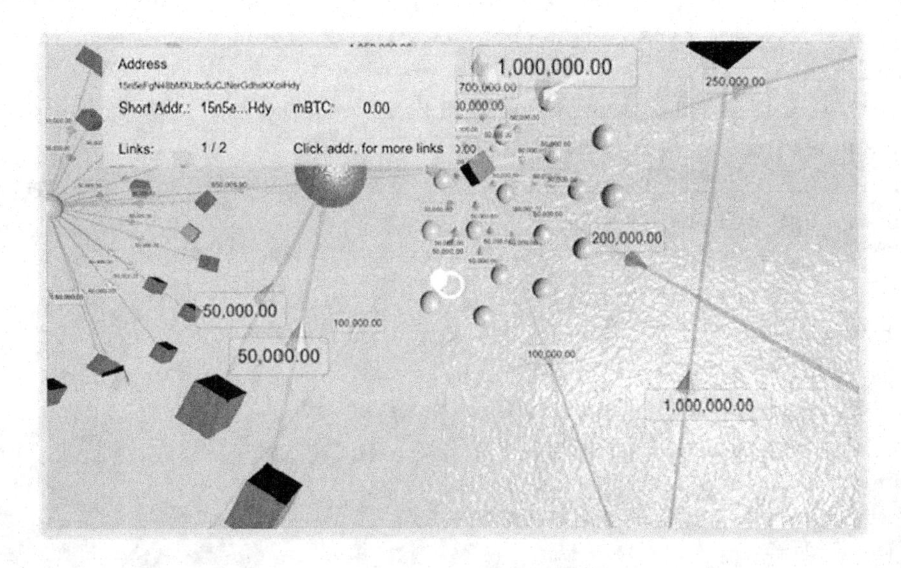

Figure 6. Block chain 3D view of IOT Transactions

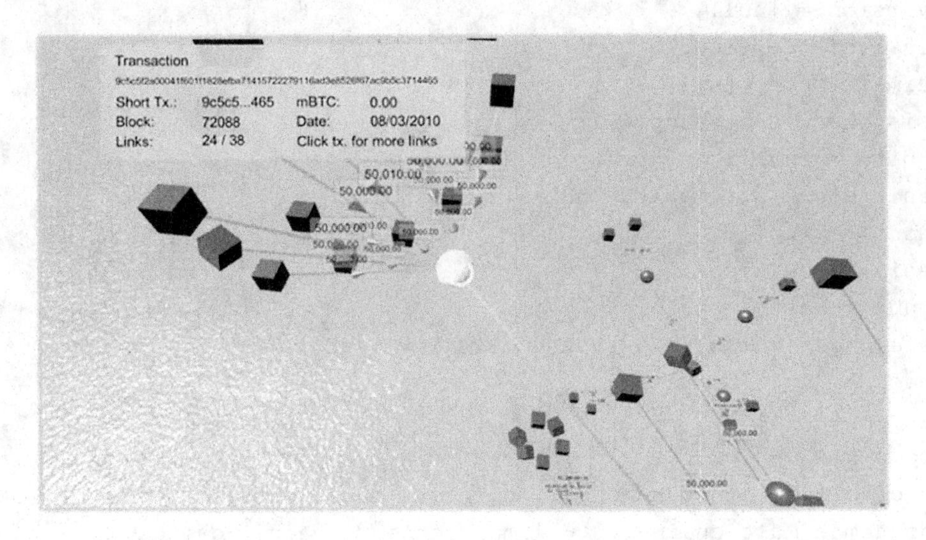

FUTURE WORK

The block chain IOT devices of different make and different protocol adherence need a clear data model like 'tangle' is to be designed. The framework of applications to package the device data across multiple network protocols to single node communication in the block chain. The devices should be digital ledger enabled. The validity of transactional data and the removal of garbage on transactional data on long block chain in nodes need matured mechanism. Our future proposed work will be on the functional flow framework of pluggable devices, defining nodes, and publishing nodes, creating or connecting to block chains, forming transactional pools and creating blocks with hashing, signature on the selected transactions. The statistics on transactional ledgers in block chain applications will give the precise details of the current state of the block chain. The visualization of the statistics in a comprehensive way could be the added advantage.

REFERENCES

Alavi, A. H., Jiao, P., Buttlar, W. G., & Lajnef, N. (2018). Internet of Things-enabled smart cities: State-of-the-art and future trends. *Measurement, 129*, 589–606. doi:10.1016/j.measurement.2018.07.067

Ammar, M., Russello, G., & Crispo, B. (2018). Internet of Things: A survey on the security of IoT frameworks. *Journal of Information Security and Applications, 38*, 8–27. doi:10.1016/j.jisa.2017.11.002

Atlam, H. F., Alenezi, A., Alassafi, M. O., & Wills, G. (2018). Blockchain with Internet of Things: Benefits, challenges, and future directions. *International Journal of Intelligent Systems and Applications, 10*(6), 40–48. doi:10.5815/ijisa.2018.06.05

Atzori, L., Iera, A., & Morabito, G. (2010). The internet of things: A survey. *Computer Networks, 54*(15), 2787–2805. doi:10.1016/j.comnet.2010.05.010

Bellavista, P., Cardone, G., Corradi, A., & Foschini, L. (2013). Convergence of MANET and WSN in IoT urban scenarios. *IEEE Sensors Journal, 13*(10), 3558–3567. doi:10.1109/JSEN.2013.2272099

Biswas, K., & Muthukkumarasamy, V. (2016). Securing smart cities using blockchain technology. In *High Performance Computing and Communications; IEEE 14th International Conference on Smart City; IEEE 2nd International Conference on Data Science and Systems (HPCC/SmartCity/DSS), 2016 IEEE 18th International Conference on* (pp. 1392-1393). IEEE. 10.1109/HPCC-SmartCity-DSS.2016.0198

Byun, J. H., Kim, S. Y., Sa, J. H., Shin, Y. T., Kim, S. P., & Kim, J. B. (2016). Smart city implementation models based on IoT(Internet of Things) technology. *Proceedings of Advanced Science and Technology Letters, 129*, 209–212. doi:10.14257/astl.2016.129.41

Dohler, M. (2011). Smart cities: An action plan. *Proc. Barcelona Smart Cities Congress.*

Dorri, A., Kanhere, S. S., & Jurdak, R. (2016). *Blockchain in internet of things: challenges and solutions.* arXiv preprint arXiv:1608.05187

Dorri, A., Kanhere, S. S., Jurdak, R., & Gauravaram, P. (2017). Blockchain for IoT security and privacy: The case study of a smart home. In *Pervasive Computing and Communications Workshops (PerCom Workshops), 2017 IEEE International Conference on*, (pp. 618-623). IEEE. 10.1109/PERCOMW.2017.7917634

Fernández-Caramés, T. M., & Fraga-Lamas, P. (2018). A Review on the Use of Blockchain for the Internet of Things. *IEEE Access: Practical Innovations, Open Solutions, 6*, 32979–33001. doi:10.1109/ACCESS.2018.2842685

Gauer, A., Scotney, B., Parr, G., & McClean, S. (2015). Smart city architecture and its applications based on IoT. *Procedia Computer Science, 52*, 1089–1094. doi:10.1016/j.procs.2015.05.122

Jaag, C., & Bach, C. (2017). Blockchain Technology and Cryptocurrencies: Opportunities for Postal Financial Services. In M. Crew, P. Parcu, & T. Brennan (Eds.), *The Changing Postal and Delivery Sector. Topics in Regulatory Economics and Policy*. Springer. doi:10.1007/978-3-319-46046-8_13

Ji, Z., & Ganchev, I. (2015). A Cloud-Based Car Parking Middleware for IoTBased Smart Cities; Design and Implementation. *Sensors (Basel), 14*, 22372–22393. doi:10.3390141222372 PMID:25429416

Mitton, Papavassiliou, Puliafito, & Trivedi. (2012). *Combining Cloud and sensors in a smart city environment*. Academic Press.

Mulligan & Olsson. (2013). Architectural implications of smart city business models: An evolutionary perspective. *IEEE Communications Magazine, 51*(6), 80-85.

Park, Y., & Rue, S. (2015). Analysis on Smart City service technology with IoT. *Technology Review, 13*(2), 31–37.

Rathore, M. M., Ahmad, A., Paul, A., & Rho, S. (2016). Urban planning and building smart cities based on the internet of things using big data analytics. *Computer Networks, 101*, 63–80. doi:10.1016/j.comnet.2015.12.023

Saravanan, K., Julie, E. G., & Robinson, Y. H. (2019). Smart Cities & IoT: Evolution of Applications, Architectures & Technologies, Present Scenarios & Future Dream. In *Internet of Things and Big Data Analytics for Smart Generation* (pp. 135–151). Cham: Springer. doi:10.1007/978-3-030-04203-5_7

Saravanan, K., & Srinivasan, P. (2018). Examining IoT's applications using cloud services. In *Examining cloud computing technologies through the Internet of Things* (pp. 147–163). IGI Global. doi:10.4018/978-1-5225-3445-7.ch008

Schaffers, H., Komninos, N., Pallot, M., Trousse, B., Nilsson, M., & Oliveira, A. (2011). Smart cities and the future internet: Towards cooperation frameworks for open innovation. In *The future internet assembly* (pp. 431–446). Berlin: Springer. doi:10.1007/978-3-642-20898-0_31

Talari, S., Shafie-khah, M., Siano, P., Loia, V., Tommasetti, A., & Catalão, J. P. (2017). A review of smart cities based on the internet of things concept. *Energies, 10*(4), 421. doi:10.3390/en10040421

Chapter 10
Security Issues on Internet of Things in Smart Cities

C. Thilagavathi
M. Kumarasamy College of Engineering, India

M. Rajeswari
Sahrdaya College of Engineering and Technology, India

Sheethal M. S.
Sahrdaya College of Engineering and Technology, India

Deepa Devassy
Sahrdaya College of Engineering and Technology, India

Priya K. V.
Sahrdaya College of Engineering and Technology, India

Divya R.
Sahrdaya College of Engineering and Technology, India

ABSTRACT

Many researchers are focusing on IoT in smart cities. It invites researchers to concentrate on simplifying engineering challenges. IoT includes recognition, locating, tracking, monitoring, and management of devices in a reliable manner. There are numerous security challenges that include network security, authentication, security-side-challenge attacks, security analytics, interface protection, delivery mechanism, and system development. IoT needs security analytics to overcome a number of problems in smart cities to prevent unauthorized access. One among the security analytics is streaming analytics, which include all the real-time data streams to detect emergency situations. Threat detection and behavioral monitoring will be done after analyzing the traffic data. The aim is to analyze and predict real-time streaming data to achieve security. Different analytical tools on security will be used to obtain the optimal result in smart cities. Traffic analysis, which is treated as a real-time stream, will be applied in street and traffic lights, transportation, and parking space occupancy and so on. Large volume of data that are received from different sensors and cameras will be given as the input in order to analyze traffic in a smart cities. Intelligent traffic congestion control system will be developed in order to analyze the heavy traffic on roadside. Security in IoT is proposed, which includes encrypting and decrypting the user request, which is further to be processed by the central processing hub, in order to prevent unauthorized access.

DOI: 10.4018/978-1-5225-9199-3.ch010

INTRODUCTION

In a day to day life, many researchers are focusing on IoT in smart cities. It invites researchers to concentrate on simplifying engineering challenges. IoT includes recognition, locating, tracking, monitoring and management of devices in a reliable manner. There are numerous security challenges which include network security, authentication, security-side-challenge attacks, security analytics, interface protection, delivery mechanism and system development and so on. The idea of the Internet of things was first proposed by Kevin Ashton, co-founder of Auto-ID centre at MIT in his presentation made to Procter & Gamble (P&G) in 1999. He discovered to link objects with an RFID tag. The concept of IoT evolved as the wireless Internet became widely developed, embedded sensors grew in worldliness. The Internet of Things (IoT) has the power to modify our world. The infrastructure of IoT encompasses interconnected objects, people, systems and information resources. These objects require intelligent services to allow them to process information of the physical and virtual world or hybrid of two and react accordingly. The intelligence of Things is necessary to pervade analytics into our systems (Wu, Chiang, Chang & Chang, 2017) and applications because collecting data alone is not enough.

IoT architecture can be considered as a global network infrastructure of a three-layer system consisting of IoT enabling technologies, IoT software and IoT applications and services. These layers are a collection of numerous active physical objects, actuators, sensors, specific IoT protocols, cloud services, communication layer, users, developers and enterprise layer. IoT has immense applications and consists of many technical and non-technical disciplines such as physical connections, manipulation of data, application interfaces, regulatory issues, and cybersecurity

Importance of IoT

Internet of things provides reliability, security and economic benefits to industries, smart home systems, traffic monitoring, waste management, healthcare systems etc.

Here are some examples of IoT on industries:

- Smart transport solutions speed up traffic flows; prioritize vehicle repair schedules and lower fuel consumption.
- Intelligent electric grids connect renewable resources efficiently which improves system reliability and charge the customers based on smaller usage increments.
- Machine monitoring sensors are used to diagnose maintenance issues, predict near-term stock-outs, and prioritize maintenance crew schedules for repair. IoT can be considered on a more personal level. Connected devices in IoT are making their way from business and industry to smart cities. Consider these possibilities:
- On the way to home from work, an alert message could be sent which is received from the refrigerator when the store is nearby to buy milk.
- Home security system, which enables to remotely control the locks and thermostats, can cool down the home and open the windows, based on the preferences.
- IoT-healthcare systems help doctors to monitor and provide the required services for patients remotely. Using smartphone applications patients get guidance to take decisions suggested by the doctor

IoT Design Challenges

The main aim for an IoT system (Zanella, Bui, Castellani, Vangelista &Zorzi, 2014) is to fulfil a reliable, secure and user-friendly interconnection of things or physical objects to finally provide a certain quality of service for a maximum number of users. Some of the key design challenges are followed: interoperability of heterogeneous systems, security and privacy issues,

- **Availability:** It is the capability of an IoT system to provide the demanding services for its customers anytime and anywhere. Availability applies to both IoT hardware and software.
- **Reliability:** An IoT system must be reliable enough for the successful delivery of IoT services at different circumstances. This feature is more critical in the field of emergency response applications where long delays and data loss can end in wrong decisions. Wrong decisions can lead to unpredictable scenarios such as confusion and irretrievable damages. For example, a failure in an IoT e-healthcare system may lead to patient death.
- **Interoperability:** It is a critical feature considered during design and build of IoT services to meet the user requirements. Heterogeneous smart devices with different platforms both at software and hardware should be integrated into an IoT system.
- **Scalability:** In an IoT, system scalability addresses the systems ability to be extended by new modules (including sensor nodes, devices, services and applications) for users without affecting the quality of available services. It must be able to handle future expansion when new technologies arrive, at both the hardware and software layers.
- **Performance Management:** Deploying billions of Internet-enabled devices causes many issues for service providers to manage the fault, performance and security of the devices.
- **Security and privacy:** For IoT system, it is difficult to guarantee a high level of privacy and security for users due to its heterogeneous networks. One of the reasons for this issue is the lack of a common standard for IoT security.
- **Big Data Analytics:** Big data analysis is one of the most challenging research fields in IoT systems. A huge amount of data needs to be processed real-time which requires fast and power efficient techniques.
- **Smart Device Design:** In the coming future, there will be an increase in the number of smart devices. Every person will carry several smart devices that require power sources to achieve sensing, computation and communication tasks.

RELATED WORK

Smart HealthCare

The main objective of Smart City is to improve the livelihood of people. Smart Health Care is an inevitable part of Smart City. Compared to the traditional healthcare system, smart health care has a variety of advantages, because all the health care devices are connected through IoT. It is easy to grab the medical information from the IoT connected medical and Healthcare devices (Yeh, 2016). The different smart devices which are coming in this area are smart pulse oximeter—to monitor pulse rate, respiration rate, amount of oxygen and haemoglobin n the blood, smart watches-to track various fitness activities, smart contact lenses-developed by Google to help diabetic patients. To ensure better health care solutions, medical practitioners, health care professionals, drug research companies, pharmaceutical companies, and diagnostic centres has to adopt the smart healthcare system with proper implementations. The use of information and Communication technology (ICT) will help to implement this system easily. This smart healthcare system helps healthy and unhealthy people to come up with healthier conditions.

The initial step against our traditional Healthcare system was Electronic Healthcare Records (EHRs). This made our traditional healthcare system to maintain the data in a centralized server. Also, it helped to get the patient history from anywhere, even though it could not be a final solution to our health-related issues. Now machine learning and Artificial Intelligence plays a major role in the analysis of healthcare data (Islam, Kwak, Kabir, Hossain & Kwak, 2015). Even though Smart Health care has its own advantages, it also has to face different challenges. In smart healthcare, data will be coming from heterogeneous sources as medical devices are connected. From the data which are coming from these smart devices can be used to help the people who have bad health habits to know about their health conditions. As per our traditional healthcare system, people are not aware of their own physical conditions even they have serious diseases. But with smart health care system, it is possible to get up to date information about each citizen. So there should be a proper mechanism to handle the heterogeneous data coming from different devices. To infer the information from these data, there is a need for advanced machine learning techniques like supervised and unsupervised learning methods. The size of these data is also large, proper storage mechanism also should be considered. This heterogeneous data should be processed and integrated. Using Machine learning algorithms, it is possible to predict the future health conditions of every citizen.

Another important challenging area is the security of personal and clinical information of patients. No one will be interested to disclose their personal or clinical information (Al-Hamadi & Ray Chen, 2017). It is necessary to focus on the privacy of patient's data for which an appropriate security mechanism has to be implemented. Apart from all these, development of smart hospital system and implementation of advanced medical services and medical devices are mandatory. Smart homes and smart hospitals need to be interconnected. In order to provide proper treatments, there should be sufficient medical resources like doctors, lab facilities etc. So there should be appropriate medical resources. With this smart healthcare, doctors can also investigate the patients and the effects of their treatments.

Security in Transportation of Smart Cities

(Raya & Hubaux, 2007) focused on the security of the network which provides a detailed threat analysis and devises appropriate security architecture. The set of security protocols has been evolved in order to protect privacy and to analyze their robustness and efficiency. Authors illustrated a good example for authentication mechanisms, where digital signatures viewed to be the most applicable technique in spite of their high overhead. This work mainly focuses on Communication aspects of transportation. The main advantage of this work lies in identifying the most relevant communication aspects; the major threats have been discovered. Implementing the security architecture along with the related protocols, it shows to what extent it protects privacy. Finally evaluated and analyzed the robustness of messages.

(Malandrino & Chiasserini, 2013) explained communication enabled vehicles which were interested in extracting information from web servers. Message dissemination in VANET was envisioned to enable

- News reporting
- Navigation maps
- Software updating
- Multimedia files downloading.

For efficient transmission of messages, the downloading process is considered to be an optimization problem.Thus, the authors concentrated on optimizing the content downloading speed.

(Fernando & Rafael, 2012) demonstrated Event Driven Architecture which was created for complex events that reflect certain activities of traffic congestion. Traffic congestion, or the so-called traffic jam, usually starts in a certain lane of the motorway because of several reasons, such as a crash between two vehicles, an obstacle on the road, or a bottleneck in the road infrastructure. Then, it started growing along the lane, and it could also spread to other lanes in the same direction of the motorway. After a period of time, the congestion in one of the lanes begins to dissolve, and finally, the traffic congestion in all the lanes disappears. Nevertheless, some environmental factors can cause a traffic jam appearance. For example, under certain weather conditions, such as heavy rain or fog, a slow flow of vehicles should not be considered traffic congestion, because all the vehicles have slowed down due to security constraints. Furthermore, this paper assumes that beacon messages are quickly spread along the motorway. Several works have focused on that issue.

(Llorca & Sotelo, 2010) proposed that the Vehicle-to-Infrastructure (V2I) communication system which was based on the geographic coverage area. GPRS (General Packet Radio Service) and Universal Mobile Telecommunications System (UMTS) were used to connect each vehicle with the central control unit. Authors illustrated a complete vision-based vehicle detection system for Floating Car Data (FCD) enhancement in the context of Vehicular Ad hoc NETworks (VANETs). Thus, a more representative local description of the traffic conditions (extended FCD) would be modelled. Specifically, global positioning and absolute velocities were used to detect the load and their velocities. These absolute velocities and global positioning were defined after combining the outputs which were provided by the vision modules with the data supplied by the CAN Bus and the GPS sensor. GPRS/ UMTS was used to transmit the information which merges the extended FCD in order to maintain an updated map of the traffic conditions.

(Zhao & Cao, 2008) have proposed VADD: Vehicle-Assisted Data Delivery in Vehicular Ad Hoc Networks in order to concentrate on the complication of delay tolerant applications in the sparse network. Here, carry and forward method was used with the predicted mobility pattern for which they projected some VADD (Vehicle-Assisted Data Delivery) protocols: Location-based L-VADD (Location First Probe), D-VADD (Direction First Probe), Multipath Direction MD-VADD (Multipath Direction First Probe) and H-VADD (Hybrid Probe).

(Frank, Giordano & Gerla, 2010) demonstrated TrafRoute: A different routing approach in vehicular networks. They used the TrafRoute routing protocol that involves self-selection and this self-selection which was based on position, knowledge of the road topology and node density. (Shah & Venkatesan, 2018) discussed the method on authentication of IoT devices and Server for which authors used secure valets. Typical IoT system architecture was illustrated by adding 3 major Components which includes an IoT device, and IoT SERVER AND user interface. First component

IoT devices have been made to take the responsibility of data collection to which various sensors were used for different purposes. After data collection got over, collected data need to be uploaded in the server which includes processing of data to make it usable with respect to different applications. Web and/or a mobile interface have been used for the user to interact with the server and IoT devise. Certain initial assumptions were defined by the authors such as initial valets need to have n keys of nm bits each based on the security requirements and memory constraints. Threat Model has also been defined by the authors to tackle various side channel attacks which require set of security keys rather using single key.

Authentication to IoT devices and IoT server has been done by using 3-way mutual authentication such as sending request, responding to IoT server is challenge to and from IoT device and server. Authors described an authentication mechanism which includes defining of secure valets consists of n keys with m bits, challenge-response mechanism and changing the secure valet. Security of the defined protocol has also been elaborated by the authors that include man in the middle attack, next password prediction, side channel attack, and DoS attack. Performance analysis has been done with ECC-Elliptic Curve Cryptography and simple single rotating for energy consumption. Energy consumption of different protocols has been tabulated with respect to average time to execute and energy consumed. Data memory and password prediction complexity has been summarized for various m and n values respectively.

(Dorothy, Ramesh Kumar & Sharmila, 2016) demonstrated IoT based home security through Digital Image Processing algorithms which gives an outline for automatic systems to handle the home securely. DIP was the key idea behind this automation process for which mechanism of digital image processing was included that consists of capturing the input image and producing the output image. DIP mechanism includes information mining technique which takes captured image as the input and produces the output for next section named as image segmentation. Enriched image has been given as the input for next step such as annotation of image which applies feature extraction and gives the required output. Authors used Template- based method and FFT Fast Fourier Transform with Twiddle factors method to retrieve and compare the image. Gray scale based matching and Edge-based matching were used for template –based technique and the spatial and time complexity were applied for comparing the techniques. Authors depicted that sum of absolute differences (SAD) measures would be adapted to handle the translation problem on image. Cooley-Turkey algorithm which is known to be a divide and conquer algorithm is used to divide entire DET into smaller DFT's to perform the computation recursively to speed up the computation process. Framework for home security is interpreted which includes usage of sensors, capturing image by using remote camera, storing the images in a database and connecting to smart phone. Authors expounded that linear filtering and Fourier transoms were the most fundamental techniques in digital image processing.

IoT in Smart City

(Zanella, Bui, Castellani, Vangelista & Zorzi, 2014) described Smart City systems which provided powerful, intelligent and flexible support for people living in urban areas. This goal can be achieved by integrating wireless sensor networks and available wireless communication services. The following research aims are targeted

1. Real-time high-level context-aware customized services;
2. Better living environments;
3. Improved utilization of the available resources.

The main elements of the Smart City architecture are to be smart health, smart environment, smart energy, smart security, smart office and residential buildings, smart administration, smart transport and smart industries. Heterogeneous information can be generated from the sensor nodes deployed in each Smart City domain. This provides the primary data source for heterogeneous information generation. These information's are collected using the existing communication services. For example, the use of satellite network for GPS devices, cellular services such as GSM/3G/4G for smartphones and the use of the internet for PCs and other navigation devices for raw data collection. The collected data are then processed and analyzed using various web technologies.

Architecture design of Smart City helps in assisting people in an intelligent manner. Some examples are discussed below

1. In case of road congestion, guiding a driver to take optimal route.
2. Alert can be generated to heart patients in situations where their heart rate is exceeding a threshold limit while performing an activity,
3. Sending alerts and warnings to people to buy their household items if it is in shortage such as buying food items via a Smart Fridge.

Implementation of the smart city architecture will follow the steps below.

At first the raw data are collected and processed to make them web consumable. Once the data are converted into a common format they are then semantically enriched with OWL ontologies based on the knowledge of domain experts. At the same level, the collected data are processed using the Dempster-Shafer rules to deal with the uncertainty aspects of the semantic model. The main idea is to identify activity and learn new rules that are governing an activity. The new rules learned at this level will be used in defining the knowledge of the semantic model as well as customized services. This will provide feedback to the end users (citizens) in the form of alerts and warnings.

Multi-Level Smart City Architecture

Sensors form the primary source of information generation. The raw data sensed by the sensor node are transferred to Level 1 of the Smart City architecture using communication services to perform further information processing. A detailed description of each Level is explained as follows:

Level 1: Concentrates on Data collection. Raw information generated from sensors are collected for further processing. Different formats in which heterogeneous data are collected are csv, tweets, database schemas and text messages. The collected formats are converted into common format using processing done by semantic web technologies. The next level describes the steps used in the conversion of data into a common format.

Level 2: Data processing Information gathered is summarized using semantic web technologies prior to transmission, analysis and fusion in the further levels. The main aim of this level is to convert the collected heterogeneous information into a common format, e.g. Resource Description Framework (RDF).

Level 3: Domain-specific data can be exploited based on the concepts and relationships between them. They are obtained by various data integration and reasoning Semantic web technologies. The techniques used in this level are:

Web ontology language (OWL)12 is an RDF graph used for the classification of the individual/ concepts based on the classes. It provides two different types of properties, namely the Data property and Object property, which can be used to form relationships between different classes. Once data classification is over, knowledge can be further improved with domain experts and uncertain reasoning. It is used for publishing ontologies for which Dempster-Shafer will be used here for activity recognition and learning new rules in a particular domain of course.

Communication Services

The communication medium plays a vital role in achieving the Smart City concept. Some of the communication servers used are:

3G (3rd generation), LTE (Longterm evolution), Wi-Fi (Wireless fidelity), WiMAX (Worldwide interoperability for microwave access), ZigBee, CATV (cable television) and satellite communication.

The main aim is to connect all sorts of things (sensors and IoT's) that can help in making the lives of citizens more comfortable and safer. For example, the Government sector uses cloud and communication services to obtain a better governance system. For the health sector, communication technologies can be used to connect health statistics, medication and location of the patient from a remote location thus helps to achieve a Smart Health system. Hence, with Smart City and communication technologies w can provide a more secure and convenient infrastructure for better living.

Customized Services

In the case of the vehicle and health domains,the impact of driver health parameters can be measured on driving conditions by combining sensor data. Combined health parameters like blood pressure and heart rate with vehicle status can help the driver to measure their real-time health condition, which can help in creating a safer environment for drivers. Similarly better monitoring of vehicle status can be achieved by using vehicle location, speed and volume of traffic approaching a junction.

In the case of the health care domain, information are gathered through wireless sensor networks about patient health and activity can assist the disabled person. The effect of temperature on home activities like eating, bathing, sleeping and cooking can be learned by combining the home and environment domain data. This can help in recognizing the correct activity status, which in turn can be a useful care tool for the elderly and people suffering from dementia.

In the case of the environment and administration domains, high level customized information can be derived from low-level information collected from the environment domain such as temperature and water level .When High-level customized information such as flood, earthquake, forest fire, landslide and other natural calamities is combined with city administration services, it could help in saving many lives. For industrial sector, a safe working environment for industry workers can be created by context-aware services obtained through heterogeneous data fusion. Furthermore a better, more productive and safer environment can be created for workers by continuous monitoring, recording and exploiting the surrounding sensor information from different domains such as harmful gas detection, workers' health and machine conditions in an industrial environment.

SECURITY IN IoT: PROPOSED SYSTEM

In IoT, everything is connected through the internet so security is a major concern. People are placing a huge amount of personal data in public as well as private networks. At times, it is very difficult to apply security mechanisms in IoT due to limited battery capacity and computing power. Security can be provided in IoT architecture at different levels 1) Security to software 2) Security to hardware and 3) Security to the network.

Classification of IoT Attacks

There are mainly four types of cyber-attacks in IoT. They are physical cyber-attacks, network attacks, software attacks and encryption attacks.

- Physical Cyberattacks

In this attack, the attacker is trying to steal the information from the sensors attached to the devices. The hacker can use this data to get into the system and can make any changes in the existing network. one can able to drain the battery charge of any device, and can make the node as an even dead node.

- Network Attacks

The attacker can access the entire network and get the data which is helpful for the attacker to stop many services which are currently running.

- Software Attacks

Anyone is able to make changes in the behaviour of the system by inserting malicious code which can propagate through the network causing DoS (Denial of Service) attack.

- Encryption Attacks

This kind of attacks can able to determine the type of encryption algorithms being used in the system and the keys used for encrypting the data. An attacker can make changes into keys so that the algorithm will be working differently for different data.

Security Threats in Different Layers

Let's now discuss the layered architecture of IoT and various security issues associated with it which is shown in figure 1.

Figure 1. Layered Architecture of IoT

Application Layer

Transport Layer

Network Layer

Perception layer

- Application Layer

Application layer deals with various user applications that are running in our system. This can be any smart applications for home, health care cities where the IoT is deployed. Almost all the nodes in smart cities are equipped with configurable embedded computers in it. Once a hacker gets into this system, they can steal the relevant data regarding the transportation and health care etc.

- Transport Layer

Internet protocols such as TLS or DTLS can be adapted to provide security in the transportation layer.

- Network Layer

Network layer deals with addressing and routing of packets between the smart devices in the network. The key establishment between the devices in the network should be more secure, otherwise, an intruder can easily get into our system for accessing the network.

- Perception Layer

This layer is the physical layer, where it senses and gathers information from the environment and recognizes the smart objects which are present in the network. There are many security issues associated with the perception layer. First one is the wireless signal which carries the information between the nodes present in the network. When someone knows about the range of frequencies used for transmitting the data, it may lead to physical attacks. So it is very important to ensure the confidentiality of the perception layer.

Mechanisms on Authentication and Access Control

Each object in the smart cities needs to be identified using some authentication mechanisms and allow only those devices to share the data across the network (Bohli, Kurpatov& Schmidt, 2015).

- Authentication

Password mechanism is the most popular and simple authentication mechanism, which we can see in almost all network authentications. But it has its own merits and demerits. Anyone can steal the user credentials and can log in into the system. To overcome the drawbacks of a password mechanism, now a day biometric recognition is being used everywhere. Authentication involves two properties. One is source authentication and another one is data authentication. Source authentication authenticates both the sender and the receiver whereas data authentication ensures the integrity of data. In cryptography; authentication can be achieved by symmetric encryption or asymmetric encryption. Also, the message authentication code (MAC) and the digital signature also use more precisely for authentication.

- Access Control

Access control allows only authorized user to access various resources such as sensors, file, devices or URL etc. The operating system limits access to these resources based on the user and application. Ensuring access control to the information which is shared in the system is very much important in IoT environments. Traditional access control models are inconvenient for IoT. For example, Mandatory access control, Discretionary access control, and Role-based access control are user-centric. It will not consider resource information, the relationship between the user and the resource provider and dynamic information. In large-scale IoT networks, it is not easy to check who gives access to which device.

Attribute-based access control replaces the discretionary permissions with policies based on attributes. It provides access to users based on attributes like resource characteristics and contextual information. In this model, possession of attribute can be easily altered without modifying the underlying structure.

Security Challenges in IoT

Security and privacy are the top concern for IoT developers. The following section describes the major challenges in the IoT network.

1. Secure constrained devices

The devices connected to the network should be properly segregated from the attack using firewalls.

2. Authorize and authenticate devices

Device authorization is used to determine which type of service and applications each device has the access in our network.

3. Manage device updates

Applying updates in the software running in the IoT devices is also a major challenge.

4. Secure communication

After ensuring security to the devices, the message sends through the network is also secured using some encryption algorithm.

Figure 2. Proposed system: Security in IoT of Smart city

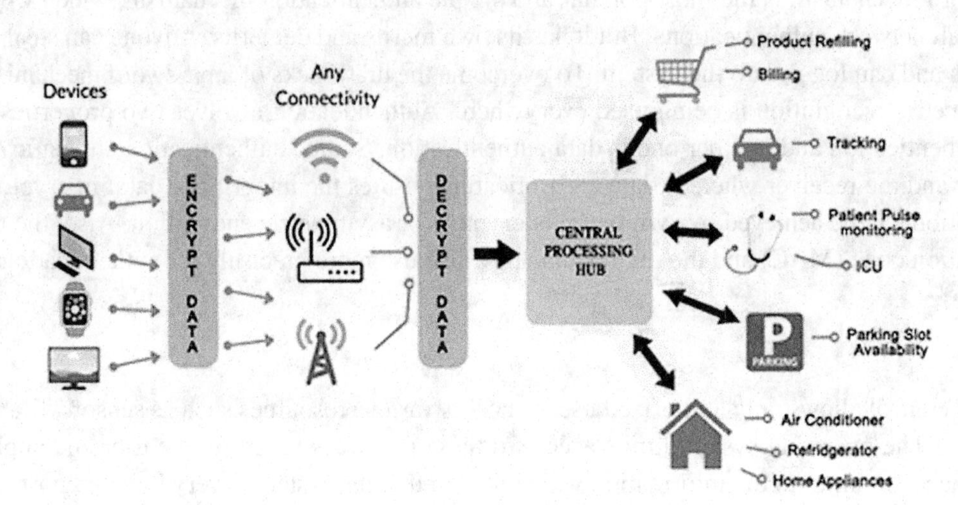

5. Ensure data privacy and integrity

Data should be transmitted, stored and processed securely in the network itself. Also, ensure the data integrity, i.e, the data has not been modified during transmission. The blockchain is the best example which offers a scalable and resilient approach for ensuring data integrity.

6. Secure web, mobile, and cloud applications

Web, mobile and cloud applications are also used to store various kinds of data in IoT. So it is necessary to provide security to these applications also.

7. Ensure high availability

IoT infrastructure provides services such as traffic control and healthcare in smart cities. To ensure high availability, these devices must be protected against cyber-attacks as well as physical tampering. IoT system should avoid a single point of failure and it is designed to be fault tolerant so that when a problem arises they can easily adapt and recover soon.

8. Detect vulnerabilities and incidents

Different strategies such as monitoring network communications and activity logs for anomalies also detect vulnerabilities and breaches in the network.

9. Manage vulnerabilities

The major challenges include which device was affected, what data or services were compromised and which users were impacted and taking actions to resolve such situations.

10. Predict and preempt security issues

Figure 3. Double Encryption using dedicated hardware device

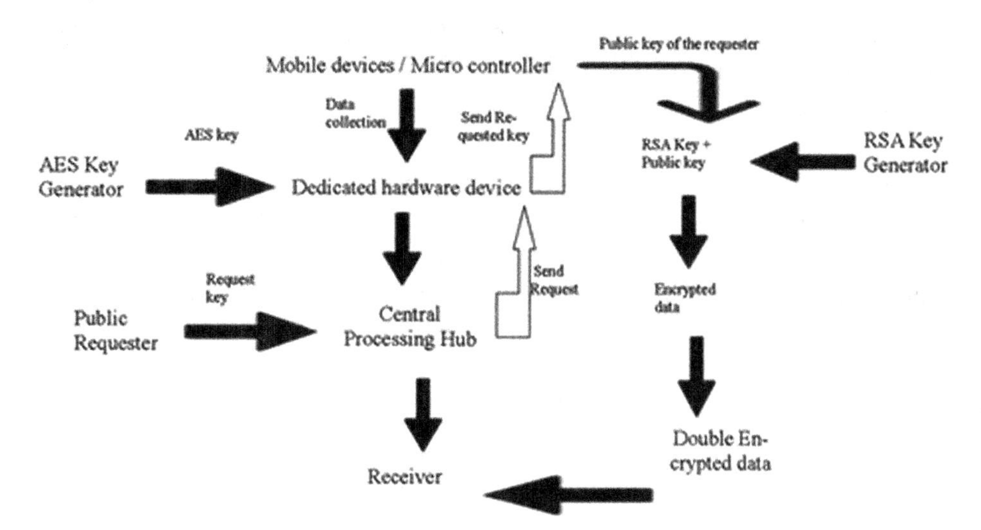

To predict the security threats, threat modelling approach can be used. There are other approaches like analytics tool to correlate events as well as applying Artificial Intelligence to adaptively adjust the security strategies applied based on the previous actions.

Smart city is highly complex network of numerous independent nodes. Nodes can be any of the peripheral devices or services that may include smart homes, smart parking, healthcare services, and smart trolleys and so on. Security in such systems is one of the major design issues associated with it, i.e., the data that needs to be transmitted between the sender and receiver nodes need to be highly secured. Hence, it is necessary to encrypt the data from the sender nodes before transmitting it via a network. Any type of transmission network may be utilised such as LAN, WLAN etc. Then, the data received from the network is to be decrypted before presenting it to the processing unit. The central processing hub, in turn, performs necessary computations and sends it to the intended receiver as already stated. If a user wants to switch on any smart home appliances like fan, AC, light then he/she needs to perform an encryption from any of the devices as shown in figure 2. Encrypted message will be transmitted to the network to apply decryption which will then be sent to the central Processing Unit so that it can be forwarded to corresponding smart home appliance and perform the action accordingly. Similarly, if it is required to monitor the heartbeat rate, blood pressure of a disabled person through our smart device, it is desired to send an encrypted message to the network which will be decrypted at the network. Then the central processing Hub will perform the necessary task. The main aim of this work is to provide the encryption and decryption in order to send the authenticated request with proper confidentiality and security. This method will not permit other unauthorized persons to make use of smart device to access smart city applications.

Fig. 3 denotes the detailed encryption and decryption operation to avoid certain attacks such as spoofing and sniffing while transmitting the data. Especially in financial and healthcare applications, sensor nodes are generating huge volume of data that needs to be secured. If encryption and decryption is done in the end devices as stated in the previous paragraph will lead to delay in processing huge volume of data which additionally consumes more power and time. It is noted that there should be a special device such as dedicated hardware device to connect to end devices to perform encryption and decryption.

Double encryption (Hussain & Abdullah, 2018; Chandu, Kumar, Prabhukhanolkar, Anish & Rawal, 2017) would be done to tackle various attacks from the secured data being accessed by unauthorized persons.

Double Encryption technique:

First level encryption

- Data collection from mobile devices / micro controller
- Dedicated hardware device perform encryption using AES for which AES key generator is used to generate AES key.
- First level encrypted data is sent to Central Processing hub to perform required operation
- If anyone requests to access central processing hub other than the authorized person then the request is sent from requester.
- Central processing hub forwards the key request to dedicated hardware device for double encryption.

Second level encryption

- Dedicated hardware device processes the request and send to mobile devices for double encryption.
- Micro controller / device will send the public key and encrypted data to RSA key generator
- RSA key generator generates RSA key which is used to perform double encryption
- First level encrypted data is to be double encrypted by using RSA which is sent to the corresponding receiver to perform decryption operation.

CONCLUSION

We are heading to a new era of ubiquity where the things can be connected at any time, anywhere and anything. One such technology is IoT which is a collection of interconnected devices that can share the information among them. These devices can be represented as nodes in a network. Nodes can be humans, vehicles or home appliances. Each node is equipped with a sensor, processor and connecting hardware. IoT is going to expand its network and its services in a more significant way in which the things connected to the internet will project to cross 20 billion in the near future. This expansion is done in two ways. First one is to expand the scope of current internet and the second one is to build a separate network from the scratch. Both techniques have its own challenges and merits.

One of the major concerns in both the techniques is that IoT cannot be considered as a single technology. It includes cloud computing, big data Analytics, networking, data science, computer vision, embedded technologies, machine learning etc. There are certain enabling technologies for IoT. It consists of RFID, nanotechnology, sensors and smart networks in which sensors and actuators are considered to be the most important enabling technology.

It is considered as a basic building block for developing smart cities or smart home. Nowadays, all the cities have become smart cities for which information technology is made as the principal infrastructure and the basis for providing essential services to residents. Services include adequate water supply, assured electric supply, automatic switching of electronic devices, monitoring heartbeat, temperature, blood pressure of a patient without the assistance of anyone else in the house.

Certain characteristics of IoT includes efficient, scalable and associated architecture, unambiguous naming and addressing, an abundance of sleeping nodes, mobile and non-IP devices, intermittent connectivity, the intelligent spark that make the product smart and smart interaction and great sensing of the physical world.

Some of the applications of IoT include street and traffic lights, transportation, parking space occupancy, infrastructure maintenance, waste management, air quality monitoring, crime detection, the architecture of the city, energy usage and distribution, traffic flow and pedestrian and bicycle needs.

IoT needs security analytics to overcome a number of problems in smart cities to prevent unauthorized access. One among the security analytics is streaming analytics which includes all the real-time data streams to detect emergency situations. Threat detection and behavioural monitoring will be done after analysing the traffic data. The main aim is to analyse and predict real-time streaming data to achieve security. Different analytical tools on security will be used to obtain the optimal result in smart cities.

Traffic analysis which is treated as a real-time stream will be applied in street and traffic lights, transportation and parking space occupancy and soon. A large volume of data that are received from different sensors and cameras will be given as the input in order to analyse traffic in smart cities. Intelligent traffic congestion control system will be developed in order to analyse the heavy traffic on the roadside. Traffic density can be calculated from real-time video streams captured by smart cameras. Captured images are pre-processed for removing noise, contrast enhancement, de-blurring to get clearer images. Then we apply various image processing techniques like region-based approaches to find the boundary of vehicles in the congested area. If the Density calculated exceeds the defined threshold value, it is noticed that the congestion is more. Further predictions on traffic can be done with various machine learning algorithms. In addition to this, QOS and security rules will be applied to overcome the security challenges when the traffic message is disseminated to the traffic control room.

Intelligent Traffic control systems will be developed by incorporating Android technology with IoT. Android apps are developed for ambulance drivers to reach their destination in time without any delay. Conventional traffic monitoring system will be replaced by using NodeMCU with advanced sensors, wi-fi module and buzzer inbuilt with traffic lights. When the ambulance is approaching the signal area, alert information will be sent to the traffic controller to clear the way to save a human life.

Security mechanisms will be applied in all the levels of the network by using techniques such as code guardian which prevents the network from security vulnerabilities, code exploits, malware and all sorts of software attacks. Authentication should be provided both at the user level and device level to provide appropriate access rights for the users and devices. Unauthorized users are restricted by limiting bandwidth and blocking techniques at the application level. Switches and access points should be equipped with smart analytical capabilities that also provide deep packet inspection capability.

In conclusion, even though IoT becomes the major booming technology, various security breaches are found due to legal, ethical, security and social concerns. In this aspect, it is necessary to know the characteristics, major concerns, security challenges, enabling technologies and analytical tools to build strong IoT models in smart cities.

Security challenges with respect to transportation, waste management, air quality monitoring, energy conservation and health care management system are going to be addressed in this chapter. The main aim is to focus all the security issues in smart cities. It becomes a major research concern in the current era.

REFERENCES

Al-Hamadi & Chen. (2017). *Trust-Based Decision Making for Health IoT Systems. IEEE Internet of Things Journal, 4(5).*

Bohli, Kurpatov, & Schmidt. (2015). *Selective Decryption of Outsourced IoT Data*. IEEE.

Chandu, Kumar, Prabhukhanolkar, Anish, & Rawal. (2017). *Design and Implementation of Hybrid Encryption for Security of IOT Data*. IEEE.

Dorothy, Kumar, & Sharmila. (2016). *IoT based Home Security through Digital Image Processing Algorithms*. IEEE. DOI doi:10.1109/WCCCT.2016.15

Fernando, T. S., & Rafael, T. M. (2012). A cooperative approach to traffic congestion detection with complex event processing and Vanet. *IEEE Transactions on Intelligent Transportation Systems, 13*(2).

Frank, R., Giordano, E., & Gerla, M. (2010). TrafRoute: A different approach routing in vehicular networks. *Proc. VECON*. 10.1109/WIMOB.2010.5645018

Hussain & Abdullah. (2018). *Review of Different Encryption and Decryption Techniques Used for Security and Privacy of IoT in Different Applications*. IEEE.

Llorca, D. F., & Sotelo, M. A. (2010). *Traffic Data Collection for Floating Car Data Enhancement in V2I Networks. EURASIP Journal on Advances in Signal Processing*. doi:10.1155/2010/719294

Malandrino, F., & Chiasserini, C. F. (2013). Optimal Content Downloading In Vehicular Networks. *IEEE Transactions on Mobile Computing, 12*(7), 1377–1391. doi:10.1109/TMC.2012.115

Raya, M., & Hubaux, J. P. (2007). Securing vehicular ad hoc networks. *Journal of Computer Security, 15*(1), 39–68. doi:10.3233/JCS-2007-15103

Riazul Islam, S. M., & Daehan Kwak, Md. (2015). *The Internet of Things for Health Care: A Comprehensive Survey. IEEE Access, 3*. doi:10.1109/ACCESS.2015.2437951

Shah & Venkatesan. (n.d.). Authentication of IoT Device and IoT Server Using Vaults. In *Computing And Communications 12th IEEE International Conference On Big Data Science And Engineering*. IEEE. DOI 10.1109/TrustCom/BigDataSE.2018.00117

Wu, Chiang, Chang, & Chang. (2017). An Interactive Telecare System Enhanced with IoT Technology. In *Pervasive Computing*. IEEE.

Yeh. (2016). A Secure IoT-Based Healthcare System With Body Sensor Networks. *IEEE Access, 4*. . doi:10.1109/ACCESS.2016.2638038

Zanella, A. (2014). *Internet of Things for Smart Cities. IEEE Internet of Things Journal, 1(1).*

Zhao, J., & Cao, G. (2008). VADD: Vehicle-assisted data delivery in vehicular adhoc networks. *IEEE Transactions on Vehicular Technology, 57*(3), 1910–1922. doi:10.1109/TVT.2007.901869

Chapter 11
Smart Parking in Smart Cities Using Secure IoT

G. Indra Navaroj
Jayaraj Annapackiam CSI College of Engineering, India

E.Golden Julie
Anna University Chennai – Regional Office Tirunelveli, India

ABSTRACT

The city is transforming into the smart city using information and communication technology (ICT), and the major role in economic development is building an infrastructure to enable greater connectivity between citizen service, energy, economics, and government. A smart city can monitor the real-world scenario in real time and support the intelligent services to both locals and travelers. Due to urbanization, people move from village to city. Increase the population in city also causes an increase in vehicles. Here, parking the vehicle securely is a challenging problem. In a smart parking system, all the devices are connected to the internet. Hackers and third parties easily access the user data or sensitive data. Smart parking system application controls the traffic, air pollution, and city functions making it easy to park the vehicle and reduce accidents. Many of the problems arise in the security and privacy of the sensitive data. In this chapter, the authors discuss security and privacy issues in smart parking systems using IoT.

INTRODUCTION

Internet of Things (IoT) devices connected to the internet. In IoT technology all the physical devices is connected to the internet. They can share the information from one to other. And also remotely monitor and control the devices, animals and Human being also monitor. This method is used to capture the theft, criminal from anywhere and monitor their activities. IoT technology is very helpful in Government and Police station also. Due to digital country city is converted into smart city. Smart city means with rising people, economy and infrastructure people moving from village to city, this is create a urbanization (Zhang et al, 2017). One important survey said in 2030 city population reach in 5 billion. So the Government have the responsibility to give the quality life, economy, living quality, smart parking. In 2020 80% of IoT devices is connected to the internet and share the private information through the network,

DOI: 10.4018/978-1-5225-9199-3.ch011

this information stored in the cloud storage. It may lead to create the problem for the citizen's. It may provide the security and privacy for the citizen's private information. This is very challenging problem. One important application of Smart city using IoT is Smart Parking System. Finding the parking spot is very challenging problem in city. Many authors proposed different parking management system and give the solution for the problem. In smart parking system is fastest growing field of IoT, and is also provide intelligent, innovative and interactive service to the human (user) using different methods and operations. All the devices are connected to the internet, heterogeneous nature, and dynamic. It may lead to or create security issues and challenges. In this chapter we investigate security issues of smart parking system using several scenarios. We investigate security vulnerability threats, Risk assessment, classified these vulnerability threats based on security objective and identified, evaluated their impact on the overall system. When city become digital, all devices connected to the internet and cloud server, people affected from serious of security and privacy attack due to the vulnerability of smart city applications (Sicari et al 2015). Day by today our cities is coved with mass number of vehicle on the road. Due to this increasing factor traffic and parking is a major issue. In smart transport and parking slot reduce the traffic using automated system. Parking slots are accurately predicted the available free slots are intimated to the driver through speaker. IoT sensors are used to reduce the traffic based on arrival and departure of vehicles. Vehicular adhoc networks (VANET) provide collusion free traffic in the busy road and also it reduces accidents.Challenges in smart Transport and Parking: In smart vehicle traffic and parking system contain mobility. Due to dynamic nature system should guide the traffic and taking final decision about parking slot is very time effective.

Internet of Things (IoT)

Internet of Things contain things(device) that have unique identities and are connected to the internet, many existing device, such as network computer or 4G enabled model phone already some form unique identities and are also connected to the internet. In year 2020 there will be a total of 50 billion devices/ things connected to the internet.

The IoTs allow people and things to be connected anytime, anyplace, with anything and anyone, ideally using any path/network and any service. They are "Material objects connected to material objects in the Internet" (Rodrigo et al, 2017). IoT is not limited to just connecting things(Device) to communicate and sharing data while executing meaningful applications towards a common user or machine goal. Evaluated from multiple technologies are wireless sensor, embedded system, machine learning, instrument control data analytics and automation. 'Things' refer daily objects or devices that communicated with other devices through internet to monitor and control the objects. Smart IoT belong to Information and Communication Technologies (ICT) applications access through IoT devices. Smart cities are developed under prevention of incidents rather than avoidances of occurrences of after the incidents. E.g., fire detection, crime prevention, floods and climate prediction.Smart city is one of the applications of IoT. It contains three parts data generation, data management, and application handling (Ankitha and Balajee, 2016).

IoT Technologies

IoT technology include RFID is known as Radio Frequency Identification, Wireless Sensor Network that is called as WSN and NFC which is called as Near Field Communication (Borgohain et al,2015).

Radio Frequency Identification

The major Technology used in IOT is Radio Frequency Identification. RFID is a Radio frequency wave, it is used for identify the object through wireless. RFID tag is easily track object from anywhere. It is support the remote access. In the frequency range car we install the RFID tag, it mainly used for finding theft car. Finding the theft car is very challenging problem for police. In RFID different range of frequency is configure. If the range of frequency is high the data transfer also high. The range is low; the data transfer from remote area is low. The RFID is contains two important components. One is RFID tag, second component is RFID transceiver.

Wireless Sensor Networks

Need for remote sensing information, gathering information and remotely control the activity we use the technology is Wireless Sensor Network (WSN).Wireless Sensor Network has cost effective, low power consumption, good in efficiency. It contain important component Sensor, Actuator, Storage. Nowadays Multimedia Sensor Network have high qualified camera so capture the image, video and audio. This is used to control and monitor the environment. It is used to collect the sensitive or private information. In later we discuss how to prevent the collected information like sensitive data.

IR Sensor Node

In Smart Parking System, sensor node place in the center of parking lot. It is sense the surrounding car parking lot information. That means empty and availability of car parking. The information capture through IR sensor, then send to the server. The server processes the information, also maintain the database, store all information then upload into webpage. The user can check the free slot information and pricing details of the particular slot, it is done by IOT technology. Without human interaction the process will be done.

Near Field Communication

This is a another core technology is used to manage the distance between few centimeters with low power and data requirements (Borgohain et al, 2015).These IoT technology is used based upon the application, and their features like range, data rate, efficiency, power consumption, energy, security.

Smart City

A smart city is an urban place that uses information and communication Technology (ICT) and major role in economic development, building an infrastructure to enable greater connectivity between citizen service, Energy, Economic and Government. A smart city can monitor the real world scenario in real time and support the intelligent services to both local surrounding and travellers like transportation, healthcare, environment, energy, home appliances.

Smart Cities Applications

In smart cities have development application enhanced in different area like Smart parking, Smart Lighting, Structural health monitoring, Surveillance, Emergency response etc (Saravanan et al., 2019). Here, we Each Smart city application is detailed.

Smart Parking

Nowadays increase the population and also increase the vehicles .So in future parking the vehicles is very challenging problem. In smart parking system give the information to the driver have the empty parking slot information. So the driver easily parking the car and also avoid the traffic and congestion. In smart parking sensor are equipped with the system and monitor and collect the information about the empty slot then give the information about the slot details to the car driver.

Smart Lighting

Smart light equipped with sensors can communicate with people and also other lights and exchange information on the sensed. Smart lighting allow lighting to be controlled dynamically, Smart light connected to the internet can be control remotely to configure lighting schedule and lighting intensity.

Smart Road

Smart road equipped with sensor can provide information on driving conditions, Travel time estimate and alert increase in poor driving condition, traffic congestion and accident. Such information can help in making the road safer and help in reducing traffic jams information sensed from the road.

Structural Health Monitoring

Structure Health Monitoring uses a network of sensor to monitor the vibration in the structures such as building and bridge. The information captured from the sensors and is analyzed to assess the health of the structure. Analyzing the data it is possible to detect cracks and leakage and any mechanical breakdown .Since Structure Health Monitoring system use large number of wireless sensor node which are powered by traditional batteries. We proposed energy such as mechanical vibrations, sunlight, and wind.

SECURITY VULNERABILITY AND RISK ASSESSMENT IN SMART PARKING SYSTEM

In this chapter we describe the security vulnerability in smart parking system. It ensures the security and privacy for an IoT system.

System Security

Information system security has three elements.

Logical Security

Logical security based on communication threats. It is used to protect the communication between two parties and two persons.

Physical Security

Physical security is also called infrastructure security. It is used to protect the information based on home appliance like physical devices. And also protect the data that means home needed information is protected from threats.

Premises Threats

Premises threats is also known as corporate and facility security. It is used to protect the people and their property within an entire data. Premises security is an access control and fire detection, environment protection etc.

Some of threats categorized in physical security threats. It give the information to the organization administrator, have the responsibility to measure threats.

Environment Threats

Natural disasters are the sources of environment threats. It is possible to assess the risk of various type of natural disaster and take suitable precaution. IoT is based on environment condition like temperature, weather that can interrupt or damage the service of information system. It may cause severe damage to the public infrastructure, and in the case of severe damage, it may take days, week, even years to recover from the severe damage.

1. In appropriate temperature and humidity.
2. Fire and smoke
3. Water Damage
4. Chemical, Radiological and Biological Hazards
5. Dust
6. Infestation

Risk Assessment

All the risk are identified first and accessed, risk assessment is used to block the threats. Documentation should have the information about the server, client, firewall, and any other device connected to network through wire and wireless device connected to the networks, once threat are identified.

Ignore the Risk

Ignore the risk means reject the risk. Found the risk is not solved by human. This type of risk is rejecting.

Accept the Risk

If the risk is identified the risk is accepted by the organizations.

Transfer the Risk

This type of risk is transfer from on organization to other organization. Otherwise move from one person to other.

ARCHITECTURE OF SMART PARKING SYSTEM

A car owner expects smart parking system as intelligent and support privacy preserved, secure. Smart parking system performs continuous parking service. Yan et al (2011) explained the Architecture of smart parking system. This Architecture contains two parts, namely Hardware Architecture and software architecture. In this section we discuss about Hardware Architecture and software architecture.

Software Architecture

In Software Architecture, Yan et al (2011) proposed four modules. That is explain below

1. **Driver Module:** Driver module is used to communicate with hardware device. It contain the following components communication driver, sensor driver for belts, communication driver for vehicle, and IFD driver is used for vehicle detection.
2. **Communication Module:** Communication module is act as a interface between sender and receiver. It performs communication operation between sender and receiver. It is used for transmit and receive between sender and receiver. Some time it support fast transmissions, error control, simplify the communication process.

Figure 1. Smart Parking Software Architecture
Gongjun yan. At el, 2016

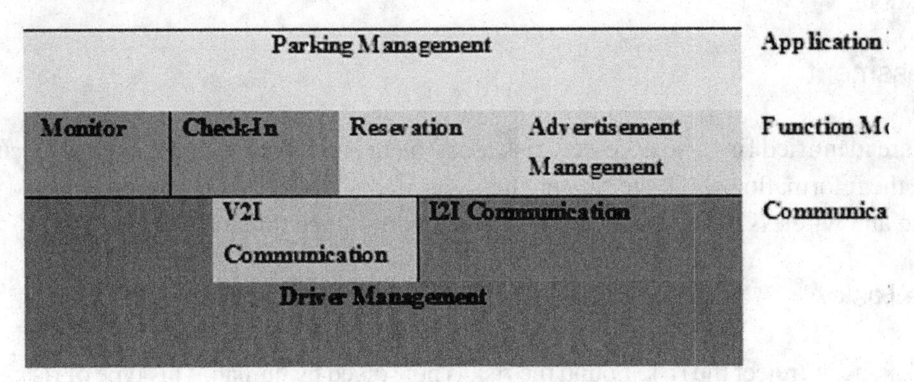

3. **Function Module:** Function Module directly communicates with the hardware device. In parking system it is a core module. It gives the thanks to the Communication module and Driver module.
4. **Application Module:** Application module is an important module in this architecture. It manages all operation of the parking system. It supports account management, maintenance management, and operation management.

Hardware Architecture

In hardware Architecture contains wireless transceiver and a simple processor. Wireless transceiver and simple processor are embedded in to EDR. Zig bee devices, Bluetooth devices, and infrared devices used for security. Important component of hardware architecture are wireless transceiver, parking belts, Infrared Device (IFD) and Control Computer (Yan et al, 2011).

1. **Wireless Transceiver** is used to transmit and receiving parking information. It is work in Wi-Fi wireless network and Wireless Lan (Wlan) network.
2. **Infrared Device (IFD)** is used for finding the parking spot is free or occupied. The function is based on the lighting scheme have blue and red light. A blue light identified the parking spot is full. A red light is identified the parking spot is available.
3. **Control Computer** is control all the operation and activity.

Components of Smart Parking System

The IoT is support for enabling object and places in physical world are converted in to the digital world. In digital world it represent the software all of the work done by automatically. Nowadays people would like to uses in smart parking, which is used to remotely monitor and interact with the physical world using the software. In smart parking system have three kind of Devices.

Sensor

Sensor is a simple or complex device, which is used to capture or sense the data from the environment or surrounding. It is used to convert the physical properties into electrical signal. Sensor devices include necessary conversion process that is convert temperature in to electrical signal.

Actuator

Actuator is a simple or complex Device that is used to convert the electric signal into physical property.

Tag

Tag as a Device is a Radio frequency identification tag. It is used for general identity. The reader device operating on a tag is typically a sensor and an actuator combined in case of writable RFID tags.

SMART PARKING WITH VARIOUS COMBINED TECHNOLOGY

An IoT Based Secure E-Parking System

In urban area population is increased so vehicle is also increase. Here traffic increase in urban area. So we need to developed automatic parking system for reduce traffic, time, fuel, air pollution .Author clearly concentration to find the improper parking vehicle and collection of parking fees. IoT based E-parking system have a combination of integrated component called parking meter (PM). It identified some issues

1. Real time detection of improper parking.
2. Estimation of each vehicle's duration of parking lot usage.
3. Automatic collection of parking charge

This system provides parking management solution and parking lot reservation system and named as parking meter (PM) based E-parking (PM-EP).

Architecture

The E-Parking system architecture consist of the important components namely are Parking meter, A WLAN or WI-FI integrated workstation, local parking management server, WI-FI Access point (Sadhukhan, 2017).

Figure 2. Networkarchitecture of E-Parking system
Pampa Sadhukhan et al., 2017

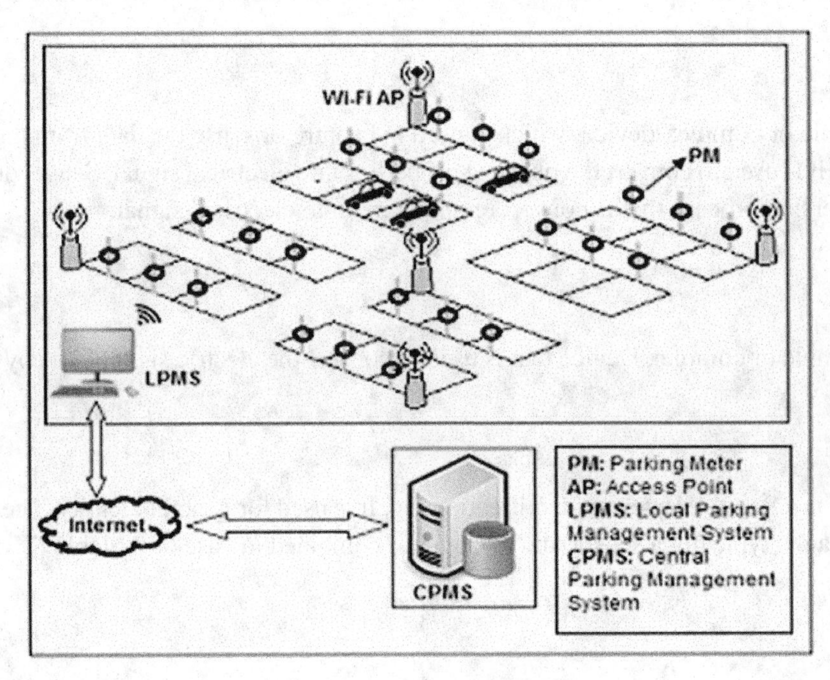

Network Architecture of E-Parking System

Each parking lot contain parking meter, it is placed in the middle of the parking lot. Parking Meter contains an ultrasonic sensor device used to find and monitoring the parking vehicle and identify the status of parking lot. The vehicle driver improper parking the vehicle in the parking lot the alarm IC module give a warning sound to the admin for alert the improper parking vehicle. Camera used for capture the image of vehicle license plate and wireless is used to communicate with parking management server. Parking Meter also contains smaller solar cells to charge the batteries. Parking Management server contain GSM it is used to send SMS to the citizens and the site officer.

Software Architecture

Software Architecture contain four module

1. Parking Lot monitoring system
2. Local Parking Management System (LPMS)
3. Central Parking Management System
4. Parking Availability Information and reservation

SECURE SMART PARKING SYSTEM USING IoT ELLIPTIC CURVE

Chatzigiannakis et al. (2016) proposed IoT elliptic curve for secure parking vehicle. The driver or user share the private information major issues and challenge arise due to user private information For security and privacy reasons the user not share their personal information . If the user shares her private information like location, car number through internet it create major issues. So the author avoided user personal information. The user can not to share their private information.

Figure 3. Block Diagram of Parking Meter (PM)
Pampa Sahukhan et al., 2017

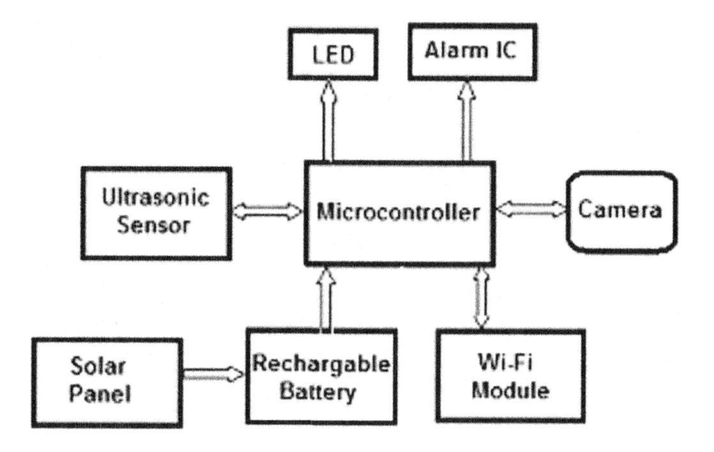

Approach

Existing system use the written code in C++. It is platform dependent. The user can change the platform also rewrite the code. But the proposed approach implemented security scheme and the cryptographic protocol can be compiled in IoT system without change the code. The proposed system is addressed the following issues.

- **Platform Independent** code can be compiled and run on a different platforms, our code is platform dependent configuration.
- **OS Independent** code can be compiled and run on different OS. The system based on C, C++, nesc, Android and ios, Linux based system.
- **Exchangeability** without affect the existing code, the components are exchange general components and platform dependent and independent components are simultaneously used.
- **Cross-Layer Algorithm** used the existing algorithm for complex one. The algorithm can be implemented out of other algorithm user can use the existing algorithm functionality.

Scalability and Efficiency

Proposed system runs heterogeneous hardware platform and different OS. Smart parking system increase many issues based on the privacy of users sensitive data and confidentiality of the user or private data. The user give the location information, that location information is communicate to the other and it is

Figure 4. Architecture of Smart Parking system
Ioannis Chatzigiannakis et al., 2016

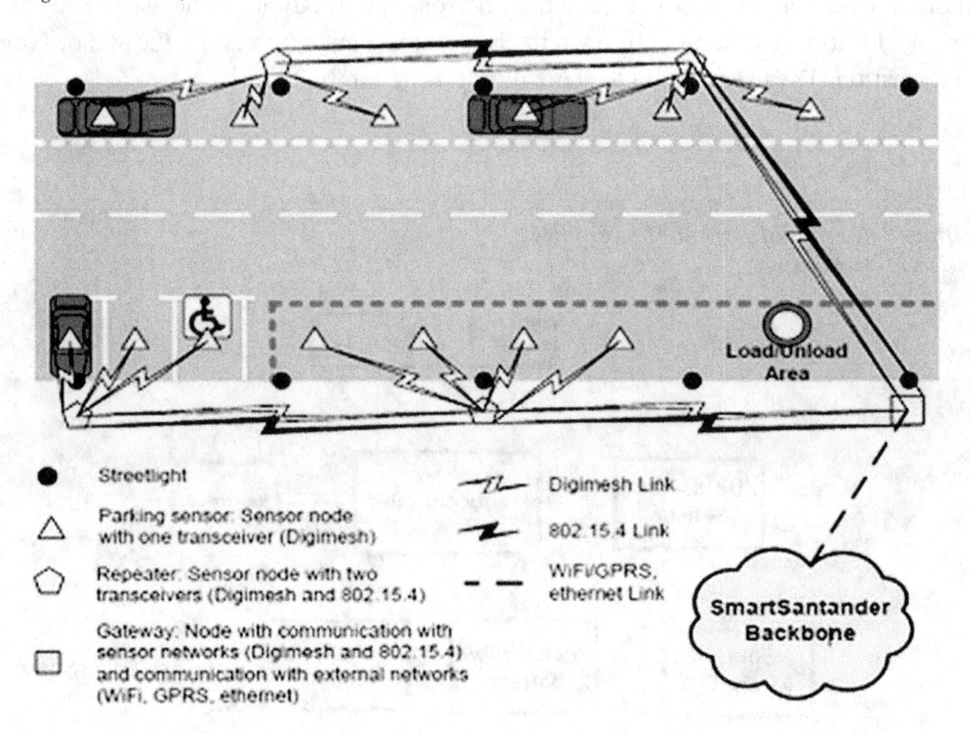

stored in the central data base. We clearly know that such private information needed to be secured. SPS is monitoring the status of the parking spaces. Many IoT devices is connected the 802.15.4 for communication purpose. Others use crowd sourcing technique use the citizen's smart phone. Thus connecting device establishing and security mechanism is based on cryptography concepts and challenges is implementing by cryptographic mechanism for IoT device privacy and security (Pyrgelis, 2016).

So the author introduced the new approaches that avoid the need of private information. Main goal is i) it protected the citizen personal information and also there is no need for storing confidentiality information i.e., Citizen's personal information ii) it keeps code maintenance cost. The elliptic curve version of Diffie Hellman problem is applied on both cloud server system embedded system. Author proposed Zero Knowledge Proof algorith322-7OWQ m worked the concept on the Elliptic Curve Cryptography algorithm to protect the citizen's privacy.

1. Smart parking services
2. Privacy preserving outdoor parking management
3. Zero knowledge protocols for smart city IoT infrastructure

Smart Parking Services

In this system offers easy to parking the car and improve the quality, convenience and choice. Smart parking services provide optimize parking space usage, improve the efficiency of parking operation and help to reduce the traffic. Smart parking services is provided all the facility like parking information, parking lot locations and risk the location information leak to the cloud server. It supports only the local interaction for the proper operations; totally avoid the confidentiality location information. Our proposed system is completely avoids the private or sensitive information. It may lead to many problem like theft and misuse the vehicle and cybercrime. Thus extract such information potential success of failure and

Figure 5. Physical Network layout of Smart Parking
(Ioannis Chatzigiannakis et al., 2016

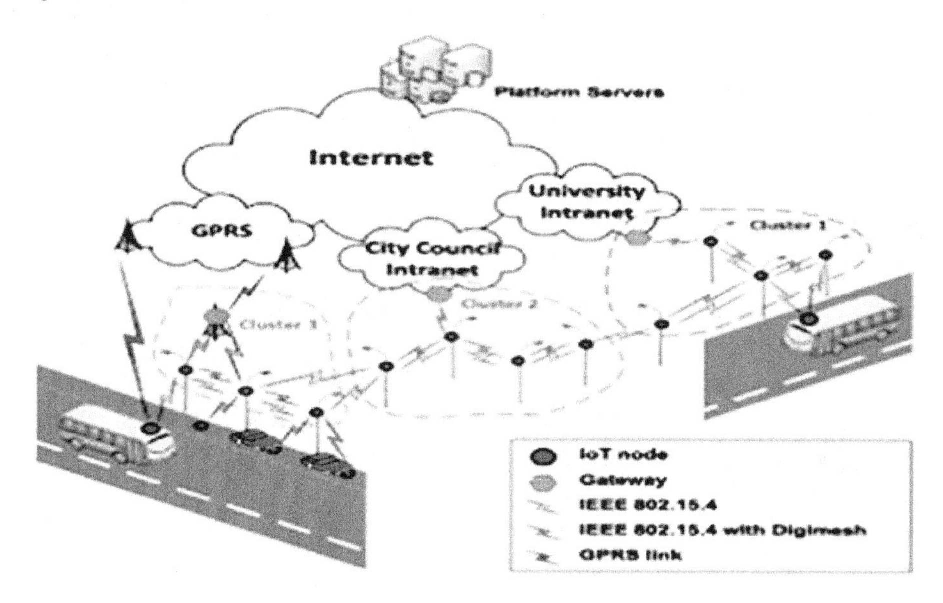

increase the cost. It give the solution for privacy preserving in the sense i)sensor embedded in the parking places ii) mobile application iii) city authorities interact.

Privacy Outdoor Parking Management

Smart parking system is allowing the citizens to park the vehicle without the need of sensitive private information (address of the user). It avoids power consuming encryption technique to protect the data. It avoids transmitting over wireless or untrusted network service information. We also completely avoid strongly sensitive private information. An SPS allow different security crypto-mechanism be interchanged and federated. This application uses a zero-knowledge protocol to improve the possession of an information. No sensitive information is transferred in the network and guaranteed the confidentiality of the private information. In this application secure database is maintain by city authority .The database contain the citizen private information. The DB is not connected to the Smart Parking System and internet clod server. Issuers have the authority to operate and maintain the secure database. A citizen is believed the issuer to protect the confidentiality of their data.

1. Citizen enrolled in the system
2. Parking in a restricted access zone
3. Integrated services

Zero Knowledge Protocols for Smart City IoT Infrastructure

Zero knowledge protocols algorithm allows the user cannot give any details of personal information as a proof. Zero Knowledge Protocol contains two important components one is a prover and another one is a verifier. Prover is to maintain the knowledge of a user's secret, while collecting no detain of customer's information, verifier conveying this knowledge to others. Zero knowledge protocol is suitable of smart parking system. Based on this works the system classified as interactive proof system and non-interactive proof system. Interactive proof system prover and verifier exchange multiple message, which keep the secret verifier is accept or reject the proof. A ZKP obey the two properties one is complete and sound. A proof is complete means the protocol succeeds, there are two types of elements an honest prover and an honest verifier. Sound if the dishonest prover have the detail proof of customer information but is negligible. Proof of knowledge algorithm has the zero detail information property of the particular customer.

SECURE SMART PARKING SYSTEM USING EMBEDDED

The author proposed secure parking the vehicle safety in a paid parking system. Finding parking space creates a challenge problem. User loss the time and fuel, create traffic for finding parking spot. The authors give the solution for parking the vehicle in a secure manner. The proposed architecture contain the components are Microcontroller, RFID Reader, RFID tag, and IR sensor. This paper explain how securely parking vehicle using gate model.

1. **Atmel Microcontroller** is support 8-bit microcontroller with 8Kbyte of in system programmable flash memory, large performance, and very low power. Microcontroller is a flash type. It is used to

Figure 6.

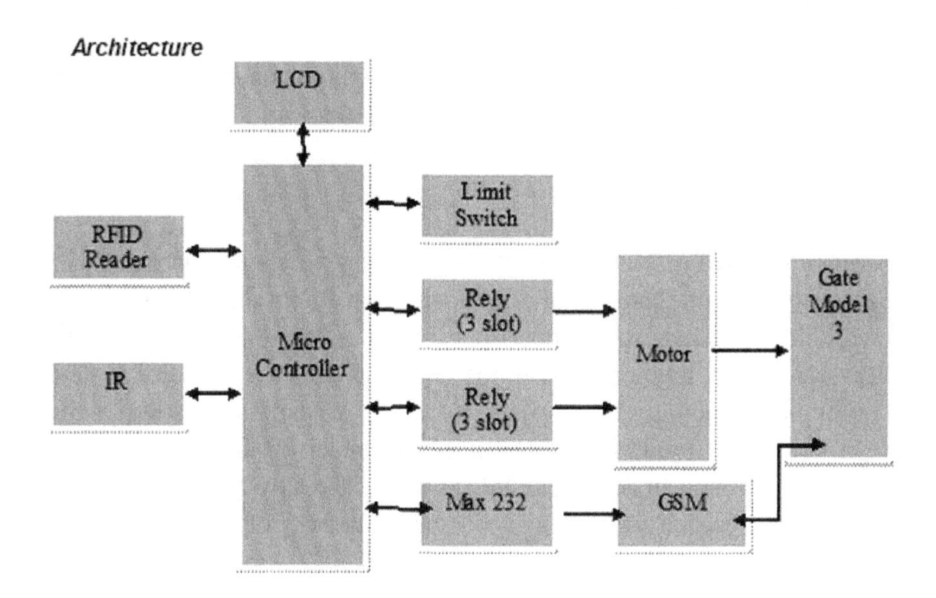

interface with RFID Reader. It is used to scan the RFID tag to scan the amount. Atmel feature are 8K bytes of flash, 256 bytes of RAM, 32 I/O lines, Three 16 bit timer/counter, two data pointer, watch dog time.

2. **RFID Reader is** an important IOT technology. It is used to identify the object automatically. RFID contain the tag, transponder, it is attached to the object.
3. **IR Transmitter and Receiver** is a type of LED device is used to emit infrared ray. The transmitted signal is passed to the IR Transmitter.
4. **Dc Motor** is an AC synchronous electric motor
5. **LCD Display** is a character based LCD. LCD interface with various micro controllers, programming, various interface.
6. **Relay Driver** is a switch worked with electromagnetic. It is used to control power supply with one circuit to another. It controls the device like lamb or electric motor.

RFID tag is used to find the amount. IR sensor is used to search the empty space in parking site. In this embedded system used to secure the every parking slot using gate system. A LED indicates the available parking site and capacity of parking area. RFID is used to indent the driver details and amount is known through the scan the user ID. Microcontroller is known the correct amount is paid then only it allow the driver to park the vehicle. The car driver registers the parking spot through the SMS. The SMS receive and transmitted through GSM module (Bharathidasan, 2017). If the parking spot is available the car driver is to allow for register. The driver pay the amount for particular parking spot, gate is open for the driver. The driver park the vehicle by the use of microcontroller. Gate is opened and closed operation is performed by the relay. Gate system is used to perform the following operations the users not register, parking lot is full, user have low amount in the account. When the driver leave the parking lot RFID tag is used to find the IN time and OUT time difference, then total parking time is calculated the remaining amount is debit from the vehicle driver account.

A SECURE AND INTELLIGENT PARKING SYSTEM

Intelligent parking system, park the vehicle as a security and intelligently. Yan et al, (2011) proposed reaction scheme for secure parking vehicle. This is categorizes as

1. Advertisement Publishing

Parking spot information is advertised to the public through the wireless network. This information is encrypted through encryption mechanism for communication. Information contains the parking spot information, the capacity of parking spot, empty spot information. The important component of parking spot is wireless transaction, topology. In parking area transceiver is called base station. That base station is used for communication. It acts as a data centre. It is transmit the parking information to the vehicle. The information is encrypted using private key; at the receiver side the information is decrypted using the public key. So no one can access the information. The advertisement information contains the location of parking spot, capacity of parking spot, and availability of parking spot (Yan et al, 2011).

2. Reserve Parking Lot

The drivers have the smart phone; can receive the information from the base station. The information is XML format. The driver orders the parking spot. The message is encrypted using private key. The order consist of the following information vehicle.;l.-0..;'s electronic license plate, reserved parking spot number, a timestamp, transaction expiry time.

3. Cancel Transaction

The driver want to cancels the parking spot, it is available in the intelligent parking system. This process is closely related to the reservation process. Driver sends the customer information to the base station. Base station receives the information. The system collects the penalty amount from customer for some period of time for the reserve the spot.

4. Payment System

Payment system is used for collect the parking fees and improper parking penalty fine from the driver. Parking fees based on the entry and exit parking time.

INTELLIGENT PARKING MANAGEMENT

Smart parking system is an intelligent parking system. The model of the parking management is birth and death stochastic process. This process is used to predict the revenue .Birth process is entering the vehicle in to the parking spot. The birth time is noted by the parking management. Death process means the vehicle is leave the parking spot, the time is noted by parking management. It is easy to find how much time the vehicle occupies the parking spot. Traffic detector is used to find the birth time and death time.

The author proposed two parking class model, which is based on the services. The namely are Economy Class Model and the Business Class Model. Economy Class Model (ECM) is low cost and low quality service, the size is larger. Business Class Model (BCM) is high quality size, cost is expensive and the size is low compare to ECM

Find the status of parking slots, use four case Parking management is based on the parking lot information, electronics, empty and occupied slot and clearance of improper parking vehicle. If the parking management work is done very well it avid the malfunctions. In smart parking system author used secure wireless network and sensor communication. Parking system as intelligent system, it supports the novel security and privacy infrastructure. In this system provide the privacy of the driver and security of the private information by using encryption and decryption approach.

A PROTOTYPE FOR IOT BASED PARKING SYSTEM

An automatic number plate recognition algorithm can be identified car number plate and we can manage the parking and fee collection for parking vehicle. But this algorithm is not suitable for our country car number plate that means our country number plate is different from foreign country number plate. Our author Gandhi and Rao (2016) introduced RFID technology for replaced number plate recognition technique. RFID is easy to track the object from remote area like car, people etc. RFID technology and wireless sensor network is apply in parking management system, it detect nearest parking site. NFC

Figure 7. Prototype of IoT based car parking management
Baratam, M. Kumar Gandhi et al., 2016

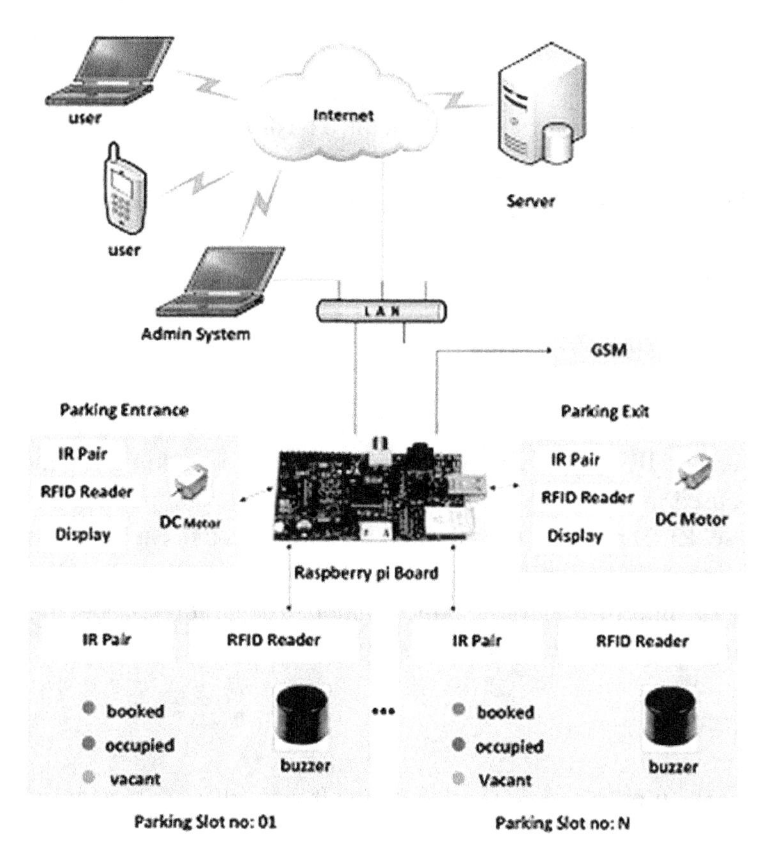

technology is support for developed the secure parking management system. Author also developed the online application (mobile application) to park the vehicle securely. Vehicle size is also varying from one car to other car. So AMR sensor is used to detect the vehicle size for securely park the car without create any damage for nearby vehicle (Gandhi and Rao, 2016).

Methodology

A prototype is used for parking the vehicle in better, flexible and secure way. The proposed infrastructure having a Raspberry ii board, it is work like a computer but the size is small. It act as a server. It is very cheap easily purchase in market. It is replace the central processing and do all the work in this simple debit card. In this system access control and communicate the things remotely.

The architecture contain four components

1. Online booking
2. Parking entrance system
3. Parking exit system
4. Parking management

The major problem of vehicle car driver is finding the parking spot in the parking area. So this system used to give the solution for the user problem user download the smart application, then parking the vehicle in secure way.

1. Online Booking

In city due to rush hour finding parking area is difficult. The driver is book the parking slot in advance using a prototype for IoT based parking system. User having mobile application and internet connect easy to find the correct parking slot. Database gives available parking slot information. User satisfied the parking are, parking slot then booking the allotted slot or available slot from the parking lot and complete the payment in advance.

2. Parking Entrance System

Parking entrance system worked with the following components such as IR sensor, DC Motor, LCD Display and a RFID reader. IR sensor is used to identify the car for parking, DC motor is used to open the gate, and LCD is used to display the status of parking vehicle and parking slot allotment. This is updated to the database. RFID reader is used to find the car details like car number plate. Number plate is detected by the RFID reader, and then compare to the theft list is given by police station. If find the car as a theft car give the information to the police station. The car driver booking the slot the gate is open then the car went into the correct parking slot.

3. Parking Management

This system is support the user to park the car correctly. If wrongly park the car we collect the penalty from the user. So each slot has one IR pair, one RFID reader, a buzzer, three LEDs. The vehicles enter into the particular parking slot. IR sensors detect the car, LED red signal ON and LED green signal OFF. Now RFID reader, read RFID tag, it match the RFID tag with the database then park the car. Otherwise buzzer is ON. If the buzzer is ON the car driver understand wrongly park the car.

4. Parking Exit System

Parking exit system consists of the major components IR sensor, DC motor, LCD display, and a RFID reader. The user exit the car from the parking site LED green signal ON and LED red signal OFF. RFID reader detects the car details and shows the fees for payment. The users want to pay the parking charge fee in online or offline method. Then the gate is open the driver leave the parking slot.

RFID technology is mainly used to analyze and detect the car is theft car or not. Then only allow the car to park otherwise give the information to police station. In this smart parking system IPV6 protocol is used for internet connection the device connect to the digital world. We use IPV6 protocol for solve the IPV4 address problem. So the user or car driver gives the permanent address to this protocol. Sometime user parking the car wrongly or damage the nearby car. This system gives the solution for this problem. The user park the car securely and user friendly.

SECURE SMART PARKING MANAGEMENT SYSTEM USING WSN, RFID, IoT

Nowadays vehicle numbers increase in the urban area. Parking the vehicle securely is difficult problem for the car owners. It leads to increase the fuel consuming, air pollution, driver frustrate and traffic congestion. The user used GPS and good mapping may lead to congestion and traffic. Author use the recent technology to solve the SPS issues. It supports real time monitoring, controlling and management parking lots status. It also supports the e-payment solution, which support optimized operation, boost productivity, saves in cost and resource (Abdulkader et al, 2018).

Figure 8. Layers of Architecture
Omar Abdulkader et al., 2016

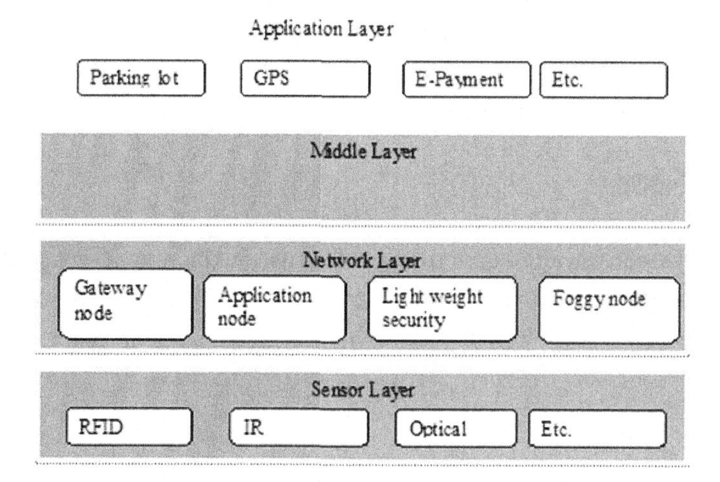

Architecture

Proposed architecture consists of four layers namely sensor layer, network layers, a middle layer and application layer.

- **Sensor Layer** is detect and determine the parking lot information such as status, capacity, no of slots etc.
- **Network Layer** is used to collect the information from sensor and transmit to the cloud based service.
- **Middle Layer** is used to support the user to get the parking lot information in real time and efficient way; it is used by the vehicle owner and car driver.
- **Application Layer** is support parking reservation and e-payment for the users.

The proposed system models have the following components.

1. Slot sensor
2. A secure and smart routing gateway
3. Smart aggregation node
4. Cloud-based service
5. Mobile Application

Slot Sensor

Edge sensors are connected to the Adhoc network. It is used to broadcast the information based on parking area to mitigate and control, manage packet flooding. The edge sensors cannot able to pass the information to gateway. The Adhoc network is used to pass the information. In parking slot contain RFID reader is used to detect the parking lot information such as empty or occupied.

A Secure and Smart Routing Gateway

Slot sensor is used to collect the parking lot information then provide quick parking the vehicle and controller. The collected information will directly send to the cloud sever. The information is encrypted the elaborated for security purpose. And also each sensor node is authenticated. Here author use light weight encryption mechanism to secure the message.

Smart Aggregation Node

Proposed system model we can use aggregation device. It is used to increase network throughput, speedup entire system transaction, decrease latency. It is used to increase the system scalability and covering wide area for smart parking solution and management.

Figure 9. Architecture of Parking System
Ndayambaje Moses et al., 2016

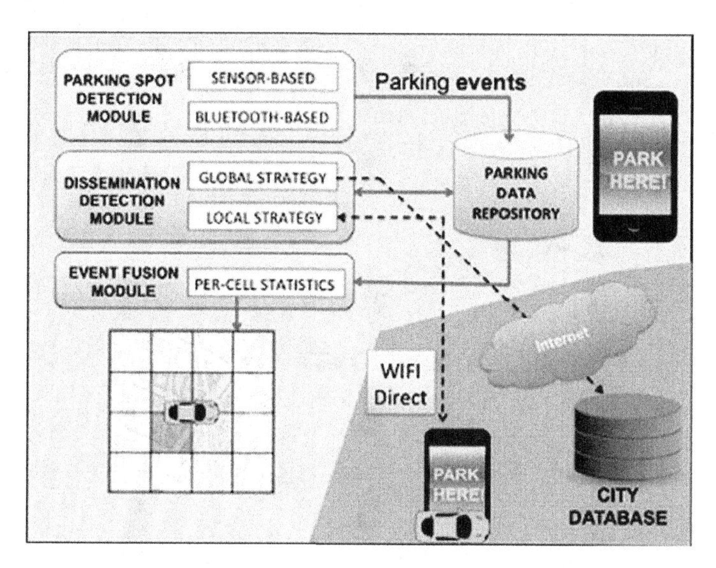

Cloud-Based Service

Proposed smart parking system is connected to the traffic agencies to support efficiency report car detection and report about stolen. Proposed model is like a sink. Collected data from various nodes and provide smart parking monitoring, controlling and give suitable solutions for its customer.

Mobile Application

Mobile application will be developed and install in user smart phone. Application will provide a service like detect parking site, reservation and payment.

Secure smart parking management system to provide cryptographically based solution for parking issues. Foggy development is support security and efficiency of smart parking issues. It is used to view parking lot status and traffic status. Author introduced the integration of WSN, RFID, Adhoc network, IoT for smart parking system. It is used to find the solution of a lot of complicated issues in SPS. It also provided strong reliability, availability, efficiency but economically feasible, low cost. These recent technology are used to security detect, monitor, controlling the SPS.

SMART PARKING SYSTEM FOR MONITORING VACANT PARKING

Author proposed car parking using smart SMS serving then allots the particular slot. It is easy to find the empty slots. This method is very simple and easy to communicate the car owner and the smart applications. In smart parking system is used for detect the vacant parking space. Vacant parking place is unused parking place. In This Parking System Used RFID technology. The RFID technology is contain both RFID tag and RFID reader. When a car parks or leaves the IPA parking lot, the RFID reader is

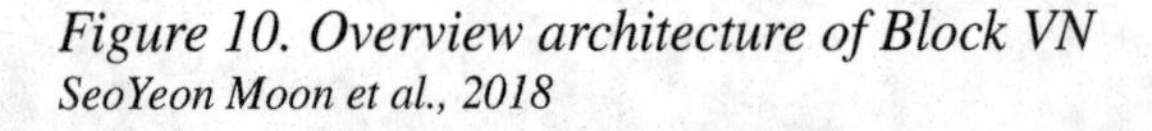

Figure 10. Overview architecture of Block VN
SeoYeon Moon et al., 2018

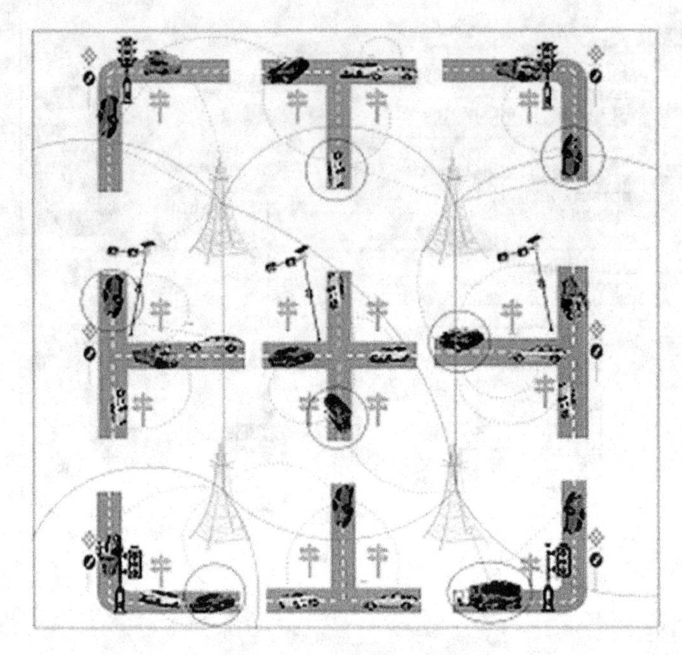

used to find which car is parking in the parking lot and which car is leaving from the parking lot. And action is detected by magnetic loop before all send the information to the unit controller .The car status information is updated by nit controller. In this architecture, used simple mathematical equation for create parking system. But this is not suitable for large scale parking system (Moses and Chincholkar, 2016).

BLOCK-VN: A DISTRIBUTED BLOCKCHAIN BASED SMART PARKING SYSTEM

Sharma et al (2017) proposed Block –VN model for solve the security issues in the smart city using IoT. It is based on the blockchain, which allow the development of automatic vehicle management in a distributed manner. In the architecture the controller is connected in a decentralized manner and provide some services. So easily achieve the high performance and scalability of the system. It is used to support the consumer to machine and machine to machine trusted free service and distributed manner and security and shared record of all information.

Benefits of the Blockchain Technique

In digital payment system blockchain is support the secure transaction from one authorize person to other authorize person. Some benefit to the beneficiary of a blockchain exchange,

- Transparency
- No risk of fraud
- Low or no exchange costs

Figure 11. Block-VN model architecture
Seo Yeon Moon et al., 2018

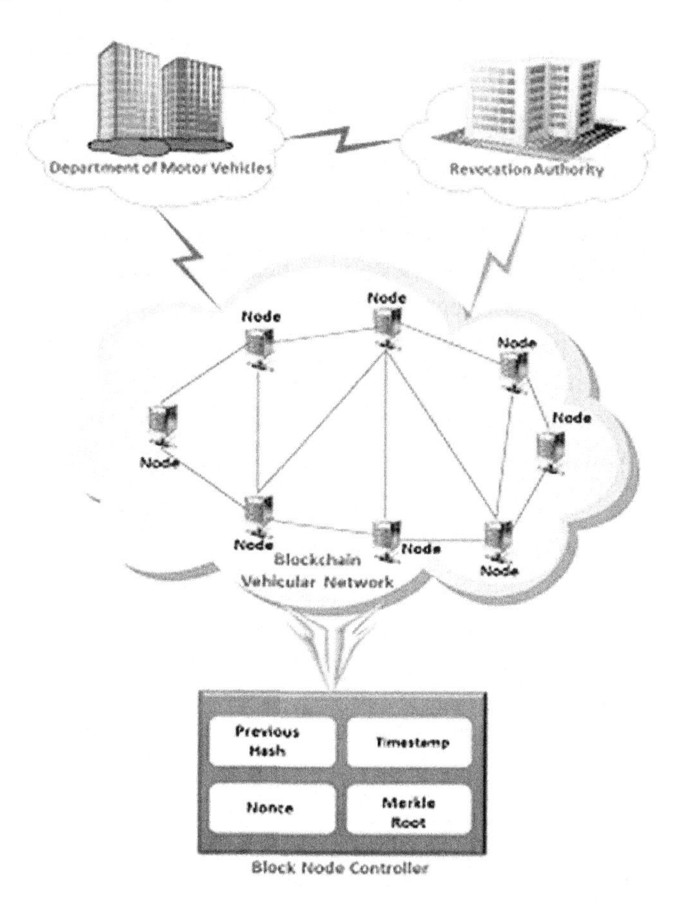

- Transactions almost instantaneous
- Network security
- Financial data assurance

Block: VN Model Architecture

Block VN Model Architecture used a blockchain technique for security and privacy of smart parking system. The author introduces the concept of smart pay that is used to pay the amount through on line mode like e-wallet. Smart share is used to share the riding information and also share the empty slots information. This architecture contains the blockchain controller, it is used to control and monitor the transaction. It contain previous hash pointer, Timestamp, Nonce, Merkle route that have perform the own service based on the blockchain controller (Sharma et al, 2017) blockchain controller work out the individual level compute and process, share based on the distributed manner and also share the information to other node. Each communication will be made in secure way we using public key cryptosystem and RSA Algorithms. Proposed system to support Privacy and security, Fault tolerance, Management of service content, Distributed operations In Smart parking system vehicle are arranged in a decentralized manner.

ADVANTAGES OF SMART PARKING SYSTEM

In parking system used to monitor the parking lot, parking lot having 16 parking spots. The parking lots include a payment station for driver to pay for the each and every parking spot after they park their car. The parking lots also contain an electronic road sign on the side of the street that display the number of empty spots. The car owner or driver, if they interest for smart parking, download the smart phone application. These applications inform them about the availability of a parking spot before they even drive on the street where the parking lot is located. Different type of sensor is used to capture and monitors the information about the empty and available spots in nearby parking lot.

Then the parking system has a payment station. The driver or owner booking a particular spot pays the fee for particular spot. The parking charge based on the time spends for the spot. Car size also varies so use the sensor device to sense the car size and choose the particular spot. In parking system have the payment station to check if a recently parked car owner actually pay the parking fees. It is used to detect the when someone park without paying.

Table 1. Comparison of different IoT parking systems

S.No	Technology	Advantage	Disadvantage
1	An IOT based secure E-parking System	1.Parking the car easily through suitable GUI 2.It support E-payment 3. Easily find the improper parking vehicle.	1. Minimize the utilization of parking space. Some parking space are unused. 2. Lack of efficiency. 3. Not concentration on damage the nearby vehicles.
2	Secure Smart Parking System using IOT Elliptic Curve	1. It is used to securely park the car 2. Reduce air pollution, traffic congestion, fuel consumption. 3. Zero knowledge protocol is used for privacy and security.	1.Lack of communication layer 2.Provide some valuable technology used for IoT Hardware 3. Not contrite on network technology.
3	Secure smart parking system using Embedded	1. Securely park the vehicle using gate system. 2.IR sensor is used to find the availability of parking area 3. RFID tag is find the car details.	1.Not concentrate platform hardware 2. Low quality of resource 3. Need high speed net connection
4	A secure and intelligent parking system	1. Support the security through wireless ns sensor network. 2.Increse space utilization 3. It support to increase the user experience.	1. Need maintenance work 2. Lack of confidentiality information 3.User may damage nearby vehicle`
5	A prototype for IOT based parking system	1.user not damage the nearby parking cars 2. User access the parking information remotely. 3.This system support for find the theft cars`	1.Not analysis the efficiency 2. Not discuss the power consumption. 3. Not support the power consumption.
6	Secure smart parking management system using WSN, RFID, IOT	1. Support cyber security and privacy in smart parking system. 2. Avoid traffic. 3. Smart parking facility securely and efficiently.	1.To measure computational cost 2. Energy consumption of smart parking for each WSN and IoT device. 3. Interference issues with RFID.
7	Smart parking system for monitoring vacant parking	1.Realtime parking navigation 2. Intelligent anti-theft protection. 3. To save parking time.	1.It is not suited for large scale parking system 2.Not concentration for payment 3.maintain large amount of database
8	Block-VN: a distributed blockchain based smart parking system	1. Secure transaction from one authorized person to another person. 2. No risk and easily find the fraud` 3.Low or no exchange costs .	1. Peer to peer communication. 2. Transparency. 3. Expensive

COMPARISONS

See Table 1.

CONCLUSION

In Smart city have many of applications like home appliances smart parking, smart irrigation, weather monitoring system, etc. Nowadays smart parking system is very popular. It is very useful for the car owner or driver to parking the vehicle without tension. In this chapter discuss the security and privacy issues in smart parking system. But security is very challenging problem. Also discuss the few of major issues in different articles. How to secure the user personal and private information, differentiate the theft car by the persons and give the information to the police station, easily find the empty slot nearby parking lot. Then compare the different technique used by the author from different article.

REFERENCES

Abdulkader, O., Bamhdi, A. M., Thayananthan, V., Jambi, K., & Alrasheedi, M. (2018, February). A novel and secure smart parking management system (SPMS) based on integration of WSN, RFID, and IoT. In *2018 15th Learning and Technology Conference (L&T)* (pp. 102-106). IEEE.

Ankitha, S., & Balajee, M. (2016). Security And Privacy Issues in IoT. *SCIREA Journal of Agriculture*, *1*(2), 135–142.

Borgohain, T., Kumar, U., & Sanyal, S. (2015). *Survey of security and privacy issues of internet of things.* arXiv preprint arXiv:1501.02211

Chatzigiannakis, I., Vitaletti, A., & Pyrgelis, A. (2016). A privacy-preserving smart parking system using an IoT elliptic curve based security platform. *Computer Communications*, *89*, 165–177. doi:10.1016/j.comcom.2016.03.014

Gandhi, B. K., & Rao, M. K. (2016). A prototype for IoT based car parking management system for smart cities. *Indian Journal of Science and Technology*, *9*(17).

Khanna, A., & Anand, R. (2016, January). IoT based smart parking system. In *2016 International Conference on Internet of Things and Applications (IOTA)* (pp. 266-270). IEEE. 10.1109/IOTA.2016.7562735

Kumar, J. S., & Patel, D. R. (2014). A survey on internet of things: Security and privacy issues. *International Journal of Computer Applications, 90*(11).

Moses, N., & Chincholkar, Y. D. (2016). Smart parking system for monitoring vacant parking. *Int. J. Adv. Res. Comput. Commun. Eng*, *5*(6), 717–720.

Rehman, S. U., Khan, I. U., Moiz, M., & Hasan, S. (2016). Security and privacy issues in IoT. *International Journal of Communication Networks and Information Security*, *8*(3), 147.

Roman, R., Zhou, J., & Lopez, J. (2013). On the features and challenges of security and privacy in distributed internet of things. *Computer Networks*, *57*(10), 2266–2279. doi:10.1016/j.comnet.2012.12.018

Sadhukhan, P. (2017, September). An IoT-based E-parking system for smart cities. In *2017 International Conference on Advances in Computing, Communications and Informatics (ICACCI)* (pp. 1062-1066). IEEE. 10.1109/ICACCI.2017.8125982

Saravanan, K., Julie, E. G., & Robinson, Y. H. (2019). Smart Cities & IoT: Evolution of Applications, Architectures & Technologies, Present Scenarios & Future Dream. In V. Balas, V. Solanki, R. Kumar, & M. Khari (Eds.), *Internet of Things and Big Data Analytics for Smart Generation. Intelligent Systems Reference Library* (Vol. 154). Cham: Springer. doi:10.1007/978-3-030-04203-5_7

Sharma, P. K., Moon, S. Y., & Park, J. H. (2017). Block-VN: A Distributed Blockchain Based Vehicular Network Architecture in Smart City. *JIPS*, *13*(1), 184–195.

Shin, J. H., & Jun, H. B. (2014). A study on smart parking guidance algorithm. *Transportation Research Part C, Emerging Technologies*, *44*, 299–317. doi:10.1016/j.trc.2014.04.010

Sicari, S., Rizzardi, A., Grieco, L. A., & Coen-Porisini, A. (2015). Security, privacy and trust in Internet of Things: The road ahead. *Computer Networks*, *76*, 146–164. doi:10.1016/j.comnet.2014.11.008

Thangam, E. C., Mohan, M., Ganesh, J., & Sukesh, C. V. (2018). Internet of Things (IoT) based Smart Parking Reservation System using Raspberry-pi. *International Journal of Applied Engineering Research*, *13*(8), 5759–5765.

Yan, G., Yang, W., Rawat, D. B., & Olariu, S. (2011). SmartParking: A secure and intelligent parking system. *IEEE Intelligent Transportation Systems Magazine*, *3*(1), 18–30. doi:10.1109/MITS.2011.940473

Zhang, K., Ni, J., Yang, K., Liang, X., Ren, J., & Shen, X. S. (2017). Security and privacy in smart city applications: Challenges and solutions. *IEEE Communications Magazine*, *55*(1), 122–129. doi:10.1109/MCOM.2017.1600267CM

Chapter 12
A Conceptual Framework for the Design and Development of Automated Online Condition Monitoring System for Elevators (AOCMSE) Using IoT

M. S. Starvin
University College of Engineering Nagercoil, India

A. Sherly Alphonse
Anna University Chennai – Regional Office Tirunelveli, India

ABSTRACT

The reliability of an elevator system in a smart city is of great importance. This chapter develops a conceptual framework for the design and development of an automated online condition monitoring system for elevators (AOCMSE) using IoT techniques to avoid failures. The elevators are powered by the traction motors. Therefore, by placing vibration sensors at various locations within the traction motor, the vibration data can be acquired and converted to 2D grayscale images. Then, maximum response-based directional texture pattern (MRDTP) can be applied to those images which are an advanced method of feature extraction. The feature vectors can also be reduced in dimension using principal component analysis (PCA) and then given to extreme learning machine (ELM) for the classification of the faults to five categories. Thus, the failure of elevators and the consequences can be prevented by sending this detected fault information to the maintenance team.

INTRODUCTION

In smart cities, the safety and privacy of people are improved by the collection and utilization of data to upgrade the lifestyle of the residents by using low-cost sensors and other wifi-enabled technologies. The IoT devices can collect and provide data at real-time so that the managers of the smart cities can

DOI: 10.4018/978-1-5225-9199-3.ch012

function in an efficient manner. In smart city, the data are collected from public populated areas such as hospitals, libraries, and schools etc for overall monitoring. The information collected is communicated through the network to the other devices in the smart city via the Internet of Things (IoT). In a smart city, the authorities can monitor the day-to-day happenings of the city from the information collected from various vital locations to enhance the quality of living, traffic regulations and resource consumption. The major advantage of this system is that the responses for the critical events can be done in a real-time. Smart city concept is implemented in various developed nations such as Singapore, United States of America, China etc. and now it is proposed in India similar to cyberville, flexicity, knowledge-based city, and ubiquitous city. The objective of the smart city is to provide hassle-free, safe environment to the people and to provide high-speed internet to the common people.

Certain applications in which smart cities with IoT devices can be used are Street and traffic lights monitoring, Parking maintenance, Crime analysis, Waste management, Air quality checking, Energy usage, Transportation and Traffic flow monitoring. The automated elevator monitoring provides superior security, better safety, and smaller costs. Most multi-storied buildings have elevators but less security. This affects the security of the tenants to a hazardous degree. Elevator monitoring systems prevent illegal visitors as well as inform the faults of the elevators to the managers of smart cities for avoiding critical failures (Heather K, 2010).

The smart cities have multi-storied buildings, where the elevators are essential for the vertical transportation between different floors. The elevator (or lift) is a type of vertical transport equipment that efficiently moves people or goods between floors (levels, decks) of a building, vessel or other structure. The elevators are generally powered by electric motors that either drives cables, hoist, or pump hydraulic fluid to raise a cylindrical piston like a jack. There are four types of maintenance activities for an elevator namely corrective maintenance, preventive maintenance, predictive maintenance, and prescriptive maintenance. Corrective maintenance activities are done after failure. Preventive maintenance is based on the expected lifetime of the parts. Prescriptive maintenance is based on the farsighted prediction of the failure. Predictive maintenance is condition-based maintenance that prevents frequent visits. Predictive maintenance can be applied very effectively in case of a smart city.

Figure 1 shows the schematic diagram of machine lifts which is widely used in commercial applications. These traction elevators work on the principle of see-saw where the car is raised and lowered by traction steel ropes rather than pushed from below. The ropes are attached to the elevator car, looped around a sheave and connected to an electric motor. When the motor turns one way, the sheave raises the elevator; when the motor turns the other way, the sheave lowers the elevator. Typically, the sheave, the motor, and the control system are all housed in a machine room above the elevator shaft. The ropes that lift the car are also connected to a counterweight, which hangs on the other side of the sheave. The counterweight and the car are perfectly balanced. In gearless elevators, the motor rotates the sheaves directly. In geared elevators, the motor turns a gear train that rotates the sheave. Nowadays, some traction elevators are using flat steel belts instead of conventional steel ropes. Flat steel belts are extremely light due to its carbon fiber core and a high-friction coating and do not require any oil or lubricant.

The elevators have different capacities like 450kg - 1,150 kg, 1,150kg-1,500kg, 1,500kg-2,000kg and others. The Miconic 10 elevator by the Schindler Company has keypads that can sort and cluster the passengers for a particular elevator based upon the floor number entered by the passenger which saves much of the time needed to reach a particular floor (Nick A, 2010). A certain type of elevators can move in the vertical as well as the horizontal direction which avoids the walking across floors (Heather K, 2010). Also, the overall energy consumed by the elevators can be reduced by a regenerative braking system.

Figure 1. Schematic diagram of Mechanical Elevator (Traction)

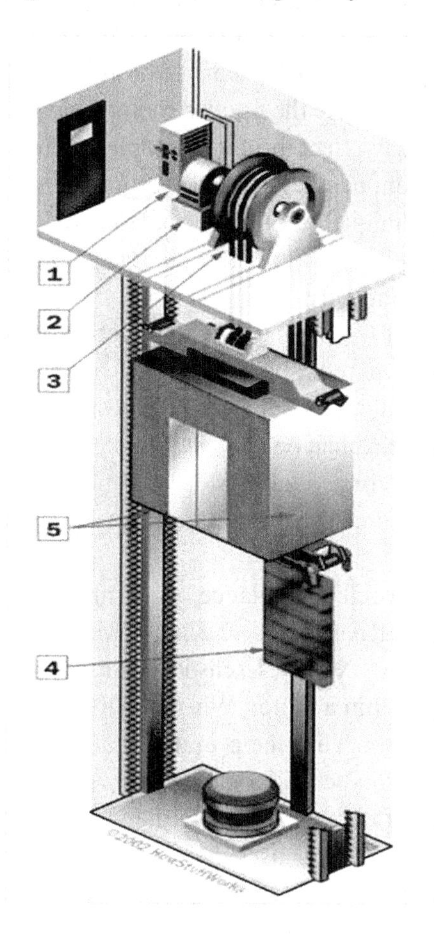

Basic Components

1. Control System

It controls the entire elevator functions (start, stop, acceleration speed, retardation etc)

2. Electric motor

The motor is used to rotate the sheave.

3. Sheave

It is connected with the motor which has a pulley with grooves around the circumference to grip the hoist ropes. So when the sheave rotates, the ropes move too.

4. Counter Weight

It hangs on the other side of the sheave by ropes and it weighs about the same as the car filled to 40-percent capacity.

5. Guiding rail

It is placed along the sides of the elevator shaft to keep the car and counterweight from swaying back and forth. It also works with the safety system to stop the car in an emergency.

When a tiny device that can collect and transmit the performance details using the Internet of Things (IoT) is attached to the elevators, the predictive maintenance of elevators becomes easier. Also, by using the Artificial Intelligence and machine learning techniques, the flaw can be correctly identified and can be rectified before the elevator breaks down. The service engineers can also get a periodic alert that enables them to perform the maintenance operations on time. This helps to perform the maintenance operations smarter and avoids the inconvenience faced by the users when the elevator is out-of-service. This method reduces the cost and energy, as the experts can avoid their travel to the elevator locations for a periodic checking. Also, based upon the information obtained by the periodic alerts, any part that needs replacements can be purchased in advance. The cost needed for the certification process and the raw materials of elevators is very high (Wowk, V. 2005). The elevators are subjected to very high wear and tear, and in case of any repair, the replacement of the parts is money-consuming. Also, the ball bearings mounted along with the motors in the elevators are subjective to wear and tear causing cracks in the ball bearings and these bearings are also susceptible to flaking, rusting, peeling and corrosion.

The traction motors result in failures due to bearing malfunctions and noisy bearings. The examination of the bearing vibrations during and after an operation is one of the primary tasks of maintenance and it should be done periodically to avoid any failure in elevators. Damage to the bearing can be avoided at an early stage by analyzing the vibrations of the ball bearing. A degree of the damage is calculated from the amplitude and the frequency of the vibrations produced. Also, the rotation of the unbalanced weight around the machine axis causes vibration, which leads to wear and tear of the motor. When the axes of a motor are misaligned it causes vibrations that are also responsible for the damage of the motor. The looseness of the ball bearings also causes vibrations that further damage the traction motor. Certain repairs may include replacing the entire device, while some others may affect the entire people's movement in a building. In the case of predictive maintenance, the problem can be predicted earlier and the cost of replacement and maintenance can be reduced. When this technique needs to be implemented on more than one elevator, more data needs to be streamed and analyzed. A cloud-based platform is very essential at this stage to perform the analysis qualitatively and quantitatively.

In this chapter, a conceptual framework that uses the vibrations of the parts within the traction motor for estimating its fault and planning its maintenance through remote monitoring using IoT has been designed. This work analyzes the various techniques available in literature and proposes the best design for a remote monitoring system for elevators using IoT as AOCMSE that can be very well implemented in a smart city. Vibration directly affects the running of a traction motor as it affects its operations and behavior. The information that can be obtained from the vibrations is more reliable. Also, through vibration changes, faults can be identified at the early stage.

The vibrations occur as a result of many conditions, such as weight imbalance, shaft misalignment, wear and looseness of ball bearings (Fluke corporation. 2018, Khan, S. A., & Kim, J. M. 2016). Sometimes, when an unbalanced weight is rotating around the machine axis due to casting faults, as the machine speed increases, the unbalancing effect causes vibrations within a motor. When a ball bearing is loosened or if it is damaged, vibrations are caused in a traction motor. Thus the overall vibration of a traction motor that determines its fault is calculated as the vibrations of its secondary parts. The vibration sensors can be placed on the input and output bearing housings to effectively calculate the vibrations. From the vibration signals obtained through the sensors, the faulty part that causes the vibration can be identified using machine learning method. The vibration signals are converted to 2D grayscale images, on which feature extraction technique is applied. The feature vectors thus obtained are reduced in dimension using Principal Component Analysis (PCA) (Jolliffe, I. 2011) and given to a machine learning technique for multi-classification of faults. Here, five types of classification for faults can be done. They are as weight imbalance, shaft misalignment, weariness of ball bearing, looseness of ball bearing and no fault. Then this fault information can be sent to the maintenance teams, who can plan for sufficient maintenance. Because of this method, the traction motor which is the vital part within the elevator, is monitored effectively using IoT.

This work contributes a good design for the remote monitoring of the motor within the elevator that includes:

- Four vibration sensors for sensing the vibrations of motors in elevators.
- MRRDTP feature extraction for 2D grayscale images constructed from vibration signals and reduction of random noise.
- PCA dimension reduction method that reduces dimensions of features.
- ELM for accurate and rapid classification.
- Raspberry Pi and IoT based communication.
- Four types of fault detection in traction motor.
- Instant communication with the maintenance teams.
- Computationally very efficient.

Section 2 describes the various techniques that have been proposed in the literature in relation to the vibration analysis, remote monitoring and IoT. Section 3 describes the architecture. Section 4 gives details about the workflow and the proposed methodology of AOCMSE. Section 5 explains the resource requirements for establishing AOCMSE. In section 6 the expected outcomes are summarized. Section 7 concludes the chapter and discusses future enhancements.

LITERATURE SURVEY

Different technologies have been reported in literature for implementing a remote monitoring system to improve the security and to enhance the efficiency of elevators. Various manufacturers design their own monitoring system for their elevators. Certain works have reported the already existing web technologies for designing a remote monitoring system as described by Liu et al (2014). Devices are installed inside the elevator room for monitoring the voltage and current. If there is any abnormality, then the sensors will send an alert to the devices that are placed both inside and outside the elevator car. This work monitors only the power failures due to abnormal voltage and uses accelerometers and infra-red sensors to monitor the comfort levels of the passenger. The sensors send the information to the processor which acts as the brain of the remote monitoring system and decides the elevator status and then sends the information to the remote monitoring system.

The Remote Condition Monitoring (RCM) by Olalere et al. (2018) is helpful for proactive maintenance. Vibration of the machine helps in finding the machine condition for prediction of faults. It uses IoT for Data Acquisition. It has sensors and Arduino microcontrollers, installed. The set-up monitors the conditions with email service. The data was analyzed and notifications created. This enables faster maintenance and prevents the breakdown of elevators. Uusitalo et al. (2018) used Raspberry Pi 3 Model B. The sensors are in TI's plug-in module are connected via I2C to the Raspberry. The system uses Marvel mind navigation system. It connects to Internet via. TUT-wireless network. Cloud server was used to save data. The experiments were done in a simulated elevator system. The features in the prototype performed well in accuracy.

Sahoo and Dhas (2018) proposed the monitoring of defective bearings through measurement of vibration signature. The vibration signature is used for fault detection. This vibration signature will be corrupted with noise. This noise needs to be removed for accurate fault detection. Adaptive noise control-based filtering techniques are used in this approach. This approach used time and frequency analysis to detect the faults. Random forest and J48 classifiers were used.

Yao et al. (2011) designed a common signal processing device for the remote monitoring and a Mono-Chip Computer (MCU) is used for the processing work. This device successfully analyses, monitors and eliminates the failures of the elevators. This work states that the elevator monitoring center can be built and be used for a building, for a district or for a county. The work also states that such periodic monitoring can reduce the down time and can also help the trapped passengers inside the elevators by a quick response and an immediate repair.

You Zhou et al. (2018) proposed a remote monitoring system for elevators based on the Internet of Things (IoT). The system performed various operations like monitoring, fault identification, alarm and also maintenance. The overall process involved in a remote elevator monitoring system involved sub-processes like gathering information, it's transmission, and processing. So the entire process was grouped under three layers like the perception layer, network layer, and application layer. The perception layer was responsible for collecting all types of information including the sensor details, the object identifier information and the scanner details. The network layer processed the details and sent the information to a communication network that is connected to the internet. The application layer is responsible for providing the information needed by the user. This method monitored the status; generated alarms in case of emergency, there was daily supervision, data recall, and data analysis. This method was also very effective in preparing the reports and also in pre-diagnosis.

Cho et al. (2016) designed and developed an elevator monitoring system for marine elevators using machine learning techniques. In the case of marine elevators, the manual monitoring of the elevators is very difficult because of their voyage. NMEA 2000 network technique is used for the effective monitoring of the marine elevators using the big data. Supervised learning technique is used for the prediction. The overall accuracy needs to be improved. This model monitored the elevators, analyzed the data and prepared the reports. Logging gateway predicts the fault using a model for prediction based on the data in the server.

Suárez et al. (2018) developed a remote monitoring model using Microsoft Azure to avoid the trapping of users inside the elevator by sending messages to the technicians about the fault. The application was implemented using cloud computing. The maintenance company receives the reports from the elevators by the App through IoT and hence takes necessary actions that reduce the events of users getting trapped inside elevators in case of some failures. They used a back-end application installed on the cloud for the software installed on the mobile phone.

Wang et al. (2015) proposed a model where the User Datagram Protocol (UDP) was used as it handled the high traffic of data. This technique also used thread to handle the multiple queues created as the data arrived from multiple elevators. The system consists of three modules i.e., data acquisition, network module and the remote monitoring center. There was a view-controller module in the remote monitoring center that gave details about the current operation. Yi (2016) has proposed an elevator remote monitoring system using big data. The data is processed at three layers namely the collecting layer, the analyzing layer, and the service layer. The collecting layer is responsible for collecting the internal and external information from the elevators. The internal information of the elevators is the information regarding the original state of the device, the testing information and the state of operation. The analysis layer analyses the safety, establishes the elevator safety record and also checks the elevator stability by intelligence analysis. The service layer is responsible for providing various services to the user. It also provides various services for the remote monitoring system. It provides services for decision support, risk analysis and early warning. The cloud platform that was utilied has three parts namely risk information collection, risk information analyzing, and risk monitoring.

Giagopoulos et al. (2018) have done a dynamic and structural integrity analysis of the elevators under real time-loads. The authors used optimization techniques for design purpose to develop light weight structures. They proposed a mixed computational – experimental analysis procedure for the dynamic and structural integrity analysis of a complete elevator system. This system was tested using worst case loading scenario in an industry-based elevator system. This elevator system was designed as a super-structure that was divided into sub-structures like chassis and cabins. Then Finite Element (FE) model of the sub-structures was developed. Then under real-time operating conditions, the accelerations of at the connection points of the chassis with the gears were calculated. Then the FE model was verified and updated. The interface acceleration values are used in the FE model and identified the maximum stress developed in the super-structure.

Yang et al. (2017) proposed a theoretical model and also an experimental verification of the vibrations of the elevator ropes and building. They were done by Hamilton's principle. The equations that describe the vibrations were formed using the kinetic and potential energies. A mini simulator was used and the results were checked. Zhang et al. (2018) proposed a model for monitoring the comfort of the elevators using the smart phones. International Organization for Standardization ISO2631-1997 standard was used to monitor the comfort level. The sensors installed in the smart phone monitor the elevators that are connected to a communication network. This method eliminates lots of errors and also makes use of the passengers to monitor the comfort level of the elevators. Orion cc software for the mobile phone was designed for the purpose of monitoring. This software monitors the comfort level by calculating the acceleration. It checks the accelerations along the all three possible directions. The horizontal vibrations are related to the smoothness of the rail which actually increases with the speed. This method is very economical and also very convenient.

Huang and Yu (2016) used directional antenna transmission technology to monitor the safety of elevators. This system has various modules for sensors, micro-controllers, display, transmission to antenna and monitoring system. The application has two versions; one for the laptops and another for the mobile phones. The alignment system of the antenna is made automatic in order to improve the rate of transmission and the quality. Dynamic Matrix Control (DMC) algorithm was used for the alignment of antenna. It can send the status information continuously and can also alert the elevator teams in case of any failure. This makes the maintenance process easy and also the installation process of the applications is also very easy.

Göksenli and Eryürek (2009) carried out failure analysis on a fractured elevator drive shaft. Chemical, micro-structural and the mechanical properties of the shaft are analyzed and the structural analyses are carried out using FEM method. Fatigue safety factor and the Endurance limit are calculated for the drive shaft and found that the change in radius of curvature causes the change in stress distribution and also the reason for the failure. Flores et al. (2008) proposed that induction motors play an important role in elevators. The normal inspection procedures cannot be used to identify the misalignment of elevators due to certain building movements caused by earthquakes. This work uses the inductor motor itself as a sensor to detect the fault. The guides of the elevator suffer from side movement, the increase of the distance between the elevator guides and the decrease of the distance of distance between the elevator guides which causes side way shakes. The increase of the distance between the guides may trigger the overload protection of the elevator system and this may also cause a disaster and increases the friction, wear and tear. Motor Current Signature Analysis (MCSA) is used to identify the faults in bearings, gear shafts etc. The increase of friction causes a torque variation within the induction motor. The Root Mean Square (RMS) plot was used to identify the malfunction within the induction motor. In this method, the induction method functions as a transducer for detecting the faults in preventive maintenance.

Shi and Xu (2018) have grouped the microprocessors, sensors, remote controls, and speech recognition technology as a remote elevator monitoring system using internet technology. They have designed an intelligent elevator that works based on the voice guide. The infrared sensors are used for seamless docking. When the elevator door is closed the voice guide system starts its operation. The infrared system and the voice guide enable the fully automatic control of the elevator system. The remote monitoring ensures the safety and reliability of the system. The infrared technique avoids unwanted openings by the passengers.

Esteban et al. (2016) implemented a method that has subsystems that are both mechanical and electrical. The observer that is based on Kalman filter monitors the speed of the elevator. Then, five performance measures that were based on the accelerations recorded are used for evaluating the performance of the elevators. Jung et al. (2017) proposed that the vibration sensors can be attached to the core of the equipment like motors and tubes to monitor the status of the equipment even when it is running. As the data read from the sensors are very large, they have proposed an algorithm for reading the vibrations. They have calculated the Remaining Usefulness Lifetime (RUL) through which the time of replacement can be calculated. They have used vibration sensor based on MEMS accelerometer which has several advantages over piezoelectric accelerometer like easy installation, lighter weight, low cost and operation on battery etc. The vibration of a pump is monitored in this approach. The vibrations are collected from the path between the motor to the exterior suction connector. The samples collected from different sensors are from different time periods. There is also noise in the samples collected from the sensors. The calculated lifetime varies depending on the age of the machine and the installation process. Besides that, the devices can also be grouped into classes.

The accelerations collected are taken from three directions. If more than one sensor is used, the vibration data acquired can be severely misaligned. So, in this work, only one sensor is used. Root Mean Square (RMS) and Power Spectral Density (PSD) can be used as features. RMS contains information about the vibrations and PSD has information about the subtle vibrations. Normalization of the features is also done. Labels are given according to the running status of the device. Here, four human labels are used as Vibrations of new machines, Unsatisfactory, Satisfactory and Severe damage. The starting and ending analysis period is set automatically by the system or by the system administrator. Data preprocessing is done to remove the noise and unwanted readings. In this method, it has been stated that peak harmonic distance can be used as features. Random Sample Consensus (RANSAC) algorithm can be used for regression.

Caesarendra and jahjowidodo (2017) reviewed various feature extraction techniques for the vibration analysis of slowly rotating bearings. The largest Lyapunov exponent (LLE) impulse factor, approximate entropy and margin factor are some of the feature extraction techniques used for the effective monitoring of ball bearing condition in literature. The analysis of vibration signals is done in the time domain, frequency domain, and time-frequency domain. The scalar descriptors from time-domain are not consistent. The frequency domain based features do not consider the amplitude which is related to rotation. A lot of time-frequency based information extraction techniques already exist in the literature. Signals, when processed as 1D does not represent the relation between the time and the frequency. If raw signals are used they are computationally inefficient. The 2D signals achieve better accuracy while diagnosing the fault. Also when the amplitude is converted to the intensity of the pixels and 2D grayscale images are created, the feature extraction techniques can be applied. From this feature vectors are generated. The dimension reduction techniques can be applied to the feature vectors to reduce the dimension further. Then the features of reduced dimension can be given as input for fault diagnosis using any multi-classification technique for classification. Lots of texture-based feature extraction techniques available in literature for biometric-images, and other imaging applications can also be used for this fault detection process using 2D images of vibration signals.

Uddin et al. (2014) proposed that the vibration signals can be converted to 2D grayscale images, on which the texture-based feature extraction techniques can be applied. The Support Vector Machine can be used for classification and the Principal Component Analysis (PCA) is used for reducing the dimension of the features. This approach used One Against All (OAA) SVM for performing the multi-classification.

Good accuracy of 100% is attained when the Gaussian RBF kernel is used for the classification. Khan and Kim (2016) proposed an approach where the slices of the vibration signal are stacked to create the 2D grayscale images. The features are extracted from these images using Local Binary Pattern (LBP) from which the histograms are used for classification. Compared to the other methods, the methods that involve image processing and machine learning concepts demonstrated more accuracy.

The usage of feature extraction methods and classification techniques helped in a better classification of faults. Compared to the other techniques for feature extraction, the Maximum Response-based Directional Texture Pattern (MRDTP) performed better in literature, as stated by Alphonse and Dharma (2017) and was also superior to LBP for extracting the features. In that paper, a Generalized Supervised Dimension Reduction Technique (GSDRS) was used to reduce the dimension of the feature vectors. The technique used RBF-based Extreme Learning Machine for classification. When the biometric feature extraction techniques like LBP can perform well in extracting the features from the 2D grayscale images of vibration signals as in literature, MRDTP can perform better in the multi-classification of fault diagnosis in motors.

Classification techniques like Support Vector Machine (SVM) and deep learning have been used in literature for classifying the feature vectors. Wan et al. (2015) collected the characteristics of vibration signal energy and the domain characteristics to construct a feature vector which is then given as input to Least Square-SVM for performing the classification with Cross Validation (CV). Shaheryar et al. (2017) proposed that different faults indicate different types of failures and are avoided by monitoring the vibrations. This work studied the best feature extraction technique for a slowly degrading ball bearing. Different feature extraction techniques for extracting the statistical and signal-based features are analyzed in this work. Deep learning with an unsupervised method of learning features has been used in this work. They have proposed Multi-channel Convolutional Neural Network with a stack of Denoising Auto-Encoders (MCNN-SDAE) for classification.

Tissera and McDonnell (2016) proposed that Extreme Learning Machines with the Supervised stack of Auto-encoders achieve good accuracy and has consumed lesser time for classification. This is because they have combined the characteristics of deep learning with ELM. So instead of using SVM and deep learning techniques for classifying the features as in literature, it is better to use this type of ELM, as this elevator monitoring system has to be implemented with real-time responses. ELM with the Supervised stack of Auto-encoders can achieve better accuracy at very less amount of time for classification.

ARCHITECTURE

The figure 2 depicts the system architecture of the AOCMSE for a single system. The vibration sensors can be installed in the bearing house of the elevator. Up to four sensors can be used to sense the four different types of vibrations depending upon the type of fault within the traction motor. The vibration sensors can be controlled by the sensor management server through the communication protocol TCP/IP. The vibration sensors can be monitored at a proper schedule of time for every 10 sec. The gateway component in Raspberry Pi 2 combines all the sensor data. The data analytics component stores the data. It then converts the data to 2D images and creates the feature vectors using MRDTP. It can also predict the type of fault using ELM. The web application component then sends the report to the mobile or the e-mail of the maintenance team who can take further action.

The overall procedure of AOCMSE is as follows:

Figure 2. System architecture

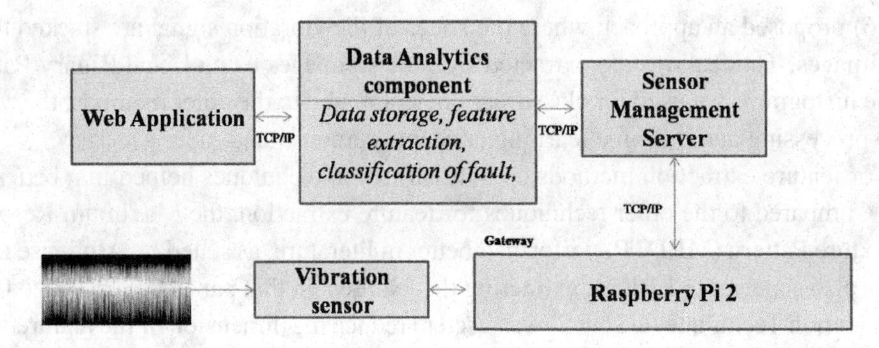

1. Acquiring the vibration data of the traction motor within the elevator
2. Converting the vibration signals to 2D images
3. Extracting the features using the MRDTP technique
4. Reducing the dimension of the feature vectors using the Principal component Analysis(PCA) (Jolliffe, I. 2011) technique
5. Predicting the type of fault using the Extreme Learning Machine
6. Sending the type of fault to the maintenance team

Design of the System

The complete workflow of the system is described in figure 3. The four MEMS vibration sensors are placed inside the bearing houses of the motor. During initialization, the sensors send messages to the Sensor Management Server and get its initial configuration data for the subsequent measurements. As many as four sensors can be used for obtaining the data. Hence, the collected vibration data is fed to the Raspberry Pi. The Raspberry Pi collects all the sensor data. Then this sensor data is converted to 2D images. These 2D images will be having random noise which contaminates the 2D image because of the conversion process. Then using MRDTP feature extraction, the noise can be eliminated and the feature vectors can be obtained. The feature vectors can then be classified using ELM to find the type of fault of the motor. The training data that are collected before for a period of two years and their fault types can be used as training data in ELM. This training data is used for classifying the fault when a new set of vibration data is collected. The information is then sent as reports to the maintenance team.

Sensors and the Communication Protocols Used

Four wireless vibration sensors can be used. They communicate using the IEEE 802.15.4 protocol. The piezoelectric sensors use the piezoelectric effect to sense the vibrations and forces. The advantages of the piezoelectric sensors are as follows:

- They are available in the desired shape.
- They are smaller in size.
- They have the rugged construction.
- They have a better frequency response.
- Their phase shift is negligible.

Figure 3. Workflow of the system

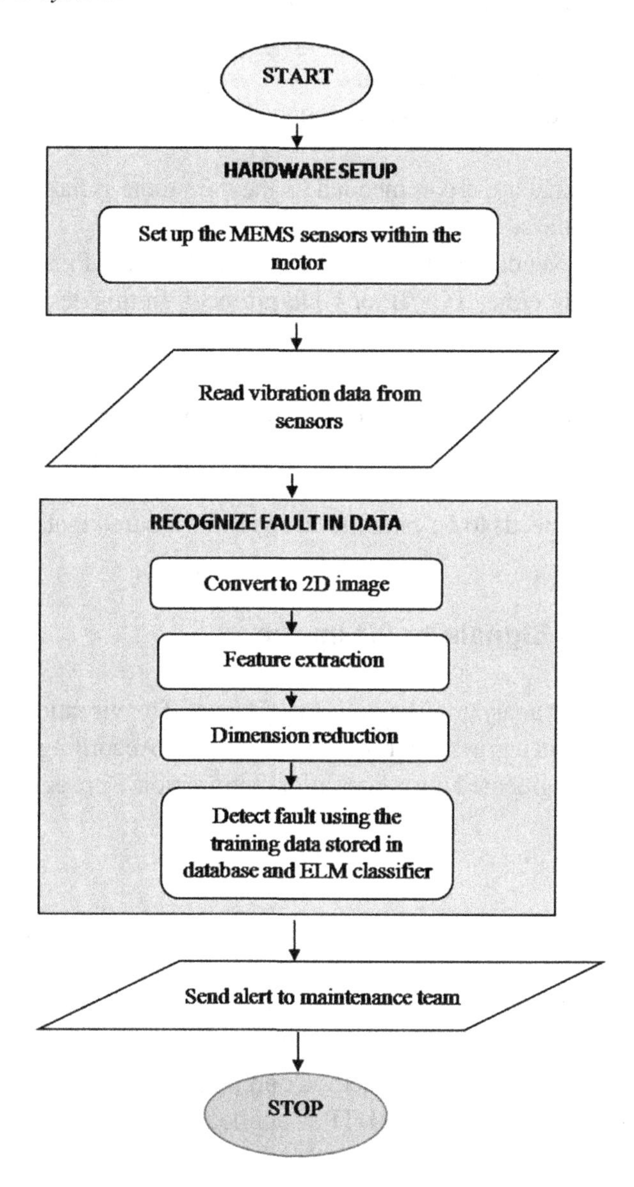

The disadvantages of the piezoelectric sensors are as follows:

- They are used for measuring dynamic readings
- They have high sensitivity to temperature
- Certain crystals used in the sensors are soluble in water and also get affected by humidity

The MEMS sensors have various advantages like:

- It minimizes energy.
- It also consumes low energy and low cost.

- It is also easier to change the parts.
- It has better accuracy, sensitivity and it's also highly reliable.
- They can be used in most of the applications.
- In case of any critical condition, alarms are produced.

In this design, MEMS sensors are recommended as they are more reliable and they are used more in real time applications nowadays.

The sensor management server can communicate with the Raspberry Pi 2 for controlling the sensors and for receiving the data using either TCP/IP or UDP protocol. In this design the TCP/IP protocol is chosen because of the following advantages

1. In TCP the data arrives at the destination at the same order in which it was sent.
2. It also guarantees the delivery.
3. TCP more reliable when compared to UDP.
4. In TCP as the connection needs to be established first, it is assured that the data reaches the destination compared to UDP.

Conversion of Vibration Signals to 2D Image

The power spectrum is used for analyzing the fault frequencies. The vibration signals are divided into slices and then combined to form a grayscale image as in figure 4. According to Khan and Kim (2016), the vibration signal should be segmented into slices after rectification. For segmentation, the slice length is taken as

$$l_v = round\left(\frac{60f_n}{\omega}, 3\right) \tag{1}$$

f_n is the frequency at which the signal is sampled and ω is the speed at which the shaft rotates. If the block size on which the feature extraction method applied is 3x3 then the quotient of eq. (1) needs to be rounded off to 3. Here, in our design MRDTP is applied on 3x3 block. So, eq. (1) can be used as such for calculating the length of the vibration signal slice. Certain numbers of these slices are stacked together to form a 2D image as in figure 4.

The 2D image size is also a multiple of 3x3 so that no portion of the signal is lost. Then the feature extraction techniques that are used in various image processing applications can be used for extracting the features. In literature, LBP has been used for extracting the features from this kind of 2D vibration images. In this design, we have proposed to use MRDTP that mainly focuses on extracting features based on both texture and shape information within the 2D images. It has also achieved good accuracy over LBP in literature. When used in this design it will achieve good accuracy while extracting the features. The procedure of finding the dominant edges in MRDTP and eliminating noise are the advantages of MRDTP over LBP. This helps in achieving good accuracy while classifying the faults based on the features found in 2D vibration images. The procedure of extracting features is explained in detail in section 3.4.

Figure 4. Conversion of a vibration signal to 2D image

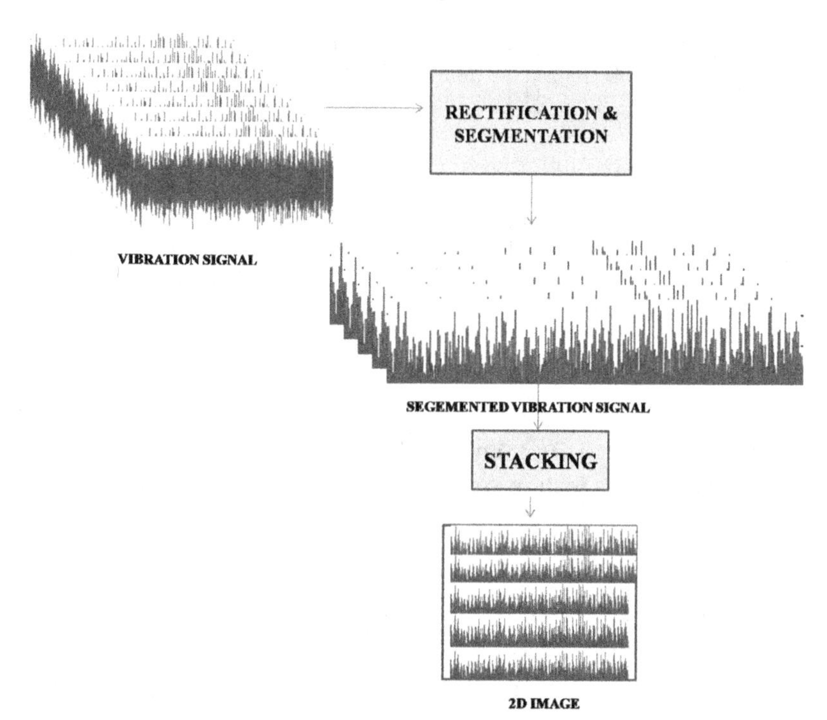

Feature Extraction From 2D Images

The 2D image obtained as in figure 4 is converted into a grayscale image. By analyzing the textures within the image we can find the variations in the vibrations. For different kinds of faults, the textures observed are different in these 2D images. The MRDTP approach has achieved superior results when applied on grayscale images and when compared with the other feature extraction methods. It extracts the information depending upon the high- frequency information within the grayscale image. It also has very low computational complexity. The feature extraction method MRDTP has the following steps as in figure 5.

Figure 5. MRDTP feature extraction procedure for 2D images

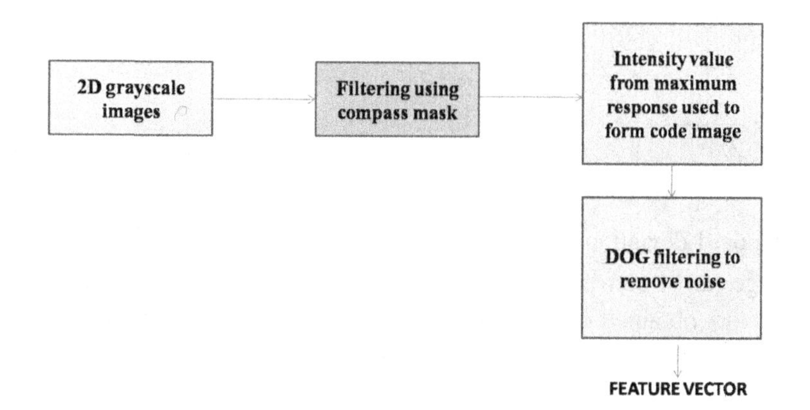

Figure 6. Kirsch mask

$$\begin{bmatrix}-3 & -3 & 5\\-3 & 0 & 5\\-3 & -3 & 5\end{bmatrix}\begin{bmatrix}-3 & 5 & 5\\-3 & 0 & 5\\-3 & -3 & -3\end{bmatrix}\begin{bmatrix}5 & 5 & 5\\-3 & 0 & -3\\-3 & -3 & -3\end{bmatrix}\begin{bmatrix}5 & 5 & -3\\5 & 0 & -3\\-3 & -3 & -3\end{bmatrix}\begin{bmatrix}5 & -3 & -3\\5 & 0 & -3\\5 & -3 & -3\end{bmatrix}$$

K_1 K_2 K_3 K_4 K_5

$$\begin{bmatrix}-3 & -3 & -3\\5 & 0 & -3\\5 & 5 & -3\end{bmatrix}\begin{bmatrix}-3 & -3 & -3\\-3 & 0 & -3\\5 & 5 & 5\end{bmatrix}\begin{bmatrix}-3 & -3 & -3\\-3 & 0 & 5\\-3 & 5 & 5\end{bmatrix}$$

K_6 K_7 K_8

The 2D images that are obtained are convoluted with Kirsch masks. Then from the obtained responses, edges are used to create the feature vectors. The Kirsch masks find out the edges within the 2D images using the non-linear process. The Kirsch compass masks $\{K_1, K_2 \dots K_8\}$, are based on the North, South, North East, South East, South West, North West, East and West directions (Kirsch, R. A. 1971). After the convolution with the Kirsch masks eight responses namely, $\{I_1, I_2 \dots I_8\}$ are obtained. The rotation using angle 45° is done for creating the eight masks as in figure 6.

When the eight directional masks are applied to images, eight different filtered responses are produced. But if the feature vectors are produced using all the responses, it will result in a high dimensional feature vector. Therefore, in MRDTP only the maximum response value is taken, which includes only the high-frequency information or only the most prominent edges. $R(x,y)$ gives the resultant information after considering only the prominent responses among all the eight responses as in eq. (2).

$$R(x,y) = \max\left(I_j(x,y) \mid 1 \le j \le 8\right) \tag{2}$$

Here, $R_j(x,y)$ is the response obtained at a position (x,y) when convoluting with a mask $K_{j,}$ $1 \le j \le 8$. Also, the resultant information creates a code image. Even though the prominent edges are considered, the code image will be still having some random noise. The Difference of Gaussian (DoG) filter is created by the subtraction of the Gaussian filter with low variance from a Gaussian filter with high variance. When this DOG filter is convoluted with the code image it removes all the noise acting like a band pass filter as in eq. (3).

$$DoG\left((x,y);w,\sigma\right) = \frac{1}{2\pi\sigma_a^{\ 2}} e^{-\frac{x^2+y^2}{2\sigma_a^{\ 2}}} - \frac{1}{2\pi\sigma_b^{\ 2}} e^{-\frac{x^2+y^2}{2\sigma_b^{\ 2}}} \tag{3}$$

where σ_a is the standard deviation used in DoG that is higher than σ_b. The micro-patterns existing within this code image can be converted to feature vector after dividing into grids and histogram formation. The feature vectors obtained can be further reduced in dimension using PCA (Jolliffe, I. 2011). Then, by using a machine learning technique, the type of fault within the motor can be predicted.

Classification Technique for Predicting the Four Types of Faults

Some of the machine learning techniques used for classification purpose is Relevance Vector Machine (RVM) (Tipping, M. 2003), AdaBoost (Adaptive Boosting) (Viola, P., & Jones, M. 2002) and Support Vector Machine (SVM) (Hearst, M. A et al, 1998). RVM is based on Bayesian inference and probability-based classification. RVM does not need the parameters for optimization. AdaBoost uses a group of machine learning algorithms. It is very sensitive to noise. SVM is a supervised machine learning algorithm, and it is the most commonly used one for facial expression recognition. It can be used for both binaries as well as multi-label classification. It can also be used for regression. It takes much time for training and testing. Deep learning (LeCun, 2015) requires lots of training data and also consumes much amount of time. ELM (Huang & Siew 2005, Huang, 2012) has very high speed compared to SVM. It is capable of the multi-classification of the faults and regression. In this design, ELM with the Supervised stack of Auto-encoders (Tissera & McDonnell 2016) is suggested because of its speed and accuracy as the AOCMSE system has to be implemented in real-time environments. Conventional methods identify the defect frequencies based upon the shaft speed that causes unavoidable changes in the defect frequencies. The analysis of these frequencies can be done by the analysis of the particular vibration signals that consumes huge amount of time and cost. In the proposed approach data is grouped based on the speed of the shaft for fault diagnosis.

The proposed method converts the time domain-based vibration signals into 2D images and then classifies them using ELM classifier based upon the textures and shape of the signal in the images. The histogram information from the code images created by MRDTP is used by the classifier to classify the type of fault by detecting the micro-patterns and the frequency distribution of the micro-patterns are represented using histograms. The histogram information helps to classify an image. The micro-patterns can be an edge, a line, a corner, or a flat area in the image according to the observed vibration signal. The micro-patterns in the 2D images created by the vibration signal vary according to the type of fault. The MRDTP is very useful in identifying the micro-patterns and eliminating the noise as per the article by Alphonse et al. (2018). Thus, the MRDTP identifies the micro-patterns that help to predict the type of fault using ELM classifier. In this article the proposed work predicts four types of faults as well as the no fault condition. The shaft speed changes the texture of images. One cycle of the signal creates one line of pixels in the image. The increase in number of cycles and the type of fault creates a distinct pattern in the image. After predicting the type of fault using the proposed method, the information is sent to the maintenance team through IoT. After predicting the type of fault, the information is sent to the maintenance team through IoT.

NEEDED RESOURCES FOR AOCMSE

Four MEMS sensors need to be placed within the traction motor that has the following configuration as in Table 1.

These sensors capture the vibrations within the motor. The sensor collects the vibrations depending upon the acceleration rate. The sampling rate of vibrations, as well as the sleeping state period and active state period of the sensors, are decided by the sensor management server. A gateway component is needed with the Raspberry Pi 2 that groups all the vibration signals within a certain time period as scheduled by the sensor management sensor. Also, an analysis component is needed by the Raspberry

Table 1. Details of sensor

Details	MEMS Sensor
Power	3mW
Noise density	4000μg
Size	0.2x0.2x0.05 inch
Accelerate range	100 g
Resonance frequency	22kHZ

Pi 2 that has the storage to store the sensor data, a subcomponent that analyzes the data and can generate reports. A web-based application is also needed for sending the reports to the maintenance team. All the communications are controlled using TCP/IP as in figure 2. Some battery or power source must be needed for this AOCMSE system to operate. Non-wired sources like solar cells can also be used. Wearable IoT devices, GPIO pins on the Raspberry Pi supply 3.3V or 5V are also needed.

Processors are required for processing the software and memory is needed for storing the data. Depending upon the communication mode chosen by the maintenance team with the device cellular networks, RFID, Bluetooth, or Low Power wide area network (LPWAN) technologies like SigFox LoRa, or NB-IoT, Universal Asynchronous Receiver Transmitter (UART), Ethernet, Controller Area Network (CAN) protocol, zigbee protocols or wifi can be used. The user interface will be needed by the maintenance team. Output devices like LEDs, speakers, actuators like motors, screens, pneumatic linear actuators, solenoids, and wireless sensor networks are also needed. Raspberry pi has the powerful processor, it is small and portable. Its raspbian OS can develop a program using python language. Raspberry pi 2 has SoC: Broadcom BCM2836 (CPU, GPU, DSP, SDRAM), CPU: 900 MHz quad-core ARM Cortex A7 (ARMv7 instruction set),GPU: Broadcom Video Core IV @ 250 MHz.

Raspberry is easily compatible with all IOT applications. MySQL server can be used as a local database. Cloud storage can be used in case of large data. Python supports both the hardware and software implementations with Raspberry Pi. Python implementations perform well for Image processing, Artificial intelligence and Machine Learning.

EXPECTED BENEFITS

This design is expected to result in the following benefits:

- Fast performance
- Time and cost-effective
- Better safety in elevators
- Fully automated and continuous monitoring of elevators
- Scaling of the system to many elevators
- As the best techniques are used in the various processing steps of the design, this will result in the best accuracy, which is very essential in a real-time application
- Periodic alerts to the maintenance team

CONCLUSION

In this chapter, the different concepts, methodologies and techniques that have been used in literature have been discussed. This chapter is useful for the future researchers to get a good insight into the elevator monitoring system. Also, this chapter proposes a unique design for the condition monitoring of the traction motor within the elevators which plays a vital role in the safe operation of the elevators. This design ensures a safety monitoring at real time. This system uses four sensors within the elevator motor for sensing the vibrations. These vibrations are analyzed and reports are sent to the maintenance teams. This saves the time and money of the maintenance operators. Also, as reports are sent even when the elevator is running, it is ensured that the elevator is under continuous monitoring using IoT. This proposed design uses image- based feature extraction technique that uses MRDTP which has low computational complexity and good accuracy when compared to other feature extraction techniques in the literature. This design proposes to use ELM, which has also achieved high accuracy at less amount of time as seen in various works of literature. The usage of ELM is very ideal to be used in real-time applications. This work can also be extended and scaled up for monitoring more than one elevator in a smart city.

REFERENCES

Alphonse, A. S., & Dharma, D. (2018). Novel directional patterns and a Generalized Supervised Dimension Reduction System (GSDRS) for facial emotion recognition. *Multimedia Tools and Applications*, *77*(8), 9455–9488. doi:10.100711042-017-5141-8

Caesarendra, W., & Tjahjowidodo, T. (2017). A review of feature extraction methods in vibration-based condition monitoring and its application for degradation trend estimation of low-speed slew bearing. *Machines*, *5*(4), 21. doi:10.3390/machines5040021

Cho, Y. W., Kim, J. M., & Park, Y. Y. (2016). Design and Implementation of Marine Elevator Safety Monitoring System based on Machine Learning. *Indian Journal of Science and Technology*, *9*(S1).

Esteban, E., Salgado, O., Iturrospe, A., & Isasa, I. (2016). Model-based approach for elevator performance estimation. *Mechanical Systems and Signal Processing*, *68*, 125–137. doi:10.1016/j.ymssp.2015.07.005

Flores, A. Q., Carvalho, J. B., & Cardoso, A. J. M. (2008, September). Mechanical fault detection in an elevator by remote monitoring. In *Electrical Machines, 2008. ICEM 2008. 18th International Conference on* (pp. 1-5). IEEE. 10.1109/ICELMACH.2008.4800064

Fluke corporation. (2018). *An introduction to machinery vibration*. Retrieved from https://www.reliableplant.com/Read/24117/introduction-machinery-vibration

Giagopoulos, D., Chatziparasidis, I., & Sapidis, N. S. (2018). Dynamic and structural integrity analysis of a complete elevator system through a Mixed Computational-Experimental Finite Element Methodology. *Engineering Structures*, *160*, 473–487. doi:10.1016/j.engstruct.2018.01.018

Göksenli, A., & Eryürek, I. B. (2009). Failure analysis of an elevator drive shaft. *Engineering Failure Analysis*, *16*(4), 1011–1019. doi:10.1016/j.engfailanal.2008.05.014

Hearst, M. A., Dumais, S. T., Osuna, E., Platt, J., & Scholkopf, B. (1998). Support vector machines. *IEEE Intelligent Systems & their Applications*, *13*(4), 18–28. doi:10.1109/5254.708428

Heather, K. (2010). *This elevator could shape the cities of future.* Retrieved from https://money.cnn.com/2016/05/03/technology/maglev-elevator-smart-city/index.html

Huang, G. B., & Siew, C. K. (2005). Extreme learning machine with randomly assigned RBF kernels. *International Journal of Information Technology*, *11*(1), 16–24.

Huang, G. B., Zhou, H., Ding, X., & Zhang, R. (2012). Extreme learning machine for regression and multiclass classification. *IEEE Transactions on Systems, Man, and Cybernetics. Part B, Cybernetics*, *42*(2), 513–529. doi:10.1109/TSMCB.2011.2168604 PMID:21984515

Huang, Y., & Yu, W. (2016). Elevator Safety Monitoring and Early Warning System Based on Directional antenna transmission technology. *Electronics (Basel)*, *19*(2), 101–104.

Jolliffe, I. (2011). Principal component analysis. In *International encyclopedia of statistical science* (pp. 1094–1096). Berlin: Springer. doi:10.1007/978-3-642-04898-2_455

Jung, D., Zhang, Z., & Winslett, M. (2017, April). Vibration Analysis for IoT Enabled Predictive Maintenance. In *Data Engineering (ICDE), 2017 IEEE 33rd International Conference on* (pp. 1271-1282). IEEE. 10.1109/ICDE.2017.170

Khan, S. A., & Kim, J. M. (2016). Rotational speed invariant fault diagnosis in bearings using vibration signal imaging and local binary patterns. *The Journal of the Acoustical Society of America*, *139*(4), EL100–EL104. doi:10.1121/1.4945818 PMID:27106344

Kirsch, R. A. (1971). Computer determination of the constituent structure of biological images. *Computers and Biomedical Research, an International Journal*, *4*(3), 315–328. doi:10.1016/0010-4809(71)90034-6 PMID:5562571

LeCun, Y., Bengio, Y., & Hinton, G. (2015). Deep learning. *Nature, 521*(7553), 436.

Liu, X. K., Chen, Y., & Yu, H. N. (2014). Research on web-based elevator failure remote monitoring system. *Applied Mechanics and Materials*, *494*, 797–800. doi:10.4028/www.scientific.net/AMM.494-495.797

Nick, A. (2010). *Smart Elevators Bring You There Faster & More Efficiently*. Retrieved from https://www.triplepundit.com/2010/04/smart-elevators-bring-you-there-faster-more-efficiently

Olalere, I. O., Dewa, M., & Nleya, B. (2018). Remote Condition Monitoring of Elevator's Vibration and Acoustics Parameters for Optimised Maintenance Using IoT Technology. In *IEEE Canadian Conference on Electrical & Computer Engineering* (pp. 1-4). IEEE. 10.1109/CCECE.2018.8447771

Sahoo, S., & Das, J. K. (2018). Bearing Fault Detection and Classification Using ANC-Based Filtered Vibration Signal. In *International Conference on Communications and Cyber Physical Engineering* (pp. 325-334). Academic Press.

Shaheryar, A., Xu-Cheng, Y., & Ramay, W. Y. (2017). Robust Feature Extraction on Vibration Data under Deep-Learning Framework: An Application for Fault Identification in Rotary Machines. *International Journal of Computers and Applications*, *167*(4).

Shi, D., & Xu, B. (2018, June). Intelligent elevator control and safety monitoring system. *IOP Conference Series. Materials Science and Engineering*, *366*(1), 012076. doi:10.1088/1757-899X/366/1/012076

Suárez, A. D., Parra, O. J. S., & Forero, J. H. D. (2018). Design of an Elevator Monitoring Application using Internet of Things. *International Journal of Applied Engineering Research, 13*(6), 4195–4202.

Tipping, M. (2003). *U.S. Patent No. 6,633,857*. Washington, DC: U.S. Patent and Trademark Office.

Tissera, M. D., & McDonnell, M. D. (2016). Deep extreme learning machines: Supervised autoencoding architecture for classification. *Neurocomputing*, *174*, 42–49. doi:10.1016/j.neucom.2015.03.110

Uddin, J., Kang, M., Nguyen, D. V., & Kim, J. M. (2014). Reliable fault classification of induction motors using texture feature extraction and a multiclass support vector machine. *Mathematical Problems in Engineering*.

Uusitalo, J. (2018). *Novel Sensor Solutions with Applications to Monitoring of Elevator Systems* (Master of Science thesis).

Viola, P., & Jones, M. (2002). Fast and robust classification using asymmetric adaboost and a detector cascade. In Advances in neural information processing systems (pp. 1311-1318). Academic Press.

Wan, Z., Yi, S., Li, K., Tao, R., Gou, M., Li, X., & Guo, S. (2015). Diagnosis of elevator faults with LS-SVM based on optimization by K-CV. *Journal of Electrical and Computer Engineering*, *2015*, 70. doi:10.1155/2015/935038

Wang, X., Ge, H., Zhang, W., & Li, Y. (2015, September). Design of elevator running parameters remote monitoring system based on Internet of Things. In *Software Engineering and Service Science (ICSESS), 2015 6th IEEE International Conference on* (pp. 549-555). IEEE. 10.1109/ICSESS.2015.7339118

Wowk, V. (2005). *A Brief Tutorial on Machine Vibration*. Machine Dynamics, Inc.

Yang, D. H., Kim, K. Y., Kwak, M. K., & Lee, S. (2017). Dynamic modeling and experiments on the coupled vibrations of building and elevator ropes. *Journal of Sound and Vibration*, *390*, 164–191. doi:10.1016/j.jsv.2016.10.045

Yao, Z., Wan, J., Li, X., Shi, L., & Qian, J. (2011). The Design of Elevator Failure Monitoring System. In *Advances in Automation and Robotics* (Vol. 1, pp. 437–442). Berlin: Springer. doi:10.1007/978-3-642-25553-3_54

Yi, X. (2016, December). Design of Elevator Monitoring Platform on Big Data. In *Industrial Informatics-Computing Technology, Intelligent Technology, Industrial Information Integration (ICIICII), 2016 International Conference on* (pp. 40-43). IEEE. 10.1109/ICIICII.2016.0021

Zhang, Y., Sun, X., Zhao, X., & Su, W. (2018). Elevator ride comfort monitoring and evaluation using smartphones. *Mechanical Systems and Signal Processing*, *105*, 377–390. doi:10.1016/j.ymssp.2017.12.005

Zhou, Y., Wang, K., & Liu, H. (2018). An Elevator Monitoring System Based On The Internet Of Things. *Procedia Computer Science*, *131*, 541–544. doi:10.1016/j.procs.2018.04.262

Chapter 13
An IoT-Based Earthquake Warning System for Smart Cities

Suja Priyadharsini S.
Anna University Chennai – Regional Office Tirunelveli, India

Ramalakshmi S.
Anna University Chennai – Regional Office Tirunelveli, India

ABSTRACT

Earthquakes are the most common natural disasters that occur in India. An earthquake warning system minimizes damage and saves countless lives. A seismic wave analysis helps develop an early warning system. The bigger the earthquake, the stronger the shaking. Hence, magnitude determination is critical to developing an earthquake early warning system. The chapter deals with detecting earthquake magnitude by identifying the individual magnitude of earthquakes. An early warning system can be effectively implemented by the proposed method, along with high-end processors and the IoT (internet of things), which has the ability to collect and transfer data over networks with no manual intrusion. The proposed early earthquake warning (EEW) system can be used to support the development of smart cities so earthquake-prone zones are made less susceptible to disaster.

INTRODUCTION

Natural disasters are sudden events caused by environmental factors that threaten lives, property and the ecosystem. Every year, they kill and maim people and damage property. Earthquakes and floods strike anywhere on earth, often without warning. In India, earthquakes are commonly occurring natural disasters. They are among the most deadly natural hazards and frequently cause surface faulting, tremors, liquefaction, landslides, aftershocks and/or tsunamis. Nearly59% of India's landmass is subject to moderate and severe earthquakes that strike without warning. Earthquakes in India are caused by the movement of the tectonic plates. A tectonic plate is a massive, irregularly shaped slab of solid rock (WHO – Technical Hazard Sheet, 2018)http://www.who.int/hac/techguidance/ems/earthquakes/en/.

DOI: 10.4018/978-1-5225-9199-3.ch013

The internal structure of the Earth has three major layers:

- Crust
- Mantle
- Core

The outer layer of the Earth is called the crust/lithosphere, which is very thin when compared to the other layers, and is broken into several large pieces called plates. The mantle consists of the upper mantle and the lower mantle. The plates of the Earth move because of the movement of the mantle. The center of the Earth is its core, which is nearly twice as dense as the mantle. The pattern of earthquake occurrences reveals that earthquake activity has occurred in a number of different earthquake belts (Stein &Wysession, 2003).

Earthquakes are happenings experienced during sudden movements of the Earth's crust. The upper part of the mantle, called the asthenosphere, is composed of liquid rock and lies under the Earth's crust. The plates of the Earth's crust which "float" on top of the asthenosphere layer are forced to shift as the outpouring molten material below moves. A huge amount of energy is released in the form of waves as these plates shift and interacts with each other. Though earthquakes can take place anywhere on the planet with little or no warning, a large number of severe earthquakes occur near plate boundaries, as the plates converge (collide), diverge (move away from another), or shear (grind past one another).Earthquakes can also be triggered by means of moving rock and magma within volcanoes. During the collision, divergence and shearing of plates, large sections of the crust can fracture and move back and forth to disperse the energy released. This "shaking" is felt as a sensation during an earthquake (Mooney, 2002).

Figure 1 shows the earthquake belt, which is a narrow zone on the earth's surface around which a majority of earthquakes take place.

Figure 2 shows the nine major plates: North American, Pacific, Eurasian, African, Indo-Australian, Australian, Indian, South American, and Antarctic. The Pacific Plate is the largest at 39,768,522 square miles. The Whole of India lies on the Indian Plate, whose relative motion is at a velocity of 5cms per year in the north-northeast direction (Srivastava, 2015).

Figure 1. Major Seismic Belts of the Earth
(Source: Geology Universe)

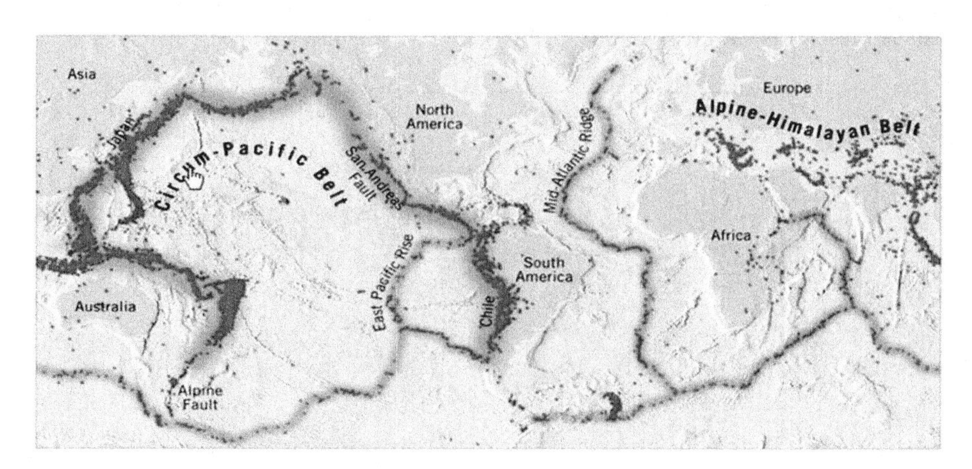

At the subduction zones, two tectonic plates meet and one slides beneath the other, curving into the mantle (the layer underneath the crust). The plates move in different directions and strike against their neighbors at boundaries. The huge forces thus generated at plate boundaries cause volcanic eruptions and earthquakes. An earthquake is a vibration of the Earth's surface caused by waves coming from a source of disturbance inside the earth (Oskin, 2015).

As per the Ministry of Earth Sciences Report, the country's earthquake-prone areas have been recognized on the basis of scientific inputs relating to seismicity, the number of earthquakes that occurred in the past, and the tectonic setup of the region. Based on these inputs, the Bureau of Indian Standards [IS 1893 (Part I):2002], has grouped the country into four seismic zones, viz. Zones-II, -III, -IV and –V as shown in Figure 2. Of these, Zone V is seismically the most active region, while Zone II is the least. Zone V comprises the entire northeast of India, parts of Jammu and Kashmir, Himachal Pradesh, Uttaranchal, the Rann of Kutch in Gujarat, parts of North Bihar, and the Andaman & Nicobar Islands. Zone IV comprises the remaining parts of Jammu and Kashmir, Himachal Pradesh, the National Capital Territory (NCT) of Delhi, Sikkim, the northern Parts of Uttar Pradesh, Bihar and West Bengal, parts of Gujarat, small portions of Maharashtra near the west coast, and Rajasthan. Zone III comprises Kerala, Goa, the Lakshadweep Islands, the remainder of Uttar Pradesh, Gujarat, West Bengal, parts of Punjab, Rajasthan, Madhya Pradesh, Bihar, Jharkhand, Chhattisgarh, Maharashtra, Orissa, Andhra Pradesh, Tamil Nadu and Karnataka. Zone II covers the rest of the country.

An earthquake is a noticeable shaking of the Earth's surface, resulting from a sudden release of energy in the Earth's crust that creates seismic waves. The pattern of earthquake occurrences around the world reveals the difficulty involved in predicting the next quake strike in terms of the time of its occurrence (Williams, 2016).

Apart from earthquakes caused by the rubbing of tectonic plates, human activity is a contributory factor, and includes the injection of fluids into deep wells for waste disposal, the secondary recovery of oil, and the use of reservoirs for water supplies (Buchanan, 2014).

Geologists have discovered that injecting waste fluids into the earth's crust could trigger seismic activity, with the constant drilling and pumping in the earth's crust culminating in tectonic action. Natural gas fields, especially hydrocarbon recovery, play a major role in manmade earthquakes (Buchanan, 2014).

Figure 2. Indian Seismic Zoning Map
(Source: National Institute of Disaster Management)

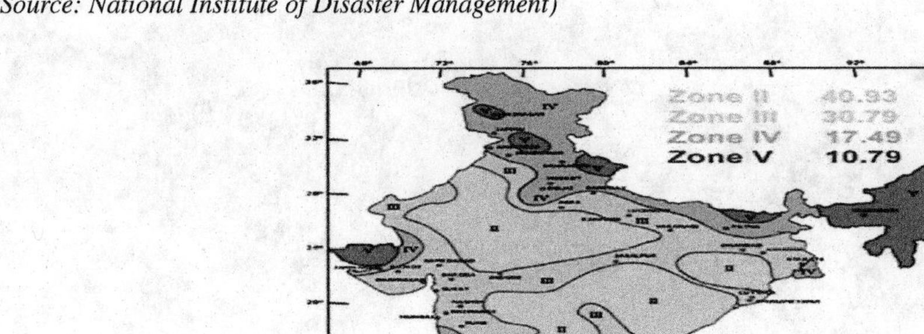

An extensive analysis was carried out by an earthquake research team from the Institute of Biophysics, China, to analyze animal behavior before the occurrence of a quake (Srivastava, 2015). The underlying reason behind the analysis is the fact that animals have a unique consciousness that humans do not, and can sense imperceptible changes in their surroundings.

- Prior to an earthquake hit, animals and birds seem unsettled.
- Fish, likewise, have an exclusive sensory receptor that can detect variations in water pressure. (Lakshmi et al., 2014).

Given the movement and rubbing of tectonic plates, an enormous amount of energy is released in the form of elastic waves (Regmi, 2017). These elastic waves are vibrations that originate from the point of initiation of the rupture and propagate in all directions through the rocks. These waves are said to be seismic waves or earthquake waves. Two types of waves, body waves and surface waves are released as a result of the elastic stress waves generated by a quake. (Erika, 2012).

As the name implies, body waves are waves that can travel through the Earth's inner layer. Body waves are faster than surface waves and reach the seismic station first. Body waves are broadly classified into primary (P) waves and secondary (S) waves (Erika, 2012).

When an earthquake occurs, body waves are the first to reach the seismic station and hence the first to be recorded. They are also called push-pull or longitudinal or compressional waves because they are propagated by the alternative compression and dilation of the ground in the direction of the propagation. These waves are fast and capable of travelling through solid, liquid and gas. Primary waves travel at a speed of 8-13 kms/second. They help detect the onset of an earthquake before the most destructive S-waves arrive. (Erika, 2012).

Secondary waves are relatively slower than primary waves and are capable of travelling only through a solid medium at a speedof 5-7 kms/second. They are also called transverse or shear waves, where the ground is displaced perpendicular to the direction of the propagation. They are severe waves which release huge masses of energy and are responsible for the destruction associated with an earthquake. Most Earthquake Early Warning (EEW) systems are operated by calculating the arrival times of P- and S-waves (Erika, 2012).

The question of "when" an earthquake will occur has not yet been answered. Research in this area is ongoing, and earthquake prediction is a challenge. Despite research on the miscellaneous facets of earthquakes, researchers have failed to turn out an efficient model for quake prediction (Wiemer, 2015).

In the operation of a seismic early warning system, rapid and accurate determination of the potential for damage from an earthquake is a critical component (Kanamori, 2005). Earthquake Early Warning (EEW) systems provide a few much-needed seconds, to as much as a few tens of seconds, prior to the onset of damaging ground motion and thus mitigate its effects(Lockman, 2005).The strength and damage of an earthquake depends upon its magnitude, location, intensity, and the frequency content of ground motions (Kanamori, 2005). Consequently, estimating the magnitude and source location of a quake is crucial in EEW systems and their efficiency depends on the epicentral distance (Ochoa et al., 2017).

Recently, Indonesia was struck by a powerful 7.5 magnitude earthquake on September 28, 2018. The tremor was centered 78 kilometres north of the city of Palu, the capital of the Central Sulawesi province. The severe earthquake struck Central Sulawesi Island at a shallow depth of some 10 kilometres (six miles). According to the U.S Geological Survey, this particular area of Indonesia has seen about 15 earthquakes with magnitudes larger than 6.5 over the last century. The powerful earthquake caused a tsunami that killed at least 844 people. An alert was declared shortly after 6 p.m., warning of potential waves of up to three meters, but cancelled at 6.36 p.m. However, the agency said that the alert had been raised only after the tsunami hit. Indonesia's warning system consists of seismographic sensors, buoys, tidal gauges and the GPS, and was finished in 2008. However, it was in no way able to predict the scale of the tsunami that occurred, which reached a height of 20 feetin the city of Palu, south of the earthquake's epicenter (Singhvi, 2018).

A tsunami warning system consists of seismographic sensors designed to record earthquakes, and their strength and location. The National Meteorology and Geophysics Agency issuesa warning, based on their severity, using text messages and sirens (Singhvi, 2018).

The design and configuration of a seismic alert system has become the need of the hour to minimize losses brought on as a result of a major quake (Lockman, 2005).

The efficiency of an EEW system depends upon how fast it processes seismic signals. An earthquake excites P- and S-waves, and the time required to provide an early warning depends on the time difference between the primary and secondary waves. S waves carry major destructive energy. The performance of an EEW system will be greatly undermined if the focus is on processing secondary waves. Such an analysis takes time, and a warning issued thereafter is unlikely to reach the respective community in time. The P-wave, which is the primary wave, is faster than the S wave and reaches the seismic station first (Kanamori, 2005). Separating P- and S-waves from the seismic signals is complicated. Commonly, a polarization analysis is applied to filter P- and S-waves and determine the onset time of P-waves (Wang et al., 2017). The polarization analysis includes certain laborious procedures and calls for specialized knowledge of polarization parameters such as the rectilinearity degree and the angle of incidence. Hence, the proposed work concentrates on the seismic record that comprises only P-wave information, eliminating the need for filtering by means of a polarization analysis. Primary wave records are adequate enough to classify earthquake magnitudes for an early alert system. An IoT- based technologies are to be incorporated to issue an efficient early alarm to the respective areas through mobile operators, communication service and regional headquarters for the support of smart city development.

A REVIEW OF LITERATURE

Odaka et al. (2003) proposed a method for quickly estimating the epicentral distance and magnitude of an earthquake from a single seismic record. The core of early earthquake detection and warning systems is the estimation of the epicentral distance and magnitude. The authors introduced a method for calculating the epicentral distance from a single seismic record. The empirical magnitude – amplitude relation was employed to determine the magnitude within a given short time interval after the P-wave's arrival. The amplitude at the beginning of the initial P-wave motion is usually quite small when compared with the later maximum amplitudes of P-and S-waves. The zero shift / DC components were removed from the data by subtracting their mean value, following which the logarithm of the absolute values was considered and the ground motion records displayed in the logarithmic scale. This method introduced a function with the form Bt.exp(-At), where A and B are unknown parameters, determined in terms of the least square method from the initial part of the waveform envelope. The function indicated that the logarithm of B is inversely proportional to the logarithm of Δ, where Δ represented the epicentral distance. Also, the parameter B was used to determine the earthquake magnitude and maximum amplitude of the signal within the short span of the time window. An empirical formula was implemented by the researchers for determining the magnitude - that is, $M = a \log S_{max} + b \log \Delta + c$, where S_{max} is the maximum resultant amplitude of the seismic signals, a,b,c – constants. The researchers inferred that the slope, indicated by the parameter *B,* depends chiefly on the epicentral distance and not on the magnitude of the earthquake.

Roueff et al. (2006) propounded estimating the polarization parameters of seismic waves in order to separate the polarized waves. The separation was done by applying an oblique polarization filter (OPF) in a time-frequency representation of the data. The OPF enhanced the wave separation and filtered the waves by applying the phase shift and rotation parameter operators. The problem with this method is that it requires knowledge of the polarization parameters of each wave, as well as a time-frequency representation of seismic waves.

Benbrahim et al (2007) discussed the discrimination of seismic signals, given that each wave holds a lot of information. This paper concentrated on two classes of seismic signals. To discriminate between seismic classes, the work involved procedures such as representing the seismic phases, data dimensionality reduction, and classifying seismic data. This work used the modified Mexican har wavelet for the representation process. A combination of random projection and principle component analysis was used for dimensionality reduction, based on the result of the time-frequency representation. The classification of seismic phases was done using a multilayer perceptron network. The limitations of this study are that the algorithm applied for dimensionality reduction is complex, and building a network for classification is hard.

Kanamori& Wu (2008) attempted to determine the strength of an earthquake from the initial P-wave by using a ground motion period parameter and a high-pass filtered vertical displacement amplitude parameter. Here, incoming strong motion acceleration signals were converted to ground velocity and displacement. When a trigger occurred, the parameters were computed. The magnitude and onsite ground motion intensity were estimated from the ground motion period and the peak ground motion velocity was estimated from the peak displacement. Information from the initial part of the P-wave is used to provide a warning.

Kurzon et al. (2014) implemented a method for the detection of P and S phases using singular value decomposition (SVD) analysis. The analysis employed three component seismic records based on a real-time iteration algorithm. The SVD algorithm was applied to the three-component data to separate the P and S phases and facilitate detection. For the detection process, the algorithm characterized both post-processing detectors and real-time detectors. The parameters used for the detectors were singular values, their time derivatives, and the cosine incidence angle. This method processed the filtered data, and the P and S phases were separated and detected by analyzing the parameters mentioned. The detectors were applied to the records, and they performed three times better than traditional detectors. The drawback of the system is that it cannot be applied to real-time records, and can only be used to preprocess known events that include S-waves.

Alphonsa A., & Ravi G. (2016) proposed an IoT based earthquake warning system using Wireless Sensor Networks. The wireless sensor network is a group of sensors which are spatially dispersed to monitor physical environmental conditions. The sensors positioned in the earth's surface to record seismic waves that radiates outward from the epicenter. In this work, vibration sensors were used to detect the ground shakes. The seismic signals which were recorded by sensors were then transferred to the gateway through ZIGBEE transmitter. The gateway consists of ZIGBEE receiver which receives the signal from sensors. Then the signals are processed by using LABVIEW software, as it provides an interface between IoT cloud and alert signal. The alert messages which include time, location and other parameters are passed to public through the smart phones by means of IoT. In addition, a GSM module is also included in the system to provide an alarm message to the people who do not have smart phones. The GSM module sends alert message to the nearby base station and from the base station, the message can be reached to the authorized numbers.

Bhardwaj et al. (2016) developed a multiple parameter-based earthquake early warning (EEW) system by considering parameters such as the maximum predominant period, average period, peak displacement, cumulative absolute velocity, and root sum of squares cumulative velocity (RSSCV). A combination of those parameters was used to develop an algorithm that issued warnings. Each parameter performs a unique task, and a combination of those pave the way for issuing alarm by setting the threshold. The RSSCV parameter was used to finds the onset time of the P-wave while other parameters ascertained the magnitude at different time windows. Different combinations of the above said EEW parameters have been tested in the design of an accurate warning system. The result reveals that the multi-parameter-based EEW system performs better than individual parametric-based systems.

Zambrano et al. (2016) developed an earthquake early warning system (EEWS) based on IoT (Internet of Things) technology. It implemented a three-layered hierarchical architecture to obtain real-time seismic information. The system used IoT technologies like the Sensor Web Enabled framework (SWE) and the Message Queue Telemetry Transport (MQTT), supported by a wireless sensor network, to make it capable of delivering an effective warning during major disasters. The observed seismic information was processed by an intermediate server where the seismic event was detected. The detected events were passed to the control center where an actual warning was distributed to aid centers through the MQTT. This Early Warning System provides a warning, through smart phones, of up to 12 seconds during the maximum seismic peak in the epicenter.

Ochoa et al. (2017) proposed a method to determine the local magnitude of an earthquake, based on the Support Vector Machine Regression (SVMR) algorithm. In this algorithm, an exponential function of the waveform and its maximum value were considered as input regression parameters. The relationship between the maximum peak of the wave for a selected time window and the local magnitude of the earthquake was considered as well. Consecutive maximum peaks were selected and the linear regression performed by applying regression parameters to the selected peaks. Thereafter, ten-fold cross validation was applied to miscellaneous signal time window parameters.

Wang et al. (2017) developed a technique using two filters to separate the P- and S-waves of seismic phases using a sliding time window. The filters were developed by considering the degree of rectilinearityand the vertical angle of incidence of the particle motion of the waves. The arrival times of the seismic phases were picked using the short-term average to long-term average (STA/LTA) ratio by finding the characteristic function. The phase onset was determined, based on the Akaike Information Criterion (AIC), by properly windowing the seismic phases. This technique was tested on the earthquake in (Wenchuan) China for time window sizes of 0.2s, 0.5s, and 1s. The performance of the seismic phase detection technique depends on the proper selection of a seismic window size.

Babu., Naidu., & Meenakshi. (2018) proposed a scheme for earthquake alarming system based on IoT. In order to detect the ground vibrations, the work employed GY – 61 analogue sensor using NODEMCU ESP8266 board. GY – 61 DXL335 is a triple axis accelerometer which has full sensing range of +/- 3g with low noise and power consumption. MQTT protocol is used to support Node Microcontroller Unit which is an open source firmware. The Lua scripting language is used for the firmware. The sum of square variations is calculated for every 48 ms to find the earthquake. If the measure exceeds the threshold value, the channel is set to the value of 2. Otherwise the channel value is set by zero. The channel value

2 implies strong earthquake, so that the alert is generated and warning about the severity of earthquake is also provided through SMS and E-Mail.

Karaci A. (2018) developed an earthquake warning system to announce an audible alert for the people in the earthquake environment. Based on IoT applications, the warning system is developed to provide an alarm when earthquake occurs. At the time of earthquake, the vibrations of the earth are measured with the help of Inertial Measurement Unit (IMU) sensor and piezo vibration sensor. These sensors are connected on the microcontroller and tested for the threshold values. If the values of the sensors exceed the threshold then the warning system is triggered, subsequently an alarm is generated and provided through Tweeter by using Wi-Fi module. A tweet is sent to the people about the severity of the earthquake by using IoT ThinkSpeak analytics platform. The buzzer is also triggered to sound at the moment of earthquake occurs. In addition to that, GSM/GPRS module is used to send SMS about the warning.

Li (2018) developed a P-wave discriminator using a generative adversarial network and random forest algorithms. The network was trained to learn the characteristics of P-waves. It was then used as a feature extractor, and the random forest classifier trained with a number of earthquake waveforms. This classifier categorizes P-waves and noisy signals and reduces false triggers in the EEWS.

METHODOLOGY

The task of an effective earthquake warning system can be divided into three subtasks.

They are:

Task 1: Identifying the magnitude of an earthquake
Task 2: Making a decision about the warning, and
Task 3: Issuing an earthquake warning.

The proposed work concentrates on finding the magnitude of an earthquake using P-waves and applying artificial intelligence. It discusses the results obtained and offers an overview of the role of the IoT in implementing an early earthquake warning system.

Figure 3 describes the overall architecture of the proposed Earthquake Early Warning System using the IoT, which is a modern technological computing concept that embeds a slew of devices/things via the internet (Zambrano, 2016).It is able to access and control the physical world by connecting physical

Figure 3. The Earthquake Warning System using the IoT

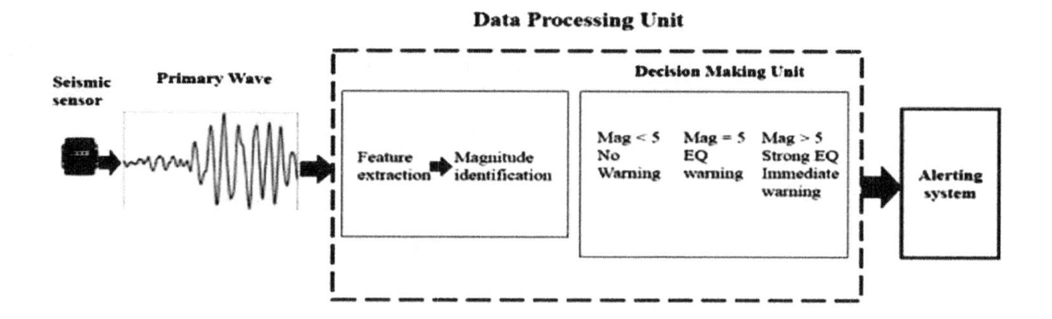

devices to the internet. The process of finding the magnitude (M) of an earthquake is vital, since the harshness of a quake depends entirely on its magnitude (Ochoa, 2017).

The performance or accuracy of the proposed system in terms of magnitude identification greatly depends on how perfectly the system classifies earthquakes, given that each magnitude has unique characteristics and distance of impact. So then, magnitude classification is an important task. The proposed system categorizes the magnitude of earthquakes using machine learning (ML) techniques.

The dataset for analyzing earthquake sizes is obtained from the IRIS (Incorporated Research Institutions for Seismology) website. The dataset consists of earthquake signals of magnitudes between 2 and 9 on the Richter scale, and includes quakes that occurred in the various countries highlighted in Figure 4.

In the proposed work, time series data recorded by the three-component seismograph is employed for the analysis. The seismograph includes three separate elements which record the movements of seismic waves in three different directions: up-down, north-south and east-west.

P-waves travel upward and tremble the ground in the direction in which they move, so they have a strong vertical component. S-waves are shear waves which tremor the ground in the direction perpendicular to their movement, and therefore their movements are recorded by the east-west component. Surface waves shake the ground back and forth, and up and down, and their movements are recorded by the north-west component of the seismogram (Earthquakes – British Geological Survey).

In the proposed system, seismic waves recorded by the vertical component of the seismogram, which contain maximum information about the primary waves, are used for the analysis. The P-wave is considered because it is the fastest wave and carries too little energy to cause devastation (Erika, 2012).

The severity of the shaking is based on the energy carried by S-waves. The S-wave is the slowest wave, and the time gap between P- and S-waves ranges from anything between a few seconds to a few minutes. Since the seismogram records P-waves first, the processing system gets the information earlier, making it easier for the system to process the data faster and provide an alarm.

Figure 4. Countries considered for analysis in the IRIS Earthquake Dataset

Feature Extraction

After obtaining the time series seismic data, features are extracted from input signals. Features are parameters which define the characteristics of seismic waves.

The following are temporal features extracted from the signals for classification:

Mean

It is the average value of a signal, calculated by summing all the (N) signal samples ranging from 0 to N-1 and dividing the sum by N.

$$Mean,= {}^{1}\!/_{N} \sum_{i=0}^{N-1} xi \tag{1}$$

where

x_i represents the ith sample and N the total number of samples.

Mean Square Error (MSE)

It is defined as the mean of the square of the difference between the estimated and actual value of the signal, and is a performance measure used for the reconstruction process (Elder, 2006).

$$Mean\,Square\,Error = \frac{1}{N}\sum_{j=1}^{N}\left(y_j - \widehat{y_j}\right)^2 \tag{2}$$

where

N represents the total number of samples, y_j the actual value, and $\widehat{y_j}$ the estimated value.

Kurtosis

It is defined as the second and fourth order moments, which is the ratio between the fourth central moment of a distribution and the fourth power of the standard deviation of the distribution (Cretu & Pop, 2008).

$$K = \frac{E[(X_i - \mu_x)]^4}{{\sigma_x}^4} \tag{3}$$

where

σ_x represents the standard deviation and μ_x the mean value.

Skewness

It is the ratio between the third moment about the mean and the third power of the normal distribution (Cretu & Pop 2008).

$$S = \frac{1}{\left(N-1\right)\sigma^3}\sum_{i=1}^{N}(X_i - \mu_x)^3 \tag{4}$$

where

μ_x represents the mean value and σ_x represents the standard deviation.

If the skewness value is negative, it indicates that the data is skewed left. For a positive value of the skewness, the data is skewed right (Liang et al., 2016).

In addition to temporal features, the proposed system also employs the spectral features described below:

Spectral Centroid

A measure which defines the spectral shape of a frequency spectrum is said to be a spectral centroid. It details spectral characteristics and is calculated as the weighted mean of the frequencies present in the signal.

The spectral centroid is the weighted average of the spectral frequency. The Fourier transform with the magnitudes is considered the weight necessary to determine the spectral centroid. The sharpness of high-frequency spectrum content is represented by the centroid. High centroid means spectra in high frequency range (Patilet al 2012)

$$SC = \frac{\sum_{k=0}^{N-1} X\left(k\right)F\left(k\right)}{\sum_{k=0}^{N-1} X\left(k\right)} \tag{5}$$

where

$X(k)$ represents the weighted frequency value and $F\left(k\right)$ the centre frequency of the bin.

Spectral Flux

It is defined as the average of the difference between the power spectra of two consecutive frames. The frame-to-frame variation in the spectral shape is measured by the spectral flux (Patil et al., 2012)

$$SF = \sum_{k=-N/2}^{\left(\frac{N}{2}\right)-1} H\left(X\left(n,k\right) - \left(X\left(n-1,k\right)\right)\right) \tag{6}$$

where

$X(n,k)$ is the kth frequency bin of the nth frame and $H(x)=\frac{x+|x|}{2}$ represents the half-wave rectifier function.

Spectral Roll-Off

The 85 percentile of the power spectral distribution is said to be a spectral roll-off. It measures the spectrum shape and produces higher values for high frequencies.

$$\sum_{n=1}^{N} M_t[n] = 0.85 \sum_{n=1}^{Rt} M_t[n] \tag{7}$$

where

M_t [n]is the magnitude of the Fourier transform at frame t and frequency bin n.

In all, 10 features are extracted in the proposed system, effectively differentiating between seismic waves in terms of magnitudes. Following the feature extraction process, the signals are subjected to the classification process using a machine learning technique.

Machine learning (ML) is a subset of artificial intelligence (AI). The focus of ML is to make computers learn automatically, identify patterns and make decisions accordingly, without human intervention. The proposed system incorporates machine learning techniques based on supervised learning algorithms.

Classification

For the classification of earthquake magnitudes, the proposed system employs a random forest classifier which builds a forest randomly, using a number of trees to handle the missing values. More number of trees in the forest gives high accuracy in the classification performance.

The random forest classifier is an ensemble-based classifier which clusters weak trees to build a strong learner for improved predictions. As the name implies, the process of root and leaf node selection is performed randomly. A random selection of features is carried out by the random forest algorithm to build a tree instead of using the entire features available (Breiman, 2001).

The random forest classifier is a bagging technique, not a boosting technique. Bagging refers to bootstrap aggregation. The random forest classifier clusters multiple estimates together and makes a prediction, depending on the majority vote from individual decision trees (Breiman, 2001). On the other hand, boosting is a procedure which improves performance by grouping weak learners (Kanamori, 2008). The selection of a root node is the first step in forest creation. The root node of a tree is any one of multiple features under consideration. Trees are then built by a random selection of features. Node-splitting is performed by evaluating the performance of the trees in a branch. The prediction is based on the voting of each decision tree (Breiman, 2001).

In the proposed system, the random forest classifier is used to find the magnitude of an earthquake using only P-waves. The P-wave signals of the earthquake, within magnitudes of M2 to M9, are given as inputs to the classifier. Training segments and features are input into the training algorithm, and the output classifier predicts the label of the new image segments. The classifier then finds the magnitude of the given seismic signal.

The next stage of the proposed system is decision making. Thesystem makes a decision to provide a warning, based on the magnitude identified by the previous classification stage.

The distance for providing an alarm for earthquake will vary for different magnitudes of the quake.

Low-magnitude (between 2 and 4) quakes are considered too low to cause major long-distance destruction, and are not felt by people. Consequently, the decision-making system decides against issuing a warning to people far from the epicenter.

If the predictive system finds an earthquake of magnitude 5 or 6, it decides to provide a warning near the epicenter. Magnitudes 5 and 6are medium earthquakes and cause short-distance destruction near the epicenter, and a warning must be issued to people closest to the epicenter.

High-magnitude, high-energy earthquakes cause long-distance destruction in terms of property and loss of human lives, and their impact on the environment is serious. Consequently, it is necessary to have a decision-making system that efficiently issues an early alarm to the entire region so as to safeguard human lives and cut losses. Based on the decision made by the decision making system, warnings will be issued to mobile operators, communication services, and regional heads through mobile and web applications.

The proposed system provides only an insight into the IoT (Internet of Things)-based earthquake warning system, and needs high-end processors and sensors to strengthen the early alarm system for major earthquakes.

RESULTS AND DISCUSSION

The development of an earthquake early warning system is a major task that warns people of impending destruction caused by an upcoming earthquake. In this work, such a system is proposed, based on the magnitude of the earthquake, which is determined by processing and analyzing seismic signals. The dataset for analysis has been obtained from the IRIS data services.

For the proposed EEW system, the data that is considered comprises all that follows one second after the onset of the P-wave The proposed work employs earthquake seismic signals of magnitudes 2 to 9 on the Richter scale. The data pertains to earthquakes that occurred in countries such as Japan, Taiwan, Mexico, India, the Fiji Islands, and Indonesia. The proposed system employs a dataset which consists of 1143seismic signals and includes 100, 104, 209, 204, 100, 217, 100, and 109 signals of magnitudes 2, 3, 4, 5, 6, 7, 8 and 9 respectively. The dataset includes earthquake signals of quakes in recent times, such as the ones in Indonesia on September 28th 2018, Iceland on 9th November 2018, and China on 10th November 2018. Seismic signals recorded by the vertical component of the seismograph are considered for processing to determine the magnitude.

The characteristics of the input signals were examined by extracting their temporal and spectral features. These features enable the classifier to categorize earthquake magnitudes effectively and enhance its magnitude classification performance. The signals, after feature extraction, were classified by the random forest classifier. The performance of the classifier is evaluated for different sets of training and testing inputs, as well as for single-test input signals with different magnitudes.

PERFORMANCE METRICS

Evaluating the performance of the classification algorithm in terms of identifying the magnitude of earthquake signals is essential.

Accordingly, the following performance measures are used

- Accuracy
- Specificity
- Sensitivity

In the proposed method, the multi-class classification of earthquake magnitudes ranging from 2 to 9 is carried out using the random forest classifier.

The performance measures of the classifier for different training and testing inputs are computed through the confusion matrix is shown in Table 1.

Classification Accuracy

The measure of the ratio between the correct predictions over the total number of samples is evaluated.

$$Accuracy = \frac{Number\ of\ correct\ predictions}{Total\ number\ of\ predicted\ instances} \tag{8}$$

$$Accuracy = \frac{TP + TN}{TP + TN + FP + FN} \times 100\% \tag{9}$$

where

- **True Positive (TP):** It represents both the actual and predicted values that are true, and is counted as the number of signals correctly predicted with respect to their magnitude. For example, a signal of magnitude 2 is classified correctly as a magnitude 2.
- **False Negative (FN):** This is an instance where the actual value is true but predicted as false. It is counted as a number of signals of a particular magnitude wrongly predicted as another magnitude signal. For instance, a signal of magnitude 2 is misclassified as a magnitude X (X represents magnitudes of 3 to 9).
- **True Negative (TN):** This is a case where the actual and predicted values are false. It is counted as a number of signals of another magnitude correctly predicted as another magnitude. For instance, a magnitude X signal is classified as a magnitude X.
- **False Positive (FP):** This is a case where the actual class is false but predicted as true. It is counted as a signal of a magnitude X that is wrongly classified as a particular magnitude signal. For instance, a signal of magnitude X is wrongly classified as a magnitude 2.

Sensitivity

Also called the True Positive Rate, it is a measure which relates to the proportion of positive data points that are correctly considered positive, with respect to all positive data points.

$$Sensitivity = \frac{TP}{TP+FN} \times 100\% \tag{10}$$

Specificity

Otherwise called the True Negative Rate, it is a measure which depicts the proportion of negative data points that are falsely considered positive with respect to all negative data points.

$$Specificity = \frac{TN}{TN+FP} \times 100\% \tag{11}$$

The performance analysis of the classifier tabulated in Table 1is computed using equations 9, 10, and 11 by substituting the values of the TP, FP, TN, and FN. Statistical parameters (TP, FP, TN, and FN) are calculated using the confusion matrix of the classifier. A sample confusion matrix for dataset 1 (247 training inputs and 896 testing inputs) is tabulated in Table 2.

Table 1. A performance analysis of the classifier for various training and testing inputs

Dataset	Number of Inputs		Performance Metrics		
	Training Inputs	Testing Inputs	Accuracy	Sensitivity	Specificity
1	247	896	98.8	95.3	99.2
2	322	821	98.71	95.11	99.21
3	536	607	98.9	96.18	99.38
4	662	481	98.59	93.95	99.14

Table 2. Confusion Matrix (Training Inputs = 247 & Testing Inputs = 896)

Magnitudes	M2	M3	M4	M5	M6	M7	M8	M9
M2	81	3	2	0	0	0	0	0
M3	3	84	1	2	0	1	0	0
M4	0	0	165	4	0	2	1	0
M5	0	0	4	160	1	1	0	0
M6	0	1	0	1	69	2	0	1
M7	0	1	0	4	0	158	0	1
M8	0	0	0	0	0	1	70	1
M9	0	0	0	0	0	2	0	71

Table 3. Values of TP, FP, FN, and TN for the 247 Training and 896 Testing inputs

Magnitudes	TP	FP	FN	TN
M2	81	3	5	777
M3	84	5	7	774
M4	165	7	7	693
M5	160	11	6	698
M6	69	1	5	789
M7	158	9	6	700
M8	70	1	2	788
M9	71	3	2	787

A sample of the statistical parameters calculated from the confusion matrix for the 247 training and 896 testing inputs is shown in Table 3.

The proposed earthquake warning system takes 5 seconds to compute seismic signal parameters and make a decision about the warning. The distribution of early warnings relies on the fact that seismic waves travel more slowly than data transmitted over a telecommunications system. Ground tremors caused by S-waves travel at an average of 2-5kms/second compared to the electromagnetic signals used in telecommunication, which travel close to the speed of light. This allows electronic warnings to reach residents before the arrival of more violent seismic vibrations, giving people extra seconds to prepare. In addition, the proposed system has had a successful run when tested with the recent data of (September2018) 7.5 earthquake struck Indonesia. The system showed its reliability in identifying its magnitude as well as those of other quakes.

FUTURE RESEARCH DIRECTIONS

The proposed idea needs to be evaluated by integrating sensors with an earthquake magnitude identification module and high-end processors. For its effective implementation, a suitable IoT protocol has to be developed so an instant warning can be issued to people through proper networks.

CONCLUSION

In the proposed work, identifying earthquake magnitudes using P-waves is detailed, and an overview of an IoT-based earthquake early warning system presented.

Today, plenty of research is being undertaken to develop IoT-supported smart cities for an efficient, sustainable and reliable environment. The principal objective of a smart city is to advance the quality of the life of its citizens by utilizing technology to foster economic growth. Assorted sectors such as transportation, waste management, water distribution, energy management and health care have benefited greatly from IoT technologies. The Internet of Things lets people link up to the internet to receive and communicate information one of the main features of smart city development is to render regions less susceptible to disasters. At present, there is a real need for a reliable earthquake warning system that makes places less vulnerable to earthquakes and prepares people better for the fallout. At the present time, there is a real need for a reliable earthquake warning system that makes places less vulnerable to earthquakes and enable the people less susceptible to the effect of earthquake.

The present work discusses the development of an IoT-based earthquake early warning system that supports smart city development.

The proposed work primarily concentrates on identifying earthquake magnitudes for such a system. The issuance of an alarm for an earthquake depends on the magnitude identified from the primary waves of seismic signals, since they travel faster than secondary (severe) waves and reach the seismic station first. P-waves are not as powerful as S-waves and help detect the onset of a quake before destructive S-waves set in. Primary waves of seismic signals within the range of magnitudes 2 to 9 have been processed by extracting the temporal and spectral features of these signals. The accurate estimation of earthquake magnitude sizes plays a key role in implementing an earthquake warning system. In this work, a machine learning-based classifier named the random forest is employed to classify the magnitude of the earthquake signals and has carried out their magnitude identification well The performance of the classifier is investigated by evaluating performance metrics such as accuracy, sensitivity and specificity. The average accuracy, sensitivity and specificity yielded by the random forest classifier in earthquake magnitude identification are 98.8, 95.5 and 99.3 respectively.

This chapter provides an insight into the development of an early earthquake warning system with the support of the IoT. In this work, the system proposed for Earthquake warning is based on identifying quake magnitudes using P-waves with a machine learning technique. For effective real time implementation, the proposed system requires high-end sensors, processors and suitable IoT protocols.

REFERENCES

Alphonsa, A., & Ravi, G. (2016). Earthquake early warning system by IoT using Wireless sensor networks. *2016 International Conference on Wireless Communications, Signal Processing and Networking (WiSPNET)*.10.1109/WiSPNET.2016.7566327

Babu, Naidu, &Meenakshi. (2018). Earthquake Detection and Alerting Using IoT. *International Journal of Engineering Science Invention, 7*(5), 14-18. Retrieved from www.ijesi.org

Benbrahim, M., Daoudi, A., Benjelloun, K., & Ibenbrahim, A. (2007). Discrimination of Seismic Signals Using Artificial Neural Networks. *International Journal of Computer, Electrical, Automation, Control and Information Engineering*, *1*(4).

Bhardwaj, R., Sharma, A. L., & Kumar, A. (2016). Multi – parameter algorithm for Earthquake Early Warning. *Geomatics, Natural Hazards & Risk*, *7*(4), 1242–1264. doi:10.1080/19475705.2015.1069409

Bhargava, N., Katiyar, V. K., Sharma, M. L., & Pradhan, P. (2009). Earthquake Prediction through Animal Behavior: A Review. *Indian Journal of Biomechanics*, 7 – 8.

Breiman, L. (2001). Random forests. *Machine Learning*, *45*(1), 5–32. doi:10.1023/A:1010933404324

Buchanan R. C., Newell, D. K., Evans, S. C., & Miller, D. R. (2014). Induced Seismicity: The Potential for Triggered Earthquakes in Kansas. *Kansas Geological Survey, Public Information Circular, 36.*

Cretu, N., & Pop, M. (2008). Higher order statistics in signal processing and nanometric size analysis. *Journal of Optoelectronics and Advanced Materials*, *10*(12), 3292–3299.

Earthquakes – Technical Hazard Sheet – Natural Disaster Profile, Humanitarian Health Action, World Health Organization. (2018). *Earthquakes - Why do we have earthquakes: British Geological Survey*. Retrieved from http://earthquakes.bgs.ac.uk

Elder, C. Y. (2006). Mean-Squared Error Sampling and Reconstruction in the Presence of Noise. *IEEE Transactions on Signal Processing*, *54*(12).

Geology Universe. (2018). Retrieved from: https://geologyuniverse.com/3-major-seismic-belts-of-the-earth/

Kanamori, H., & Wu, Y.-M. (2005, June). Rapid Assessment of Damage Potential of Earthquakes in Taiwan from the Beginning of P Waves. *Bulletin of the Seismological Society of America*, *95*(3), 1181–1185. doi:10.1785/0120040193

Karaci, A. (2018). IOT-Based Earthquake Warning System Development And Evaluation. *Mugula Journal of Science and Technology*, *4*, 156-161. doi:10.22531/muglajsci.442492

Kurzon, I., Vernon, F. L., & Rosenberg, A. (2014). Real-Time Automatic P and S Waves Singular Value Decomposition. *Bulletin of the Seismological Society of America*, *104*(4), 1696–1708. doi:10.1785/0120130295

Lakshmi K. R., Nagesh, Y., & Krishna, V. M. (2014). Analysis on Predicting Earthquakes through an Abnormal Behavior of Animals. *International Journal of Scientific & Engineering Research*, *5*(4).

Li, Z., Meier, M. A., Hauksson, E., Zhan, Z., & Andrews, J. (2018). Machine Learning Seismic Wave Discrimination: Application to Earthquake Early Warning. *Geophysical Research Letters*, *45*(10), 4773–4779. doi:10.1029/2018GL077870

Liang, X., Zhang, H., Tingting, L., & Gulliver, A. (2016). A Novel Time of Arrival Estimation Algorithm based on Skewness and Kurtosis. *International Journal of Signal Processing, Image Processing and Patten Recognition*, *9*(3), 247–260. doi:10.14257/ijsip.2016.9.3.22

Lockman, B. A. (2005). Single-Station Earthquake Characterization for Early Warning. *Bulletin of the Seismological Society of America*, *95*(6), 2029–2039. doi:10.1785/0120040241

Mooney, W. D., Prodehl, C., & Pavlenkova, N. I. (2002). Seismic Velocity Structure of the Continental Lithosphere from Controlled Source Data. International Handbook of Earthquake and Engineering Seismology, 81A.

National Institute of Disaster Management. (n.d.). Retrieved from: http://nidm.gov.in/safety_earthquake.asp

Ochoa, L. H., Nino, L. F., & Vargas, C. A. (2017). *Fast magnitude determination using a single seismological station record implementing machine learning techniques*. Geodesy and Geodynamics. doi:10.1016/j.geog.2017.03.010

Odaka, T., Ashiya, K., Tsukada, S., Sato, S., Ohtake, K., & Nozaka, D. (2003). A New Method of Quickly Estimating Epicentral Distance and Magnitude from a Singe Seimic Record. *Bulletin of the Seismological Society of America, 93*(1), 526 – 532.

Oskin, B. (2015). What is a Subduction zone? *Planet Earth, Live Science.* Retrieved from https://www.livescience.com/43220-subduction-zone-definition.html

Patil, H. A., Baljekar, P. N., & Basu, T. K. (2012). Novel Temporal and Spectral Features Derived from TEO for Classification Normal and Dysphonic Voices. In S. Sambath & E. Zhu (Eds.), *Frontiers in Computer Education. Advances in Intelligent and Soft Computing* (Vol. 133). Berlin: Springer. doi:10.1007/978-3-642-27552-4_76

Patil, H. A., Baljekar, P. N., & Basu, T. K. (2012). Novel Temporal and Spectral Features Derived from TEO for Classification Normal and Dysphonic Voices. In S. Sambath & E. Zhu (Eds.), *Frontiers in Computer Education. Advances in Intelligent and Soft Computing* (Vol. 133). Berlin: Springer. doi:10.1007/978-3-642-27552-4_76

Regmi, J. (2017). Rupture Dynamics and Seismological Variables in Earthquake. The Himalayan Physics, 6-7, 96-99.

Roueff, A., Chanussot, J., & Mars, I. J. (2006). Estimation of polarization parameters using time-frequency representations and its application to wave separation. *Signal Processing*, *86*(12), 3714–3731. doi:10.1016/j.sigpro.2006.03.019

Singhvi, A., Saget, B., & Lee, C. J. (2018). What went wrong with Indonesia's Tsunami warning system. *Asia Pacific, The New York Times*. Retrieve from https://www.nytimes.com/interactive/2018/10/02/world/asia/indonesia-tsunami-early-warning-system.html

Srivastava, H. N. (Ed.). (1983). *Earthquakes Forecasting & Mitigation*. National Book Trust.

Stein, S., & Wysession, M. (2003). *An Introduction to Seismology, Earthquakes and Earth structure*. Blackwell Science, Inc.

Wang, Z., & Zhao, B. (2017). Automatic event detection and picking of P, S seismic phase for earthquake early warning and application for the 2008 Wenchuan earthquake. *Soil Dynamics and Earthquake Engineering*, *97*, 172–181. doi:10.1016/j.soildyn.2017.03.017

Wiemer, S. (2015). Earthquake Statistics and Earthquake Prediction Research. Institute of Geophysics, Zurich, Switzerland.

Williams, M. (2016). What is an Earthquake? *Universe Today, Space and astronomy news*. Retrieved from https://www.universetoday.com/47813/what-is-an-earthquake

Yamaski, E. (2012). What We Can Learn from Japan's Early Earthquake Warning System. *Momentum*, *1*(1), 2.

Zambrano, M., Perez, I., Palau, C., & Esteve, M. (2016). Technologies of Internet of Things applied to an Earthquake Early Warning System. *Future Generation Computer Systems*. doi:10.1016/j.future.2016.10.009

KEY TERMS AND DEFINITIONS

Confusion Matrix: A matrix that illustrates the correct and incorrect predictions in classification of classifier for each class.

Epicentral Distance: An epicenter is a location on the earth surface directly above the hypocenter where the earthquake originates. The distance from epicenter to any interested point is said to be as epicentral distance.

Machine Learning: An artificial intelligence technique that makes the system to learn automatically with the help of statistical techniques.

Magnitude: A quantitative measure of the actual size of the earthquake.

Richter Scale: A numerical scale used for measuring the severity of an earthquake.

Seismic Waves: The waves of energy released when two tectonic plates are rubbed against each other and the energy releases in the form of elastic waves.

Seismic Zone: Seismic zone is a geographical area that has high probability of seismic activity.

Tectonic Plate: A lump of gigantic, asymmetrically shaped solid rock. Most of the seismic activity takes place at the boundaries of the tectonic plates.

Chapter 14
Design and Prototyping of a Smart University Campus

Vincenzo Cimino
University of Palermo, Italy

Simona E. Rombo
University of Palermo, Italy

ABSTRACT

The authors propose a framework to support the "smart planning" of a university environment, intended as a "smart campus." The main goal is to improve the management, storage, and mining of information coming from the university areas and main players. The platform allows for interaction with the main players of the system, generating and displaying useful data in real time for a better user experience. The proposed framework provides also a chat assistant able to respond to user requests in real time. This will not only improve the communication between university environment and students, but it allows one to investigate on their habits and needs. Moreover, information collected from the sensors may be used to automatically identify possible anomalies in the available spaces of the campus, facilitating this way the planning actions necessary to solve them.

INTRODUCTION

It is well known that the "city" represents a suitable context for the design and development of solutions based on Information and Communication Technology (ICT), which make accessible the challenges for a sustainable technological advancement toward a "smart" planning and organization of resources and data. On the other hand, we have witnessed the revolution of the digital age, which has radically changed the way of living by introducing not only new ways to collect information, but also new techniques for their analysis and processing. Nowadays anyone can access the internet from any point or location, at home or on the road, with the possibility of processing functions for any need, as defined in the "*ubiquitous computing*" paradigm. Ubiquitous computing can be defined as a new computing strategy such that computers are implicitly integrated into the daily actions of users, who are frequently in contact with electronic equipment such as personal computers, smartphones, bracelets, refrigerators and even

DOI: 10.4018/978-1-5225-9199-3.ch014

glasses. At the base of the ubiquitous computing we find new network infrastructures, operating systems, sensors, microprocessors, all supporting the idea of surrounding ourselves with computers and software constantly active to provide assistance based on daily activities and according to our interests (Pradhan et al., 2014). This idea is further confirmed by the technological development of Internet, which is easily accessible thanks to special components from any object able to communicate with people and other machines, creating this way a network of objects called the "Internet of Things" (IoT). Data produced by users and their devices, in the IoT field, and by the digital urban infrastructures, in the smart city area, therefore require solid technologies that provide a great deal of information, in order to detect, aggregate and extract novel information useful to support decisions and to improve the experience of each user.

Ubiquitous computing is therefore a basic concept in order to understand the notion of *smart environment* which, according to (Weiser et al., 1999), is "a physical world that is richly and invisibly interwoven with sensors, actuators, displays, and computational elements, embedded seamlessly in the everyday objects of our lives, and connected through a continuous network." While the smart city is a very complex and complete smart environment, smaller and simpler examples are those of hospitals, universities, shops, etc., which are part of a city in their turn.

In this chapter, the design and prototyping of a smart university campus is described. The proposed framework is based on the data supplied by the users and provides useful services for the integration and mining of these and other data retrieved from sensors. The main goal here is to support the "smart planning" of the university environment, improving the management, storage and mining of the involved information. To this aim, a platform is proposed which allows for interaction with the main players of the system, generating and displaying in real time useful data for a better user experience. The proposed framework provides also a chat assistant able to respond to user requests in real time. This will not only improve communication between the university environment and the students, but above all it will be useful to know their habits and needs. Information collected from the sensors also allow to automatically identify anomalies in the campus, making easier to perform actions for their solution.

BACKGROUND

Middleware and Technologies

In order to define a good infrastructure, IoT needs to be supported by a middleware that allows consumers and application developers to interact in a user-friendly way, despite the differences in each user's perspective of IoT systems (Boman, Taylor & Ngu, 2014). The literature reveals a variety of strategies, models, and frameworks in order to define a Smart Campus environment. It is possible to distinguish two main factors for their design:

1. Communication protocols;
2. Storage and persistence technologies.

As for the first point, the main aim is to identify a protocol suitable for the transport and reliability of information within an area. Among the most common protocols on the Internet is *HTTP (HyperText Transfer Protocol)*, which is based on the request/response pattern. Within the Internet of Things, several "binary" protocols are emerging, based on the "publish/subscribe" paradigm and guaranteeing a low byte transfer. A well known protocol among them is the *MQTT (Message Queue Telemetry Transport)* protocol. Other technologies are also used for small network areas, such as ZigBee, LoRa and *Bluetooth* (Van Merode, Tabunshchyk, Patrakhalko & Yuriy, 2016). As for the storage and persistence technologies, the choice depends mainly on the type of data to be considered. In particular, relational databases represent the most widespread system for creating databases through the use of tables and relationships between them in *SQL* (Structured Query Language), but recently also non-relational databases have been proposed, also called *NoSQL*, which stands for "NotOnly SQL", because their peculiarity is not in the query language, but in the way heterogeneous and unstructured data are related, while maintaining a certain compatibility with SQL. The implementation of these latter technologies requires not only the development of a solid back-end side, but also the use of local resources and continuous maintenance of the system. For this reason, cloud infrastructures have been created allowing to delegate most of the work and speed up of the process of developing web applications and online platforms. Among the most complete is the Google Firebase platform, often used in the mobile and home automation. In the Literature several efforts have been provided in this direction, using the service to manage notifications between users' devices (Heryandi, 2018), create mobile applications (Jain, Garg, Bhosle & Sah, 2017) or support the storage of real-time information (Anggoro, 2018).

Internet of Things and Cloud Computing

Among the most recent ICT technologies in the Smart City and IoT fields, another key factor is the use of the "information cloud" to which specific tasks related to the applications are delegated. Cloud Computing is a source that provides software and hardware resources (such as mass storage for storing data or computational units for data processing) whose use is offered as a service by a provider. The resources requested from the supplier are assigned automatically and quickly, in such a way to be completely transparent to the user, and bringing enormous benefits such as the drastic reduction of costs (no hardware or software needs to be purchased), the speed of service (offered in a self- service and ondemand), global scalability (dynamically optimized resources), productivity (there is no need for machine maintenance), reliability (thanks to backup measures and data mirroring).

The types of services offered by the Cloud can be summarized in three main categories:

- **SaaS (Software as a Service):** Used for the distribution of Internet applications. With this service, cloud service providers host the software application and the underlying infrastructure, dealing with maintenance, such as machine upgrades and security patches. End users use this application by connecting to the Internet generally from a web browser on the phone, tablet or PC.
- **PaaS (Platform as a Service):** Refers to cloud services that provide an on demand environment for the development, testing or distribution of software applications. It is used to allow developers to create mobile apps for mobile devices, without having to deal with the configuration or management of the underlying server infrastructure.
- **IaaS (Infrastructure as a Service):** It is the basic service which is offered, and consists of a complete infrastructure including its own hardware and software. It provides also virtual machines, storage resources, networks and operating systems.

Smart Planning

In contrast to other smart campuses proposed in the Literature which mainly aim at optimizing the energy consumption (Popoola, Atayero, Okanlawon, Omopariola, & Takpor, 2018; De Angelis, Ciribini, Tagliabue, & Paneroni, 2015), here a different approach is followed. In particular, a in-depth study is provided on how, over time, many factors related to actions and behavior within the ecosystem can be used to: (1) optimize in an intelligent way actions in daily routine, (2) improve the management of areas and classrooms within the campus, and (3) suggest useful information according to a system of recommendations. In order to deal with these three issues, a suitable data analysis is required.

PROJECT OF A SMART CAMPUS

Relying on the notions of innovation and "intelligent" development, inspired by the Smart City model, the research presented here focuses on how it is possible to use heterogeneous data sources for a better understanding of the world around us, particularly in the university environment.

The idea is to make the university experience more compliant with the new technologies and make it enjoyable to the university environment (students, teachers, collaborators), so as to be informed in real time on the basis of their current needs and/or have an overview of the statistics on specific subjects. As an example, a student needs to read a specific book, and the system provide a list of the open libraries and their level of occupancy. At the same time, the user may contribute to improve the system by enlarging its data collection, as better explained below.

The information, in the case considered in this chapter, has a fundamental role as it is analyzed from a double point of view: very short term and long term. Typically the data flows coming from the surrounding environment and from the user activity can help to understand certain factors of fruition of the spaces, habits and necessities. The university environment is a source of data of various kinds; and it is possible to distinguish the following types of sources:

- **Direct:** Generated by the physical control system (university card control, entrance gates, libraries);
- **Automatic:** Generated by automatic systems (university web portal, sensors, video surveillance);
- **Volunteers:** Directly provided by the users (Social Networks).

Once defined the actors of the ecosystem and the sources of data, it is necessary to outline how the environment (i.e., the university) will involve them and, at the same time, may receive information and plan, thanks to the user, a vision of the future in an intelligent way. The dialogue between the parties have to take place in a structured and digitized context, aimed at the collection of actions and decisions shared by participants.

It is worth pointing out that, in order to take into account the candidate information coming from each user's action, especially in the initial phase of the project suitable listening techniques have been used, such as sample surveys, also in electronic form, web questionnaires, supported by sections "Critiques and suggestions", etc.. The collected data are then supplemented by the messages received from the chats, forums or blogs, social networks and applications. Once the available data sources and the possible implementations have been chosen, characterizing factors have been established on which the project will be built. Figure 1 shows a block diagram representing the project, highlighting the three main blocks:

Figure 1. Block organization of the whole project

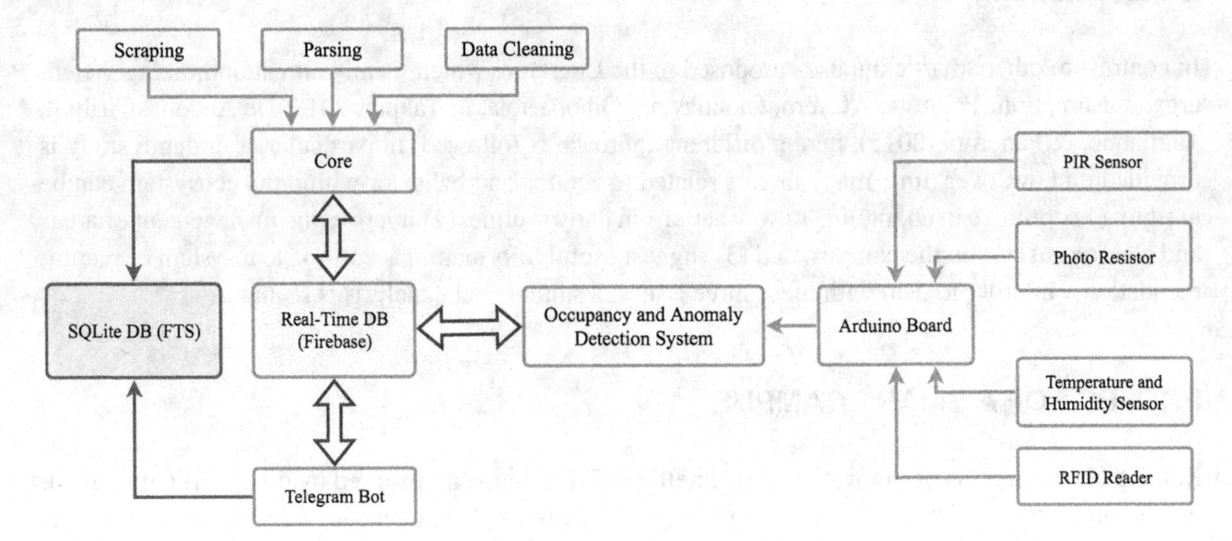

- Core of the platform,
- IoT system for Occupancy and Anomaly Detection System,
- Telegram bot.

Core of the Project

The first section of the platform is dedicated to the main component, that is the module that manages all the information obtained from the university portal, data sources, web sites and students. It was developed in NodeJS, which offers a versatile and efficient language for creating web applications and managing many requests in real-time. This block, which is central for the project, provides the functions of research, analysis and processing of information retrieved from existing sources to create a collection of departure. Moreover, it provides also procedures related to the manipulation and processing of data used in the output.

Choice of Data Sources

The first important aspect in the creation of a starting data collection is the choice of the sources from which information may be obtained. This aspect influences the data format and storage, as well as the quality and integrity of the contents also with reference to their corresponding sources. In particular, the sources used in this project are:

- **Official University Portal:** Used to obtain data on professors, classrooms, exams, educational calendar and organized events;
- **Libraries Portal:** Web portal library management within the university, used for obtain access data and library occupancy;
- **Sensor Detection:** System for classroom monitoring with Arduino, used to obtain data on humidity, temperature, brightness, movement, RFID tags;

- **Telegram Bot:** Chat for the mobile application Telegram. It uses the data defined previously and is used to obtain information about the habits and activities of the students.

Collection of Information

The data, although having heterogeneous sources among them, are collected following the memorization criteria to be manipulated using similar tools and structures. However, this choice depends on the type of data to be manipulated and its use. Consider, for example, the management of data relating to classrooms; over time they are characterized by the same attributes and values (address, floor, number of seats), except for very rare exceptions. The only value that can vary is that related to the daily employment status, which has a value dependent on time. This is the reason why different types of data persistence are used. The data used in this work can therefore be divided into two categories:

- Data for consultation (static data, not time dependent);
- Data dedicated to monitoring (dynamic data, time dependent).

The first, which are data that do not change over time, as the class name and position, the names of the professors, the names of the degree courses, have been saved in a SQLite database with a particular extension suitable for textual searches, FTS4 (Full-text search, version 4). This module performs text searches within the database quickly and efficiently even if the table contains many large documents. For example, in a database of half a million documents, the DB query will return the result in about 0.03 seconds, compared to 22.5 seconds for the query of an ordinary table. The FTS4 table for the analyzed example consumes about 2006 MB on disk compared to only 1453 MB for the ordinary table, with a slightly lower table dip-time to the detriment of FTS4. This means that, to the detriment of a higher cost in terms of space, with a table FTS4 there is an almost instantaneous result in terms of research, which in the case of a web application is essential. The remaining data, those relating to the hours of lessons and the identification of the entrance to the libraries, given their dynamic nature, are saved in a in real time database, in which there are many more updates and writings. The choose for the best database involving devices and the system fell on Firebase. Its main qualities are:

- Backend flexibility;
- Reduced development time;
- Excellent scalability;
- Offers all the benefits of the cloud, so no software configuration is included;
- Storing data as a native JSON, then what is stored is what will be provided in the future;
- Data protection as Firebase requires 2048 SSL encryption for all information exchanges;
- Data persistence in secure areas, in order to avoid negligible information loss;
- Cloud storage immediately accessible on any device or terminal;
- Cross Platform API (if using this DB with an app).

After identifying the sources from which to obtain information and the tools used for their memorization, suitable techniques will be used to retrieve the information, i.e., the web scraping operation and the parsing of CSV files, as shown in Figure 2.

Figure 2. Representation of the information gathering phase: from the selection of the sources to the collection of the raw data

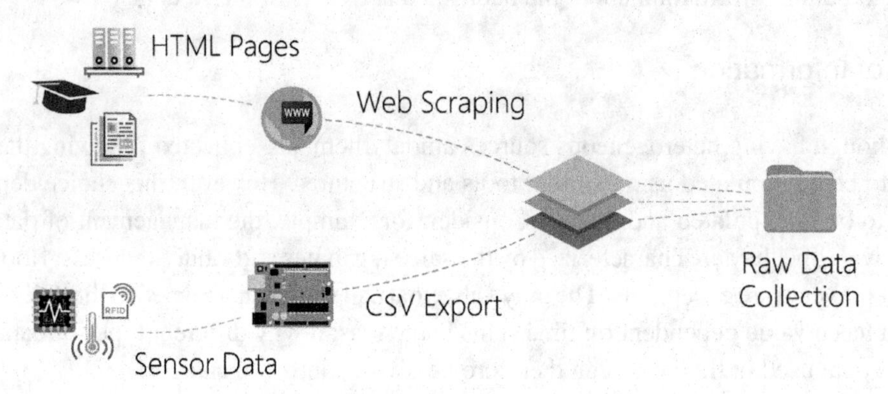

The first of these operations is dedicated to infer the useful information from the sources that provide information via web pages. In particular, the scraping phase consists on the extraction, from each visited page, of all the sections relating to events, professors, libraries, etc.. This is possible by the use of suitable NodeJS software libraries, and by selecting during the design phase the sections to be considered by the definition of regular expressions and patterns. However, it is worth to point out that this method depends on the page structure, and therefore on the rules defined by the website developer.

The parsing operation is used to separate the elements within the CSV (Comma-Separated Values) files containing the sensor information. The CSV format is a text file format used primarily for exporting a table of data. The first line of the file contains the names of the fields separated by a comma, while the second line onwards contains the data, also separated by a comma. The data produced by the prototyping cards such as Arduino and ESP8266 (used in the project) are essentially raw data, partly analog and partly digital, which are converted into numerical format to be sent to the server. Since these embedded systems do not have a large amount of memory and computing power, they are sent at time intervals of the CSV files to the server, which manages the task of performing the parsing and creating a more suitable structure.

Analysis and Processing

After defining the information collection strategies, we need to illustrate how to extract the essential data starting from the recovered blocks. This phase is necessary to standardize the structure which models the stored object.

Data obtained following the scraping phase are characterized by greater strings of text without any syntax or structure. For this reason, it is necessary to carry out a meticulous control on the data so that these are updated with the variation of time avoiding cases of duplicates or transcription errors. In this phase the contents of each string, the syntax, the format and its quality are analyzed. Figure 3 shows the idea behind this step, taking as an example the information extracted during the scraping phase of the University website. Raw data are characterized by text, links to external web sites, url to images, dates and times with different formats, places and addresses. It is therefore necessary to find a solution in order to obtain, from a simple text, useful data to be stored in a database. For example, the local date format is a discriminating factor for efficient data manipulation. Servers, client and software libraries used in

Figure 3. Data cleaning process of raw data to obtain structured data

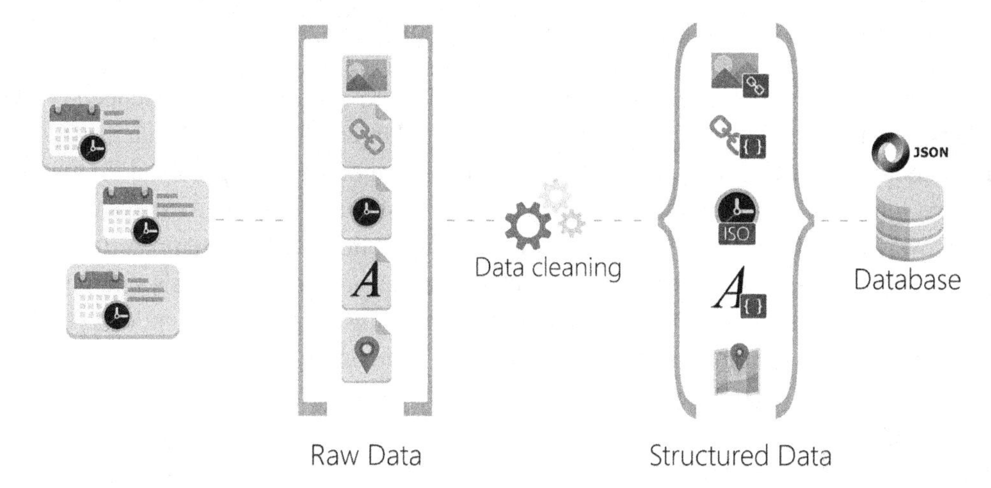

order to manage time are based on a precise syntax. This is the reason why all the dates extracted from the web pages have been converted into ISO 8601 format, an international standard for the representation of dates and times, so as to be managed with simplicity and allow better manipulation in the phase of comparisons between dates and difference in hours, days and months.

Among the main features of the information provided by the services of the platform, there is to suggest places of interest, events and planned activities. In this respect, the main problem is the lack of references on the location of the object under consideration, as the only data available is a string representing the address of the place taken into consideration. For this reason a geocoder has been used, that is a service with the ability to convert an address into a pair of values: latitude and longitude. The library used in this case is NodeGeocoder, which is based on the Google Maps service. All geographical locations are stored on the Firebase DB, in order to reduce the amount of requests only to places not yet present in the collection.

CSV files, on the other hand, represent a semi-structured data by themselves, as they contain within them a structure that delimits the fields and their attribute of origin. However, the validation control rules are missing and as data from the sensors will also be present spurious signals due to anomalies or malfunctions. Therefore, there is the need to build an auxiliary structure used to reduce the anomalies identified and prepare the data to be processed together with other structures.

Detections With IoT

The developed module that required more attention is related to the IoT. The idea is to create a monitoring system for the shared spaces within the university campus, in particular their employment and regular use based on the educational calendar. In the literature, referring to presence monitoring is a very general and very controversial problem. The type of strategy depends on the purpose of the application. It is indeed possible to distinguish a hierarchy of problems (Akkaya, Guvenc, Aygun, Pala, & Kadri, 2015):

- Presence detection;
- Presence counting;

- Occupancy tracking;
- Event recognition.

Here the problem of presence detection has been taken into account, and subsequently integrated with an anomaly detection system. This system has been implemented to make the data available in real time and to understand how the current areas management is updated and optimized. In the Internet of Things context, presence detection can be performed by a vast variety of heterogeneous raw data from sensors: microwave or infrared sensors, optical camera, microphone signals, temperature sensors. In a domestic environment it is easy to determine the presence or absence of a room's internodes, but in a university environment one needs to define some discriminating parameters in order to understand if the classroom in question is used to perform the lecture is occupied for other reasons.

Hardware Used

- **ARDUINO:** Arduino is a programmable hardware platform. Assembles an ATMega328 processor, suitable for applications in the IoT field. It is easily extensible with components of various kinds: sensors, expansion cards, network.
- **ESP8266:** Like Arduino, this is also a programmable card, which compared to the previous one is already equipped with a wi-fi module for connections to networks, therefore preferable to Arduino for IoT projects.
- **DHT11 SENSOR:** Digital temperature sensor, has the advantage of being able to detect both temperature and humidity, but the defect of being inaccurate.
- **BH1750 SENSOR:** Digital brightness sensor. This card is a simpler and more accurate version of the photoresistor.
- **PIR SENSOR:** A passive infrared sensor (PIR sensor, acronym of Passive InfraRed), widely used as motion detectors.
- **RFID READER:** RFID tag reader (Radio-Frequency IDentification), technology for the automatic identification and storage of information on objects, people, environments.
- **OLED DISPLAY:** Display with low consumption OLED technology used to display data processed by programmable cards.
- **EXTENSION FOR SD MEMORY:** Extension board for Arduino which contains a buffer battery for the timer, an SD card reader and control LEDs.
- **RASPBERRY PI:** Like Arduino, Raspberry Pi is a prototyping platform, usable for the creation of servers of various kinds, home automation, hot-spot, media center, and much else.

Initial Considerations

The initial idea is to identify, if possible, the degree of employment in a classroom, then check if it is used in a certain moment and with a certain authorization. The teaching areas are not always used for the purposes of the lesson, but remain available to the students as areas of study that are not used by the teachers. For this reason, surveys were carried out for about a week within selected classrooms to study the parameters obtained by the sensors and search for common patterns to recognize three fundamental states:

- Empty classroom;
- Classroom occupied by students;
- Classroom occupied by lesson or seminars.

The first surveys were conducted using temperature, humidity, brightness, sound and movement sensors (Pedersen, Nielsen, & Petersen, 2017). Once the acquisition has been completed, the result of each sensor is processed individually:

- **Temperature and Humidity Sensor:** This sensor maintains a constant behavior during detection, except for some spurious signals that produce noise during the measurements. The variation in temperature and humidity is minimal and is more evident when the sun illuminates through the windows. The average temperature between one room and the next is almost identical.
- **Light Sensor:** This sensor records large variations unlike the previous one. The amount of light at times when the classroom is closed (with lights off) is very different from when the classroom is used (lights on open windows). This feature is essential for determining the state in which the classroom is located. However, the data showed slight momentum during hours when the lesson was scheduled. If more attention was paid to these anomalies, it was found that during those hours a lighted-out lesson was carried out to facilitate the viewing of the displayed displays with a projector. This factor means that the amount of light can be scarce or almost zero even during the lesson. In addition, each room has a different average lighting as the latter depends on the number of windows and the position of the sun at a given time.
- **Sound Sensor:** The sound sensor has the ability to perceive sharp and fairly loud noises. During the lessons, most of the time, this kind of sounds are caused only by the opening and closing of the doors, but not by the voice of the teachers or the activity of the classroom. This makes this sensor almost unusable, completely excluding it during the next detection phase.
- **PIR Sensor:** This is the sensor that provides the most consistent information to the desired result, as it tracks the presence of movements inside the classroom with a good degree of reliability. During the moments of uselessness of the classroom there are no signs of movement, while during the beginnings of the lessons the sensor is almost always in active state. The situation changes during the longer lessons, without pause and with the teacher seated at the desk; in that case the sensor perceives very few movements and does not just enough to activate it.

In order to understand if the classrom is empty, an RFID tag sensor (Oshin, Owoniyi, Oni, & Idachaba, 2017) has been used. Typically, within the university environment, particular cards have been distributed to students that uniquely identify them. These cards are mainly used for accesses within the campus, for access to libraries, for access to the university canteen and other services dedicated to student routines. After having carried out some reading tests through an NFC reader, the verification was also made by the reader connected to the Arduino board. The acquired data is a four-byte code expressed in hexadecimal digits (UID deltag), which uniquely identify the card as soon as read. Flanking this information to the data collected so far, excluding the sound sensor, it is possible to have a complete picture to describe precisely what happens inside a classroom. It is also supposed that the cards will be recognized using their UID as authorized and unauthorized. This determines the use of the classroom by normal students or by a staff authorized to use the classroom for lessons, conferences, seminars or other types of events. A further

Figure 4. Subdivision in blocks of the sequence of detections: three types compared

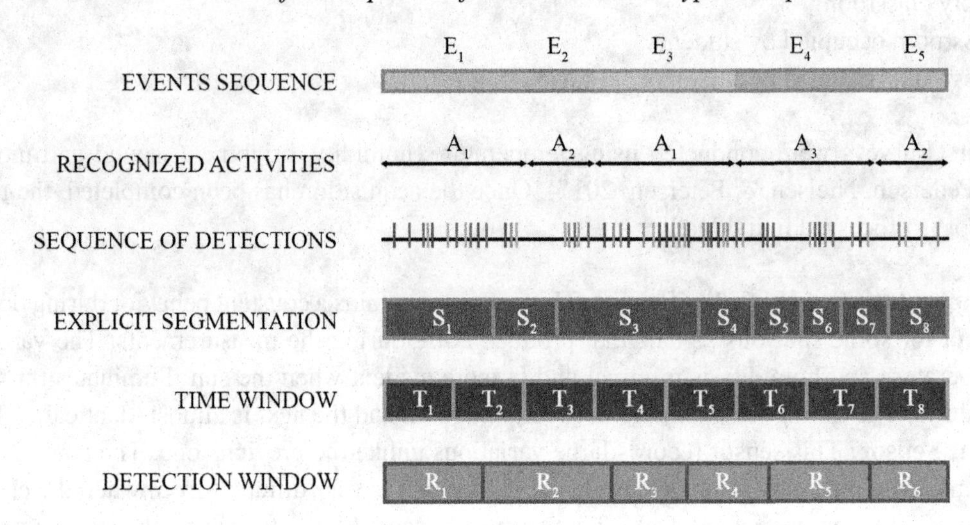

analysis is carried out on the division into blocks of the data obtained from the sensors, since operating on the entire dataset, which includes hours of detections, does not allow to obtain detailed information.

Figure 4 shows how activities are interpreted within a space such as the classroom and some techniques for dividing information are illustrated. Three different methods have been tested in the Literature (Bersch, Azzi, Khusainov, Achumba, & Ries, 2014; Chahuara, Fleury, Portet, & Vacher, 2016; Krishnan, & Cook, 2014):

- **Explicit Segmentation:** This subdivision in blocks provides a check on the frequency of the measurements made. If a change is found that is consistent with the sequence of sensor readings, then a block is terminated and a new one is generated. This method can be useful for analysis carried out on a single sensor, as blocks are defined in which the sensor behaves differently, but in the totality of the measurements it has returned the results completely in phase with what is called the sequence of the events, compromising the recognition.
- **Time Division:** This strategy offers the advantage of being able to decide the size of the time window; this choice should not be overlapped on very large values, or on values that are too small, since in the first case there is a risk of capturing a longer amount of time than the event itself, while in the second case there is a risk of having not enough data available in order to understand the type of activity in progress.
- **Subdivision by Detections:** This method takes into consideration some factors of the first method and others of the second method. The subdivision is based on the measurements made, in particular a quantity of elements is identified, which can be variable during the test phase, which define the duration of a window. This means that in a temporal portion where the sensors have few activations there will be a very long window, while in the high activity phase the windows will be very small. This leads to the same disadvantages as the first proposed method.

The technique among those proposed with a smaller amount of disadvantages seems to be the temporal division, provided that these windows are arranged in an optimal manner. To avoid problems due to synchronization there is the possibility of introducing an overlap of a small amount of time between one window and the other, in such a way as to reduce the margin of error. However, the reality taken into consideration in this project is characterized by precise times defined during the day. For lectures, seminars and events, the time is set at intervals of one hour or at most half an hour, so there is awareness of the start time and sometimes the end time of the single event.

Strategy

To manage this variety of information and detections, the strategy adopted in this elaborate consists in summarizing the data obtained by the sensors at the time windows to generate different arrays of attributes, useful for identifying anomalies. This approach is in line with the type of expected result because the types of classes to be identified are a combination of data from the sensors and can be described by statistics on the variables calculated within a portion of time data. The raw data taken into consideration, collected by an Arduino board, consists of an infrared presence sensor (PIR), a temperature and humidity sensor, a light sensor and an RFID (Radio-Frequency IDentification) reader. These output values

Figure 5. Processing of sensor data for the purpose of anomaly detection, separating the detections in time blocks Wi of dimension T

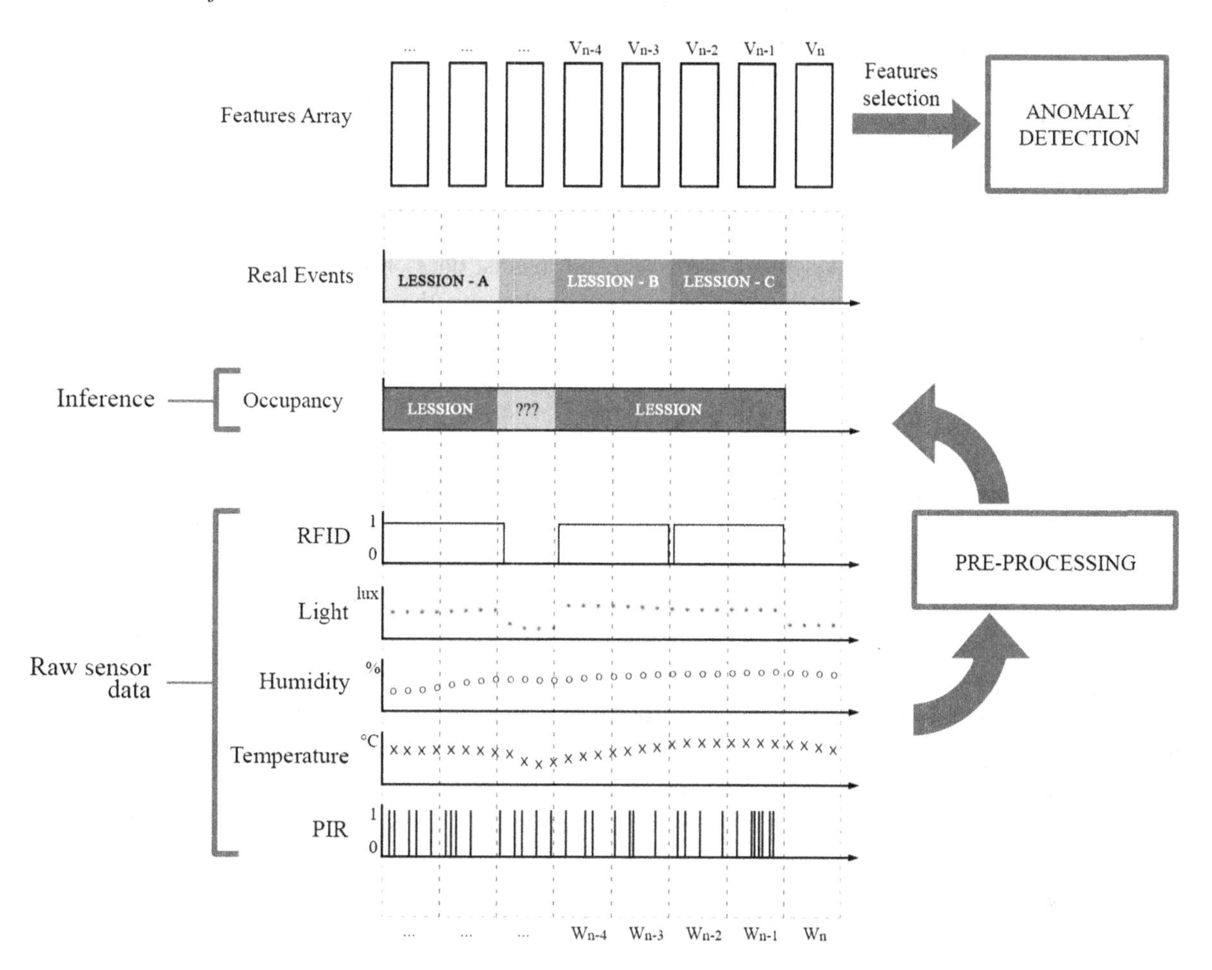

related to date and time (timestamp), irregular sampled signals (temperature, humidity and light) and status values (movement and RFID). All these data are subsequently pre-processed to obtain higher level information useful for better treatment (Tapia, Intille, & Larson, 2004; Jurek, Nugent, Bi, & Wu, 2014).

The time factor therefore allows to define a coordinated structure with real time, in which each *Wi* time window is synchronized with the beginning of each hour. With regards to the duration of the time window *Wi*, *T* has been chosen such that it was shorter than any event to be identified, and long enough in order to have significant content (as shown in Figure 5). Any event has to be long at least *T/2* in order to be detected by the sensors. Within the university, lectures and other activities typically start at a certain hour *h* o'clock, or at half past *h*. However, there may be delays due to unpredictable factors, which determine the start of lessons later than expected. To solve this problem, it was decided to operate on 15-minute time windows, thus defining a one-hour macro window consisting of four time windows. Often the operations on the raw data are difficult to be generalized, for this reason suitable statistics have been used in order to summarize the data contained in each time window (e.g., frequencies, means, etc.). For the PIR sensor the number of activations in the time interval has been calculated. For all analogue sensors (temperature, humidity and light), standard deviation and mean has been calculated. For the RFID reader, a Boolean value is assigned if the duration of the RFID tag with the reader is the majority of the total time window. To return a positive value, the id of the RFID tag needs to be registered among those allowed (assumed that this is defined by the university administrative area), in this way the classroom activity will be recognized as "authorized activity", otherwise will be recognized as "generic activity". The use of these parameters has a very low computational cost, also giving the possibility to have contextualized information within the time window and some information related to the activities within the university.

Anomaly Detection

Finally data recovered from the sensors are compared against data on the official occupation of the university areas. Occupational data consist of an array of elements that define the degree of occupation of the classroom, in particular the following types will be assigned:

- Type 0 if the classroom is empty or unused;
- Type 1 if the classroom has internal activity but is not provided by the didactic timetable;
- Type 2 if the classroom is occupied in an official manner, authorized by RFID tag.

Data from the sensors present all three values, as the motion sensor activates in the event of movement even if the RFID tag is not used, as shown in Figure 6. In the case of data coming from the University portal, they are characterized by values 0 and 2 to represent the state of inactivity of the classroom or active lesson, respectively. These data are obtained through the operation of scraping on the web pages concerning the occupational plans of the classroom. The obtained data structure contains information on places, teachers in the classroom, course names. However, these data are incompatible with those that are currently the values coming from the sensors, so a structure similar to that obtained by the Arduino board has been created. The difference between the two sources generates an object containing the anomalies, which is conveniently stored in the Firebase database. All daily anomalies are also visualized within a calendar represented by a grid, in which each element is characterized by the square of the Euclidean norm of the array.

Figure 6. Example of type assigned to a set of sensor data by the anomaly detection process

Timestamp,	Temp,	Hum,	Light,	PIR,	RFID	
20180519121455,	25.4,	20.0,	250,	0,	0	TYPE 0
20180519121458,	25.3,	20.0,	248,	0,	0	
20180519121501,	25.4,	20.1,	249,	0,	0	
...	...	...	...	...	...	
20180606143001,	26.4,	30.4,	350,	1,	0	TYPE 1
20180606143010,	26.3,	30.4,	340,	1,	0	
20180606143017,	26.5,	30.4,	351,	1,	0	
...	...	...	...	...	...	
20180607103000,	25.9,	30.4,	317,	1,	1	TYPE 2
20180607103010,	26.4,	30.4,	322,	1,	1	
20180607103020,	26.3,	30.4,	320,	1,	1	

Arduino and Raspberry Pi

From the hardware point of view, the test phase of the anomaly detection system was carried out by creating an ad-hoc network between Arduino and Raspberry Pi. The ESP8266 board, being equipped with a Wi-Fi module, transmits data autonomously, while the Arduino board requires a USB connection to the Raspberry Pi. To better understand the life cycle of the processes initiated in the client terminals and server, it is represented in the diagram flow (Harsha, Reddy, & Mary, 2017) in Figure 7.

Figure 7. Infinite cycle processes related to the client (Arduino) and to the server (Raspberry Pi)

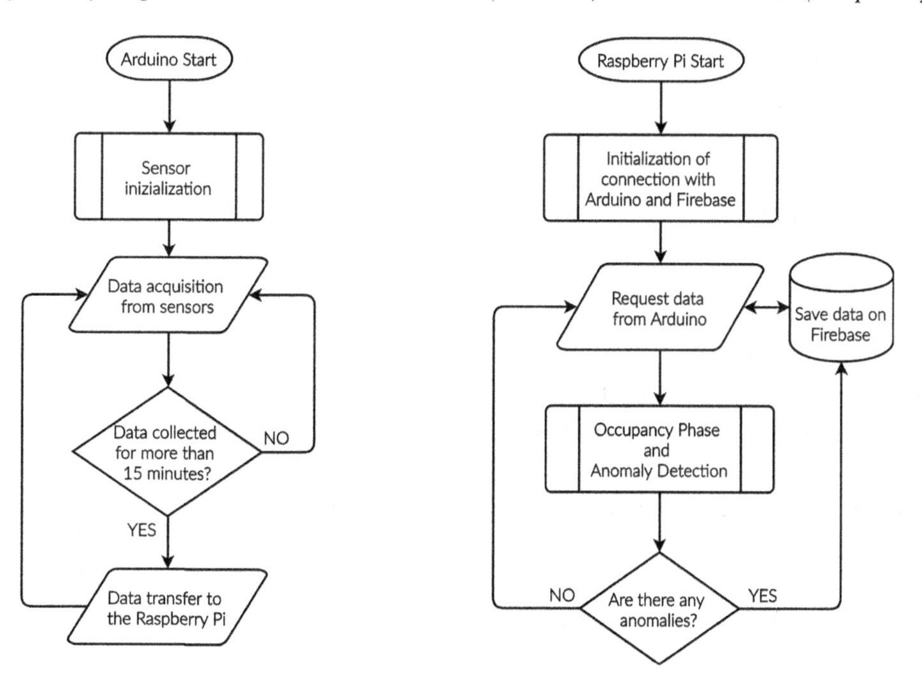

Telegram Bot

The initial target is to create a useful tool that not only allows the consultation of the data collected even on the move, but which also gives the possibility of obtaining implicit and explicit feedback on the quality of the products offered and the way in which they are used. The realtime database Firebase is reachable at any time from any mobile terminal and not, so it is used to store the data of interest in the database on the cloud in order to find information easily.

In general, at least three solutions may be considered in order to solve the problem of data harvesting:

- Development of a native mobile application;
- Development of a web app on the web;
- Development of a Telegram bot.

If a smartphone application is created, two separate applications would have to be developed in parallel (Android/iOS). On the other hand, the creation of a website would have involved the design of an additional back-end and front-end module. The creation of a bot for Telegram, instead, allows to develop only the back-end side, given that the front-end is the application itself.

Telegram is now a very popular application in the mobile and desktop environment. It is one of the best messaging applications and allows the multitude of functions available to create chatbots or simply "bots", which provide information to the user interactively. The interaction with the user is facilitated by the simplicity of the operations, allowing to write through the keyboard of the device in order to perform a conversation. Therefore, Telegram provides the chance of using APIs to interface bots and users. The use of these procedures is possible in a native way, which involves a long period of development, or in an intermediated way by libraries or "wrappers" used to design a bot from any development platform.

List of Commands

The conversation with the bot starts just like in the case of a real person, using a username in the place of a phone number. Once identified, the "*/start*" command may be used to establish a conversation. The main operations immediately available for the user and that do not require parameters are:

- **/libraries**: list of libraries and their available seats;
- **/libinfo**: overview with statistics of all libraries;
- **/events**: events organized by the University;
- **/canteen**: information related to the catering service;
- **/feedback**: sending feedback;
- **/commands**: list of bot commands;
- **/info**: general information about the bot.

By consulting the command section, the user can also view the secondary ones, which require parameters to be used correctly:

- **/professor <name>**: search for a professor by name;
- **/classroom <name or department>**: search for a classroom by name or department;

- **/course <keyword>**: search for the ID of a degree course by keyword;
- **/timetable <code>**: request for the timetable related to a degree course;
- **/exams <code>**: request for exams related to a degree course.

Table 1 provides information on the bot commands, highlighting the storage system used, the validity period of the data and the type of output offered by the single command. For example, the first implemented command was that relating to the occupation of libraries in real time. The following information is retrieved from the web portal through a scraping operation:

- Name of the library or faculty to which it belongs;
- Membership building and floor;
- Occupied places;
- Total seats.

Data have a very short validity, because the free seats inside a library can change quickly even in a few minutes. To keep the information always up-to-date, the bot queries the web portal every two minutes, keeping in memory the data read to provide it to users rather than having to perform continuous queries to the university server, overloading it with requests.

Data Acquired by the bot

The structure of the data acquired by the bot during its operation is characterized by the following sections:

- **actionsLog**: this section logs the requests made to the bot, taking into account the requested function, time and user id, collected day by day;
- **librariesLog**: the data stored in this section represent the occupation of the libraries used to inform users. They are grouped by university area and inside we find the employment data grouped by day;

Table 1. The commands of the bot and its characteristics of use: database, expiration time and output

Command Name	Database	Expiration Time	Output
libraries, libinfo	Firebase	2 minutes	Text
events	Firebase	6 hours	Text, locations and links
canteen	Firebase	1 day	Text, locations
professor	SQLite	6 months	Text and external links
classroom	SQLite	6 months	Text, locations and links
course	SQLite	Never	Text
timetable, exams	None	None	PDF Document
feedback	Firebase	None	None
info, commands	None	None	Text

- **feedback**: all information sent explicitly by users, or evaluations, comments, criticism and advice, will be collected in this section. This function is offered by the homonymous command used by the bot;
- **users**: the users section collects the information necessary to give the bot a "memory" useful for distinguishing users. While the bot is running, a user may decide to block or terminate the dialog. To avoid sending notifications to users who no longer use the bot, a structure has been created to manage even such eventualities:
    - User ID;
    - Last choice made;
    - Last command executed;
    - Timestamp of first use;
    - Timestamp of last use;
    - Bot presence flag;
    - Language used;
    - Flag activity;
    - Flag receiving notifications.

Pros and Cons for Designing and Prototyping

During the study and development, this project reached a phase of medium fidelity prototyping. This state adds a lot of details and the interaction is closer to the final product. In most cases, medium fidelity prototype is enough for the user to fully experience the final product, however there are several aspects to consider as pros and cons in the design and prototyping of the university campus.

Pros:

- Big amount of Data: the more the information, the easier it is to make the right decision;
- Tests: possibility to test the usability of the product to find problems;
- Reduced time and costs: with prototyping it is possible to detect early what the end user wants with faster and less expensive software.
- Users feedback: easy to communicate with users in order to obtain comments and positive or negative feedback.

Cons:

- Insufficient analysis: a focus on a limited prototype can distract study from properly analyzing the complete project;
- Compatibility: as of now, there is no standard for tagging and monitoring with sensors.
- Safety: there is a chance that the software can be hacked, so it is necessary to introduce many security measures as authentication system and data encryption;
- User confusion: the worst-case scenario of any prototype is users mistaking it for the finished project and they may not understand it merely needs to be finished or polished.

TESTS AND RESULTS

In this section it is firstly shown what data have been collected in their entirety, how these data have been analyzed and manipulated and how they are able to offer a new point of view on their understanding, going beyond their original meaning. The data presented will therefore have two different measures, setting time as a key factor. The proposed project has provided for the collection of data in total a period of about 7 months and are distinguished as:

- Data from the IoT system.
- Data from the Telegram bot and access to libraries.

Tests Related to the IoT System

The data collected on the classrooms by the IoT system were carried out within the University and in particular in three classrooms: Classroom A, Classroom B, Classroom C. At the end of each day, the data from the sensors were compared with the official timetable data. To identify the degree of the anomaly a matrix of differences was calculated, within which the daily anomalies will be found, as shown in Figure 8. The areas highlighted are those relating to common activities, consequently it is possible to notice a large number of anomalies. These are visible inside the structure on the right, representing the differences between the first two structures.

The anomalies detected day after day are thus collected and displayed on a web page in the form of a grid, in which each element represents the day and each group of elements represents the month in the academic year. The color of each element of the grid varies depending on the degree or intensity of the anomaly associated with that particular day.

Figure 9a shows the anomalies analyzed in classrooms A, B and C, respectively. All three have a large number of anomalies, however, as previously described, there are different types of anomalies, those due to motion sensors (with value 1) and those due to problems related to the didactic timetable (with value 2). In Figure 9b an analysis is carried out on these types: the dark bars define the amount of anomalies found in their entirety, while the light bar defines the amount of anomalies due to inconsistencies in the

Figure 8. Definition of the data structures used to identify the degree of anomaly of the classrooms

SENSOR DATA

```
{ '30-05-2018':
   {
     '7': [ 1, 1, 1, 1 ],
     '8': [ 1, 1, 1, 1 ],
     '9': [ 1, 1, 2, 2 ],
     '10': [ 2, 2, 2, 2 ],
     '11': [ 2, 2, 2, 2 ],
     '12': [ 2, 2, 2, 2 ],
     '13': [ 2, 2, 1, 1 ],
     '14': [ 1, 1, 2, 2 ],
     '15': [ 2, 2, 2, 2 ],
     '16': [ 2, 2, 1, 1 ],
     '17': [ 1, 1, 1, 1 ]
   }
}
```

OFFICIAL UNIVERSITY DATA

```
{ '30-05-2018':
   {
     '9': [ 0, 0, 2, 2 ],
     '10': [ 2, 2, 2, 2 ],
     '11': [ 2, 2, 2, 2 ],
     '12': [ 2, 2, 2, 2 ],
     '13': [ 2, 2, 0, 0 ]
   }
}
```

DIFFERENCES

```
{ '30-05-2018':
   {
     '7': [ 1, 1, 1, 1 ],
     '8': [ 1, 1, 1, 1 ],
     '9': [ 1, 1, 0, 0 ],
     '13': [ 0, 0, 1, 1 ],
     '14': [ 1, 1, 2, 2 ],
     '15': [ 2, 2, 2, 2 ],
     '16': [ 2, 2, 1, 1 ],
     '17': [ 1, 1, 1, 1 ]
   }
}
```

COMMON DATA (NO ANOMALIES)
UNCOMMON DATA (ANOMALIES)

Figure 9. Representation of the anomalies identified: (a) through a grid system; (b) through a bar chart

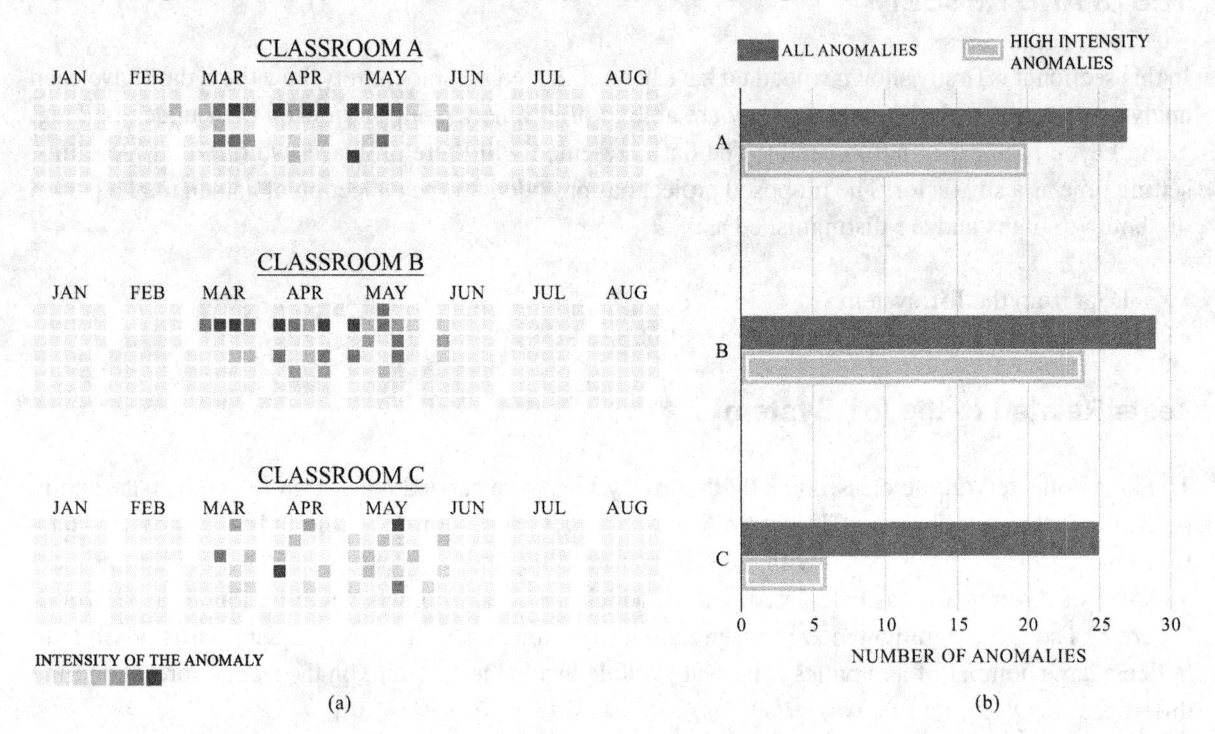

occupational plan of the classroom. From this point of view, classrooms A and B appear to have many more anomalies than classroom C, which mainly contains slight anomalies, due to the simple activity inside the classroom by the students.

An additional information that can be drawn from the displayed grid is the presence of periodicity within the structure. The presence of adjacent anomalies in a row (as in the example shown in Figure 9a) represents a periodic anomaly, since the colored squares refer to the day of the week in which it has been detected some anomaly.

Tests Related to the Telegram Bot

The experimentation phase of the bot required approximately seven months, consisting of two initial weeks of testing and feedback and subsequent publication. During this time enough data have been collected allowing to draw some conclusions about their usefulness. All data have been stored in JSON format files for better understanding and versatility. We can therefore summarize the data collected as follows:

- **Dataset 1:** Daily log of libraries occupancy status;
- **Dataset 2:** Daily log of user activities.

The first dataset contains all the data concerning the occupancy status of the libraries of the university campus. It includes a total of 200 days of acquisition, each containing an average of 250 daily surveys for each of the libraries, divided by area. Every single survey indicates the occupation at a specific moment in the library.

The second dataset is characterized by data related to the actions committed within the bot. In particular, there are all the commands typed side by side of the id of the user and timestamp of the action issued, distributed in daily collections.

The analysis of these collected data offers a manipulation of the time factor to obtain forecasts in the medium and long term. By focusing on the first dataset, observing all surveys in the space of a day, for example, it is possible to understand how quickly a library fills up or empties. However, each data varies from one library to another. In fact, it can be noted that the degree of filling of a library is proportional to the maximum number of places available and each library can also have a different closing time from the others. One factor in common, however, is the way in which the curves that represent the trend of the graphs grow and settle during the day. The tests focused on identifying the factors that the observed libraries may have in common:

- Time of maximum variation;
- Maximum peak time.

However, these values should be computed with an evaluation in the average case. We then proceed with the computation of the average of all the surveys, taking as an example a test library, called Library X. Following the definition of the average trend, it is necessary to identify from this the factor for which the library fills up during the day, or define those points where the filling speed is faster than other points of the day. To do this, the derivative of the previously identified curve has been computed. Only the positive values of the derivative curve were considered to isolate only the upward and downward peaks (consisting of negative values). The resulting curve has marked peaks in all the areas where there is an increase of presences in the library, so now it is sufficient to isolate the major peaks to define, among all, which is the moment or moments in which the flow of employment of the library is larger. Then applying this procedure to Library X, the result is that reported in Figure 10, which shows the average daily trend, its derivative and the peak of maximum turnout, which in this case coincides with the hours 9:45 am.

Figure 10. Chart regarding the average day and related derivative with the identification of the moment of maximum variation.

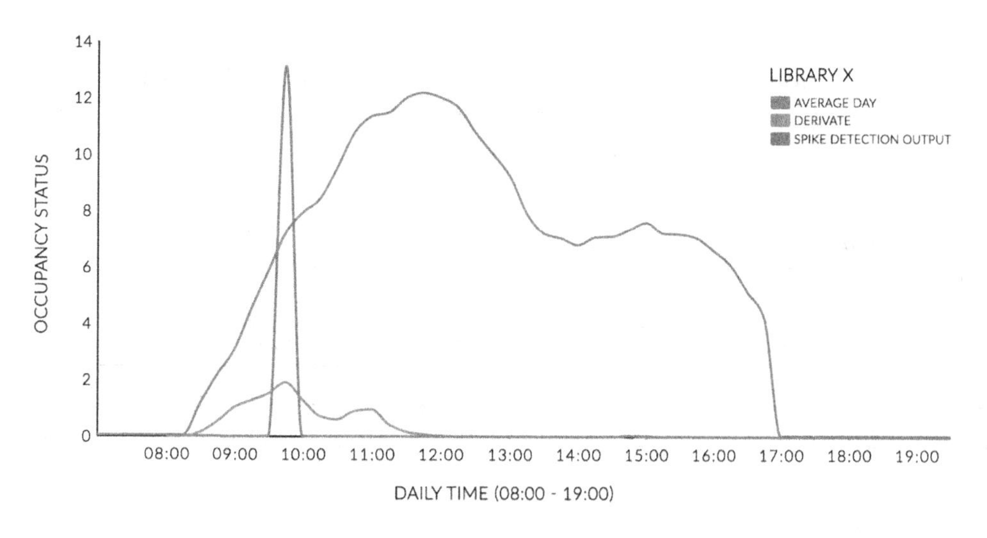

Figure 11. Box plot of the maximum peak time regarding occupancy in the test libraries

A second analysis was used to identify the time to reach the maximum daily occupation. The data used are those relating to the three test libraries: Library X, Library Y and Library Z. In order to have a more precise measurement in relation to this distribution, a box plot has been constructed that represents a detailed measure using the quartiles on the distribution of the obtained values, as shown in Figure 11.

The user activity dataset, on the other hand, contains every single activity performed inside the bot. It is therefore possible to consider this information as real statistical data on the use of the bot. The first test performed on this dataset is related to the distribution of operations performed by the users of the bot during the hours of the day.

This operation consists of a simple management of counters, one associated with each hour of the day: if an operation has been carried out at a certain time, the corresponding counter is increased by one.

Given the large amount of different operations and for a better visualization, it is possible to represent such data in relation to the type of command used in the bot in a certain time slot. To have a clearer view on this data, the keywords used in the bot have been grouped as follows:

- **Libraries:** Aggregates all the messages relating to the library areas to the library command;
- **Course:** Aggregates the functions related to the degree course (exams and timetables);
- **Campus:** Aggregates the functions related to the university (events, canteen, classrooms and professors);
- **Bot:** Aggregates the commands for bot management.

Once the representation criteria have been defined, what is said is summarized in Table 2. In most of the time slots, the most requested information is that relating to libraries. In addition, the hours with the most requests are those between 13:00 and 15:00.

Table 2. Distribution of user activities by keyword and time slots

Keyword	Time Slot						
	5-9	10-12	13-15	16-18	19-21	22-24	01-04
Libraries	44	43	112	46	6	28	5
Course	3	38	33	34	9	7	5
Campus	8	13	43	28	7	3	1
Bot	13	25	49	36	9	26	6

All the information extracted and collected from the log files can be used to identify, in an intelligent way, tips and suggestions to be notified directly to users. For example, from Table 2 it is clearly visible that the most used function corresponds to libraries. Thanks to the analysis described above, it is possible to calculate the best time to notify users in relation to the free seats in libraries, optimizing the movement within the campus.

CONCLUSION AND FUTURE DIRECTIONS

In this chapter we have proposed a platform able to aggregate, process and analyze data collected from a university campus in order to plan, in a "smart" and automated way, actions or operations in a university ecosystem. The proposed framework, thanks to the use of different data management techniques, a Telegram bot, and an IoT sensor system within the university environment, allows the solution of many problems that arise today between users and administration, such as errors on data due to poor maintenance or information retrieval on university facilities, which are often complex and not very intuitive. Some preliminary experiments showed that it is possible to obtain a better understanding of the activities and needs of the students, by analyzing already existing data. The Telegram bot has given the opportunity to directly compare student requests, analyzing types of requests, hours of use and giving the possibility to create a profiling system on which it is possible to deepen the work that, with the use of classification algorithms, will produce improvement of the whole system from the point of view of efficiency and automation.

An innovative aspect is that of improving the routine of each individual user, who will be able to manage his daily actions in an optimized manner thanks to the recommendation system and the information provided by the bot.

The topics dealt within this chapter may be considered as a starting point for the solution of other problems, which technical details are not reported here. One potential research direction is to use machine learning algorithms in order to perform user profiling and to design a recommendation system aiming to offer personalized information, possibly different for each user. Even the IoT system, thanks to the use of clustering algorithms or neural networks, could give further results beyond those provided during this study. A clustering algorithm could improve the search for specific patterns or the current system of anomaly detection, while a system based on neural networks could be useful to predict certain behaviors and further improve the management of the educational calendar.

A further step forward could be that of introducing Big Data techniques, allowing not only to better manage all heterogeneous sources, but also to introduce operations for computational optimization such as MapReduce and distributed computing. For the management of a large amount of data, these aspects are essential, as they also guarantee security and better performance on the operations.

Another factor to be considered will be the use of Open Data, already established in the public sector and in public administrations as an essential factor for a smart city, bringing benefits to citizens in terms of costs, transparency towards local authorities, urban security, public transport.

Finally, another important issue that has not been addressed in this study is the use of social networks. The data coming from these sources are full of useful information to analyze problems in a Smart City. By the analysis of tweets from Twitter, for example, important news, malfunctions and incidents may be identified in real time. By the use of images, videos, events and places of interest from other social networks, such as Facebook, it is possible to organize and improve the daily routine by suggesting activities, entertainment and important notices.

Finally, this project, currently proposed in the university field, can also be extended in a different context, such as the city, with appropriate changes to the strategies for managing data and communication protocols.

REFERENCES

Akkaya, K., Guvenc, I., Aygun, R., Pala, N., & Kadri, A. (2015, March). IoT-based occupancy monitoring techniques for energy-efficient smart buildings. In Wireless Communications and Networking Conference Workshops (WCNCW), 2015 IEEE (pp. 58-63). IEEE. doi:10.1109/WCNCW.2015.7122529

Anggoro, A. G. P. (2018). *Monitoring server room temperature remotely in real time using raspberry pi and firebase* (Doctoral dissertation). Universitas muhammadiyah surakarta.

Bersch, S. D., Azzi, D., Khusainov, R., Achumba, I. E., & Ries, J. (2014). Sensor data acquisition and processing parameters for human activity classification. *Sensors (Basel)*, *14*(3), 4239–4270. doi:10.3390140304239 PMID:24599189

Boman, J., Taylor, J., & Ngu, A. H. (2014, October). Flexible IoT middleware for integration of things and applications. In *Collaborative Computing: Networking, Applications and Worksharing (CollaborateCom), 2014 International Conference on* (pp. 481-488). IEEE. 10.4108/icst.collaboratecom.2014.257533

Chahuara, P., Fleury, A., Portet, F., & Vacher, M. (2016). On-line human activity recognition from audio and home automation sensors: Comparison of sequential and non-sequential models in realistic Smart Homes 1. *Journal of Ambient Intelligence and Smart Environments*, *8*(4), 399–422. doi:10.3233/AIS-160386

De Angelis, E., Ciribini, A. L. C., Tagliabue, L. C., & Paneroni, M. (2015). The Brescia Smart Campus Demonstrator. Renovation toward a zero energy classroom building. *Procedia Engineering*, *118*, 735–743. doi:10.1016/j.proeng.2015.08.508

Harsha, S. S., Reddy, S. C., & Mary, S. P. (2017, February). Enhanced home automation system using internet of things. In *I-SMAC (IoT in Social, Mobile, Analytics and Cloud)(I-SMAC), 2017 International Conference on* (pp. 89-93). IEEE. 10.1109/I-SMAC.2017.8058302

Heryandi, A. (2018, August). Developing Application Programming Interface (API) for Student Academic Activity Monitoring using Firebase Cloud Messaging (FCM). *IOP Conference Series. Materials Science and Engineering*, *407*(1), 012149. doi:10.1088/1757-899X/407/1/012149

Jain, S., Garg, R., Bhosle, V., & Sah, L. (2017, August). Smart university-student information management system. In *Smart Technologies For Smart Nation (SmartTechCon), 2017 International Conference On* (pp. 1183-1188). IEEE.

Jurek, A., Nugent, C., Bi, Y., & Wu, S. (2014). Clustering-based ensemble learning for activity recognition in smart homes. *Sensors (Basel)*, *14*(7), 12285–12304. doi:10.3390140712285 PMID:25014095

Krishnan, N. C., & Cook, D. J. (2014). Activity recognition on streaming sensor data. *Pervasive and Mobile Computing, 10*, 138–154. doi:10.1016/j.pmcj.2012.07.003 PMID:24729780

Oshin, O., Owoniyi, A., Oni, O., & Idachaba, F. E. (2017). *Programming of NFC Chips: A University System Case Study*. Academic Press.

Pedersen, T. H., Nielsen, K. U., & Petersen, S. (2017). Method for room occupancy detection based on trajectory of indoor climate sensor data. *Building and Environment, 115*, 147–156. doi:10.1016/j.buildenv.2017.01.023

Popoola, S. I., Atayero, A. A., Okanlawon, T. T., Omopariola, B. I., & Takpor, O. A. (2018). Smart campus: Data on energy consumption in an ICT-driven university. *Data in Brief, 16*, 780–793. doi:10.1016/j.dib.2017.11.091 PMID:29276746

Pradhan, P., Naik, A., & Patel, P. (2014). Location Privacy in Ubiquitous Computing. *International Journal of Research in Science and Technology*, 54.

Tapia, E. M., Intille, S. S., & Larson, K. (2004, April). Activity recognition in the home using simple and ubiquitous sensors. In *International conference on pervasive computing* (pp. 158-175). Springer. 10.1007/978-3-540-24646-6_10

Van Merode, D., Tabunshchyk, G., Patrakhalko, K., & Yuriy, G. (2016, February). Flexible technologies for smart campus. In *Remote Engineering and Virtual Instrumentation (REV), 2016 13th International Conference on* (pp. 64-68). IEEE. 10.1109/REV.2016.7444441

Weiser, M., Gold, R., & Brown, J. S. (1999). The origins of ubiquitous computing research at PARC in the late 1980s. *IBM Systems Journal, 38*(4), 693–696. doi:10.1147j.384.0693

ADDITIONAL READING

Batty, M., Axhausen, K. W., Giannotti, F., Pozdnoukhov, A., Bazzani, A., Wachowicz, M., ... Portugali, Y. (2012). Smart cities of the future. *The European Physical Journal. Special Topics, 214*(1), 481–518. doi:10.1140/epjst/e2012-01703-3

Cook, D., & Das, S. K. (2004). *Smart environments: Technology, protocols and applications* (Vol. 43). John Wiley & Sons. doi:10.1002/047168659X

Deakin, M. (2012). From Intelligent to Smart Cities: CoPs as organizations for developing integrated models of eGovernment Services. In City Competitiveness and Improving Urban Subsystems: Technologies and Applications (pp. 84-106). IGI Global.

Kunzmann, K. R. (2014). Smart cities: A new paradigm of urban development. *Crios, 4*(1), 9–20.

KEY TERMS AND DEFINITIONS

Back-End: Program with which the user interacts indirectly, usually through the use of a front-end application. In a client/server structure the back-end is the server.

Data Mining: Data mining is the set of techniques and methodologies that have as their object the extraction of knowledge or knowledge from large amounts of data and the scientific, industrial or operational use of this knowledge.

Feature Selection: Is the process of selecting a subset of relevant features like variables or predictors for use in model construction.

Front-End: Program with which the user interacts directly, which acts as an interface between the user and another program (back-end). In a client/server structure, the front end is the client.

Smart Campus: Environment that links devices, applications, and people to enable new experiences or services and improve operational efficiency.

Smart Device: Device that uses machine-to-machine technology to offer services that enhance the user experience.

Smart Environment: Aspect of a smart city concerning the control and monitoring of environmental factors such as pollution, waste, planning of green areas, and energy.

Smart Living: Aspect of a smart city regarding lifestyles, consumption, behavior, in the interest of greater social cohesion.

Chapter 15
Smart IoT Meters for Smart Living

Ramesh Kesavan
https://orcid.org/0000-0003-2733-4484
Anna University Chennai – Regional Office Tirunelveli, India

Pushpa Jaculine J.
Infant Jesus College of Engineering, India

ABSTRACT

Smart cities and smart villages provide technology-based, sophisticated, and better lifestyles to their citizens. Smart cities include traffic control, transport management, managing spare resources like power and water, solid waste management, e-health monitoring, infrastructure management based on internet of things (IoT) technology. IoT is a technique that combines sensors, electronic devices, information and communication technology, and software for the social wellbeing of the common man. In recent years, many IoT-based smart devices, namely smart garbage bins, automatic parking system, smart electric meters, supervisory control and data acquisition (SCADA) for water distribution, have been devised and used successfully in many cities. Mostly, smart meters are used in recording electric power and gas consumption.

INTRODUCTION

This chapter introduces the basic concept of smart meter and we will also learn about the standard architecture of smart energy meter, smart gas meter and smart water meter.

Smart Metering

Smart metering is a technology enabled businesses procedure to keep track of energy consumption, automatic meter reading collection, process the collected data and use the processed information for business development. As smart metering is an efficient and effective method for tracing resource utilization in

DOI: 10.4018/978-1-5225-9199-3.ch015

the distribution network. It acts as a decision support system for the supplier and for customers it acts as an assistant who informs and guides in energy utilization. On note of vast benefit electric, gas and water distributing companies and corporations are moving towards the smart metering concept (Lloret, Tomas, Canovas, & Parra, 2016). It is also an essential framework for establishing smart cities and smart villages.

Why Smart Meters?

Water, electricity and cooking gas are indispensable resource for life, economic well-being, and environmental integrity of the society. Even distribution of these vital resources to its citizens is a prime responsibility of the government. Decades ago the government used analog meters to record the consumption for billing purpose. Due to the advancement of science and technology the developing countries are trying to monitor and ensure the distribution along with billing using modern electronic devices. What is a smart meter? Smart meters are digital meter which replaces the traditional analog meters used in measuring and monitoring the consumption of resources like electricity, gas and water by households and industries. Smart meter embeds microcontroller, sensor, wireless and web technology to upgrade the mechanical device to an IoT based smart digital device. Smart meter prototype will provide technical solutions to monitor the distribution networking system and to ensure even distribution with nominal operational cost.

Advantages of Smart Meters

1. **Digital Display:** Most Smart meters have a digital display screen that shows the reading in their place at real time.
2. **Accurate:** As it is digital the reading are more accurate when compared with analog meters.
3. **Ease of Monitoring:** Smart meters provide provision to monitor the consumption more precisely so valuable resources like electricity and gas can be consumed in right way.
4. **Frequent Reading:** Can record reading frequently. i.e. hourly or daily based on the necessity.
5. **Automatic Reading:** Automatically transmits the reading to the customer as well as to the supplier or the distributer of the product. In case of analog meter a meter reader needs to collect record and transmit information to the administrative office.
6. **Timely Decision Making:** As the readings are collected frequently and as it is in digital format, collected information can be used for real time analytics. Based on the outcome of the analysis real time decision making can be done for the proper utilization and distribution of resources.
7. **Accurate Billing:** Transparent and accurate billing.

Application of Smart Meters

Smart Cities

As discussed earlier, smart cities use digital technology to enhance the life style of the people in the city. In the point of energy and resources conservation and preservation, adaptation of digital communication and information technology will optimize the utilization of resources. In smart cities "smart grids" play a vital role in effective distribution and conservation of electrical energy. Smart grid is the next generation of electric grids which uses digital technology to manage and monitor energy distribution.

The first and foremost step moving from traditional electric grid to smart grid is replacing the existing electromechanical meters with smart energy meters. In the same way smart meters are used to measure, process and transfer the consumption data of different resources like water and gas. Heat and cooling meters are applications that are used to measure the energy consumed to either heat or cool a centralized heating or cooling system installed building or apartment.

Farming and Agriculture

Smart metering is also an integrated part of smart farming and agriculture. Particularly smart meter will be an opt solution for the optimum usage of water in irrigation. Energy meters can call also be used for automatic function of water pump.

Automatic Meter Reading (AMR) in Developed Countries

Energy conservation strategy in developed countries has taken a multi-prong approach through pricing, mandatory conservation requirements and promoting and encouraging ownership and voluntarism. Developed nations are treating power, gas and water as an economic good. Pricing of resources is an important and effective mechanism in encouraging customers to conserve energy resources. Sensor based meter reading system was coined in early 1970's. But the initiative on automatic meter reading in developed countries got its full momentum in the past two decades. As per 2015 report, in developed countries more than 90% of the outages has been fitted with *Advanced Metering Infrastructure* (AMI) (Uribe-Pérez, Hernández, de la Vega, & Angulo, 2016).

AMI is the collection of peripherals needs to setup a complete smart meter.AMI includes hardware, software, communications, consumer energy displays and controllers, customer associated systems, Meter Data Management Software (MDMS), and supplier business systems. Figure 1. illustrates the building block of an AMI. It consists of three blocks. Data collection block is where the meter is physically placed. Data communication block is which contains the wireless communication devices like Wi-Fi, Satellite communication or GSM module to have two way communications between the data collection unit and the MDMS. The third block is the MDMS which holds the memory for storage, serve for processing the data and visualization components.

Automatic Meter Reading in Developing Countries

As developed countries are ahead in technology, infrastructure and economical status adapting to AMR is not a major problem. Although adaptation of AMR for energy and water distribution are beneficiary, in developing countries implementation of AMR faces many hurdles due to the lack of infrastructure, lack of qualified manpower, and limited financial resources. Several countries may have manpower available but lack finance. Others may have ample funds, but have a shortage of trained people. Many have neither of them. Each of the countries has its own options and constraints for future developments. In many developing countries there is also a serious scarcity of resources. In these countries the introduction of resource management and distribution using IoT practices would be an obvious option for overriding the above stated automatic distribution problems.

Figure 1. Building Blocks of AMI

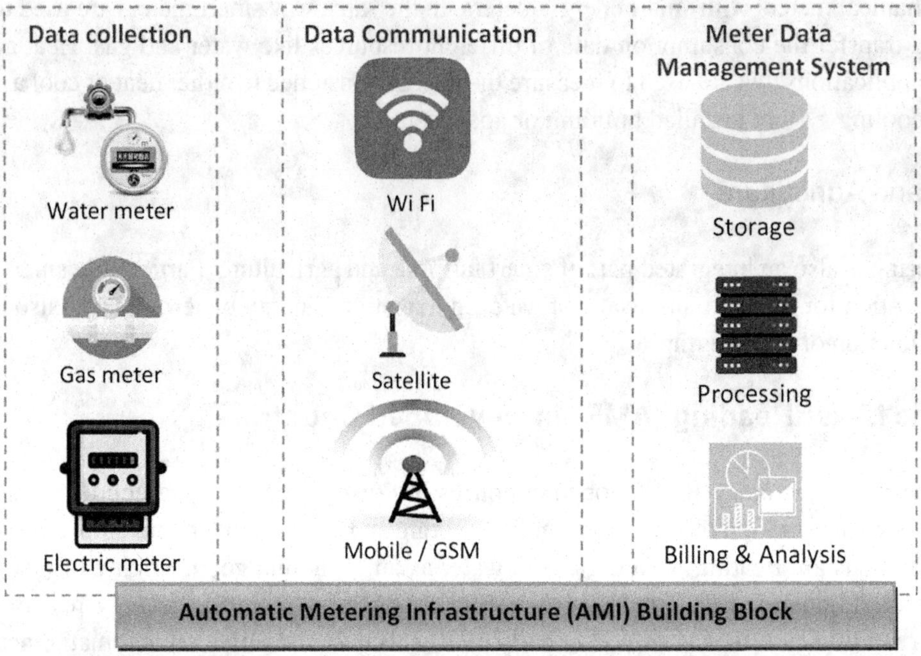

IoT in Smart Metering System

IoT is the networking of physical devices by which electronic devices can communicate among themselves. The data communication networking, data collection and remote handling and accessing of the smart metering systems are done by integrating IoT technology.

The following section describes features of a smart energy meters, smart water meter and smart gas meter.

Smart Energy Meter

Smart energy meter is an electronic device that record consumption of electric energy in intervals of time and communicates that information for monitoring, billing and also displays data on web page (Alahakoon & Yu, 2016). The system architecture of a typical smart energy meter consists of two modules namely data acquisition module and wireless transmission module as shown in figure 2. Data acquisition module consists of energy meter, current/voltage sensor and a microcontroller (Benghanem, 2010). The input power supply for the load in the premises is passed through the electric meter. The electric meter passes the current consumed or the load utilized to the current/ voltage sensor. The voltage and current sensors in the data acquisition module measure the RMS (Root Mean Square) values of voltage and current and feed them to the microcontroller, where calculations for active and reactive power are performed. The microcontroller consists of a Central Processing Unit (CPU) to process, Random Access Memory (RAM) and Read only Memory (ROM) to handle and store data, an Analog to Digital Converter (ADC) to convert the analog meter reading to digital signal for future processing and transmission. The microcontroller controls the wireless module and sends the information to database and consumer.

Figure 2. Block diagram of Smart Energy Meter

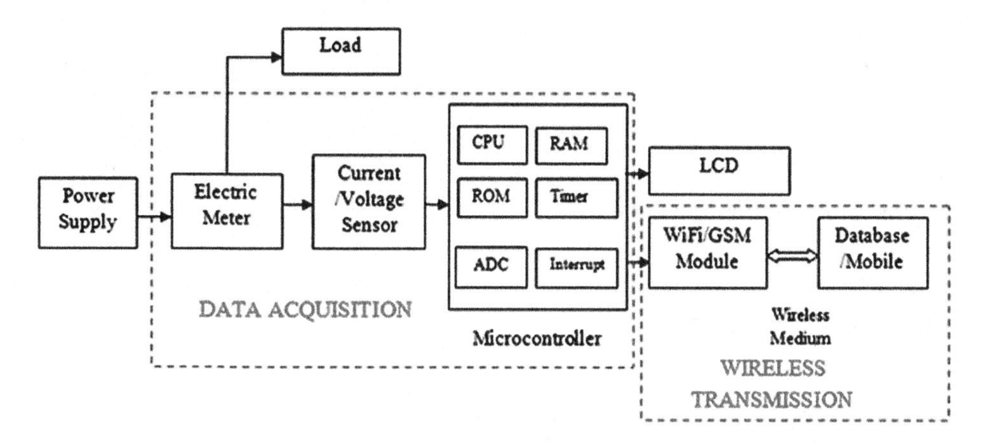

Components of Smart Energy Meter

Electric Meter

The meter which is used for measuring the energy utilized by the electric load is known as the electric meter. The energy is the total power consumed and utilized by the load at a particular interval of time.

Current / Voltage Sensor

In a smart energy meter to measure the current flow either a current sensor or a voltage sensor is employed.

Current sensor is used to measure electric current flow in a wire. Current sensor detects and converts consumed current to proportionate output voltage. The output of current sensor is usually analog voltage signal but modern sensors generate easily accessible digital signal corresponding to the current flow. The output signals from current sensor can be visualized in a digital display unit (LED or LCD) or the signal may be stored as reading for billing and further analysis.

Voltage sensor is an alternative to current sensor which also detects and measures the AC and/or DC voltage levels in a circuit. Common applications of current and voltage sensors are billing or metering, failure and power quality monitoring.

Microcontroller

As the name suggests microcontroller is the core controller of all operations in this metering system. Microcontroller is a single integrated circuit, with the following peripherals. A processing unit called CPU which ranges from 4 bit processor to 32 or 64 bit processors. Volatile RAM for data storage, ROM, Erasable Programmable Read-Only Memory (EPROM), Electrically Erasable Programmable Read-Only Memory (EEPROM) and Flash memory for programming as well as storage of the processing parameters. Bi-directional I/O pins allowing control and detection of logic state, Universal Asynchronous Receiver/ Transmitter (UART), Serial communication Interfaces, Serial peripheral interface and controller area network for system interconnect, timer, counter, Pulse Width Modulation (PWM) generator, watchdog timer, clock generator, ADC and DAC for digital conversions.

Case Study on Microcontroller Used in Smart Meter

1. Arduino UNO

Arduino is an open-source hardware/ software platform used for building multidisciplinary electronics projects. Arduino consists of a physical programmable circuit board (a microcontroller) and Arduino software IDE (Integrated Development Environment) for programming, which can be executed in a normal computer.

There are many varieties of Arduino boards for different purposes or objective. The common components of major Arduino's are

- **USB Power:** Used to power the Arduino UNO using USB cable.
- **Barrel Jack Power:** for DC wall supplies.
- **Pins** (5V, 3.3V, GND, Analog, Digital, PWM, AREF)
- **GND:** Used to ground the circuit.
- **5V & 3.3V:** 5V pin supplies 5 volts of power and the 3.3V pin supplies 3.3 volts of power.
- **Analog:** These pins read the signal from an analog sensor and convert it into a digital value.
- **Digital:** These pins can be used for both digital input and digital output.
- **PWM (Pulse-Width Modulation):** These pins act as normal digital pins.
- **AREF:** Analog Reference.
- **Reset Button:** Restart any code that is loaded on the Arduino
- **Power LED Indicator:** To indicate about the ON / OFF mode of the system.
- **TX RX LEDs:** Transmitter/ Receiver LEDs, indicates whether Arduino is receiving or transmitting data
- **Main IC:** Integrated circuit.
- **Voltage Regulator:** Controls the amount of voltage that is let into the Arduino board

2. PIC

PIC (Peripheral Interface Controller) is another familiar smallest microcontroller which can be programmed to carry out different tasks in electronics based industries. PIC microcontroller can be programmed using Circuit-wizard software. Its architecture comprises of CPU, I/O ports,A/D converter, memory organization (RAM& ROM), timers, counters,serial communication, interrupts, oscillator and CCP (Capture/ Compare/PWM) module and protocols like Serial Peripheral Interface (SPI), UART, Controller Area Network (CAN) which are used for interfacing with other peripherals as shown in Figure 3. The advantages of using this microcontroller include low power consumption, high performance, support hardware and software tools such as simulators, compilers, and debuggers.

3. Raspberry Pi

Raspberry Pi is a series of small single-board computers with a dedicated memory, processor, and a graphics card for output through HDMI. Though various versions of Raspberry Pi are available, the basic building block of all models features a Broadcom system on a chip (SoC) with an integrated ARM-compatible, CPU and on-chip GPU. The raspberry pi board comprises of the following hardware.

Figure 3. Architecture of PIC microcontroller

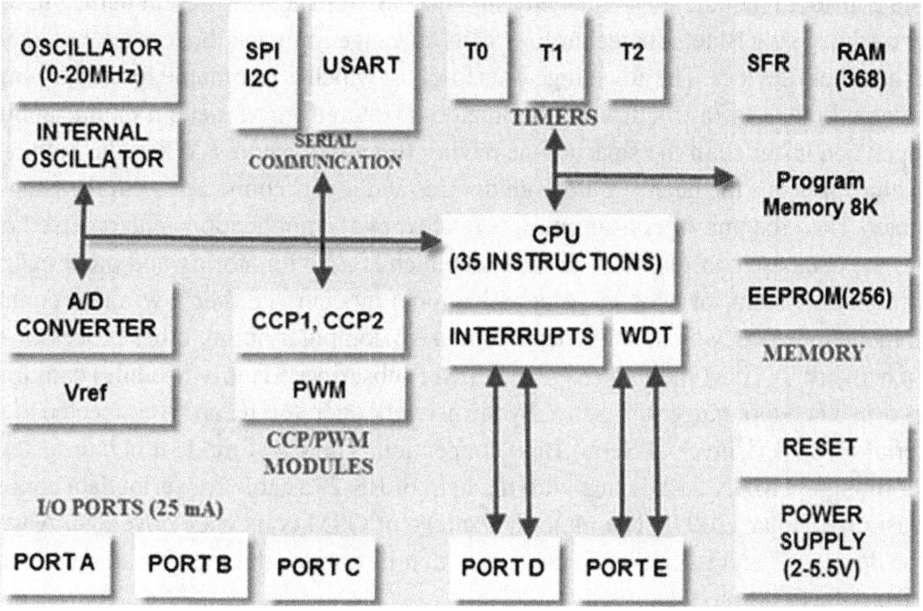

- **Memory:** A program memory (RAM) of size 256MB or 512MB
- **CPU:** CPU carries out the instructions of the computer through arithmetic and logical operations. Raspberry pi uses ARM11 series processor.
- **GPU (Graphics Processing Unit):** Raspberry board comes with Broadcom video core IV chip for graphical processing.
- **Ethernet Port:** The Ethernet port is used as the base for all communication.
- **GPIO Pins (General Purpose Input & Output Pins):** GPIO acts as an interface between other all components in the system.
- **XBee Socket:** For wireless communication purpose.
- **Power Source Connector:** To connect external power source.
- **UART:** A serial input & output portused to transfer the serial data.
- **Display:** Outputs can be displayed either using HDMI or Composite video.

Data Processing Module

The second and prominent module which makes this digital device as smart device is the data processing module. This module comprises of communication module, database to store and a web application or a mobile application for billing and analysis purpose. The Communication module transmits the current consumption data from the microcontroller to the database using wireless communication systems. Some of the commonly used Wireless Communication Systems are Bluetooth, Wi-Fi, Infrared (IR) wireless communication, satellite communication, and Zigbee etc.

Bluetooth technology is a high speed low powered wireless technology link that is designed to connect phones or other portable equipment together. Wireless signals transmitted with Bluetooth cover short distances, typically up to 30 feet (10 meters). The Bluetooth module can receive and send information wirelessly from microcontroller to a mobile application or a computer. Communication between micro-controller and Bluetooth module is established using the USART module present in the microcontroller. The limitation while using Bluetooth technology is its coverage limit and the reading has to be collected manual with handheld devices. The advantage is its low cost which is affordable for developing countries with the available limited infrastructure. For a Bluetooth featured smart meter a mobile application has to be developed and installed in the smartphone having Bluetoothfeatures. When the mobile Bluetooth is turned on, the app scans the nearby Bluetooth devices and it gets connected with the Bluetooth module in the meter. Then the unit of consumption is fetched by the application. The received data is then transferred to the database console through internet which is used for storing and processing the data.

A GSM (Global System for Mobile communications) modem is either a wireless communication module or a modem device, which can be used to make a computer or any other processor to communicate over a network. A GSM modem requires a SIM (Subscriber Identity Module) card for operation and operates over a network range subscribed by the network operator. It can be connected to a computer through a serial or USB (Universal Serial Bus) connection. The GSM modem is interfaced to the microcontroller through a MAX-232 device with the help of RS-232 cable for serial data communication (Islam & Wasi-ur-Rahman, 2009). The major advantage of GSM is its wide coverage. A web application has to be developed and maintained such a way that it receives the meter reading sent in the form of message from the GSM module.

IR Communication is a medium range electromagnetic wave based communication system for transmitting and receiving data between two portable or fixed devices. Wireless Local Area Network (LAN) communication system is a diffuse point IR based communication system. To fetch data from IR module integrated meters a mobile application has to be developed and installed in the Smartphone or a Personal Digital Assistant (PDA) having IR features.

Selection of wireless communication technology depends upon the type of application and the financial implication towards the installation of the meter.

Mobile Application

A mobile application, most commonly referred to as an app, is a type of application software designed to run on a mobile device, such as a Smartphone or tablet computer. Mobile application for smart metering is designed in a way that it receives data from the microcontroller through a wireless module and stores it in a permanent database. It should also have a Graphical User Interface (GUI) for easy processing and visualization of the meter reading. There are many tools available for developing mobile apps for different mobile device operation systems like Android, Windows, IOS.

Web Application

Web application is another interface which receives meter reading through internet and store in a remote web server using web technology. The microcontroller transmits the received data to the web application in real time or in non-real time. Web application acts as a mediator between the resource provider and the consumer by inheriting the infrastructure of the internet. Using the web applications the billing, online payment and remote monitoring of the meter can be done easily.

Smart Water Meter

Among all natural resources, water is an indispensable resource for life, economic well-being, and environmental integrity of the society. The water scarcity issue is intensifying primarily due to variation in rainfall and rise in consumption but principally due to poor water distribution and monitoring system. A survey states that by 2025, 1.8 billion people will be living in countries with water stressed situations. Both developed and developing countries are working on solutions to meet out the basic water need of their citizens. One among the initiative is adapting technology enabled water distribution for effective water resource management. In particular, usage of smart water meter will lead to effective monitoring and controlling of water distribution (Meseguer & Quevedo, 2017).

Smart water meter is an electronic device that records water consumed by individual consumer daily and communicates the water consumption detail to supply board for monitoring and billing. This handy smart device can be fitted at different households and industries to measure the water flow and to record the water consumption. Bluetooth/GSM enabled feature in the device transmits the quantity of water consumed to the mobile application with the bill collector. The flow rate is transferred and stored to a central server for the further processing like billing and planning.Smart water network solution improves the longevity and reliability of the water network by measuring, collecting and analysing a wide range of network area.

Smart Water Meter Architecture

The methodology used for smart water distribution management is shown in figure 4 (Pushpa Jaculine, Ramesh, 2018). The system to be installed at the inlet of water supply line comprises of three modules namely, Wireless enabled smart meter, mobile application and web application. The smart meter consists of water flow sensor, wireless module and microcontroller. The water flow sensor in the water distribution pipeline is connected to microcontroller which is used as the core controller. The volume of water used is detected by a water flow sensor. The microcontroller receives the usage data from flow sensor. The wireless module is connected to the controller for wireless transmission of flow rate in real time.

The overall block diagram of a typical water monitoring system is given in Figure 5. It comprises of three main components namely the main controller, Smartphone application / web application and database. The water flows through the distribution pipe and enter into the flow sensor of individual houses

Figure 4. Architecture of Smart Water Management

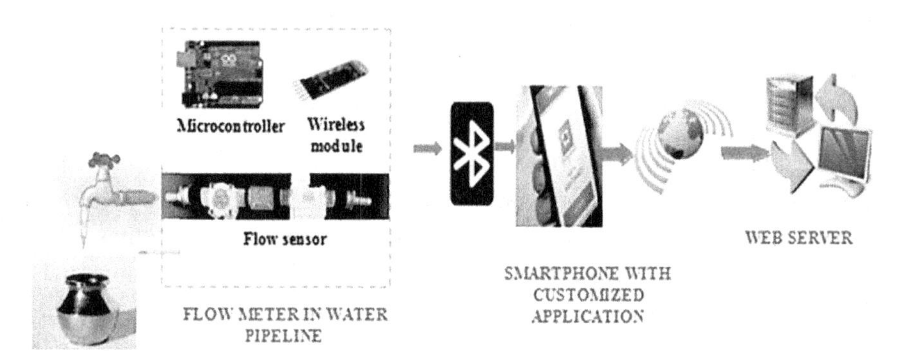

Figure 5. Block Diagram of Smart Water Management System

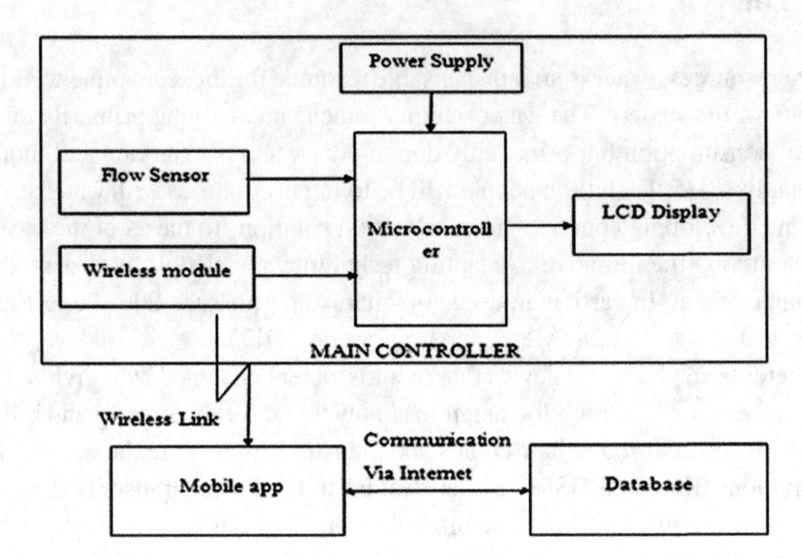

or industries. The flow sensor calculates the volume of water passes through the pipe with the help of microcontroller by using pulse width modulation technique and stores the flow rate.

Hardware Requirements

The smart water meter replaces the energy meter and the current / voltage meter in smart energy meter with a flow sensor.

Flow Sensor

Flow sensor is used to measure the water flow of different households or industries. The sensors are solidly constructed and provide a digital pulse each time when a particular amount of water passes through the pipe. The output can easily be connected to a microcontroller for monitoring water usage and calculating the amount of water consumed. The water flow sensor is positioned in line with water supply pipe line and the water flow through pipe pushes the rotor vanes of the flow sensor. It uses a pinwheel sensor to measure how much liquid has passed through it. The pin wheel has a little magnet attached, and there's a hall-effect magnetic sensor on the other side of the plastic tube that can measure how many spins the pinwheel has made through the plastic wall (Cominola, Giuliani, Piga, Castelletti, & Rizzoli, 2015).

Smart Gas Meter

Piped gas network is a basic facility integrated in smart cities proposal which provides uninterrupted supply of cooking gas and natural gas for domestic use and commercial purpose. Piped gas network needs a meter to measure the gas usage on real time. Smart Gas meter is yet another special digital flow meter which measures the volume of fuel gases used at residential buildings, commercial, and industrial outlets. It should measure all fuels includes natural gases and Liquefied Petroleum Gas (LPG). The major

Figure 6. System architecture of a smart gas meter

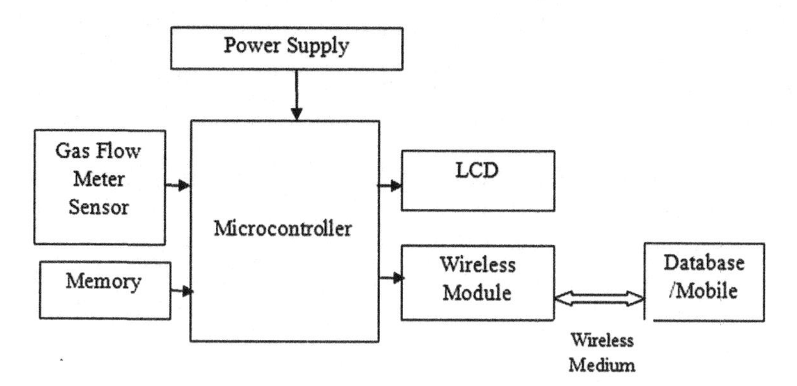

challenge in measuring gas fuel is its volume differs with change in pressure and temperature. To overcome this issue, most gas meter measures the volume of gas consumed irrespective of its temperature and power. Total gas consumption is calculated by the gas flow meter sensors, and information passed to both the distributer and the consumer.

The system architecture of a smart gas meter is shown in figure 6. The basic module of the system is similar to water metering system. Since the physical property of the gas differs from water we need a unique gas flow sensor in the place of water flow sensor.

As this monitoring system integrates different existing techniques and low cost components, the total amount required for developing the device is relatively small which is affordable for developing countries to setup the device in households and industries. The lifetime of the device is also large and hence it is economically feasible. The device does not need trained personnel's to handle it and it is less error prone in operation and also highly reliable. It is a device that provides transparent and effective distribution system for industries and households and also it creates awareness among the people about the resource management (Dong et al., 2017).

REFERENCES

Alahakoon, D., & Yu, X. (2016). Smart electricity meter data intelligence for future energy systems: A survey. *IEEE Transactions on Industrial Informatics*, *12*(1), 425–436. doi:10.1109/TII.2015.2414355

Benghanem, M. (2010). *RETRACTED: A low cost wireless data acquisition system for weather station monitoring*. Elsevier.

Cominola, A., Giuliani, M., Piga, D., Castelletti, A., & Rizzoli, A. E. (2015). Benefits and challenges of using smart meters for advancing residential water demand modeling and management: A review. *Environmental Modelling & Software*, *72*, 198–214. doi:10.1016/j.envsoft.2015.07.012

Dong, S., Duan, S., Yang, Q., Zhang, J., Li, G., & Tao, R. (2017). Mems-based smart gas metering for internet of things. *IEEE Internet of Things Journal*, *4*(5), 1296–1303. doi:10.1109/JIOT.2017.2676678

Islam, N. S., & Wasi-ur-Rahman, M. (2009). *An intelligent SMS-based remote water metering system*. Paper presented at the 2009 12th International Conference on Computers and Information Technology.

Lloret, J., Tomas, J., Canovas, A., & Parra, L. (2016). An integrated IoT architecture for smart metering. *IEEE Communications Magazine*, *54*(12), 50–57. doi:10.1109/MCOM.2016.1600647CM

Meseguer, J., & Quevedo, J. (2017). *Real-time monitoring and control in water systems. In Real-time monitoring and operational control of drinking-water systems* (pp. 1–19). Springer.

Ramesh. (2018). IoT based smart water distribution system. *International Conference on Applied Soft Computing Techniques ICASCT-18.*

Uribe-Pérez, N., Hernández, L., de la Vega, D., & Angulo, I. (2016). State of the art and trends review of smart metering in electricity grids. *Applied Sciences*, *6*(3), 68. doi:10.3390/app6030068

Chapter 16
IoT for Waste Management: A Key to Ensuring Sustainable Greener Environment in Smart Cities

Jaganthan Thirumal
Anna University, India

Usha Kingsly Devi
Anna University, India

Dynisha S.
Anna University, India

ABSTRACT

Smart cities incorporate information and communication technology to enhance the quality and performance of urban services. The element of smart cities includes physical infrastructure and IoT technology, which gives a framework, methodology, technology, and management solution and efficient waste handling and reduction with the assistance of software analysis tools. It provides effective environmental resource flow integration. IoT system provides a digital access to waste management. This system uses online smart monitor sensors that monitor the performance of water supply and effluent handling system utilizing a cloud-based platform. This enhances real-time planned performance and increases life-cycle equipment. This technique enhances the synergistic use of resources due to climate mitigation and adaptation for sustainable growth and this technology also uses air quality sensors across the city to collect open data platform for monitoring and reducing primary and secondary pollutants and systematically instruct the pollutant-causing sources to maintain ambient air quality.

INTRODUCTION

Now a day's cities is changing to the more complicated ecosystem due to rapid increase of India's Economy and a steady shift of strategically culture from Rural to Urban. By 2030, over 60% of the world's population will be surviving in cities. The vast urban growth results in pollution, traffic congestion, and social inequality. This modifies the urban center as a stage of convergence of social, demographic, eco-

DOI: 10.4018/978-1-5225-9199-3.ch016

nomic and environmental risks (Ramaprasad et al., 2017). In recent times, the population explosion has resulted in a number of environmental problems. The usefulness of resources in the environment is primarily damaged by pollution and causes illness to humans, plant and animal life. Depending upon the area of the environment affected and developmental activities such as construction, manufacturing and transportation not only deplete the natural resources, but also produces large amounts of wastes that lead to pollution of air, water and soil. This causes ill health and loss of crop productivity. Hence, smart systems are evolving to overcome such difficulties and concoct cities as smarter one. As a fashion label, the term smart city is used by the researchers and governments since 1990 (Ramaprasad et al., 2017). The invention of World Wide Web and wireless communication to mankind are now subjected to the most unmanageable juncture of the Internet rebellion—the "Internet of Things" (IOT) which is familiarly termed as "Ubiquitous Computing. Smart cities bring smart system using IOT for governing smart living, smart economy and smart environment for the benefit of both human and the ecosystem. The conceptualization of smart city depends on the level of development, reform, resources, and aspirations of the city residents towards their willingness to commute. Smart city primarily focuses on inclusive and sustainable development. The main infrastructure elements of smart city include sufficient supply of water, non intermittent electricity supply, sanitation that includes management of solid waste, good urban mobility and public transport, affordable housing, robust communication, information flow using connectivity, digitalization, improved governance and citizen participation, sustainable environment, safety and security of citizens, health and education. As a result of modernization, there is a patulous progress in Internet technologies and wireless sensor networks. The Internet of things (IOT), which is a newfangled archetype bedecked with transceivers, micro controllers, and suitable concordant stacks for digital communication. With a service of sensors, IOT expedite an environmental protection by keeping track of air quality, water quality, atmospheric and soil conditions.

SMART CITY

Smart Cities is an innovative city that uses ICTs to improve quality of life characterized by development in economic condition, sustainability, and a high quality of life. This innovation integrates conditions of all of its critical infrastructures through monitoring. It should engage with all the services on offer and connects people, information and city elements using new technologies in order to achieve the efficiency of urban operation and services. Also it should satisfy the needs of present and future generations with respect to social, economic, and environmental aspects Smart City is persistently dominated by new modes. As a consequence, in order to keep up with rapid development the city must be susceptible and adaptable. The modes which make smart cities possible will also put higher demands on the underlying communications infrastructure. The modes in communications technology and other related technical advances such as 5G, big data and cloud services are entered in technical modes. The applications or tools that help citizens, businesses and city administrations develop services including health care and transport in new and existing areas comes under the application mode. Further, how a city chooses to organize and procure its solutions and how these affect the city's role as the facilitator of an efficient Smart City is discussed by market model modes. The figure 1 represents the modes of smart city.

Figure 1. Modes of smart city

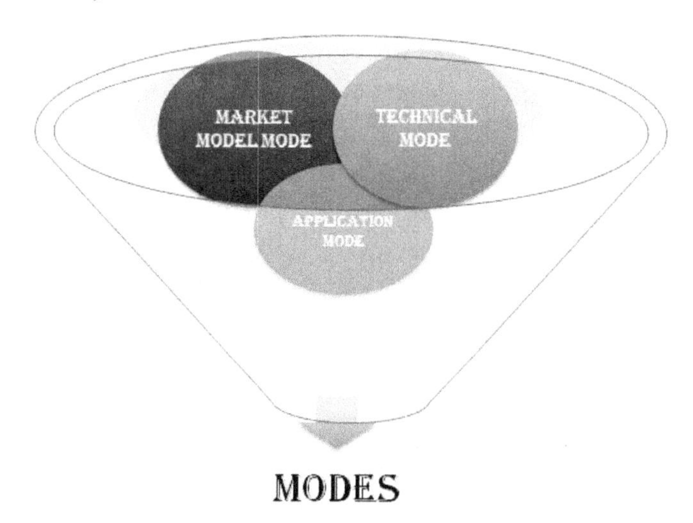

INTERNET OF THINGS

A novel technique of internet approaching is Internet of things. It refers to the interconnection of uniquely identifiable embedded computing devices within the existing Internet infrastructure. This interconnection of machine-to-machine communications (M2M) is now emerged nearly in all the sectors. It accedes to frame the way of life easier, reliable and resourceful. It is not gentle to contrive a wholesome environment because of Industries and transit of wastes. A variety of Protocols, domains, and applications is offered by advanced connectivity of devices, systems, and services. The substantial emerging strands of IOT are sensor networks and Radio Frequency Identification (RFID). It is the connectivity of devices, physical objects and buildings which enclosed with software, electronics, sensors and network congruence to gather and swap data. The fragmented market and island solution of smart Cities applications has overcome by IOT technologies and provide inclusive solutions for all cities. The successful outcome of IOT relies on systematization which contributes interoperability, compatibility, effectual operations and trust worthy on a global scale. Around 50 to 100 billion things will be electronically linked with internet by 2020. To attain eco- friendly universe and tolerable existence, the smart environmental devices assimilated with Internet of things for pursuing, perceiving and observing environmental challenges (Zeinab et al., 2017). An IOT functions in perception, network, middleware and application layers for observing environmental condition (Supriya et al., 2017). The figure 2 show the layers of IOT.

1. Perception Layer

The service of perception layer is to accumulate real time datasets for surveillance of environmental troubles.

2. Network Layer

Linkage of the systems and platforms as well as Information conveyance is the elemental operation executed by network layers. It is mainly composed of networks of access and networks of transport.

Figure 2. Layers of IOT

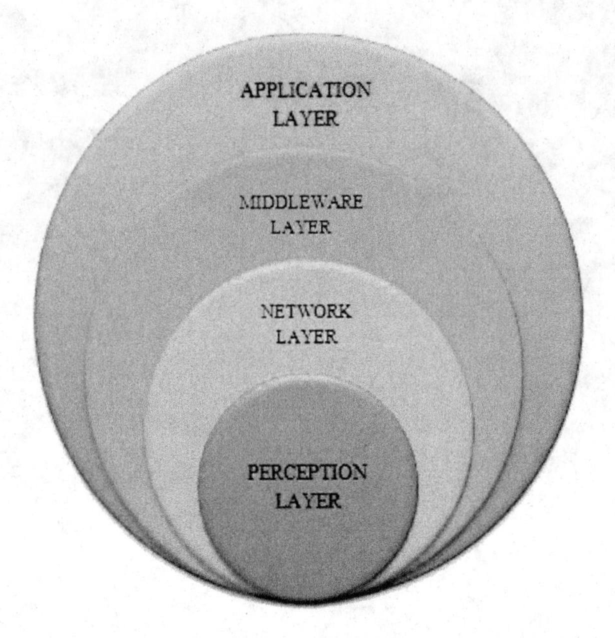

An accessed network includes Short-range wireless networks comports of the Sensors Area Network (SAN), 2G, 3G, Wi-Fi, and Zigbee.

Wide Area Networks (WAN) of wired or wireless hybrid network subsystems of ECS with wired and wireless broadband IP network cloud with web service based global network transport protocols were transport networks.

3. Middle Layer

For the manipulation of data, software, prototype and platforms and interruption among the network a firm of sub- layers called middle layer was adopted.

4. Application Layer

Applications such as caching, establishing, demonstrating and distributing the environmental intelligence procured from sensors, devices and web services were executed by using the application layer.

The IOT is said to be a broadband network since it uses standard communication protocols. Targets can be identified, measured and understood and that can change the environment and they are integrated based on geographic location. The IOT based interaction is shown in figure 3.

FRAGMENTS OF IOT

Iot gets fragmented into hardware, middleware and user end visualization (Daiwat et al., 2015). The figure 4 represent the IOT fragments.

Figure 3. IOT based interaction

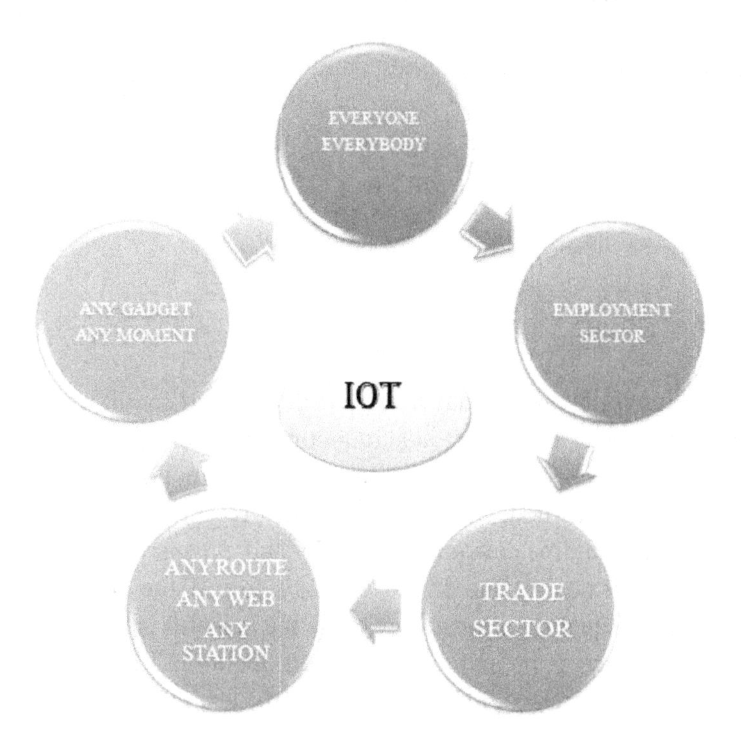

1. **Hardware:** It configures sensors, actuators, embedded devices and other communication devices.
2. **Middleware:** For cache information gathered by sensor devices a diverse contrivance necessitate arises and that can be refined by embedded devices and tools used for data modeling.
3. **User End Visualization:** Initially with statistics recognition visualization and explication appliances which can be approached on platforms which relieve the end user to keep a track of various events driven by the data collected by sensory hardware.

Figure 4. Fragments of IOT

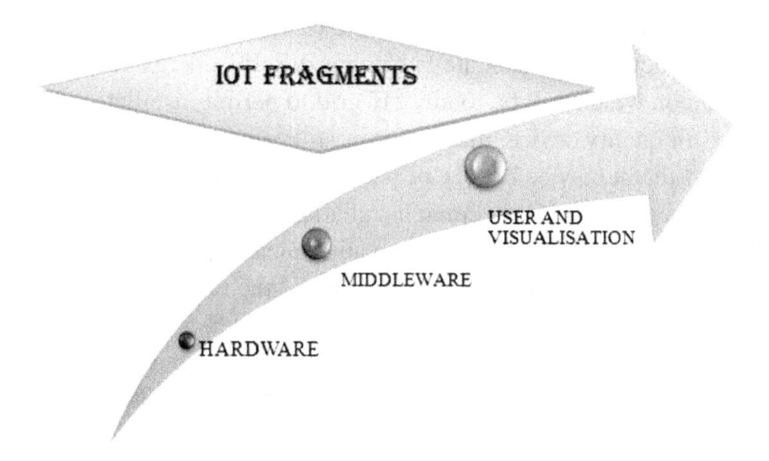

APPROACHES OF IOT IN SMART CITIES

Depending on the nature of pollution in smart cities, an IOT approach is classified into following phases

1. Air phase
2. Water phase
3. Soil phase
4. Noise phase

IOT Approach in Air Phase

An energy efficient solution for facing the day to day challenges and creation of smart environments is IOT. The congregate technologies of IOT, cloud computing, remote sensing, Geographic information system and global positioning system develops an integrated information systems to analyze the variation in climate mode. In classic method, the data loggers visit the site every time to garner the data for analyzing air pollution. This is a time consuming, lengthy and expensive process to carry out. Hence, IOT is preferred to monitor air quality due to its flexibility. In IOT the quality of air is monitored by the web server. Whenever the quality of air gets poor by the harmful gases like CO_2, smoke, alcohol, benzene and NH_3, an alarm is triggered and displays the air quality in PPM on the LCD and as well as on the web page. An IOT sensor array could be connected to the low power consumption LPC 2148 with 8 channels of analog to digital converter. An integrated TCP/IP protocol stack Wi-Fi module ESP8266 (Snehal et al., 2017) is assembled with AT commands in both client as well as service modes to detect the level of pollutant gases (Kondamudi & Gupta, 2016). The sensor elements employed to analyze the data set based on linear regression prediction with the help of machine learning algorithm to monitor the air pollution in real time and also predict the measurement in the next given interval. To measure the Co_2 and Co concentration in air MG811 and MQ7 sensors is used. Temperature can be predicted by LM35 sensors (Reshma & Ravindra, 2016). Also moisture content and alcohol can be analyzed through SY-HS 220 and MQ6 sensors. Light dependent resistor monitors the light intensity of the surrounding environment. The switching actions for AC/DC devices can be performed by the relay. Using Wi-Fi communication the collected data could be sent to the network. The figure 5 show the block diagram of air quality monitoring. It is a safer and better option to monitor the air quality periodically using IOT in the crowded areas, parks or fitness trails.

To date, the AirQ device used 3G or Wi-Fi property to transfer knowledge from the deployed sensors. Smart Sense has currently developed associate in Nursing NB-IoT energy-efficient technology and aim to evolve any with a version which will be totally off-grid to permit installation in any location. A new NB-IoT steam-powered air quality device, that performs still because the existing sensors and optimized for NB-IoT, which means that it may be battery or star steam-powered in future. This can enable the device to be put in in more locations, while keeping installation and operative prices low. The AirQ device used to optimize the answer to fulfill NB-IoT information measure needs and to re-design the devices computer code and hardware to support NB-IoT. The aim of the readying is to check the performance of NB-IoT for assortment of all the information presently collected by the AirQ device and to live the facility needs for developing battery steam-powered device within the future. The first NB-IoT pilot is currently up and running in Xanthi, Hellenic Republic with real knowledge (Smart Cities).

Figure 5. Block diagram of air quality monitoring

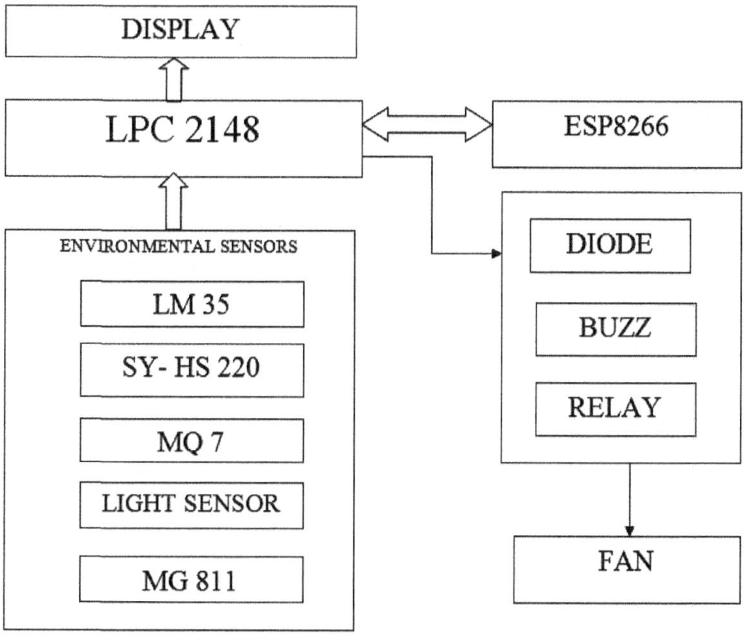

IOT Approach in Water Phase

While approaching IOT, water phase can be classified further as smart water, wastes and sewage attribute surveillance.

Smart Water Observance by IOT

By the UN forecast in 2050 over six billion people will be surviving in cities. Hence, the demand of water will be increased to 55% than present scenario. Global urbanization proposed a number of water-related challenges mainly, Supplying of clean drinking water and proper disposal of effluent. There is a lack of water near cities near cities in many regions of the globe. Hence, the water is conveyed into the cities with the long- distance pipelines or desalinated from sea water. To address these challenges cities are looking ahead for innovative technologies and the better management. As a work with mobile operators and other stack holder for the Smart water management employed by IOT technologies to afford clean and safe drinking water for the mankind. To improve metering and flow management digital technology is used in water distribution. This increases the reliability and transparency to save water in reduced costs. Physical pipe networks are overlaid with the help of data and information network to determine the anomalies (such as leaks) in real time to analyses the pressure and flow data. For instance, in India as a part of improvements to the water supply system Mumbai has installed remotely controlled smart water meters. This reduces more than 50 percentage of water leakage. An IOT based water quality monitoring, the sensor nodes are stationed along the bank of water. The water level is monitored using LV- Max Sonar- EZI sensor. The quality of water is analyzed by evaluating the degree of turbidity using SEN0189. The pH of water is analyzed using pH sensor and temperature by DHT11 sensor (Rahul &

Figure 6. Block diagram of water quality monitoring

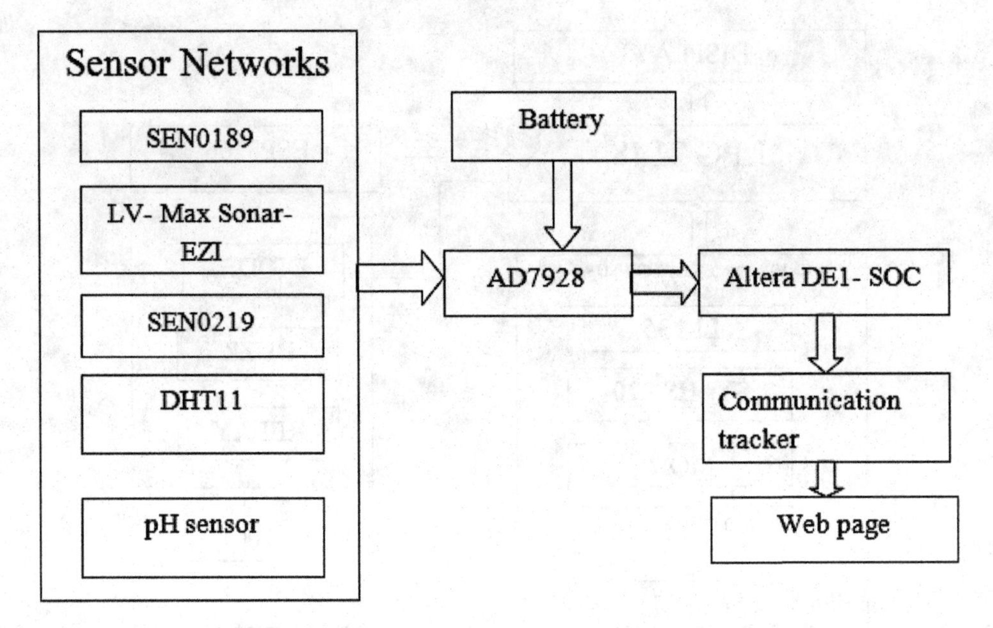

Sridevi, 2017). SEN0219 measures the concentration of Co_2 in parts per million. The measured water parameter data are collected by the sensor nodes and sent to FPGA board (Cho Zin Myint et al., 2017). The Altera DE1- SOC, Field programmable gate array containing 85K programmable logic elements, 4450K bits embedded and hard memory controllers is utilized through the entire system. The analog output from CO2 sensor and Turbidity sensor were digitized by AD7928 Analog to Digital converter. Using communication tracker the data is send to the webpage. The block diagram of water quality monitoring was represented in figure 6. Hence, by IOT water situation information is provided to customers to conserve water in real-time.

IoT technology is also economical in tracking and analyzing water use in buildings. Banyan Water, a sensible water management company, claims it's helped customers to save lots of over seven billion liters of water since its origin in 2011. The approach by putting sensors and supersonic meters that track water consumption across the building, exploitation package to investigate the gathered information and realize anomalies like leaks and overspend.

Banyan Water recently introduced variety of latest product and options geared toward meeting the requirements of business, institutional, and multi-family water users. In April the corporate launched Irrigation Insight, a flow watching and leak detection and mitigation answer that works with any existing irrigation system. The company's latest technology, Smart Rate, could be a water management tool that interprets water consumption into greenbacks spent, giving users valuable information to assist manage budgets and attain property goals. Smart Rate works in tandem bicycle with Banyan Water's initial technology giving and good Irrigation. The newest water management dashboards on Banyan Water Central, the company's internet platform, show a variety of product-specific key performance indicators that remodel water information into unjust intelligence. Information square measure summarized at each the portfolio and property levels, permitting users to grasp however and wherever water is being consumed over time. The addition of the Smart Rate technology to Banyan's dashboards permits a selected breakdown of water-related prices and additionally provides price projections supported water use behavior.

Banyan additionally recently free variety of package tools specifically tailored to the requirements of landscapers and irrigation technicians. Designed to be used on mobile devices whereas at properties, these tools facilitate landscapers take a look at, troubleshoot and verify performance of irrigation zones to confirm the system is in correct operating condition that could be a key element of semi permanent water potency. Banyan additionally recently enlarged its operations into 5 new states: CA, Missouri, Kansas, North Carolina and Tennessee ("Banyan Water's Data-Driven", 2015).

Waste Governance by IOT

At faster than that of urbanization, generation of waste gets increased day by day. It is difficult for Cities to find the source, and separate them. The usage of different kinds of waste that potentially returned to a life cycle of a consumer and affects the health of human. The main inability of waste management is the prediction of waste and picked up. Trucks are often sent to collect waste, when bins are full. Smart waste management systems using IOT provide the means to reduce the waste and manage the flow of waste from source to disposal. In which, waste released into the surroundings is monitored, collected, transported for processing, and then finally to recycle or disposed off. Besides it is used to convert a waste into a resource and create closed-loop economies. Through web page using transmitter module the trash bins are monitored. The wet wastes are decomposed by blades and motors. Hence, 50% of the transportation gets reduced when the wet wastes are crushed. With the use of servo motor, these wet wastes are converted into manure and helps the crops to grow. In rainy seasons, if the trash bins are not closed the water gets filled into it and creates unhygienic conditions. Hence, to avoid such situations, rain sensors are used in trash bins which implements the opening and closing mechanisms. In smart cities, the intelligent bins become feasible one in the collection of waste. Smart waste management should be intended as the simple installation sensors on bins, and it should cradle integrated planning strategy customized for recovery of resources and efficiency with in economic circular frame work. Smart collection bin works with IR sensor TSOP1738 that indicates the weight and different levels of garbage in the dustbins and also the weight sensor gets activated in order to send its output when its threshold level is crossed. The data received by the microcontroller ARM LPC2148 based on a 32/16 bit (Navghane et al., 2016) is then transmitted to the receiver section through transmitter module. The Figure 7 shows Block diagram of waste governance monitoring

Figure 7. Block diagram of waste governance monitoring

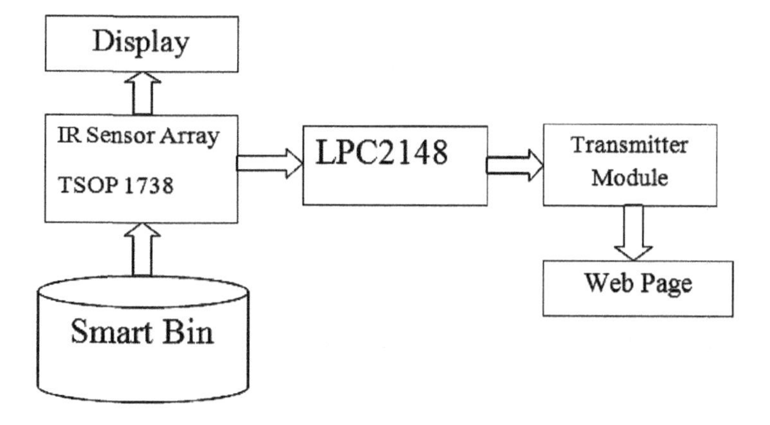

Sewage Observance by *IOT*

When the stratum of sewage in manhole crosses a particular threshold, it is logged by the ultrasonic sensor. The perceived database is upgraded by the universal module and conveys the warning to a server through Wi-Fi module. The coordinates of a manhole are consigned by GPS module. The data is analyzed by the supervisor and assigned the manhole for cleaning. When cleaners log into the android application, they will be disclosed about the location, direct route and statistics of blocked manhole. After flushing the manhole, cleaners update the position of cleaned manhole to the database. To safeguard this whole module from sewage, it is lying within a robust and waterproof IP67 rated case. A small hole located on the lid of the manhole through which magnetic charging port antennas will be extended upward to the lid surface for the superior GPS monitoring and fine wireless connectivity (Akshay et al., 2017). The model of sewage observance is given in figure 8.

IOT Approach in Soil Phase

In soil phase the approach of IOT is categorized into two sectors as follows.

1. Agricultural sector
2. Landslide sector

Agricultural Sector

The demand for food is increasing in terms of quantity and quality. It embossed the entail for intensification and industrialization of the agricultural sector. To ensure global food security, agriculture sector should be more efficient and resilient The "Internet of Things" (IoT) offers a solution towards the modernization

Figure 8. Model of sewage observance

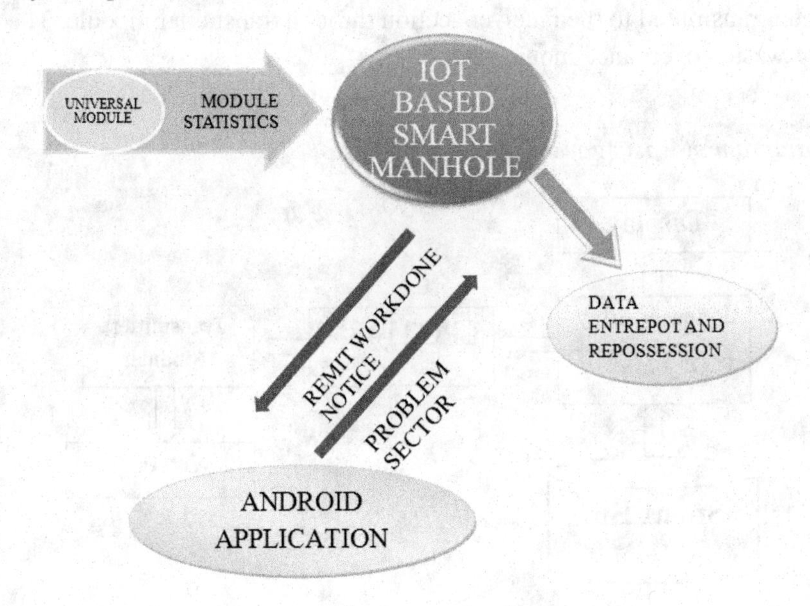

of agriculture. An IOT technique in agriculture amalgamates farmland, agricultural machineries, and fresh agricultural products with the chips, broadband network and database systems. Where the agricultural sensor information such as temperature, humidity, pressure, gas concentrations and vital signs, agricultural products attribute information along with agricultural location information are collected in the Information/ Sensor collection layer using: two-dimensional code labels and readers, RFID tags and readers, cameras, GPS sensors, terminals, cable networks, sensor networks and wireless networks. The agricultural information acquired through Sensor layer is collected and summarize for processing in transport/ network Layer. It transmits and processes the data with the fusion of Internet network and telecommunication, network management center, information center and intelligent processing centers. For agriculture, transport Layer is the nerve center and cerebra of Internet of Things. The collected data is then analyzed and processed in application layer by combining IOT and Agricultural Market intelligence (Patil et al., 2012). Also the documentation of agro product is Orthodoxly compiled by manual and bar code for agricultural supply chain. This concludes in blunders, reprieves, and inadequate information for chain supply. IOT on the agricultural supply chain can manage and streaks the trait of agricultural products by fusing agricultural supply chain with farmer's investment. The intelligence of production, dispensation, security and quality of the agricultural supply chain unified by RFID technology and cloud computing. It incorporates the IOT with farmers and makes persuasion of the holistic agricultural supply chain. It processes the hypothesis information of each facet including the production, endowment, depot, transference and marketing (Xiaohui & Nannan Liu, 2014).

Landslide Sector

A landslide occurs due to construction without proper engineering, removal of vegetation cover and deforestation on slopes. This results in death and injury of peoples and animals. For saving the lives and assets from devastating landslide can be monitored, forecasted and warned using IoT (Jadhav et al., 2018). The relative humidity, wind speed, rainfall, temperature and wind vane are procured from micro meteorological node base. By using AM2315 sensors I^2C, temperature and relative humidity are measured. At different depth, the moisture content in the soil is determined using ground node with the help of soil moisture sensor 200SS. Barometric sensor BMP280 and a weather meter spark fun are used to measure wind speed, wind direction and rainfall. The data from the sensors is transmitted from both nodes by multi protocol LoPy. The movements of landslides were measured by MPU 9250 accelerometer, compass and gyroscope. Finally, collected information is forwarded to cloud architecture. For the crumbly environment, the protocol uses IP67 enclosure for reliable performance and to protect effects of immersion up to 1 meter deep (Meryem et al., 2018).

IOT Approach in Noise Phase

In urban areas, the noise pollution is mainly by increase of vehicles and traffic. Last few decades the vital issue in big cities is the traffic management. Due to the rapid population growth the usage of vehicles increases and causes major traffic congestion. By using network communication between server and hardware module the traffic signals can be monitored to avoid traffic congestion (Nikita et al., 2016). The vehicle mounted board and the regional computing unit are placed in the system. To send and receive the radio waves, vehicle mounted board has the transmitter and receiver. Regional computing units are placed in workstations to receive data from vehicles and transmit signal using RFID. The end users can

Figure 9. Block diagram of noise pollution monitoring

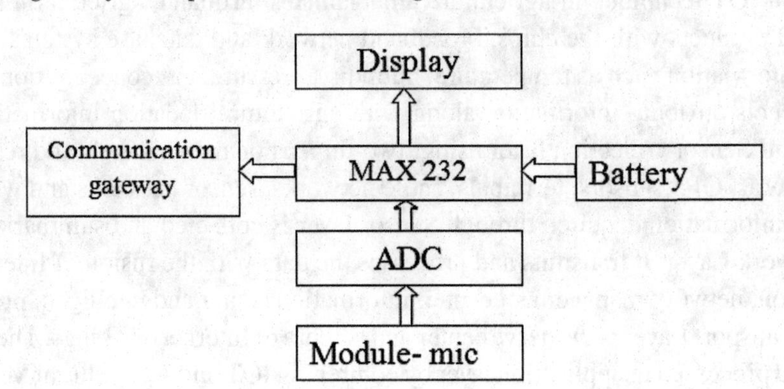

interact with wired or wireless connection of their Smartphone's with embedded system using GSM module (Mahesh, 2016).

The communication gateway and sound level sensors were used for monitoring the noise pollution using IOT. The figure 9 show the block diagram of noise pollution monitoring. Communication module is connected to the MAX 232 microcontroller and it interacts with microcontroller through the transmitter and receiver ports Wi-Fi module. From the remote location Wi-Fi module access the data and display it into the LCD system connected to the micro controller. The sound level is constantly measured by this system and reported to online server over IOT. The data gets processed when the sensors interact with microcontroller and transmits over the Internet. A sound sensor module –mic is used to measure the sound levels. The data obtained from this sensor is in analog form which is converted to its equivalent digital form. From the module, analyzed data is sent to any location and that can be fetched using a laptop /mobile by providing Wi-Fi details and IP address of the website connected to the internet (Sarika et al., 2017) .

CONCLUSION

The succeeding era of cityscape outgrowth is the progression of an intelligent city to enhance the efficacy, assurance and authenticity of an established city. By 2020, urban cities will begin to impose smart cities with an eco- friendly environment. By 2030, all the major cities in the world are on their way to becoming clever cities and existing cities will persist to ripen and embrace newfangled competences. Smart cities are not a query of whether but an immediate method of probe. It is not easy to create the healthiest environment because of industries and transportation wastes. To shape the city into smart, the environment needs a smart technology for monitoring and management the pollution. This can be accompanished by using the IOT technique for seeking, perceiving and supervising environmental pollution for the eco-friendly and sustainable life.

REFERENCES

Banyan Water's Data-Driven Water Management Solutions Save Businesses And Organizations 1.75 Billion Gallons Of Water To Date. (2015, September 10). Retrieved from https://www.prnewswire.com/news-releases/banyan-waters-data-driven-water-management-solutions-save-businesses-and-organizations-175-billion-gallons-of-water-to-date-300139938.html

Bhalerao, Ghosh, Mhatre, Vadgaonkar, Wajge, & Shinde. (2017). IoT Based Smart Waste Management System. *IJCTA*, *10*(8), 607–611.

CitiesS. (n.d.). Retrieved from https://www.gsma.com/iot/smart-cities/

Deshmukh, Surendran, & Sardey. (2017). Air and Sound Pollution Monitoring System using IoT. *International Journal on Recent and Innovation Trends in Computing and Communication*, *5*(6), 175–178.

El Moulat., Debauche, Mahmoudi, Ait Brahim, Manneback, & Lebeau. (2018). Monitoring system using Internet of Things for potential landslides. *Proceedings of the 15th International Conference on Mobile Systems and Pervasive Computing (MobiSPC 2018).*

Lakshminarasimhan. (2016). Advanced traffic management system using Internet of Things. *IEEE Journal,* (1), 1-9.

Mohammed, & Ahmed. (2017). Internet of Things Applications, Challenges and Related Future Technologies. *International Journal World Science News*, *67*(2), 126–148.

Myint, Gopal, & Aung. (2017). Reconfigurable Smart Water Quality Monitoring System in IoT Environment. *Proceedings of the IEEE/ACIS 16th International Conference on Computer and Information Science (ICIS).*

Navghane, S. S., Killedar, M. S., & Rohokale, V. M. (2016). IoT Based Smart Garbage and Waste Collection Bin. *International Journal of Advanced Research in Electronics and Communication Engineering*, *5*(5), 1576–1578.

Padwal & Kurde. (2016). Long-Term Environment Monitoring for IOT Applications using Wireless Sensor Network. *International Journal of Engineering Technology, Management and Applied Sciences*, *4*(2), 50–55.

Patil, V. C., Al-Gaadi, K. A., Biradar, D. P., & Rangaswamy, M. (2012). Internet of Things (IoT) and Cloud computing for Agriculture: An overview. *Proceedings of the third national conference on Agro-Informatics and Precision Agriculture.*

Reshma, Mundhe, & Dabhade. (2016). *Environmental Monitoring For IoT Applications Based On Wireless Sensor Network*. Academic Press.

Sai Ram, K. S., & Gupta, A. N. P. S. (2016). IoT based Data Logger System for weather monitoring using Wireless sensor networks. *International Journal of Engineering Trends and Technology*, *32*(2), 71–75. doi:10.14445/22315381/IJETT-V32P213

Snehal, Shinde, Karode, & Suralkar. (2017). Review on - IOT based environment monitoring system. *International Journal of Electronics and Communication Engineering and Technology*, *8*(2), 103–108.

Tanappagol & Kondikopp. (2017). IoT Based Energy Efficient Environmental Monitoring Alerting and Controlling System. *International Journal of Latest Technology in Engineering Management & Applied Science*, *6*(7), 83–86.

Tendulkar, Sonawane, Vakte, Pujari, & Dhomase. (2016). A Review of Traffic Management System Using IoT. *International Journal of Modern Trends in Engineering and Research*, *3*(4), 247–249.

Vyas, Bhatt, & Jha. (2016). IoT: Trends, Challenges and Future Scope. *IJCSC, 7*(1), 186-197.

Wang & Liu. (2014). The application of internet of things in agricultural means of production supply chain management. *Journal of Chemical and Pharmaceutical Research*, *6*(7), 2304–2310.

Chapter 17
Farming 4.0:
Technological Advances That Enable Smart Farming

Saravanan Radhakrishnan
VIT University, India

Vijayarajan V.
VIT University, India

ABSTRACT

Generally, the rate of technological advancement is increasing with time. Specifically, the technologies that are the building blocks of Farming 4.0 are now advancing at a rapid pace never witnessed before. In this chapter, the authors study the advances of major core technologies and their applicability to creating a smart farm system. Special emphasis is laid on cost of the technology; for, expensive technology will still keep small farmers at bay as major population of farmers inherently are new to technology, if not averse. The authors also present the pros and cons of alternatives in each of the subsystems in the smart farm system.

INTRODUCTION

Farming, one of the oldest occupations of mankind, has undergone several refinements all through the history of human civilization, but had never been perfected though. Farming in itself is a conglomeration of multiple sciences like Botany, Zoology, Chemistry, Geology and Meteorology. Traditional farming practises involved leveraging knowledge on these disciplinary fields and applying them on agricultural field (pun!) subjectively. Though these scientific disciplines are at the core of farming, technological aids come from other disciplines of science. Until last decade most of the technological aids came in the form of mechanical engineering which without doubt did contribute primarily to mechanize most of the hard work that demanded muscle power.

In the past decades, powered machinery had replaced many farm tasks erstwhile performed by humans or animals. The current era of farming partially incorporates multiple aspects of automation

DOI: 10.4018/978-1-5225-9199-3.ch017

through computer technology. For example, semi-automated farm irrigation system irrigates the farm with minimal manual intervention as required, by employing timers and solenoid valves.

The next revolution in this sector, appropriately termed as Farming 4.0, is a culmination of several complex sciences and technologies and has immense potential to revolutionize farming. Farming 4.0 would certainly unleash the true potential of this amalgamation of technologies to come up with innovative solutions to farming problems with the objective of doing more with less. In other words, Farming 4.0 would increase the quantity and quality of farm produce and reduce the application of nutrients and pesticides. To be more precise, it attempts to optimize fertilizer and pesticide application by identifying which section of the field needs what inputs and how much and applying the same accordingly. This is called Precision Farming and the technologies that make this possible are Variable Rate Technology, Wireless Sensor Network, Drone technology, Computer Vision, Artificial Intelligence, Robotics, Cloud Computing, Big Data, etc. Strictly speaking, many of these are cross disciplinary themes and very much lie at the heart of Smart Farming.

The technology that lies at the forefront of making the paradigm shift from traditional open field cultivation to closed, protected and controlled cultivation is Green House Technology. Moreover, sensors, microcontrollers and actuators are increasingly being used in protected cultivation to bring in a degree of automation wherein each of the above said parameters may be precisely controlled to create optimal growth conditions for the crops. Further, in this chapter, we will explore how technology can help increase the efficiency of some of the major farm activities in the growth stage of crop management like:

- Water management
- Nutrient management
- Pest management
- Crop Stress management

The objective of the chapter is twofold:

- Help farmers realize that technology has arrived at their doorsteps (farm gates?!) and it is just waiting for them to invite with both hands into their farms
- Get the new researchers up to speed on the technological advances in smart farming, so as to help them steer further researches in the right direction

Table 1. Interplay of various Technologies as part of Farming 4.0

Functional Areas	Farm Activities	Technologies Enabling Automation	Interface for Farmers
Water management	Borewell motor operation, Irrigation valve controls	Sensors, Robotics, Cloud, AI/ML	ChatBots, mobile apps, web interfaces
Nutrient management	Crop monitoring, Deficiency identification, Fertigation	Computer vision, Imaging, Sensors, Drones, Cloud, AI/ML	
Pest management	Pest monitoring, Disease identification, Pesticide application	Computer vision, Imaging, Sensors, Drones, Cloud, AI/ML	
Crop stress management / micro-climate maintenance	Crop monitoring, stress identification, Actuating Shade-nets, Foggers, Misters, Fans, Side-curtains, etc.	Computer vision, Imaging, Sensors, Drones, Robotics, Cloud, AI/ML	

The tabular column shown in Table 1 explains the various farming activities during the growth stages of the crops and the technologies involved in those activities towards betterment of crop health management. Nevertheless, technological aid is just not limited to the above, but also with other farm activities during the sowing/planting stage and even during harvesting (plucking flowers and fruits) or post-harvest stages (packing, price/demand forecast, etc.) as well. That said, the scope of this study is limited to the advances and affordability of various technologies during growth stages of crop management.

Typically, the technology life cycle has four distinct stages:

1. Research and development
2. Ascent
3. Mature
4. Decline

We shall assess these fast emerging and disruptive technologies that are changing the farm landscape on the below parameters:

- Stage in the technology lifecycle (R&D, ASCENT, MATURE, DECLINE)
- Economic viability to use in farms (LESS VIABLE, MODERATELY VIABLE, HIGHLY VIABLE)
- On-going research works, potential innovations (LOW, MODERATE, HIGH, VERY HIGH) and
- Future research directions

In this chapter we dive deep to find how these staggering new technologies are coming together as part of Agriculture 4.0 and how surprisingly closer their applications are to the farmers.

The organization of the next sections is as follows:

- Green House Technology as the frontrunner in modernizing farming sector
- Variable Rate Technology as the encompassing technology that facilitates managing variations accurately to grow more produces using fewer resources.
- Sensor Technology, advances in sensor technology, wireless communication technologies and their applicability to farming
- Computer Vision, Multispectral/Hyperspectral/Thermal imaging and their maturity level
- Drone Technology, UAV and their affordability
- Artificial Intelligence / Machine Learning techniques and the ones best suited for crop health management
- Cloud/Fog/Edge Computing, various IoT platforms and the ones most suited for farming
- ChatBots as the crucial link between farmers and the technologies

Variable-rate technology facilitates varying the quantity of water, fertilizer, pesticides, etc., for different thus reducing input costs and maximizing output. In this chapter, we'll study the recent research breakthroughs in VRT and their feasibility in deploying them in the fields.

Advances in Sensor technology have made sensors as commodities. Not only that sensors are becoming cheaper and cheaper, but the power needs of these sensors also have come down drastically as they are now equipped with low-power electronics for sensing and transmitting the signals. We will explore sensors, WSN topologies in the context of energy efficiency and security and also explore how innovative geo-mapping techniques like What3Words will contribute to WSN.

Drone technology is becoming very rampant these days. Agricultural drones are already finding their applications in farms to monitor crops. The results of combining drone data, sensor data and vision computing to create strategies for the farm is far-fetched. Let us explore the maturity levels of this sophisticated technology in the context of small scale farming.

Robotics is undoubtedly a game changer in Farming 4.0. But how prepared are they in rolling up their sleeves and get their hands dirty in the farm? Will they be dexterous enough to discharge the duties in the farm with manifold productivity? Even if they are, how expensive would it be for a small scale farmer to own them? Does the increased productivity justify the cost? May be, recent research has revealed the use of these tiny genies in the farms. Let us explore more in this chapter.

Cloud Computing was only seeing little progress in the past few decades but are now exponentially growing thanks to the rapid increasing processing power and storage at much lower costs. We'll analyse recent research literature of the variants and suitability of Cloud/Fog/Edge computing technologies for small scale farms.

Artificial Intelligence, Internet of Things and Big Data technologies are joining forces to analyse correlations in wide variety of huge amount of structured and unstructured data acquired from multiple sources like historic weather data, soil status data, crop images, etc., to extract knowledge and provide farmers with meaningful insights and recommendations to take action to maximize crop yield. This cognitive IoT combination can further be augmented with market data to help farmers with marketing their produce. We'll study recent advances in these technologies in the context of Farming 4.0.

Chatbots which are virtual assistant that automates interactions with end users are now becoming ubiquitous. They are promising to go a long way to bridge the farmer-technology gap. Chatbots trained in local dialects through Natural Language Processing and enriched with agriculture knowledge through AI-ML would let farmers interact with the Smart Farm system in a more natural way. We will look at recent frameworks that would enable this last mile delivery of technology to the farmers.

We conclude the chapter with a recommendation of technology infrastructure requirements for Farming 4.0 at a village level that would drive policy decisions from Government as part of smart cities/villages initiative is also presented in brief.

GREEN HOUSE TECHNOLOGY IN THE FOREFRONT OF FARM MODERNIZATION

Conventionally, crops are cultivated in open fields where the sudden and drastic changes in climatic conditions like temperature, humidity and light intensity adversely impact the crops.

If it would be possible to create a suitable micro-climatic condition around the crops, then it should be possible to control the external influence due to the macro-climate on crops to a reasonably good extent, thus protecting the crops from harsh weather conditions. This is precisely what is called as Protected Cultivation. The advances in this technology have led to the creation of various types of agro-climatic zones suiting the needs of various types of crops. The physical structures to create this protected cultivation come in different types like Green House, Shade House, Glass House, Plastic House, Lath House, etc., that provide optimum growing environment for different crop types.

In addition to the fact that these greenhouses protect the crops from severe weather conditions and damages due to rain and wind, each of the micro-climatic parameters like light, temperature, humidity, etc. can also be controlled by employing various techniques like rolling benches, moveable shade nets, drop-down side-curtains, misters, foggers, fans, PAR (Parabolic Aluminized Reflector) lightings.

Parvatha Reddy (Parvatha Reddy, 2016) analyses the recent technological advances in different types of green houses, various climatic factors that need to be taken care in green house management and how various environmental parameters may be controlled to create microclimate suitable to a given crop. Shamshiri et al. (Shamshiri et al., 2018) reviews several aspects of modern green houses, Controlled Environment Agriculture (CEA) and their derivatives. While the study finds automation and technology as the solution to the increased demand in food, the paper doubts the increased profitability relative to the costs and suggests further studies in this aspect. Similar study by Kooten (Van Kooten, Heuvelink, & Stanghellini, 2008) and team also vouches that greenhouse technology has the solution not only in terms of keeping plant-harmful events at bay, but also in terms of minimizing resource usage.

Assessment Report

- Stage in the technology lifecycle: MATURE
- Economic viability to use in farms: MODERATELY VIABLE
- On-going research works, potential innovations: MODERATE
- Future research directions: Designing greenhouses appropriate for different terrains, geographies, climatic conditions and crops.

VARIABLE RATE TECHNOLOGY AND RECENT BREAKTHROUGHS TO PRODUCE MORE WITH LESS

The most important aspect of precision farming is producing more with less. In other words, resource utilization must be very precise. Thus any input like water, fertilizer, pesticide, etc. must be used at the very optimal level. Generally, farmers tend to use same quantity of resources throughout the farm. In case of precision farming, there should be a means of identifying what is needed, where it is needed and how of much of it is needed in the farm and also there should be a means of providing the exact same quantity of whatever is required at precisely the location where it is required. The technology that makes this happen is called Variable Rate Technology (VRT). Lot of studies have been happening for past couple of years to get this technology right for farming.

Some of the common approaches in VRT are:

- Manual
- Map based
- Sensor based
- Hybrid – Sensors & Maps

The complexity and thus the cost increases as one moves from manual to map based to sensor based approaches. In all these approaches, the first step is to delineate the farm into regions. Smaller the regions, higher the precision will be. Also, the shapes of the regions also play a major role. In the early stages of VRT applications, simple grids like sectors of a circle or cells of evenly divided rectangle were used as regions. In more recent times, sensible zones are created as regions based on homogenity of certain physical or chemical properties like terrain, soil structure, water retention capacity, electrical conductivity, pH value, yield map, NDVI. In manual approach, the variable rate applicator (VRA) is controlled or operated by humans based on his subjective observation. In map based approach, the crop expert will analyse the farm condition and create a prescription map which will be fed to a system that controls the VRA. In sensor based approach, the controller operating the VRA processes real-time data from sensors and determines the need at various zones and applies the exact resource and its quantity as needed at variable rate.

Very recently, with the advent of powerful drones which has GPS navigation capabilities, advanced multispectral imaging sensors, high processing power, onboard storage and appllicator lifting capabilities, hybrid approach of map and sensor based VRAs are gaining importance in precision farming. In this approach, prescription map gets generated using historical yield maps and fertility maps and fed to the drone system which also processes real time data from its image sensors and ground sensors and determines the variable rate real time and applies the input precisely throughout the farm.

Assessment Report

- Stage in the technology lifecycle: RESEARCH AND DEVELOPMENT
- Economic viability to use in farms: LESS VIABLE
- On-going research works, potential innovations: HIGH
- Future research directions: Making multitude of technologies to work in tandem to make VRAs effective, Creating standards for interoperability of related technologies

SENSOR TECHNOLOGY, ADVANCES IN SENSOR TECHNOLOGY, WIRELESS COMMUNICATION TECHNOLOGIES AND THEIR APPLICABILITY TO FARMING

Sensors are integral part of IoT eco system since they are the eyes and ears of the entire farm automation system that perceive the world and pass it on to the brain of the IoT system which is nothing but the AI/ML services hosted usually in cloud servers. As illustrated in Fig 1, the data sent by the sensors get processed by these intelligent services and the insights derived are passed back to the farms either as information to farmers or as instructions to the actuators.

Since these sensors are going to be implanted in large quantities, though on sampling basis, throughout the length and breadth of the farm to measure various parameters, they need to be economically viable, physically small and consuming less power and still wireless so that they can be deployed in farms. Sensor technology has evolved to meet these needs and is further evolving thus making the sensors smart, robust and self-healing.

Farmers now have a wide range of sensors to choose from. For example, there are so many different temperature and Humidity sensors available like LM3X, TMP3X, TMP00X, DHTXY, AM23XY, SHT3X. Similarly there is an array of sensors for monitoring rain, wind, water flow in the pipelines, soil characteristics like soil moisture, pH, salinity, electrical conductivity, nutrition (NPK), etc. Depending on the use case and the budget a particular sensor type can be decided.

Figure 1. IoT eco-system in a typical farm automation

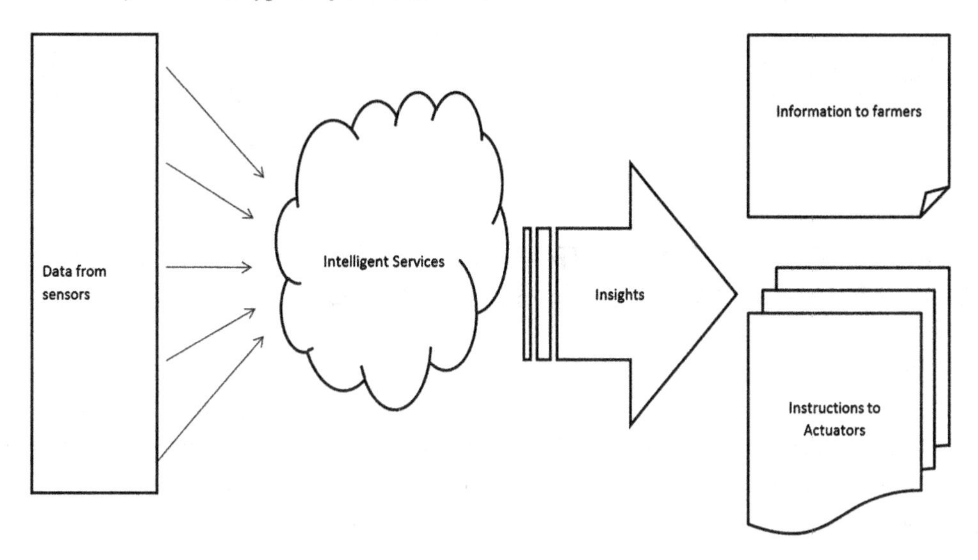

Typically farms will be in remote locations and hence may not have adequate network coverage. Hence sensors that are smart enough to take care of themselves like self-calibration and self-diagnosis are best for farms. In fact, sensors that can do data pre-processing to some extent, so as to avoid heavy loads on networks are preferred in farms. For example, if the sensor is intelligent to keep sensing data periodically, but skips communication to the central hub for certain number of times as long the data has not changed significantly will be very useful in terms of reducing load on network and cloud servers, especially when they are installed in farms in large quantities. Such smart sensors will reduce decision latency as well. Thus, these smart sensors may be entrusted with decisions not needing so much of computing power and data. Such smart sensors will reduce decision latency as well. This is the basis for Edge computing which is now taking higher priority than cloud computing in applications like precision farming.

Major researches in sensor technology are in the areas of creating intelligent sensors, power optimizations, power harvesting from ambient energy and charging the sensor batteries, etc.

For instance, a research (Hoang, Julien, Berruet, Detection, & Fdi, 2013) describes the problems associated with such sensor nodes when present in harsh environment and part of a wireless sensor network as how it would be in farms. The authors discuss approaches to identify an internal failure and take corrective action to improve both reliability and energy-efficiency. This reduces the node's vulnerability and alleviates maintenance costs.

The most popular wireless communication technologies in Wireless Sensor Networks are BLE (Bluetooth Low Energy), ZigBee and WiFi. While Wi-Fi is the fastest among these and has the maximum communication range, the energy consumption is also the highest among these and hence not much suitable for Farms. The next fastest among these is BLE which has relatively lower energy consumption. But ZigBee is the lowest energy consumption technology, has a low latency and low duty cycle and is also the cheapest, making it ideal for farms where hundreds or thousands of sensors are connected in Mesh network which allow nodes to pass data through multiple paths. The most recent addition to this is LPWAN technologies, briefly explained later in this chapter under Cloud Computing topic.

Assessment Report

- Stage in the technology lifecycle: MATURE
- Economic viability to use in farms: HIGHLY VIABLE
- On-going research works, potential innovations and future research directions: VERY HIGH

COMPUTER VISION, MULTISPECTRAL / HYPERSPECTRAL / THERMAL IMAGING AND THEIR MATURITY LEVEL

Conventionally, unhealthy plants are identified by the symptoms on flowers and leaves like abnormal colour, shape and texture. More often than not, such symptoms become visible to naked eye only after a certain stage of the infection. At this stage, it typically becomes very difficult to reverse the effects, at least is much more difficult than addressing it at an earlier stage. Computer vision systems and spectral imaging technologies let farmers identify unhealthy plants right at the early stage of infection – much earlier than when they would develop visual symptoms.

There was a comparison study (Xu, Riccioli, & Sun, 2016) to identify if hyperspectral imaging technology would fare better than Computer Vision Systems when it comes to differentiating, salmon fish that are organically farmed from those that are conventionally farmed. Their research concluded that hyperspectral imaging indeed is superior to Computer Vision Systems.

Although the cost of acquiring hyperspectral images is typically high, hyperspectral remote sensing use is increasing for monitoring the health of crops due to its higher bang for the buck. This is due to the fact that the technology enables detection of diseases much earlier than is otherwise possible thereby enabling the reversal of the damages and thus reducing the losses. A researcher (Ahmadi. P, 2017) was able to demonstrate the same by moving the detection of basal stem rot disease much ahead in the lifecycle.

Researchers (Moghadam et al., 2017) described the significance of hyperspectral imaging technology in identifying infected plants. In order to differentiate infected plants from healthy plants, they employed machine learning techniques for training the classifiers which were fed with hyperspectral imagery data.

Spectral imaging adds another dimension to the usual optical imaging, thus creating a dataset called data cube that contains both spatial and spectral data. In Hyperspectral imaging light is divided into hundreds of small bands that enable lot more details to be captured. The presence of the spectral content in the pixels in spectral imaging helps detect even subtle colour changes on foliage. Drones mounted with a hyperspectral sensor can identify anomalies from well above the canopy level of the crops thus facilitating farmers to detect even minor problems in their crops and take immediate corrective actions.

This wonderful technology hitherto economically unviable to farmers is now becoming cheaper thanks to the recent initiatives to reduce the costs and hence this technology seems heading towards the farm! Until the cost of this technology becomes affordable, there is a cheaper alternative called multispectral imaging. In multispectral imaging the light gets divided into small number of bands, usually in single digit or in tens. Compared to hyperspectral imaging, multispectral imaging is less efficient in agricultural sector. Nevertheless, given the cost effectiveness and commercial viability of this technology, multispectral sensor technology is finding increased use in Precision Farming. Vegetative indices of the field can be created using multispectral imaging camera drones. These vegetative indices and prescription maps thus created from aerial multispectral imagery data can be very useful in optimizing resources like water, pesticides and fertilizers.

Figure 2. Sample NDVI calculation for healthy and unhealthy/stressed plant

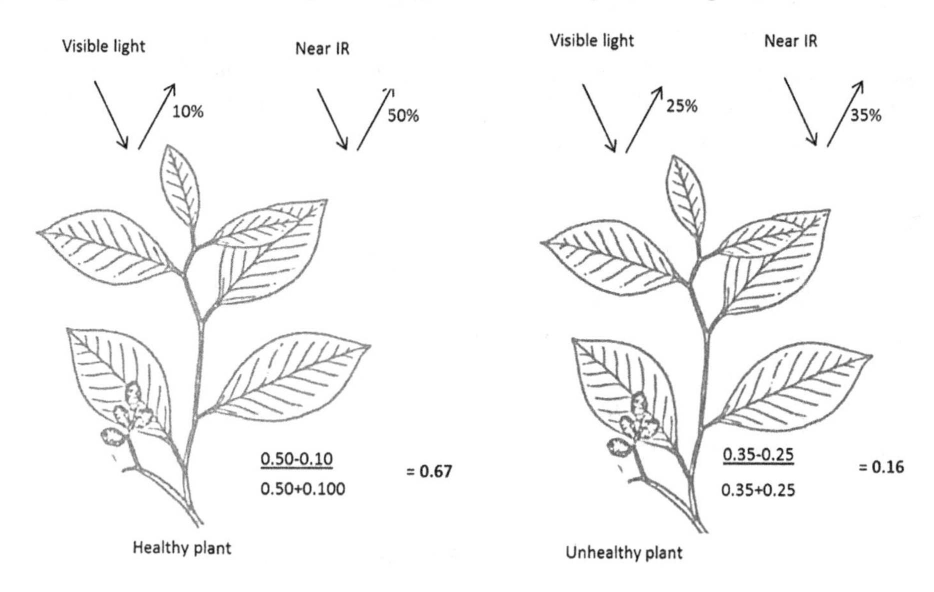

Thermal Imaging is another technology that is much cheaper than hyperspectral imaging, but is very promising in farming. Thermal imaging technology uses thermal infrared band and relies on the thermal energy radiation from objects. Thermal Imaging Cameras are far cheaper than hyperspectral cameras and they are predominantly useful in irrigation management though they have wider applications in crop management including but not limited to disease detection and yield forecasting. For instance, healthy plants absorb more radiation in visible spectrum for photosynthesis and reflect more radiation in the Near-IR while unhealthy/stressed plants absorb relatively less radiation in visible spectrum and hence the difference in their reflectance between visible and Near-IR radiation is reduced. Making use of this fact, an index called NDVI (Normalized Difference Vegetative Index) is obtained, which is a powerful metric to identify stressed plants even before the symptoms start showing up to naked eyes.

As illustrated in Fig 2, a healthy plant absorbs more visible light for photosynthesis while a stressed/ unhealthy plant reflects more visible light, due to its inability to do normal photosynthesis making use of visible light. Hence the NDVI for these two sample plants are 0.67 for healthy plant and 0.16 for stressed plant. NDVI is calculated using the below formula:

NDVI = (N-V)/(N+V), where N is Near-IR (%) reflectance and V is visible light (%) reflectance

The algorithm calculates the index by comparison of the reflectance intensities. Higher the index, healthier the plant is. There are now cameras that capture Near IR in addition to the standard RGB light. In fact, there are devices that calculate the NDVI with the imagery data captured using such onboard cameras and send this NDVI dataset over network. This becomes even more powerful when such modules are fitted to drones, one can get time series of healthiness index of crops. The next section explains more about this.

These technologies are reaching maturity stage and many companies have started developing automated detection and analysis technologies using spectral imaging. When it comes to analysis of hyper/ multispectral data, the parametric techniques like functional statistics and simple/multiple regression are not efficient. Hence non-parametric techniques like Partial Lease Square, Principal Component Analysis will need to be used. One of the researches (Whetton, Hassall, Waine, & Mouazen, 2017) demonstrated using Principal Component Analysis on hyperspectral imagery data of Barley and Wheat crops and classified fungal diseases like yellow rust and fusarium head blight accurately.

The development of spectral imaging technology is progressing rapidly and more breakthroughs are expected. This would allow farmers to choose camera models adjusted and best optimised for their farms.

Assessment Report

- Stage in the technology lifecycle: between ASCENT and MATURITY
- Economic viability to use in farms: LESS VIABLE
- On-going research works, potential innovations: VERY HIGH
- Future research directions: Bringing down costs of wonderful multispectral imaging and vision technologies suitable for agriculture

DRONE TECHNOLOGY, UAV AND THEIR AFFORDABILITY

Drones are the newest additions to precision farming. A technology once thought to be useful and constrained only to military surveillance has found increased applications in other fields and reached the farms as well. The Unmanned Aerial Vehicle (UAV) technology may sound very high-tech and very expensive, but in reality they are becoming cheaper and cheaper since the components used in drones like gyros, accelerometers, radio receivers, GPS, etc. are being mass produced. The processors used in drones are also becoming incredibly powerful.

Agricultural Drones come in various forms and features and a farmer can choose one that best suits his budget, farm type and size. They can be as simple as accommodating a digital camera to take still pictures and steered / controlled through remote control or as sophisticated as fit with sensors and actuators that can detect needs of individual crop segments and apply / spray fertilizers and sprays precisely how much ever is required for the crop segment, making Variable Rate Application/Technology a reality. This type then becomes an amalgamation of drones and robotics and is still way too expensive to enter into a small scale farm.

Further advances in drone technologies enable small and lightweight multispectral / thermal imaging sensors to be fit in drones and these relatively inexpensive drones with such advanced sensors and imaging capabilities help farmers do better crop health management thus reducing crop mortalities and increasing yields.

As illustrated in Fig 3, drones provide below three additional dimensions to crop health diagnosis:

1. Height: Seeing a crop from above canopy level would reveal problems like pest and fungal infestations generally not so apparent at eye level
2. Hyper/multi spectral images: Drones fit with hyper/multispectral imaging cameras have the ability to capture information in a wide spectrum range including IR and visual range thus facilitating identification of unhealthy plants that otherwise is not distinguishable to human eyes.
3. Capture Frequency: Drones can analyse the field at much higher frequency like multiple times in a day which otherwise is a laborious process. The data thus obtained during periodic intervals which when combined to create a time series animation will reveal problem areas in the farm that warrants immediate attention.

Figure 3. Additional three dimensions provided by drone views

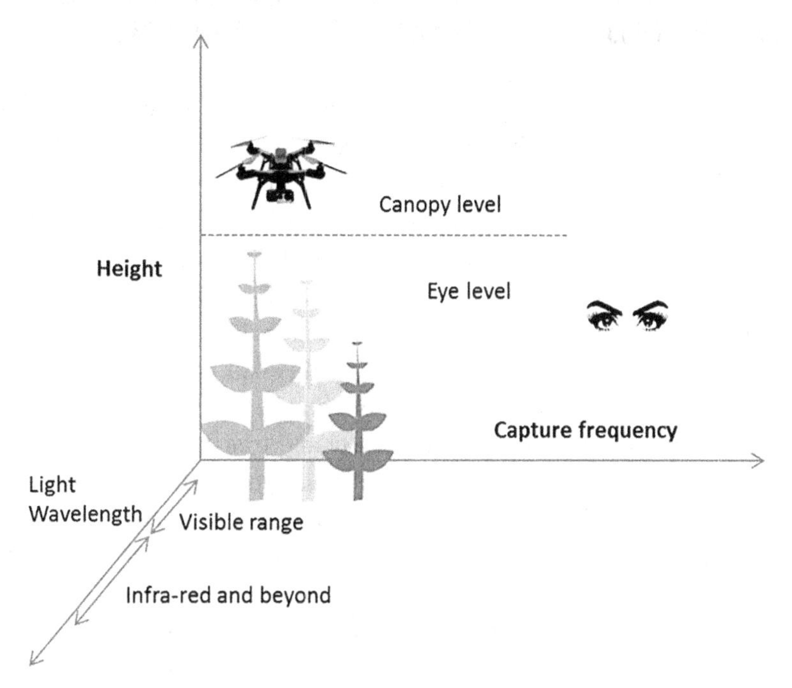

Veroustraete (Veroustraete, 2020) in his research in the year 2015 tried to validate the tall claims of drone advocates on the applications of drones in precision agriculture like crop health monitoring, irrigation monitoring, mid-field weed identification, variable-rate fertility. He concluded that though drones had limited practical applications in agriculture as of 2015, they do have potential to extend their applications to lot more use-cases as more and more researches take place in this technology.

More recently, Mogili and Deepak (Mogili & Deepak, 2018) et al in their work in 2018 reviewing the application of drones in farming, provided a detailed technical analysis of various models of UAVs applicable in agriculture on a couple of parameters like cost, technology, load carrying capacity, flight range, limitations, etc. and proposed a system describing a drone that does crop monitoring using onboard multispectral camera, analyse the images to identify zones needing pesticide spray and the drones with sprinkler system auto navigates with the GPS coordinates and sprays the pesticides on the identified zones. Their conclusion is that drone applications in precision agriculture is still not mature, but has a lot of potential for making a positive impact in precision agriculture with advances in research and development of the core technology and their applications in agriculture.

Assessment Report

- Stage in the technology lifecycle: ASCENT
- Economic viability to use in farms: LESS VIABLE (especially for small farms)
- On-going research works, potential innovations: VERY HIGH
- Future research directions: Bringing down cost, increasing their versatility in terms of farm applications and reducing the need for technical knowledge (eg: auto take-off & landing, auto navigation)

ARTIFICIAL INTELLIGENCE / MACHINE LEARNING TECHNIQUES AND THE ONES BEST SUITED FOR CROP HEALTH MANAGEMENT

Artificial Intelligence / Machine Learning services running in cloud servers are cost effective solutions for crop health management, especially in the areas of detecting and classifying unhealthy plants.

Researchers (Ramcharan et al., 2017) explored the application of transfer learning for training a deep convolutional neural network for identification of three diseases and classification of two pest damages. They achieved an overall accuracy of 93% using their best trained model.

Moreover, researchers (Abdullahi, Sheriff, & Mahieddine, 2017) could demonstrate that the use of CNN representations for the problem of estimating plant health on a maize plantation produced an average prediction accuracy of 99.58%. Until this point, this had been the best result achieved. That said, modelling a CNN for training requires millions of data, heavy computational power and complex mathematics which takes a long time to generate results and transfer learning takes off the last layer of a trained network from the neural network. Himatu et al. applied this technique by using CNN that saves what it learns in weights and fine tuning the learned weights from existing architecture. The last layer of an image is mapped on to the new image i.e. the existing CNN with the existing layer is used to train the database.

CNN has the ability to learn generic features and the features can be used with classifiers to solve most machine vision challenges. In this process, the activations from the last layer are used as features after removing the last layer from the trained CNN. By doing this, the last layer becomes a feature extractor rather than a classifier. Research has shown that this approach can be used for a dataset with a small number of images which was a challenge previously in producing accurate results. It is also reported to outperform both the fine tuning and training from the scratch approach by producing results with greater accuracy with both small and larger datasets.

It is evident that artificial intelligence is gaining rapid momentum in the area of farming. Since it is getting popular with farmers, AI will truly become the super brain behind farm automation and having a say in most aspects of farming, be it for killing weeds, controlling the micro-climate, monitoring evapo-perspirations of plants, detecting plant anomalies, growing healthy crops or producing better yield.

Another technology which is being explored with great enthusiasm by researchers and corporates alike to introduce into the farm is Robotics. One of the aspects of robots that is still being perfected to let robots enter into the farms is the dexterity in which these little machines handle the farm produce like flowers and fruits. True, farm robots are still very expensive. But technology is hard at work and farm robots are becoming less expensive day by day due largely to the unbelievable advances in the components used in robotics. Once it achieves the economies of scale, farm robots will become affordable to farmers. As of now, Robotics is somewhere between *R&D* and *Ascent* stages as far as its application in Farming is concerned.

Assessment Report

- Stage in the technology lifecycle: R&D(Robotics)/ASCENT(AI)
- Economic viability to use in farms: LESS VIABLE(Robotics)/VIABLE(AI)
- On-going research works, potential innovations: VERY HIGH
- Future research directions: Bringing down cost, increasing their dexterity in terms of handling farm produces and reducing the need for technical knowledge (eg: voice activated robots with AI, Deep Learning applications for monitoring and control of farm activities)

CLOUD/FOG/EDGE COMPUTING, VARIOUS IOT PLATFORMS AND THEIR SUITABILITY FOR FARMING

As the wireless sensor network is proving to be practical in farms and starts sending raw data to the central hub, there needs a lot of heavy lifting to be done by these intelligent services to process these data and make meaning out of them to send derived information to farmers and instructions to actuators and that is precisely where the need of Cloud Computing comes. Cloud provides the necessary speed, storage and computing power to perform complex analysis of the data collected from the farm. In fact, additional data from weather stations can also be used during this analysis to derive more meaningful information that would be helpful for the actuators and farmers.

Microsoft Azure IoT Suite, Amazon Web Services IoT platform, IBM Watson and Google Cloud IoT Core are some of the leading cloud based IoT platform/frameworks that also contain rich set of Machine learning services.

Microsoft Azure contains IoT concentrator and advanced machine learning services. The AWS IoT platform provides an extensive SDK to create connected devices and data transfers between them through varied protocols like MQTT and HTTP. IBM Watson provides a model builder that facilitates a fully automated data processing and model building interface to start processing data, preparing models, and deploying them into production. Google Cloud IoT Core is a platform that provides complete solution to collect, process, analyse and visualize data from geographically dispersed IoT devices. ThingWorx is another platform that facilitates faster development of solutions through its simple interface. In addition to this, there are lot more open source IoT platforms claiming to provide end-to-end solutions for specific domains. While these are slightly pricier than the aforementioned frameworks, they make implementing farm automation solutions lot more easily. The authors of this chapter do not advocate one platform or framework over the other.

Typical problems associated with cloud, however, are the network coverage / internet speeds in rural areas, data privacy, security and cost of cloud computing resources. To overcome these problems, one solution could be to have the local server or IoT Gateway node that aggregates sensor data to do some of the processing and send encrypted data to cloud in a controlled manner to do the complex AI based analytics utilizing the ML frameworks and services. As already discussed in Sensor Technology section, pushing some of the decisioning logic upstream to the IoT gateway (nodes where the data consolidates) and/or every further upstream to the sensors (edge nodes where the data originates) themselves will reduce data load on network and cloud servers and most importantly speed up decisions. The former is what is termed as Fog computing and the latter is called Edge computing. Edge and Fog computing are finding increased importance in Precision farming, especially to overcome aforementioned problems with network coverage in rural areas where generally farms are located. In this context, most recent long range wireless communication technologies like LoRa are finding significance in farms. LoRa, being one of the Low Power Wide Area Network (LPWAN) technologies meant for low bandwidth data transfer for long range and finding faster adoption, is best suited for Farming 4.0. Tapashetti et al. (Tapashetti & Shobha, 2018) studied the application of LoRa in precision agriculture and concluded that LoRa does facilitate faster modernization of agriculture.

Erstwhile, fog computing lacked the power to train NN models and hence could not be leveraged for use-cases needing AI decisions. Now, the advent of MCUs capable of embedding AI in the microcontroller would allow pre-trained NN models to be deployed in the Fog nodes. This takes Fog computing to the next level. For instance, MicroController manufacturers like ST are coming up with toolkits like

STMCube32.AI that can convert Neural Networks into optimized code to run on their ST32 microcontrollers. It is expected other microcontroller manufacturers also would follow suit and come up with similar capabilities which would make Fog computing to be the best bet for Farming 4.0

Assessment Report

- Stage in the technology lifecycle: MATURE
- Economic viability to use in farms: MODERATELY VIABLE
- On-going research works, potential innovations: VERY HIGH
- Future research directions: LPWAN technology applications in farming

CHATBOTS: THE CRUCIAL LINK BETWEEN FARMERS AND THE TECHNOLOGIES

One of the applied technologies that benefited largely from the maturity of Artificial Intelligence is ChatBot technology where conversational virtual assistant powered by AI can interact with farmers in a natural way using Natural Language Processing.

The maturity of Bot technology has led way to many vendors creating Communications Backend-as-a-Service, Bot building frameworks and hosting platforms, Cognitive Services.

There is a lot of buzz in every industry to provide conversational interface for their customers/users. But no much exploration has been done to create a more intuitive interface for farmers that would understand their terminologies and dialect and converse with them more naturally.

Generally farmers are new to technologies and hence such technology as Chat Bots will play a major role in connecting them truly to the IoT eco-system and letting them leverage the true potential of technology. Farmers will be more comfortable interacting with the automation system through an intelligent chatbot trained with farming domain knowledge than through mobile app or web interfaces.

Mostaço et al.(Mostaço, Costa de Souza, Campos, & Cugnasca, 2018) have tried to prove this concept by implementing a chatbot called AgronomoBot, using Telegram API and IBM Watson Conversation. They used ML and NLP to extract intentions and entities from queries of the users so that meaningful responses are generated using data acquired from Wireless Sensor Networks in wine production environment.

Mohit Jain et al.(Jain et al., 2018) took this one step further and created a conversational agent using cloud technologies and leveraging huge volume of Kisan Call Center (KCC) data that was made available for public by the Government to build knowledge base for the conversational system. Through a user study with about 30 potato farmers from rural India, the research team could prove that all common questions relating to farming activities from farmers could satisfactorily be answered by a chat bot in an intuitive and natural way, thus eliminating the need for farmers to be literates, leave alone being educated.

Assessment Report

- Stage in the technology lifecycle: MATURE
- Economic viability to use in farms: HIGHLY VIABLE
- On-going research works, potential innovations: LOW

- Future research directions: Building knowledge base with agricultural expertise for chatbots, NLP with vernacular language and local dialects of farmers in different regions, User experience and usability study of chatbots with farmers.

CONCLUSION, PERSPECTIVE AND BEYOND

In this chapter we reviewed the maturity levels, economic viability, ongoing researches and possible research directions of some of the prominent technologies that make Farming 4.0. The levels of many of the technologies have reached *Maturity* stage and some of them are still in *Ascent* stage. Nevertheless, technology has definitely arrived at the farms. With technologies like sensors, drones, imaging, intelligent monitoring systems and robots making into field trials, Agriculture 4.0 is becoming a reality. Moreover, with continued research and development in these and other related technologies from the perspective of farmers, there is no denying that the next phase of ultra-precision agriculture has commenced. Now, as part of the Smart Village initiative, the Government may consider including the below infrastructure proposal to cater to the farming sector that make up a major part of smart village. This summarizes most of the technologies discussed above in the context of farms in Smart Villages.

There is no point in talking about implementing a smart village without a recommendation of dedicated technology center located at each village that would provide all the technology support to the village.

Figure 4. Typical model of a LoRa based Smart Village

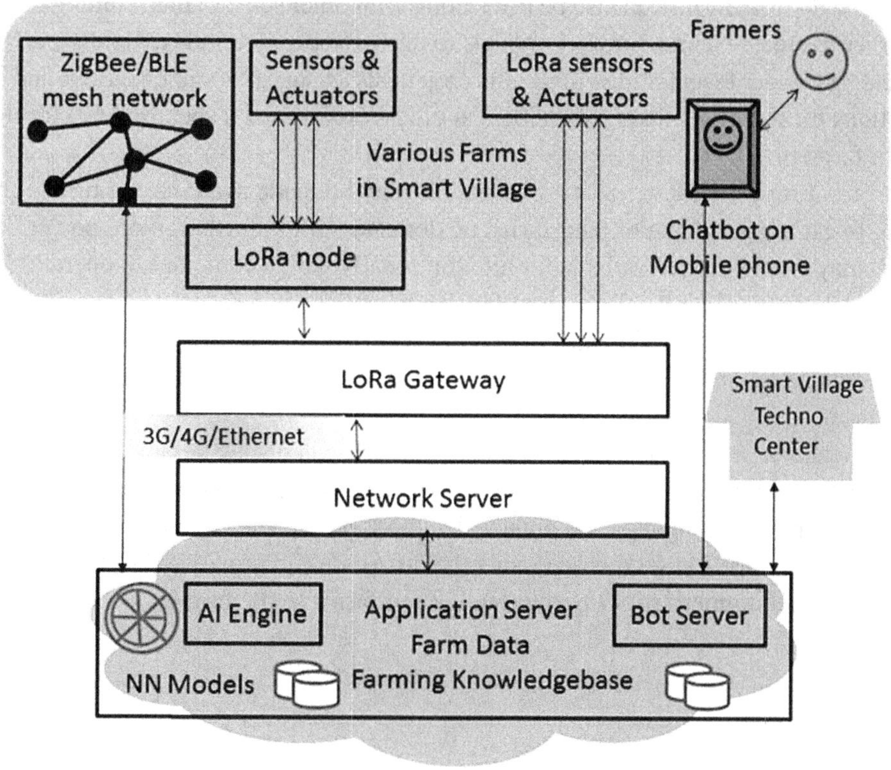

As this chapter is focusing on farming 4.0, the discussion about this proposed Smart Village Technology Centers (SVTC) will be centered on those proposed activities that would enable smart farms run by smart farmers.

In the below recommendations for policy makers, though the authors mention LoRa as the communication technology for smart villages, other Low Power Wide Area Network technologies like SigFox and the commercial NB-IoT may also be considered as deemed appropriate.

Fig 4 illustrates the typical model of such a Smart Village where LoRa technology coexists with other communication protocols. In this model, while it is still possible to get the often discussed Bluetooth, Wifi, ZigBee WSNs in farms connected to the central application server hosted in Cloud and managed by SVTC, the focus is on LoRa technology. The primary reason for this recommendation is LoRa's open source eco-system, long range (in kilometres) and low power consumption (battery life in years), thus connecting farms (which generally lack network coverage of any sort) in villages very easily together to the central application server through the network server. This is achieved by strategically placing the LoRa gateways in the village so as to get complete LoRa coverage.

The proposed SVTC established by Government may include but not limited to the following services and capabilities:

1. Ensuring LoRa coverage for the entire village by establishing LoRa gateways at strategic locations in farms.
2. Subsidize LoRa nodes and sensors to capture farm data and send to the SVTCs.
3. Getting an aerial survey of all farms using UAV/drones maintained by these SVTCs, data of which Government can use as inputs while framing policies and decisions.
4. The aerial data from farms can also be used along with other sensory data coming into the application server to create Neural Network models to derive vegetative index, yield predictions, market price, etc. at a wider level for the village and crop health, usage recommendations and optimization suggestions for water, fertilizer, pesticide and other resources at a specific level for the individual farms to farmers.
5. UAV/drones fitted with devices like IR cameras may be made available for hire to those farmers wanting to get imagery data of their farms on demand. Going further, more powerful UAVs with sprayers may also be made available for hire (optionally with technicians to operate them) to make best use of Variable Rate Application of fertilizers and pesticides.
6. Though little far fetching at this stage, the SVTCs may train small neural network models specific to individual farms and push them on to Fog nodes like STM32 MCUs, subsidized and deployed at each farm.
7. Providing chatbot service as an app to farmers which when they deploy it on their mobiles can converse naturally with chatbots to get most of the common questions relating to agriculture answered in their native language and dialect. The bot service when running in the SVTC alongside the intelligent application servers accessing the farm data can provide very intelligent and specific responses and recommendations pertaining to their farms to the farmers.

REFERENCES

Abdullahi, H. S., Sheriff, R. E., & Mahieddine, F. (2017). Convolution neural network in precision agriculture for plant image recognition and classification. *2017 Seventh International Conference on Innovative Computing Technology (INTECH),* 1–3. 10.1109/INTECH.2017.8102436

Ahmadi, P., Technology, A., & Agriculture, F. (2017). *Early Detection of Ganoderma Basal Stem Rot of Oil Palms Using Artificial Neural Network Spectral Analysis*. Academic Press.

Hoang, V., Julien, N., Berruet, P., Detection, A. F., & Fdi, I. (2013). *On-line self-diagnosis based on power measurement for a wireless sensor node*. Academic Press.

Jain, M., Kumar, P., Bhansali, I., Liao, Q. V., Truong, K., & Patel, S. (2018). *FarmChat : A Conversational Agent to Answer Farmer Queries*. Academic Press.

Moghadam, P., Ward, D., Goan, E., Jayawardena, S., Sikka, P., & Hernandez, E. (2017). *Plant Disease Detection using Hyperspectral Imaging*. Academic Press.

Mogili, U. M. R., & Deepak, B. B. V. L. (2018). ScienceDirect ScienceDirect Review on Application of Drone Systems in Precision Agriculture. *Procedia Computer Science, 133*, 502–509. doi:10.1016/j.procs.2018.07.063

Mostaço, G. M., Costa de Souza, Í. R., Campos, L. B., & Cugnasca, C. E. (2018). *AgronomoBot : a smart answering Chatbot applied to agricultural sensor networks*. Academic Press.

Parvatha Reddy, P. (2016). *Sustainable Crop Protection under Protected Cultivation. Sustainable Crop Protection under Protected Cultivation*. Academic Press. doi:10.1007/978-981-287-952-3

Ramcharan, A., Baranowski, K., Mccloskey, P., Ahmed, B., Legg, J., & Hughes, D. P. (2017). *Deep Learning for Image-Based Cassava Disease Detection*. Academic Press. doi:10.3389/fpls.2017.01852

Shamshiri, R. R., Kalantari, F., Ting, K. C., Thorp, K. R., Hameed, I. A., Weltzien, C., … Shad, Z. M. (2018). *Advances in greenhouse automation and controlled environment agriculture : A transition to plant factories and urban agriculture*. Academic Press. doi:10.25165/j.ijabe.20181101.3210

Tapashetti, S., & Shobha, K. R. (2018). Precision Agriculture using LoRa. *International Journal of Scientific & Engineering Research, 9*(5), 2023–2028.

Van Kooten, O., Heuvelink, E., & Stanghellini, C. (2008). New developments in greenhouse technology can mitigate the water shortage problem of the 21st century. *Acta Horticulturae*, (767), 45–52. doi:10.17660/ActaHortic.2008.767.2

Veroustraete, F. (2020). The Rise of the Drones in Agriculture. *Cronicon, 2*(2015), 325–327.

Whetton, R. L., Hassall, K. L., Waine, T. W., & Mouazen, A. M. (2017). ScienceDirect Hyperspectral measurements of yellow rust and fusarium head blight in cereal crops : Part 1 : Laboratory study. *Biosystems Engineering, 166*, 101–115. doi:10.1016/j.biosystemseng.2017.11.008

Xu, J., Riccioli, C., & Sun, D. (2016). Comparison of Vis or NIR Hyperspectral Imaging and Computer Vision for Automatic Differentiation of Organically and Conventionally Farmed Salmon. *Journal of Food Engineering*. doi:10.1016/j.jfoodeng.2016.10.021

Chapter 18
Cost Effective Smart Farming With FARS-Based Underwater Wireless Sensor Networks

E. Srie Vidhya Janani
Anna University Chennai – Regional Office Madurai, India

A. Rehash Rushmi Pavitra
Anna University Chennai – Regional Office Madurai, India

ABSTRACT

Smart farming is a key to develop sustainable agriculture, involving a wide range of information and communication technologies comprising machinery, equipment, and sensors at different levels. Seawater, which is available in huge volumes across the planet, should find its optimal way through irrigation purposes. On the other hand, underwater wireless sensor networks (UWSNs) finds its way actively in current researches where sensors are deployed for examining discrete activities such as tactical surveillance, ocean monitoring, offshore analysis, and instrument observing. All these activities are based on a radically new type of sensors deployed in ocean for data collection and communication. A lightweight Hydro probe II sensor quantifies the soil moisture and water flow level at an acknowledged wavelength. The freshwater absorption repository system (FARS) is matured based on the mechanics of UWSNs comprised of SBE 39 and pressure sensor for analyzing atmospheric pressure and temperature. This necessitates further exploration of FARS to complement smart farming. Discrete routing protocols have been designed for data collection in both compatible and divergent networks. Clustering is an effective approach to increase energy efficient data transmission, which is crucial for underwater networks. Furthermore, the chapter attempts to facilitate seawater irrigation to the farm lands through reverse osmosis (RO) process. Also, the proposed irrigation pattern exploits residual water from the RO process which is identified to be one among the suitable growing conditions for salicornia seeds and mangrove trees. Ultimately, the cost-effective technology-enabled irrigation methodology suggested offers farm-related services through mobile phones that increase flexibility across the overall smart farming framework.

DOI: 10.4018/978-1-5225-9199-3.ch018

INTRODUCTION

Wireless Sensor Networks (WSNs) comprises large number of sensor nodes capable of sensing, processing and transmitting information into a large scale sensing area without a predesigned network (A.Rehash Rushmi Pavitra & E.Srie Vidhya Janani 2017). Sensors communicate with another part of networks via wireless interface. Wireless networks well suits real time applications as acoustic detection, forest fire detection, environmental monitoring, military surveillance, inventory tracking, medical monitoring, process monitoring and smart spaces etc (Ian F.Akyildiz 2005). Energy efficiency seems to be critical in UWSNs. On the other hand, less than ten percent of the entire ocean volume is being investigated while the remaining area is still not explored.

Recently, researches on UWSNs primarily focus on underwater wireless medium modeling and end-to-end communication (Mohammad Sharif-Yazd et. al. 2017). The sensor node collects the information from the underwater surroundings and transfer to the sink nodes positioned on the water surface; finally sink node moves the data to the Base Station (BS) (Almir Davis & Hwa Chang 2012). Although, UWSN simulate some standard properties of the terrestrial wireless sensor network, Radio Frequency (RF) signal is widely used as wireless transmission media. However, RF is not suitable for underwater transmission due to immediate attenuation. Thereby, the acoustic signal is employed in underwater environment with range and speed is (1km, 1500 m/s) (Ian F.Akyildiz 2005). Underwater sensor network encounters specific challenges like: acoustic signal transmission, limited signal speed in comparison to electromagnetic waves. On the other hand, acoustic signal transmission greatly complements for dynamic network topology where nodes move on water irrespective of distance which seems to be impossible with electro- magnetic waves due to its high power absorptions. Also, time synchronization is crucial with relevance to underwater environment because of spread delay and velocity of sound.

Underwater Autonomous Vehicle (AUV) offers an additional perspective through aerial monitoring capability (Russell B.Wynn et. al. 2014). Implementation cost of an AUV is high; this necessitates cost effective long range Sensor Buoy System (SBS) for monitoring atmospheric temperature and pressure in underwater environment. Subsequently, Mangroves are the tropical forest systems which are most probably intensively exploited and degraded by human action. Over the last 25 years, the mangrove lands have been reduced severely, turning this land into pasture land or soil without any use. Mangroves are object of many research projects focused on their conservation. This kind of trees is very useful in natural disasters such as tsunamis and they are often used for raising shrimp and fish culture. But nowadays, the population of mangroves is dramatically decreasing because of the increment of salinity in water. The critical salinity level in mangroves depends on each species but generally salinities higher than 16 parts per million make impossible to develop the new individuals.

Over the last 70 years the world population has tripled and by the year 2020 an exponential increase in population around 9 to 10 billion is expected. Food and Agriculture Organization (FAO) predicts that global food production will rise by 70% to meet estimated demand over 2020. Usually, rapid population growth, resource constraints and climate change encountering feeders need technology facilitated irrigation. In turn, application of modern ICT (Information and Communication Technologies) enabled smart farming offers capital-intensive and hi-tech system of growing hygienic food ensuring a sustainable environment. This in turn necessitates an FARS based UWSN for monitoring the crop field with distinct sensors (humidity, temperature and soil moisture) thereby automating the irrigation process. Hence, farmers can monitor the field conditions from anywhere, anytime through mobile phones which is highly efficient in comparison to the conventional approaches for smart farming. Hence, it is highly necessary to build a cost effective framework for smart farming with underlying UWSNs thereby addressing the mentioned challenges.

Seawater which is available in huge volumes across the planet should find its optimal way through irrigation purposes. Freshwater differs from seawater with concentration in terms of salt and other ions (sea water has an average salinity of 3.5%); where 74.5% of freshwater prevail in the form of glaciers. This necessitates a huge investment in water consumption for domestic purposes such as drinking water; irrigation for agriculture etc. thereby increasing the implementation cost of an irrigation system. Subsequently, physical irrigation systems suffer due to insufficient water resources. Hence, an optimum irrigation is to be facilitated to the farm lands with relevance to their growth conditions and environmental conditions. Without technology enabled agriculture, it is highly crucial to identify the ideal plant growth conditions as soil moisture, water level, atmospheric temperature and humidity. On the other hand, UWSNs which inherently complements for automated irrigation in smart farming, has its own issues. The major challenges of UWSNs are: acoustic signaling with limited bandwidth, path loss incurred on an acoustic channel, extraneous noise in sensor node communication.

Characteristics with Terrestrial Sensor Networks

The main differences between terrestrial and underwater sensor networks are as follows:

- **Cost:** The terrestrial sensor nodes are conventional to become progressively economical, underwater sensors are expensive appliances. Hence, it is exclusively due to the more complex underwater transceivers thereby hardware protection is necessary in the extreme underwater environment.
- **Deployment:** In general, terrestrial sensor networks are densely deployed; in other hand the deployment is deemed to be sparser in underwater environment, due to the cost involved and several challenges seems to be very crucial in underwater environment.
- **Topology:** The network topology of the terrestrial sensor networks can be static or low dynamic. In a consequence, topology is high dynamic due to continual movement of nodes by water current.
- **Power:** The power desired for acoustic underwater communication is greater than the terrestrial radio communication due to scattered higher distance and to more complex signal processing at the receiver need to remunerate the impairment of the medium. Further, underwater network require higher battery capacity requirement.
- **Memory:** Terrestrial sensor nodes have very limited storage capacity. Whereas, underwater sensor nodes require more data catching as a result of intermittent channel specification.
- **Spatial Correlation:** With relevance to terrestrial sensors are regularly correlated. Further, correlation seems to involve overhead in terms of higher distance and limited bandwidth in underwater network.
- **Network Components:** The main components of the terrestrial sensor networks are sensor nodes, sinks, actuators and base station. Besides, underwater network involves underwater sensor nodes, AUV and onshore base station.

Underwater Sensor Node Architecture

The common internal components of an underwater sensor are depicted in Fig. 1. It comprises of a controller, memory, sensor, acoustic modem and power supply.

Figure 1. Underwater sensor node architecture

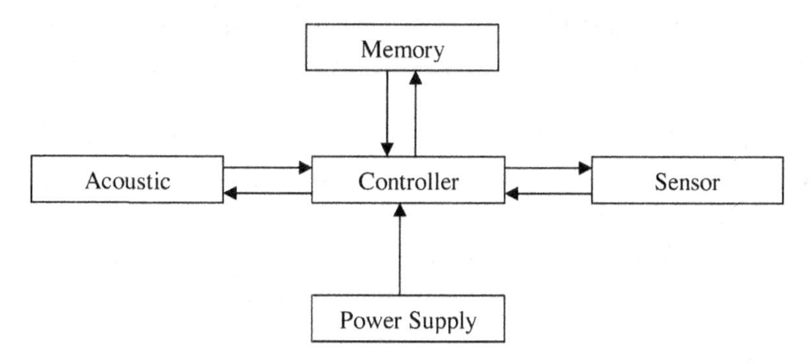

The components of underwater sensor are described in detail as follows

- **Controller:** The controller collects data from the sensor, process data and stores it in the memory. Microcontrollers are low power device which perform one task and run one specific program. The information is stored in ROM (read-only memory) generally does not change.
- **Memory:** The memory component is used to store intermediate sensor readings, data from other nodes.
- **Sensors:** It can gather a physical quantity data at standard interval, the turn around the data into signal and ultimately send the signal to base station. Typical examples for such underwater sensors include temperature sensors, density, salinity, acidity, chemical, conductivity, pH, oxygen, hydrogen, turbidity and so on. Subsequently, Sensor node is also capable of identifying their neighbor nodes to form a network.
- **Acoustic Modem:** Device is used to exchange data between individual nodes. The objective is to generate a signal that can be transmitted easily and communication media can be radio frequencies, optical signal and acoustic signal. Acoustic modem is indicated in Fig. 2.
- **Power Supply:** The power supply has significant impact on their performance. It supplies electric power at regular time period.
- **UWSNs Applications**
- **Water Quality Monitoring:** With a combination of sensors such as chemical and salinity sensors address the quality of water based on water salinity proportion. It manages salinity level by closely monitoring the water surface.

Figure 2. Acoustic modem

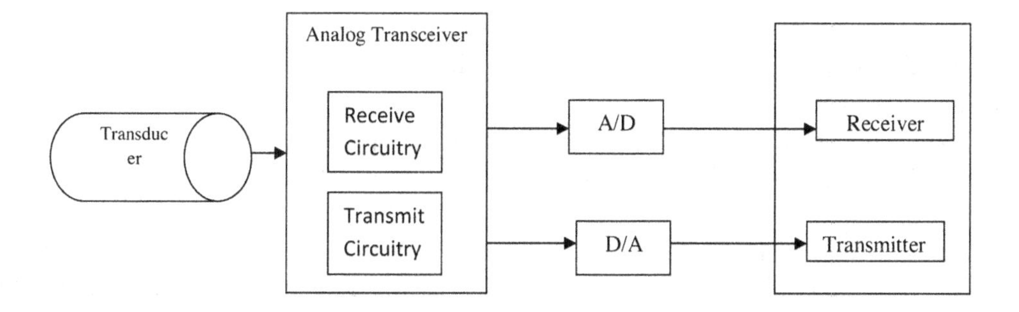

- **Undersea Explorations:** Large number of minerals is presented in underwater environment which is required to be explored such as oil and gas. Underwater sensor are used to detect underwater oilfields or reservoirs, determine routes for laying undersea cables and assist to examine of an important minerals.
- **Disaster Prevention:** Evidently, the natural disasters are irresistible. Sensor nodes read the seismic activity from remote locations and provide appropriate tsunami warnings to coastal areas or to analyze the issues of underwater earthquakes.
- **Assisted Navigation:** Underwater environment is extremely divergent, unexplored, random, and heavy with increasing depth. Underwater sensors can be used to determine hazards on the seabed, locate dangerous rocks or shoals in shallow waters, mooring positions and submerged crashes.
- **Military Surveillance:** UWSNs employed to assist military applications thereby different sensors deployed in underwater to identify various aspects of military applications. Different sensors such as cameras, sonar imaging and metal detectors integrated with AUV are used to measure the underwater mines, securing ports, and submarines and are also used for monitoring and surveillance. This technology allows enables an economic solution to protect marine forces.

Existing Literature for UWSNs

Intelligent agriculture through LoRa based approach over eight sensors incorporated with multi function sensor components has been introduced (Yi-Wei Ma & Jiann-Liang Chen 2018) Collected information in the agricultural field transmits data to remote computer through LoRa network. In a contrast, sensor house are positioned to protect against differing environmental aspects ensuring a feasible form of the network.

ZigBee (IEEE 802.15.4) is high-level communication protocols; data transmission over sensor nodes towards the coordinator using Xbee has been performed (G.Sahitya et. al. 2017). Sensors are associated to the arduino which isolates a set of data into individual node value.

Agricultural Environment Monitoring Systems (AEMS) employ nodes with fixed sensors confined for agricultural observing (Mohamed Rawidean et. al. 2017). Also, the sensor data forward the information through cloud gateway thereby defining Software as a Service (SaaS), Platform as a Service (PaaS) and Infrastructure as a Service (IaaS) for accurate monitoring and controlling of agricultural environment.

Enhanced ShockBurst protocol with customized-driven applications for real time assessment is suggested (Argyris Theopoulos et. al. 2018) to efficiently validate the system performance with F-test analysis. The research emphasizes on MultiCeiver mode that can broadcast the information with automatic packet assembly and timing, automatic acknowledgment (ACK) and retransmissions of packets for hydroponics cultivation.

Optimum Productivity by implementing low cost SmartNode for precision agriculture has been proposed (Juan M et. al. 2017). ZigBee network topology seems to monitor productivity, environment temperature and soil moisture through Kriging estimation algorithm which is further presentable in a regression analysis database.

Sensor equipped Unmanned Aerial Vehicle accomplish at optimal speed in order to achieve highest network throughput (Prusayon Nintanavongsa & Itarun Pitimon 2017). Also, random way point model form mobility pattern.

Agriculture environmental monitoring is identified to be better in terms of smart micro- needle and nitrate sensor (Brendan O'Flynn et. al. 2017) with Security in Agriculture, Food and the Environment

(SAFE) system for temporal and spatial analysis of environmental nutrients. The system not only measure the level of nutrients over farm field but also provides real-time information of the growing conditions which ensures a potential for optimized resource utilization in a precision agriculture application.

Water distribution for overall network has been proposed depending on linear interpolation method (Manish Bhimrao Giri & Dr. Ravi Singh Pippal 2017). Sensor nodes are located at the root of crops to reads the moisture level of soil in the active root zone. Primarily, system endorses to evaluate the water distribution level with multiple crops.

With relevance to the closed-loop system sensor nodes seem to monitoring and controlling the soil moisture together alerting the end users focused by hardware requirements namely end device, router and gateway (Anindita Mondal et. al. 2017). The prototype of this model was implemented and tested which increases the reliability of the network with dynamic data transmission.

An optimal resource along with enhanced cultivation over collection of significant information is analyzed to be better in terms of sensor nodes deployment has been made feasible through Software de Gestion (Jorge Granda-Cantu˜na et. al. 2018). Therefore this technique necessitates a minimum of ten samples and ten iterations for distinct soil types.

To determine the sparse and dense region by deploying Mobile Sinks (MSs) over extended range for UWSN has been proposed (Arshad Sher et. al. 2017). Sensor nodes are positioned and static sink is located at the mean of the network field. The research emphasizes on i) sparsity-aware energy-efficient clustering (SEEC) that involve hello packet based on its position. On the other hand, sensor nodes forward hello packet within their transmission range together enclosing the hop count from a sink thereby sparse and dense region are evaluated over multiple path and forward data packets at different regions. Also, Cluster Head (CH) nodes are chosen based on two parameters, residual energy and low depth. ii) To ensure an energy efficient communication to maximize the network coverage is accurately derived in circular sparsity-aware energy-efficient clustering (CSEEC) which in turn divided into concentric circle of equal parts. Further the approach seem to prevent energy hole formation. iii) circular depth-based sparsity-aware energy efficient clustering (CDSEEC) has two phases: lower semi-circle and upper semi-circle phase. During lower semi-circle phase CHs never broadcast their data packet to the static sink; in a contrast nodes send data to MSs. Moreover, sensor nodes at the closest proximity to the static sink transmit data based on depth threshold value which seems to be substantial network lifetime and stability period.

A Smart Irrigation Decision Support (SIDS) system is suggested (Yousef E. M. Hamouda 2017) to efficiently forecast irrigation time based on fuzzy logic controller. Also, logic rules are selected just as crop and soil conditions as a consequence of output parameters seems to be efficient water utilization.

BEER is localization free routing protocol, where mobile sinks are positioned into rectangular region (Junaid Shabbir Abbasi et. al. 2017). Flow of mobile sinks is in clockwise order. To determine the direction of mobile sink, nodes forward hello packet to the neighboring nodes that fall within their transmission range together enclosing the node ID and sink coordinates. On the other hand, energy levels of sensor nodes are evaluated thereby comparing the transmission distances to fixed transmission range. Based on this the choice of communication is done in terms of direct or multi-hop communication. Therefore this technique identifies itself to be relevant for terrestrial network due its distance constrained communication.

Dynamic data transmission through flooding based approach over sender nodes towards the surface sink has been introduced (Tariq Ali et. al. 2014). The flooding zone is involved to avoid the overflow on the total network. DVRP supports broadcast communication mechanism. Consequently, node collects the same data packet over multiple-path and forward data packets at different times. Thereby, priority queue mechanism focuses to reduce the forwarding nodes and manipulate the number of multiple paths. DVRP enhances the packet delivery ratio with continuous transmission. In a contrast, priority queue mechanism increases communication latency which is not suitable for underwater communication.

Optimum energy resources by avoiding direct communication over extended range for UWSN has been proposed (Nadeem Javaid et. al. 2016). Sensor nodes are extended across a ring sector and sink node is located at the mean of the network field. Neighboring nodes are evaluated by location aware strategy; node finds its neighbor then relevant communication is desired to deliver the data from source to destination. Forwarder nodes are chosen through route establishment phase in other hand node communication takes place in data transmission phase. Therefore, this technique increases congestion and delay which adds over time complexity.

Multiple transmissions with equivalent energy utilization for overall network has been performed; depend on table driven routing protocol (Ayaz M, Abdullah A & Jung LT 2010). The research emphasizes on energetic courier node that can broadcast the hello packet message with exact time to the ordinary nodes. Primarily, courier node requires more power to transmit the data towards surface sinks. Hence, this approach finds courier node is high-priced sensor and performance is also not stable.

Agriculture System (AgriSys) monitors the environment to manage its adequacy (Aalaa Abdullah et. al. 2016). Fuzzy control system act as a human operator thereby shift the input-output relationship in order to predefined set of member function and rules. Therefore, this method increases productivity.

With relevance to the stochastic geometry based capacity research analytic (Xia Li & Dongxue Zhao 2017), a set of nodes are organized into clusters. The cluster members periodically transmit their data to the corresponding cluster head. Cluster head node aggregates the data and then transmits the concurrence data to the sink node. Primarily, transmission is done in terms of limited bandwidth over carrier frequency thereby interference involve exclusive in cluster head nodes. In consequence, the research addresses the signal to interference ratio (SIR) to avoid the ambient noise on the total network. Subsequently, it is proven that transmission capacity outperforms outage probability in terms of optimal network node density.

Sensors connected to the arduino and zigbee seem to expand the probability of collecting the data (G.Sahitya et. al. 2016). Primarily, the system evaluates the crop yielding rates which enhance the maximum cultivation to the end user.

Exploring Sustainable Agriculture

The above sections distinctly related the current performance on smart agriculture using WSN; with reference to the analysis there is no attempt for seawater based irrigation for sustainable agriculture, this research contributes frequent stimulating research highlights as follows:

- The proposed SBS demonstrated here show advantages of easy fabrication, low cost, high sensitivity and compatibility of temperature monitoring in large scale environment.
- The research attempts to exploit UWSNs for efficient utilization of seawater increases production, resolve chemo-physical policies and socioeconomic constraints as crop production monitoring.
- Smart farming increases productivity which in turn ensures environmental and agricultural sustainability.
- To determine the high network throughput attained with the facilitation of SBS in underwater environment.

UWSNs Limitations

Underwater Wireless Communication (UWC)

UWC is entirely diverse from the terrestrial network communication in various aspects facing challenges as ambient noise, multipath, transmission loss etc.

Ambient Noise

The ambient noise (N) is inferior on the deployment surrounding, probably attains from diverse sources (turbulence (N_t), shipping (N_s), waves (N_w) and thermal noise (N_{th})) (Junaid Shabbir Abbasi et. al. 2017), from equation (1). Power Spectral Density (PSD) of the four noise components in decibel that relative to micropascal per hertz as a function of frequency (f) in kilohertz (KHz).

$$10 \log N_t (f) = 17 - 30\log f$$

$$10 \log N_s (f) = 40 + 20(s - 0.5) + 26 \log (f) - 60 \log (f + 0.03) \quad (1)$$

$$10 \log N_w (f) = 50 + 7.5w^{1/2} + 20 \log f - 40 \log (f + 0.4)$$

$$10 \log N_{th} (f) = -15 + 20 \log f$$

Where s is the shipping activity factor, $0 \le s \le 1$, and w is the wind speed in metre per second and noise density is denoted as $N(f)$. The overall PSD of the ambient noise is

$$N(f) = N_t(f) + N_s(f) + N_w(f) + N_{th}(f) \quad (2)$$

Attenuation

Attenuation or path loss attained in underwater wireless communication over a transmission distance d for a signal frequency f, is expressed in equation (2).

$$A(d, f) = A_0 d^k a(f)^d \quad (3)$$

Where A_0 is a unit normalizing constant, absorption coefficient is represented in $a(f)$ and k is denoted as spreading factor. Thereby, absorption coefficient can be achieved using Thorp empirical equation, as described in equation (4).

$$10\log a(f) = \frac{0.11f^2}{1+f^2} + \frac{44f^2}{4100+f^2} + \frac{2.75f^2}{10^4} + 0.003 \quad (4)$$

Propagation Delay

The fundamental necessitate of acoustic signals in underwater environment seems to be high propagation delay that is comparatively 2 × 105 times moderate than the electromagnetic propagation in terrestrial sensor networks. The acoustic propagation delay is extremely changing which in turn uncertain on particular aspects that incorporate water temperature, acidity, depth, and salinity.

Limited Bandwidth

The acoustic signal bandwidth is shortened to a limited dimension due to differing factors that comprise high force of sound at lower frequencies and wide channel penetration by higher acoustic frequencies. Also, acoustic systems operate over 30 KHz.

High Transmission Power

High transmission power is recommended to generate the acoustic signal. Hence, it increase network load thereby energy efficiency is highly critical for underwater environment.

Bit Error Rate

High bit error rate obtains as by reason of connectivity loss, multipath interference and random motion of water current characteristics of the acoustic medium.

Intermittent Connectivity

Provisional loss of connectivity is occurred due to dense area and inaccuracy of position prediction. Hence, acoustic signal does not attain or concentration of the signal is ideally low.

UWSNs Communication Mechanism: Radio Frequency Signal vs. Acoustic Signal

UWSN is a network of self-powered sensor nodes and Autonomous Underwater Vehicle (AUV) deployed underwater to perform collaborative tasks using acoustic links. Thereby, acoustic signal is employed underwater to propagate over a range of 1 km at a rate of 1500 m/s. The sensor node collects the information from the underwater surroundings and transfers to the sink nodes positioned on the water surface; finally sink node moves the data to the Base Station (BS). Although, UWSN simulate some standard properties of the terrestrial wireless sensor network, Radio Frequency (RF) signal is widely used as wireless transmission media. However, RF is not suitable for underwater transmission due to immediate attenuation (Ian F. Akyildiz et. al. 2005). Underwater sensor network encounters specific challenges like: acoustic signal transmission, limited signal speed in comparison to electromagnetic waves. On the other hand, acoustic signal transmission greatly complements for dynamic network topology where nodes move on water irrespective of distance which seems to be impossible with electro- magnetic waves due to its high power absorptions and delay spread.

UWSNs: Trends and Challenges

UWSN has its own characteristics; underwater acoustic channels are unique (Salvador Climent et. al. 2014). Ultimately, terrestrial network specification is not feasible for UWSN.

Physical Implementation Limitations

Radio or optical communication ensures a long-distance communication with higher bandwidths; thereby finding their way with terrestrial network applications. On the contrary, water attenuates and scatters waves of all electro-magnetic frequencies, making acoustic waves preferable for underwater communication.

Medium Access Control and Resource Sharing

Sharing communication resources among nodes can be performed in multi-user systems. In wireless sensor networks, the frequency spectrum is inherently shared and interference needs to be properly managed. Efficient sharing of resources among the communicating nodes is performed through different methods to separate the signals coexisting in a common medium. Subsequently, resource sharing is facilitated for underwater networks by focusing its shortcomings as longer delays incurred with acoustic channels, frequency-dependent attenuation, propagation delay suffered by acoustic signals and bandwidth constraints of acoustic hardware. An efficient MAC mechanism for UWSN is achieved by exploiting the acoustic signals which can be deterministically separated in time or frequency.

Necessity of Gateway in Smart Farming: Low Power Wide Area Network (LPWAN)

LPWAN ensures wide area coverage to low power devices at the expense of low data rates. Standard developing organizations (IEEE, IETF, 3GPP, ETSI) and multiple industrial alliances built individual LPWA technologies to promote new standards. LoRa (Long Range), a Chirp Spread Spectrum (CSS) type modulation ensures multiple data rates together addressing bandwidth and spreading constraints. In addition, power consumption across the network is reduced and balanced with the low power consumption ICs which expose a maximum of 10μA power consumption during communication (25mA TX, 10mA RX). LoRaWAN offers secure bidirectional symmetrical link between the CH and LoRa gateway (MN) thereby ensuring reliable data transfers to the BS well in prior connecting to end-user applications facilitated with GPS-free localization services. Subsequently, efficient LoRaWAN enables UWSN to attain an outset for both uplink/downlink data which in turn increases communication robustness with high end deployment (star of star topology). The research addresses a cost-effective commercial LPWAN is at an outlay of $2000 (approx. the rate of the sensors deployed) ultimately reducing the capital expenses (CAPEX) and operating expenses (OPEX) for both end-users and network operators. LPWAN – LoRa specifications as mentioned in Table 1.

Node Hardware: Standard and Specifications

The FARS is matured based on the mechanics of UWSNs holds radically new type of Sensor Buoy System (SBS) is deployed underwater with counter weight. The integral part of SBS involve SBE

Table 1. LPWAN – LoRa specifications

LPWAN Features	LoRa
Uplink	Data
Downlink	Data + ACK
Symmetrical Technology	Y
Payload Size	19-250 bytes
Encryption	AES-138 E2E
Open Standard	Y
Technology	CSS/FSK/OOK/GMSK

39 sensor is a high-accuracy, fast-sampling temperature (pressure optional) recorder with integrated Inductive Modem (IM) interface for long duration deployments. For guaranteed energy autonomy, SBS carries 2.5 W solar panels which assure optimum energy even under prolonged poor light conditions. The top layer of buoy system equipped with RF antenna with Beacon light offers power supply regulation and management system, a set of interfaces for accessing the sensors, a module for amplification, module for conversion (analog to digital) and multiplexing of the data read from the sensors (underwater) and a CPU (microprocessor) to centralize the whole process to implement the user-defined monitoring function.

From Sensors to Decision Making: Smart Farming Use Case

1. Sensor Buoy System

The FARS exploits the mechanics of UWSNs holding a light weight Sensor Buoy System (SBS) deployed underwater with counter weight. The vertical based SBS comprises a corrosion proofing stainless steel tube of 3 m length and a width of 25 mm. In addition, middle part of the SBS includes a steel tube comprising an elliptic shaped flotation material coated with polyvinyl chloride complementing a rigid and flexible system. Subsequently, the upper part is enclosed with 2.5 W solar photovoltaic panels which act as storage batteries with an IP-68 watertight box of measurement 12 × 12 × 7 cm; an 8 dBi polarized omnidirectional antenna (EnGenius NET-WL-ANT-008ON) and a light-emitting beacon to ensure an optimum visibility of atmospheric conditions.

Initially, sensors are deployed in the network field of 1200*1200 square meter area enclosed by porous cardboard. The board ensures an ideal humid condition which in turn promotes increase regional rainfalls. FARS is used for the real time storage of freshwater. The RO process converts seawater into freshwater. Seawater passed through a synthetic plastic polymer pipe further demineralizes undergoing an RO process through a semi-permeable reverse membrane. Freshwater from the RO process is stored in a water storage tanker. On the other hand, residual water from the RO process is exploited to irrigate salicornia seeds and mangrove trees to enrich the soil quality thereby ensuring a suitable farming condition all around the year. The architecture of FARS-UWSNs is demonstrated in Fig. 3.

Figure 3. Architecture of FARS-UWSNs

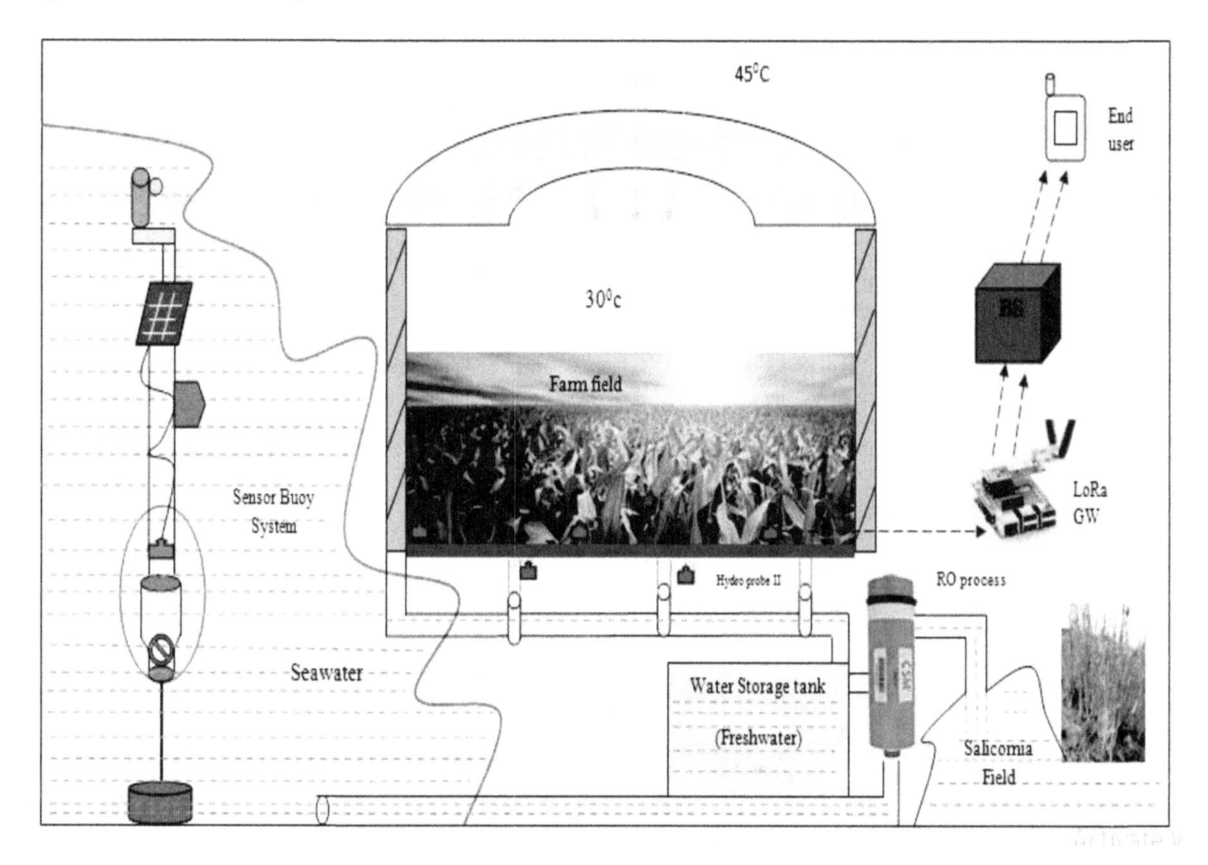

2. Offshore Data Logger

For offshore monitoring Printed Circuit Board (PCB) logger unit is deployed with integrated data storage and batteries. Fig. 4 shows the schematic overview of the data logger. The current value from the SBS transfers 4-20mA analog signal through 8 dBi Cisco aironet omnidirectional antenna (AIR-ANT2480V-N) converts into digital data and multiplexing of 16 bit value into 32-GB secure digital (SD) cards.

3. Clustering Approach

To analyze UWSNs, Hexagonal Based Clustering with Mobile Sinks strategy reduces overall network energy consumption complementing network lifetime. The Hydro probe II sensors are organized into hexagonal cells forming cluster; the node is chosen as a CH based on residual energy. If the residual energy of the node is greater than the average residual energy of an individual node ($E_R > AE_R$) and minimum distance from CM act as CH. Moreover, overall holding capacity of a CH to CM is ($0 \geq$ CM ≤ 20). Mobile Sinks is placed at left and right side of the farm field, whereas both MS1 and MS2 moves randomly to aggregate information from CH. Consequently, MS avoids direct communication from sensor to BS in order to avoid delay in the network.

Figure 4. Schematic overview of the data logger

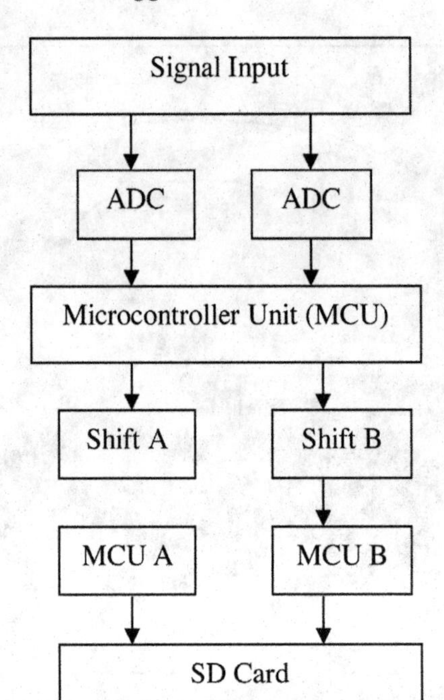

The outcome of CH selection procedure will be the set of nodes having higher calculated value of Position metric (POS) with reference to Equation (8) derived at the end of this section. HBC-MSs capitalizes on a CH selection procedure that promises every sensor node has only one CH i.e. only one node within its transmission range will be selected as CH. In this substance, proposed technique ensures that no two CHs are immediate neighbors. There might be a situation when two CHs results in same calculated value of the POS. In such situation, an optimal solution is to select a node having low path loss value among the competing set of nodes.

Initially, for each sensor node in the network, the neighboring nodes within its transmission range $txion_{Range}$ are identified.

$$ND_{v=} \sum \left\{dt(v,v')\right\} < txion_{Range}; \text{for each node in the network} \quad (5)$$

Where *v* is the source node and *v'* are the neighboring nodes; $txion_{Range}$ is the transmission range of the source node that is assumed to be 25 m; $dt\left(v,v'\right)$ indicates the distance between the source node and neighbor nodes.

Subsequently, the distance between the source node 'v' and neighbor node v' ; $dt\left(v,v'\right)$ can be derived using Euclidean distance formula

$$\sqrt{dt\left(v,v'\right) = \left(\left(x_v - x_{v'}\right)^2\right) + \left(\left(y_v - y_{v'}\right)^2\right)} \quad (6)$$

For every node ν, a maximum node degree (total number of neighbors) the node can handle is predefined to be 20. The sum of distances to all neighbors D_v is computed for every node as,

$$D_v = \Sigma\left\{dt\left(v, v'\right)\right\} for\, V' \in ND_v \quad (7)$$

Where, $dt\left(v, v'\right)$ indicates the distance between the source node and neighbor nodes that is derived using Euclidean distance formula. For every sensor node, $AVGRE_v$ needs to be computed to examine if its residual energy is greater than the average of residual energies of all its neighbors.

$$AVGRE_v = \frac{1}{ND_v} \Sigma E_v\, for\, v \in ND_v \quad (8)$$

Where, ΣE_v is the sum of residual energy of all the neighboring nodes and ND_v indicates the node degree.

Finally the position function POS, is calculated for each node with reference to Equations (5), (6) and (7) as,

$$POS_v = ND_v + AVGRE_v + \frac{1}{D_v} \quad (9)$$

Ultimately, node with the highest POS, residual energy and minimum distance act as CH. The neighbors of the selected CH become the corresponding CMs to form a cluster which is discussed in the section that follows immediately.

4. Sensor Network Topology

Consider two nodes N1 and N2 with different energy levels. CH selection process is employed to choose an optimal CH in terms of residual energy and distance from the surface level of the farm field. Assuming N1 is selected as a CH, it broadcasts hello message packet to its neighboring sensors within its transmission range. On the other hand, Cluster Member (CM) nodes in a cluster send their data to the corresponding CH.

5. Data Transmission

In FARS, after node organization and schedule generation, the consequent step is data transmission. Sensor nodes collect data and transfers to CH; CH sends acknowledgment to the MS. The sink moves towards the CH and collects data and sends to the BS medium of LoRa gateway.

6. Seawater Composition

Water inadequacy enhanced rising severe in arid and semi-arid regions across the world. However, seawater which is available in huge volumes across the planet should find its optimal way to reduce water

Table 2. Standard seawater composition

Chemical Ion	Concentration (ppm)	Total Salt Content (%)
Chloride Cl^-	19,345	55.0
Sodium Na^+	10,752	30.6
Sulfate SO_4^{2-}	2,701	7.6
Magnesium Mg^{2+}	1,295	3.7
Calcium Ca^{2+}	416	1.2
Potassium K^+	390	1.1
Bicarbonate HCO_3^-	145	0.4
Bromide Br^-	66	0.2
Borate BO_3^{3-}	27	0.08
Strontium Sr^{2+}	13	0.04
Fluoride F^-	1	0.003

demand. Freshwater differ substantially from seawater by the relative amount of salts. Table 2 shows a typical seawater composition.

7. Pressure-Driven Membrane: Reverse Osmosis (RO) Process

Seawater passed through a synthetic plastic polymer pipe of length 3000m further de-mineralizes undergoing an RO process through a semi-permeable membrane. The RO process converts seawater into freshwater. The performance of RO process recovery rate (RR) of freshwater can be expressed as

$$RR = \frac{NP}{NF} \times 100 \tag{10}$$

Where, NP and NF are the filtrate (or permeate) passing through the membrane and flow rates, respectively.

8. Nodes Deployment in the Farm Field

Hydro probe II sensors are deployed in the network field of 1200*1200 square meter area enclosed by porous cardboard. Fig 3 shows architecture of FARS-UWSNs. The board ensures an ideal humid condition which in turn promotes increase regional rainfalls. FARS is used for the real time storage of freshwater. Freshwater from the RO process is stored in a water storage tanker. On the other hand, residual water from the RO process is exploited to irrigate salicornia seeds and mangrove trees to enrich the soil quality thereby ensuring a suitable farming condition all around the year.

Ultimately, the proposed cost effective smart farming solution ensures a 10-25% lesser operational cost in comparison to the existing smart farming methodologies. The flow chart specification is demonstrated in Figure 5.

Figure 5. Flow chart of FARS

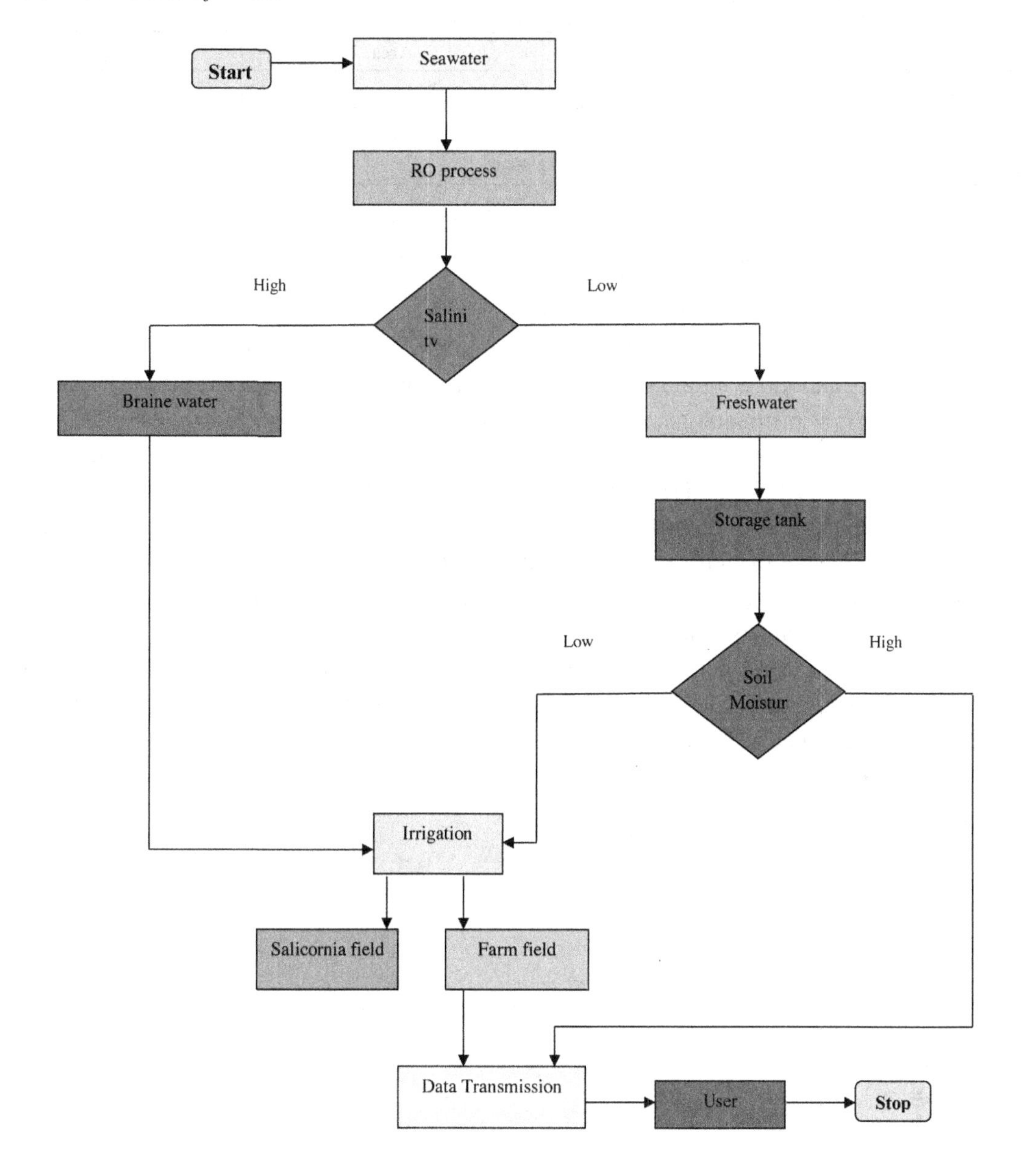

Social Relevance and Utility of the Proposed Research

The inception of smart technologies across multiple productions has been a primary evolutionary factor over the recent past. Moreover, smart farming endeavor a path towards sustainable agriculture over heterogeneity of technologies, crop production systems and networks across all aspects of the agriculture environment. Food and Agriculture Organization (FAO) predicts that global food production will experience an exponential rise by 70% to meet estimated demand over 2020. Rapid population growth, resource constraints and climate change is projected to reduce global average crops. Rather, the adoption of advanced production techniques and modern explorations can contribute to overcome this problem.

Table 3. Total mangrove area by regions

Region	Area	Global (%)
South and SE Asia	75173	41.5
Australia	18789	10.4
America	49096	27.1
West Africa	27995	15.5
East Africa and Middle East India	10024	5.5

FARS based UWSNs comprise the application of communications technology to make agricultural process easier, faster and more automated. Conversion of seawater into freshwater through RO process was enhanced as a key methodology. From the analysis it is clearly revealed that "semi-permeable reverse membrane (diameter 2.50 inches, length 40.00 inches)" is the most essential conversation of freshwater. However, soil moisture level data are continuously forwarded to BS over LoRa gateway. On the other hand, residual water from the RO process is exploited to irrigate salicornia seeds and mangrove trees. Salicornia seeds and mangrove trees are salt-tolerant forest ecosystems provides many ecological, environmental and socio-economic benefits. In turn, mangroves create unique niche that hosts rich agglomeration of species diversity and act as physical barrier to mitigate the effects of coastal disasters like tsunami, hurricanes, waves and stabilizing sediments and absorption of pollutants. Ultimately, the proposed cost effective smart farming achieves approximately 20% performance improvements in terms of crops production and network lifetime compared to the existing smart farming methodologies. Table 3 shows a total mangrove area by regions.

Figure 6. Soil moisture level in FARS

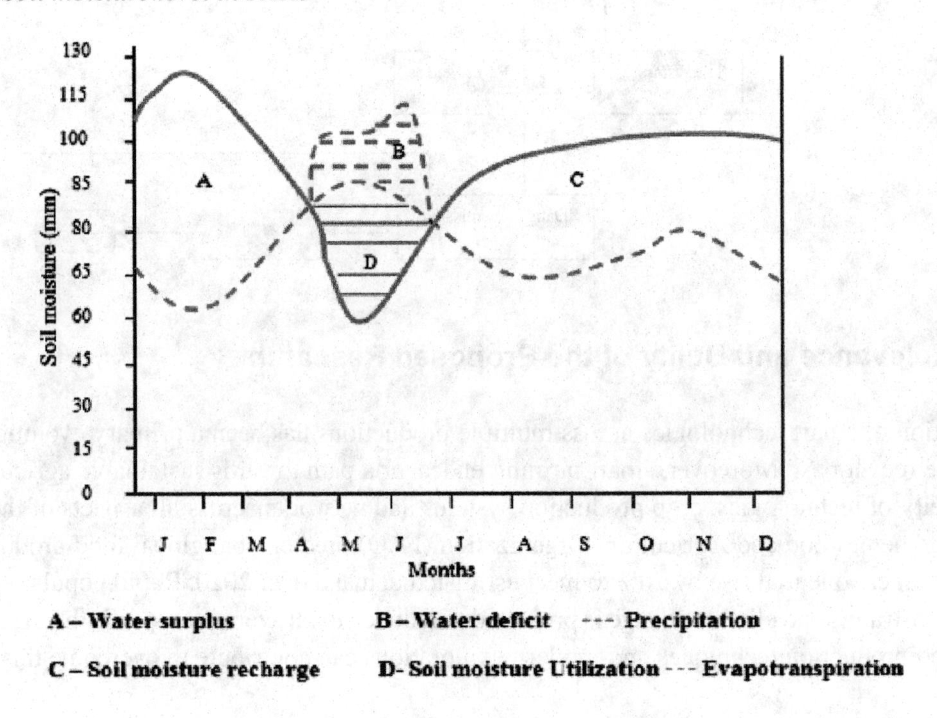

Figure 7. Productivity in smart farming

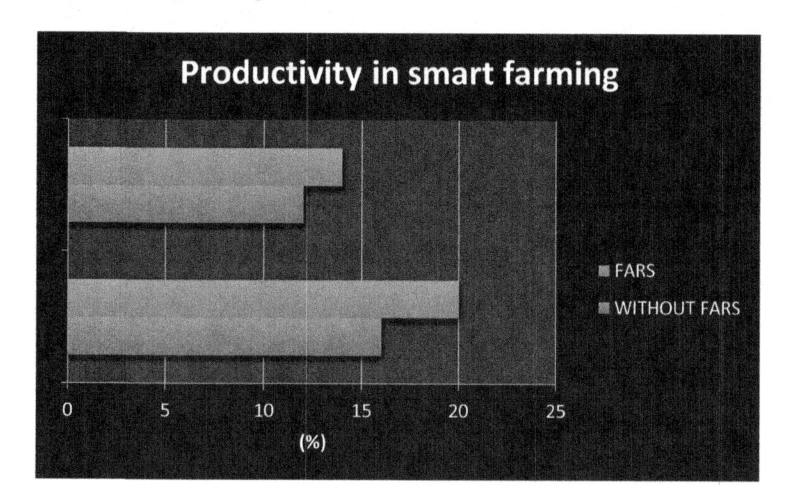

Results and Discussion

This section revealed the results of the study, Fig. 6 indicating the actual results are better consistent through FARS-UWSNs, which shows that the model is feasible during the analysis on water deficit for agronomic traits of crops (wheat, grain, cotton and rice).

Productivity

Productivity is quantified in terms of crop traceability and increased crops productivity. In Fig. 7 FARS-UWSNs seems to be convincing in terms of productivity, when compared to traditional smart farming. Since LoRa is involved in data collection task, transmission delay reduces phenomenally in FARS-UWSNs.

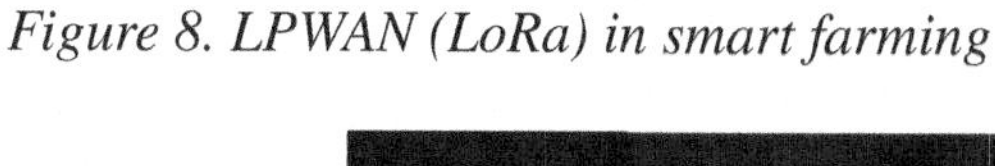

Figure 8. LPWAN (LoRa) in smart farming

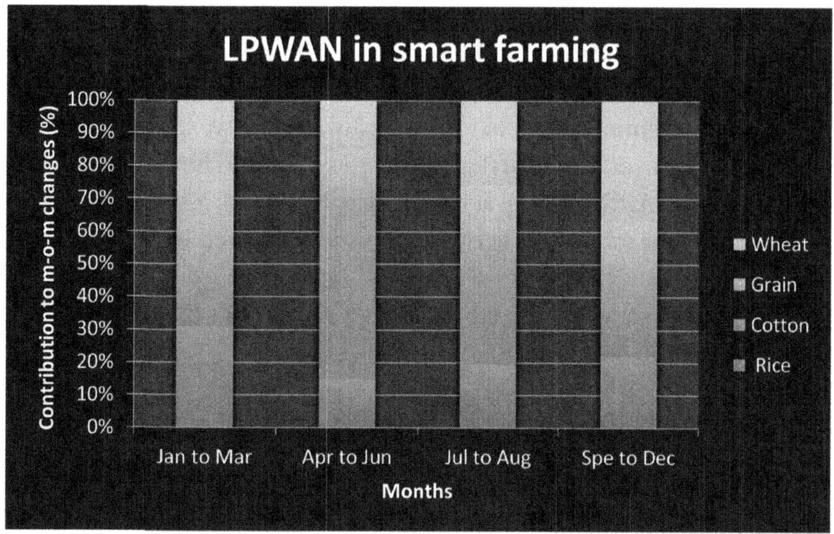

Figure 9. Network Lifetime of FARS

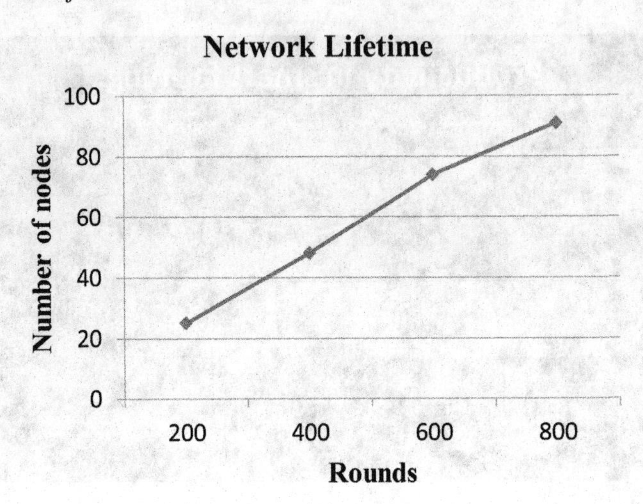

LPWAN in Smart Farming

The results of the LoRa gateway Fig. 8 shows the total amount of crops in terms of percentage at an operating temperature of 60^0c. The bar line indicates the amount of various crops in a certain months.

Network Lifetime

Network lifetime is quantified in terms of the total number of nodes alive over different ranges of common rounds. The analysis results prove that all nodes are alive after the completion of rounds. Fig. 9 shows the network lifetime of proposed FARS.

CONCLUSION

In this study, FARS based UWSNs is relevant for large scale sector in smart farming. In order to converts seawater into freshwater through RO process was enhanced as a key methodology. From the results it is clearly revealed that "semi-permeable reverse membrane (diameter 2.50 inches, length 40.00 inches)" is the most essential conversation of freshwater. However, soil moisture level data are continuously forwarded to BS over LoRa gateway; Analytical results show that significant productivity improvement is achieved by FARS strategy achieves approximately 20% performance improvements in terms of increased crop traceability and network lifetime compared to the nominal communication based data forwarding. Ultimately, salicornia seeds and mangrove trees play a very significant role in maintaining the coastal environment thereby reducing the impact of wave action and soil erosion in the coastal areas is avoided.

REFERENCES

Abbasi, J. S., Javaid, N., Gull, S., Islam, S., Imran, M., Hassan, N., & Nasr, K. (2017). Balanced Energy Efficient Rectangular Routing Protocol for Underwater Wireless Sensor Networks. *IEEE Access: Practical Innovations, Open Solutions*.

Abdullah, A., Al Enazi, S., & Damaj, I. (2016). AgriSys: A Smart and Ubiquitous Controlled-Environment Agriculture System. *International Conference on Big Data and Smart City*. 10.1109/ICBDSC.2016.7460386

Akyildiz, I. F. (2005). Underwater Acoustic Sensor Networks (UW-ASN). *BWN Laboratory*. Retrieved from http://bwn.ece.gatech.edu/Underwater/index.html

Akyildiz, I. F., Pompili, D., & Melodia, T. (2005). Underwater acoustic sensor networks: research challenges. Ad Hoc Networks, 257–279.

Ali, T., Jung, L. T., & Faye, I. (2014). Diagonal and Vertical Routing Protocol for Underwater Wireless Sensor Network. *International Conference on Innovation, Management and Technology Research*, 372 – 379. 10.1016/j.sbspro.2014.03.690

Ayaz, M., Abdullah, A., & Jung, L.T. (2010). Temporary cluster based routing for underwater wireless sensor networks. *Informat Tech (ITSim) Internat Symposin*, 1009–14.

Climent, S., Sanchez, A., Capella, J. V., Meratnia, N., & Serrano, J. J. (2014). Underwater Acoustic Wireless Sensor Networks: Advances and Future Trends in Physical, MAC and Routing Layers. *Sensors (Basel)*, *14*(1), 795–833. doi:10.3390140100795 PMID:24399155

Davis, A., & Chang, H. (2012). Underwater Wireless Sensor Networks. *Oceans*.

Giri & Pippal. (2017). Use of Linear Interpolation for Automated Irrigation System in Agriculture using Wireless Sensor Network. *International Conference on Energy, Communication, Data Analytics and Soft Computing*. 10.1109/ICECDS.2017.8389716

Granda-Cantu˜na, J., Molina-Colcha, C., Hidalgo-Lupera, S.-E., & Valarezo-Varela, C.-D. (2018). Design and Implementation of a Wireless Sensor Network for Precision Agriculture Operating in API Mode. *International Conference on eDemocracy & eGovernment*. 10.1109/ICEDEG.2018.8372346

Hamouda, Y. E. M. (2017). Smart Irrigation Decision Support based on Fuzzy Logic using Wireless Sensor Network. *International Conference on Innovation, Management and Technology Research*, 361-368. 10.1109/ICPET.2017.26

Javaid, N., Shah, M., Ahmad, A., Imran, M., Khan, M. I., & Vasilakos, A. V. (2016). An Enhanced Energy Balanced Data Transmission Protocol for Underwater Acoustic Sensor Networks. *Sensors (Basel)*, *16*(4), 487. doi:10.339016040487 PMID:27070605

Juan, M., Nunez, V., Faruk Fonthal, R., Yasmin, M., & Quezada, L. (2017). Design and implementation of WSN for precision agriculture in white cabbage crops. *IEEE International Conference on Electronics, Electrical Engineering and Computing (INTERCON)*.

Li, X., & Zhao, D. (2017). Capacity Research in Cluster-Based Underwater Wireless Sensor Networks Based on Stochastic Geometry. IEEE, 14, 80-87.

Ma, Y.-W., & Chen, J.-L. (2018). Toward Intelligent Agriculture Service Platform with LoRa-based Wireless Sensor Network. *Proceedings of IEEE International Conference on Applied System Innovation*, 13-17. 10.1109/ICASI.2018.8394568

Mondal, A., Misra, I. S., & Bose, S. (2017). Building a Low Cost Solution using Wireless Sensor Network for Agriculture Application. *International Conference on Innovations in Electronics, Signal Processing and Communication*. 10.1109/IESPC.2017.8071865

Nintanavongsa, P., & Pitimon, I. (2017). Impact of Sensor Mobility on UAV-based Smart Farm Communications. *International Electrical Engineering Congress*.

O'Flynn, De Donno, Barrett, Robinson, & O'Riordan. (2017). Smart Microneedle Sensing Systems for Security in Agriculture, Food and the Environment (SAFE). *IEEE Sensors*.

Pavitra & Janani. (2017). Low Delay and High Throughput Data Collection in Wireless Sensor Networks with Mobile Sinks. *International Journal of Scientific & Engineering Research*. doi:10.1109/IEECON.2017.8075822

Rawidean, M., Kassim, M., & Harun, A. N. (2017). Wireless Sensor Networks and Cloud Computing Integrated Architecture for Agricultural Environment Applications. *International Conference on Sensing Technology*.

Sahitya, Balaji, Naidu, & Abinaya. (2017). Designing a Wireless Sensor Network for Precision Agriculture using Zigbee. *IEEE International Advance Computing Conference*, 287-291. 10.1109/IACC.2017.0069

Sahitya, G. (2016). Wireless Sensor Network for Smart Agriculture. Academic Press.

Sharif-Yazd, M., Khosravi, M. R., & Moghimi, M. K. (2017). A Survey on Underwater Acoustic Sensor Networks: Perspectives on Protocol Design for Signaling, MAC and Routing. *Journal of Computer and Communications*, *5*(05), 12–23. doi:10.4236/jcc.2017.55002

Sher, Javaid, Azam, Ahmad, Abdul, Ghouzali, … Khan. (2017). Monitoring square and circular fields with sensors using energy-efficient cluster-based routing for underwater wireless sensor networks. *International Journal of Distributed Sensor Networks*, 13.

Theopoulos, A., Boursianis, A., Koukounaras, A., & Samaras, T. (2018). Prototype wireless sensor network for real-time measurements in hydroponics cultivation. *International Conference on Modern Circuits and Systems Technologies*. 10.1109/MOCAST.2018.8376576

Wynn, R. B., Huvenne, V. A. I., Le Bas, T. P., Murton, B. J., Connelly, D. P., Bett, B. J., ... Hunt, J. E. (2014). Autonomous Underwater Vehicles (AUVs): Their past, present and future contributions to the advancement of marine geosciences. *Marine Geology*, *352*, 451–468. doi:10.1016/j.margeo.2014.03.012

Chapter 19
Air Quality Monitoring Using Internet of Things (IoT) in Smart Cities

Gayatri Doctor
CEPT University, India

Payal Patel
Oizom Instruments Pvt Ltd., India

ABSTRACT

Air pollution is a major environmental health problem affecting everyone. An air quality index (AQI) helps disseminate air quality information (almost in real time) about pollutants like PM10, PM2.5, NO2, SO2, CO, O3, etc. In the 2018 environmental performance index (EPI), India ranks 177 out of 180 countries, which indicates a need for awareness about air pollution and air quality monitoring. Out of the 100 smart cities in the Indian Smart City Mission, which is an urban renewal program, many cities have considered the inclusion of smart environment sensors or smart poles with environment sensors as part of their proposals. Internet of things (IoT) environmental monitoring applications can monitor (in near real time) the quality of the air in crowded areas, parks, or any location in the city, and its data can be made publicly available to citizens. The chapter describes some IoT environmental monitoring applications being implemented in some of the smart cities like Surat, Kakinada.

SMART CITIES

Smart Cities can be defined in many ways. The meaning of smart cities has evolved over the years with different meanings to people who come from different areas. Although, there are different meanings of Smart Cities, whenever one uses the word 'smart' it means that there will be usage of Information and Communications Technology (ICT) or Internet in order to be able to address the various urban challenges (Mitchell S., Villa, Stewarts-Weeks, & Lange, 2013). A Smart City can be identified along six main areas, Economy, People, Mobility, Living, Governance and Environment (Giffinger, et al., 2007).

DOI: 10.4018/978-1-5225-9199-3.ch019

Smart Cities would be a driver for economic growth, improving people's quality of life by developing the local area with the use of technologies that can provide smart outcomes.

A variety of challenges are faced in today's cities. These include creation of jobs, sustaining the environment and economic growth. Understanding of Internet's contribution and application is important to planning processes of future cities. (Mitchell S., Villa, Stewart-Weeks, & Lange, 2013).

The Government of India under the leadership of its Prime Minister Shri Narendra Modi initiated in June 2015 a Smart City Mission in India with an aim to develop 100 smart cities. This Mission is an urban renewal and retrofitting program. The Ministry of Housing and Urban Affairs(MoHUA), then known as Ministry of Urban Development (MoUD), shortlisted a 100 Cities from potential smart cities identified by the State Governments. Each of the potential Smart Cities, were required to prepare a 'Smart City Proposal' (SCP) with the help of consultants and participate in a 'City Challenge'. The Smart City Proposals would contain the strategies for 'Area Based Development' (ABD) and a 'Pan-City initiative'. The Area Based Proposals could be retrofitting (in an existing built-up area of more than 500 acres), redevelopment (replacement of existing built up area of more than 50 acres) or Greenfield (vacant areas). The Pan-City Initiative would envisage application of selected smart solutions to the existing city-wide infrastructure. Citizen participation was an important aspect of formulating these proposals.

The Ministry of Urban Development (MoUD) shortlisted 20 cities in the 1st round that were declared in January, 2016. Other cities were asked to improve on their proposals for selection in the next round. It was observed that these 20 cities were mostly in eight states. Thus a special round called the 'Fast Track' was conducted for the 23 States and Union Territories to shortlist the smart cities from amongst them. 13 Cities were selected from this Fast Track round and were declared on 24th May, 2016. The 2nd round was declared on 20th September 2016 and 27 cities were shortlisted in it. 30 cities were declared in the 3rd round in June 2017, followed by 9 cities in the 4th round on 20th Jan, 2018. The last city was selected on 20th June 2018, thus totaling to 100 smart cities being shortlisted in different rounds.

A smart city comprises of core infrastructure elements like supply of sufficient water, electricity that is assured, sanitation facilities, solid waste management, public transport to ensure efficient urban mobility, housing which is affordable, especially for the poor, security and safety of citizens, namely, women, children and elderly. It would also comprise of IT digitalization and connectivity which is robust, governance that is good and uses e-Governance and participation from citizens. It must also have health, education initiatives and environments which are sustainable.

An investment of Rs.2,01,981 crore is proposed under the smart city plans by the 99 cities. Projects which are focused on restructuring a specific identified area are known as Area Based Projects and they are estimated to cost about Rs. 1,63,138 crore. Initiatives in the Smart City which pan across the city are known as Pan City Initiatives and are estimated to cost about Rs. 38,841 crores.

In order to implement the different Smart Cities Plans, a Special Purpose Vehicle (SPV) was to be setup at the city level. The SPV would be a under the Companies Act, 2013, be a 50:50 equity shareholding jointly between the Urban Local Body (ULB) and the State/Union Territory and a limited company. Each selected Smart Cities is required to setup an SPVs and implement its Smart City Proposal. The making of Detailed Project Reports (DPRs), various tenders etc has to be done by the SPV. Project Management Consultants (PMCs) will assist the SPV in creation and implementation of the projects from the Smart City Proposal (Ministry of Housing and Urban Affairs, Government of India, 2017).

ISO 37120

In order to ensure that policies get converted to practice, city indicators can be used by various people and professionals like city managers, researchers, planners, politicians to promote livable, sustainable, inclusive and prosperous cities across the globe. (ISO 37120, 2014) In order to measure the performance of cities with respect to their quality of life and sustainability, cities need indicators. These indicators are very often are not standardized or comparable across different cities, especially over time.

In May 2014, the International Organization for Standardization (ISO) announced the first international standard on city data, namely, Indicators for City Services and Quality of Life. The standard includes a comprehensive set of 100 indicators among 17 Themes. These themes include Economy, Education, Energy, Environment, Finance, Fire & Emergency, Governance, Health, Recreation, Safety, Shelter, Solid Waste, Telecommunications & Innovation, Transportation, Urban Planning, Waste Water and Water & Sanitation with core and supporting indicators.

In 2018, ISO 37120:2018 Sustainable Cities and communities – Indicators for city services and quality of life defined 100 city performance indicators among 19 Themes namely Economy, Education, Energy, Environment & Climate Change, Finance, Governance, Health, Housing, Population & social conditions, Recreation, Safety, Solid Waste, Sports & Culture, Telecommunications, Transportation, Urban local agriculture & food security, Urban Planning, Wastewater and Water with core and supporting indicators (ISO 37120:2018, 2018).

Further development of indicators to support smartness (ISO 371221) and resilience (ISO 371232) in cities is ongoing in ISO/TC 268. Maintaining, enhancing, and accelerating progress towards improved city services and quality of life is also fundamental to the definitions of both smart cities and resilient cities.

Environment & Climate Change which is one of the Themes has core indicators like Particulate Matter concentration PM 2.5, PM 10, Gas emissions from greenhouses which is measured in tonnes per capita; and supporting indicators like Percentage of areas designated for natural protection, Nitrogen Dioxide (NO2) concentration, Sulphur Dioxide (SO2) concentration, Ozone (O3) concentration, Noise Pollution, percentage change in number of native species as given in Table 1.

Table 1. ISO37120 Indicator – Environment Theme

Indicators - Core
• Fine Particulate matter (PM2.5) concentration • Particulate Matter (PM10) concentration • Greenhouse Gas emissions measured in tonnes per capita
Indicators - Supporting
• % of areas designated for natural protection • Nitrogen Dioxide (NO2) concentration • Sulphur Dioxide (SO2) concentration • Ozone (o3) concentration • Noise Pollution • % change in number of native species

Source: ISO 37120 Indicators

AIR POLLUTION

One of the major environmental problems is Air pollution, and it is affecting a large number of people. Whenever the environment or air is modified from its natural characteristic because of being contaminated by some chemical, physical or biological agent, it causes asthma and respiratory infections, more prominently in children. Air Pollution also attributes for cardiovascular diseases like chronic respiratory disorders, cancers etc which have been the cause of mortality and morbidity. Air pollution is thus one of the largest environmental health risk hazards in the world.

Keeping note of the same, that air pollution is a major health hazard, the Government of India has formed a steering committee which would discuss and deliberate the health effects of air pollution. This would be done not only in the ambient air pollution but also the household air pollution (World Health Organization (WHO) India, 2018).

For a large number of people around the world, dust, soot, sulfur oxides and ozone are a becoming a threat to their health. The World Health Organization reports that 93% children across the world breathe in air with pollution levels that are harmful, way beyond the prescribed guidelines. According to the WHO's World Global Ambient Air Quality Database, 80% of urban areas and approximately 9/10 people breathe highly polluted air as outdoor pollution is way beyond permissible standards.

India is one of the countries that have consistently terrible air quality. In 2016, air monitoring stations across the 4,300 cities were studied and inferences made on measurements and calculations obtained. WHO reported that India's cities were the maximum affected. A ranking of particulate pollution in a study of cities from the database indicated that 11 out of 12 cities with the highest levels were located in India. PM2.5 is one of the most hazardous particle which is commonly measured for air quality. The PM 2.5 yearly average reading for the city of Kanpur in India whose population is about 3 million is 319 micrograms per cubic meter. Kanpur is the most polluted, ranking 1 of the 12 cities. The one city outside India from these 12 cities is Bamenda, Cameroon. As on Oct 31st, 2018 an article indicated that the worst time of the year for air pollution in India was round the corner. The country's air quality becomes very toxic during the winter season. (Vox Media, 2018).

Robert Rohde, a lead scientist at Berkeley Earth mentions in a tweet on 31st October 2018 that it has been observed that in the past few years, during the first two weeks of the month of November, the city of Delhi in India faces the worst bout of air pollution. 2018 appeared to be on similar lines, as all the daily averages were recorded very in the 'very unhealthy' range. He has shared the particulate air pollution in New Delhi, India from 2016, 2017 and 2018 till date as seen in Figure 1 (Rohde, Particulate Air Pollution in New Delhi,India for 2016,2017,2018, 2018). Another tweet on 5th November 2018 mentioned that on that day in Delhi in the morning, fine particulate concentrations were above 500ug/m3 for the first time this season. Ideally, any values over 55ug/m3 are to be considered unhealthy, while those above 250ug/m3 are termed as acutely hazardous. Figure 2 shows the PM2.5 values in New Delhi over the last 14 days (Rohde, New Delhi - PM2.5 - Last 14 days, 2018).

The overall Air Quality Index (AQI) of Delhi was recorded at 370μg/m^3 at 11am on Sunday, according to data by the Central Pollution Control Board as mentioned in the article on 11th Nov 2018 in the Times of India (Times of India, 2018).

These days, data-driven analytics is being used in environmental policy making. The United Nations 2015 Sustainable Development Goals is requiring governments to explain how they are managing pol-

Figure 1. Particulate Air Pollution in New Delhi, India
Source: https://twitter.com/RARohde/status/1057532316170551296/photo/1

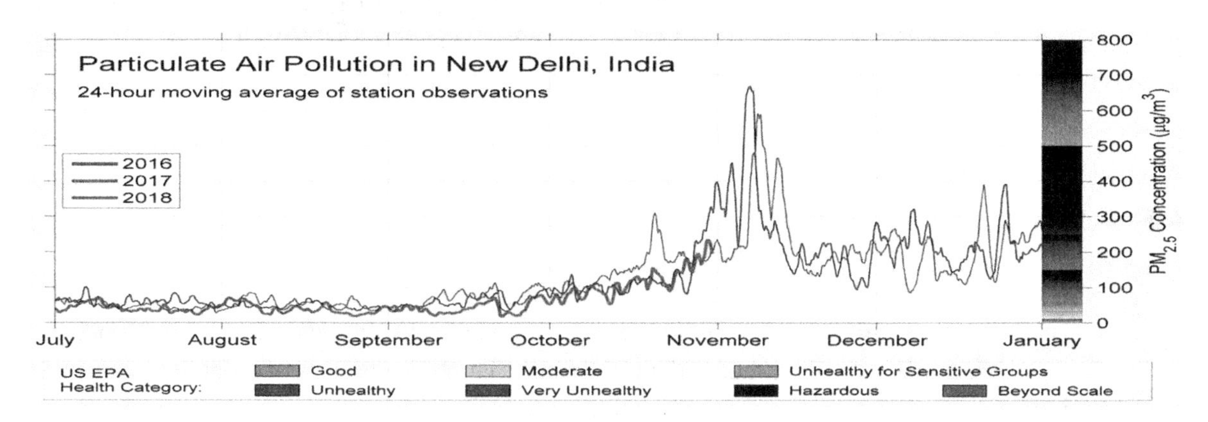

Figure 2. New Delhi - PM2.5 – Last 14 days
Source: https://pbs.twimg.com/media/DrN3iceWkAsU6vG.jpg:large

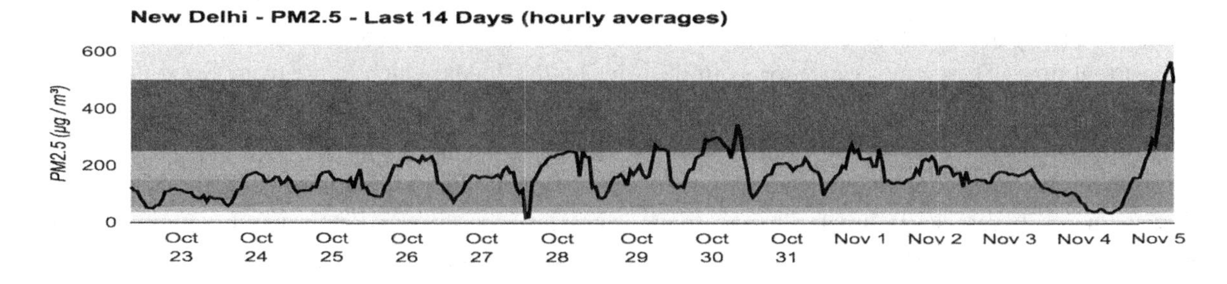

lution control and natural resource management with special emphasis on quantitative metrics that are available. It is easier to identify problems and best practices, follow trends, emphasis policy failures and successes with the help of empirical and data-driven approaches in environmental protection.

A measurement of environmental trends helps provide a basis for policy making which is effective. In 2018 the Environmental Performance Index (EPI) ranked countries (a 180 countries) on performance indicators (24 indicators) across categories (10 categories) which covered environmental health and the viability of the ecosystem. EPI gives a scorecard that indicates the countries whose environmental performance is good implying the leaders and those whose environmental performance is poor indicating the trailers. This scorecard also helps judge the level of the countries with respect to the environment policy goals that have been established.

EPI provides an insight on best practices and guidance to countries who may aspire to improve their performance and become leaders. Columbia University and Yale University have a collaboration with the World Economic Forum and are jointly producing the EPI. India ranks at 177 from 180 countries and this is an alarming situation which gives rise to a need for an awareness about air pollution and air quality monitoring (Wendling, et al., 2018).

Figure 3. Air Quality Index (India)

Good	Satisfactory	Moderate	Poor	Very Poor	Severe
(0-50)	(51-100)	(101-200)	(201-300)	(301-401)	(>401)

AIR QUALITY MONITORING

A variety of sources are responsible for the pollutants in the air. These sources affect the composition of the atmosphere and the environment. Air pollution sources and the absorption or dispertion capacity of the atmosphere determine the concentration of air pollutants. The concentration of air pollution changes from location to location, varies with time depending on topographical and meteorological conditions, thus causing different patterns in air pollution. The sources of air pollutants include vehicles, domestic sources, industries, etc are the sources of air pollution (Bhavani & Reddy, 2015).

For those citizens who have illnesses due to exposure to air pollution, the awareness of the level of air pollution on a daily basis is important. An Air Quality index has an objective of informing citizens about the quality of air and the pollutants it has like PM 2.5, PM 10, SO2 etc. This information is usually almost real time. There are six categories in the Air Quality Index which range from Good to Severe. Details of the six categories of the index are shown in Figure 3 (Central Pollution Control Board, Ministry of Environment, Forests and Climate Change, 2014).

In India, the apex monitoring agency is the Central Pollution Control Board (CPCB), Ministry of Environments, Forests and Climate Change. It has established the strategy for air quality monitoring and has adopted AQI – the air quality index system which is an established tool globally to have different categories of air quality based on the degree of the problem. Various stations which are being operated by different Pollution Control Boards across the country provide information to CPCB regarding the air

Figure 4. Pollution Information at Anand Vihar, Delhi
Source: https://app.cpcbccr.com/AQI_India/

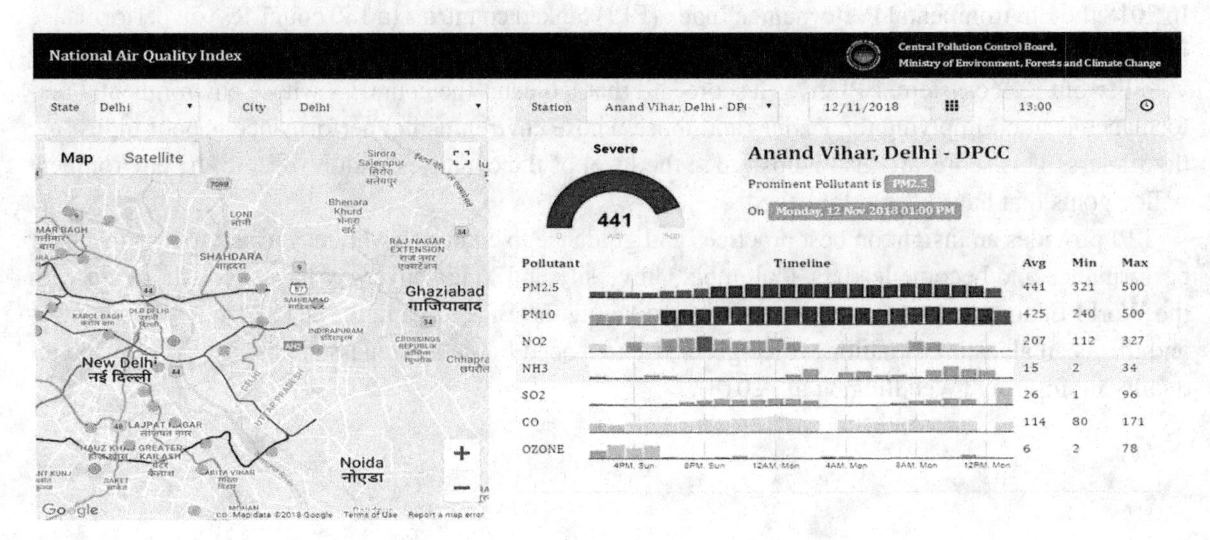

quality. CPCB also has a site available at https://app.cpcbccr.com/AQI_India/ where the National Air Quality Index data for different states, some cities and locations is available online. Pollution information at Anand Vihar, Delhi on 12th November 2018 at 1pm can be seen on the portal as shown in Figure 4.

Most Indian cities and those in developing countries are faced with the challenges of identifying a technique to monitor air pollution which is affordable as well as easy to deploy. Scalable Air-Quality Monitoring Infrastructure is not viable and sustainable due to exorbitant costs, space requirements, energy consumption, heavy operation/ maintenance costs.

INTERNET OF THINGS (IOT)

Kevin Ashton was the one who first coined the term Internet of Things (IoT) in 1999. Everyday objects used by consumers and equipment used in industries, called devices are connected on a network. These devices are managed and gather information via software so that new services can be enabled, efficiency increased and benefits can be achieved in areas of environment, safety or health. All this collectively, makes up IoT. (Goldman Sachs Global Investment Research, 2014). Defining the Internet of Things (IoT) simply means a system where sensors which are attached to or in devices or items in the physical world, use a wireless or a wired network connection to the Internet.

Smart Systems and the Internet of things are driven by a combination of:

- Sensors and Actuators- GPS sensors which use location data, cameras and microphones which act as the eyes and ears, measurement of pressure changes and temperature like sensory organs.
- Connectivity- digitized inputs available on networks (LAN, WAN, Wifi, Bluetooth, 2G, GSM, 3G, GPRS etc.)
- People & Processes- Integration of data, processes, people and systems along with these networked inputs which are combined to form bi-directional systems.

The interactions between the entities like sensors, connectivity, people and processes enable entities to create applications and services that are of new types. Activity trackers that can capture the heart rate pattern, calories spent, remotely turning on or off appliances are some devices that are already available in the market. A recent addition is connected cars that can be tracked and rented using a smartphone.

Some diverse applications which are available:

- For the Consumer at Home: Light bulbs, smoke alarms, refrigerators
- For Transport Mobility: Package monitoring, smart parking, traffic routing
- For Body Health: Equipment monitoring, remote diagnostics, bio-wearables.
- For Buildings Infrastructure: Security, Lighting, HVAC, Emergency Alerts
- For Cities & Industry: Surveillance, Air Quality monitoring, Management of Waste.

These services and connected devices create applications which are across industries or even within their own verticals. For example, transportation can be combined with smart cities, healthcare can be connected to smart homes, smart buildings and mobility. With reference to time, location & services, these integrations can become quite tightly coupled (Harbour Research and Postscapes, 2013).

There are a number of important technological changes that are the enablers of the use of IOT. These include:

- Sensors which are cost effective, Internet Bandwidth and Processing.
- Smartphones serve as remotes or hubs for connectivity to homes, cars, fitness devices, ie, they are now IoT personal gateways.
- Wi-Fi coverage is widely available these days and also available at very low prices or free.
- The use of IoT will generate large volumes of unstructured data, and thus one of its key enablers is the availability of big data analytics, to be able to analyze the volumes of data being generated.
- IPv4 is soon to be replaced by a new Internet Protocol called IPv6. This is now supported by most new networking equipment available.
- IPv4 supports 32-bit addresses, which means that it can support about 4.3 billion addresses. IPv6 supports 128-bit addresses, which means that it can support about 3.4 x 1038 addresses, which as on date are almost unlimited number of addresses and would be able to handle a very large number of IoT devices (Goldman Sachs, 2014)

In the future of smart cities, the Internet of Things (IoT) would play an important and large role. Monitoring various things like the air quality, garbage, sewer details, and the impacts can be done with the help of sensor enabled devices. Environmental monitoring applications which are IoT based, usually have sensors which monitor air and water quality, soil and atmospheric conditions, and even wildlife movement monitoring, thus can help in environmental protection.

Monitoring the quality of air in parks, public spaces, can be done by IoT devices in urban areas. This would also mean that IoT sensors would be installed at numerous locations in the city and data from these sensors would be made available to citizens. Pollution monitoring and reporting sensors have seen an emergence, with the technology developments in wireless communication technologies.

SMART CITIES AND IOT ENABLED ENVIRONMENT SENSORS

Many cities in the world have air quality sensors installed along with the existing infrastructure to monitor and track the quality of air in major areas. In 2014, Chicago installed on lampposts, a network of sensors called the Array of Things. This was developed by the Chicago department of Innovation and Technology along with the Argonne National Library. They use a technology called 'waggle chips'. Air pollutants like carbon monoxide, ozone particulate matters etc can be tracked using these sensors the waggle chips. There is a plan for monitoring volatile organic compounds (VOC) in the future with these sensors. In order to take preventive action, Chicago uses data from these sensors to be able to predict the air quality in major areas and then take some actions to prevent the same. Chicago with its open data portal provides this data to the citizens.

Barcelona Lighting Master plan is somewhat similar. There are smart lighting systems with sensors that give information about air quality to the citizens and the city authorities. Worldwide, more than 65 cities, including Los Angeles, Boston etc have benches in parks called Sofa benches. These benches have solar panels which provide electricity and via USB ports can be used to charge devices. Thus, the Sofa bench is used not only as a social space, but has sensors to record the air quality, traffic, temperature and provides a sustainable source of energy.

The Environmental Defense Fund (EDF) along with a partnership with Google is using Street View cars in order to monitor the methane levels in many cities by ensuring the cars have a methane analyzer and an intake tube. The city of Dublin in 2014, had a pilot trial, where it installed 30 bikes with air sensors in order to measure the carbon dioxide, particulate matter, carbon monoxide and smoke. MIT's Senseable City Lab, in New York City, has used cell phone data, which is anonymous; to be paired with air quality measures and it can help to determine the extent to which New Yorkers are exposed to different chemicals (Bousquet, 2017).

There are two companies Ex Machina (EXM) and Libelium, who primarily work in IoT technologies. These two companies have joined together to help Athens airports executives monitor the pollutant levels and the aircraft locations. Libelium's uses its Waspmote Plug and Sense! Sensor Platform while EXM's custom firmware and standard hardware are used. All the sensor nodes, have probes with which temperature, humidity, particulate matter, atmospheric pressure and ozone are measured. In the Airport Carbon Accreditation programme which is conducted by the Airports Council International (ACI), Athens airport has achieved 'carbon neutral' status. (Twentyman, 2017)

As low-cost sensor technologies are emerging in the market, it is possible for everyone to monitor air pollution. It is now possible to monitor air quality at spatial resolutions that was not possible with traditional monitoring systems. These new sensors available today are not only low cost, but are small, easy to use and portable and so citizens can also help by monitoring the environment.

Nine cities, namely, Barcelona, Vienna, Edinburgh, Oslo, Belgrade, Haifa, Ljubljana, Ostrava and Vitoria are participating in CITI-SENSE. People from these nine cities share information, both subjective and objective about air quality, thermal comfort and acoustic. The portal, CITI-SENSE Citizens' Observatories Central Web Portal (http://co.citisense.eu) enables citizens to access environmental information in real time. In addition to this, citizens also have a forum for discussion, debate and sharing of their personal observations. They can access information provided by the various static and portable sensors, mobile apps and varied perception surveys regarding air pollution. In order to monitor environmental quality within the city, in public spaces Citizens' Observatories have been developed in Vitoria, Spain (CITI-SENSE, 2016).

In the Indian context, Ministry of Housing and Urban Affairs (MoHUA) previously known as Ministry of Urban Development (MoUD) selected a 100 Cities for development as Smart Cities. All the cities are currently at various stages of development. Out of the 100 smart cities, many cities have considered the inclusion of smart environment sensors or smart poles with environment sensors as part of their proposals. It can be observed that cities from the first 20 smart cities like Bhubaneshwar, Pune, Jaipur, Surat, Kochi, Vishakhapatnam, Indore, Kakinada, New Delhi, Indore, Coimbatore, Bhopal all have included smart environment sensors or smart poles as a part of their proposals and are at different stages of execution.

A combined Expression of Interest for Environmental Sensors Implementation in 10 Smart Cities (Coimbatore, Madurai, Salem, Thanjavur, Tiruchirappalli, Vellore, Tirunelveli, Tiruppur, Thoothukudi and Erode) of Tamil Nadu was floated in March 2018. In addition to these cities, many others like Gandhinagar, Vadodara, Hubli-Dharwad, Nashik have also included environment sensors in their proposals and are testing their implementations.

There are various organizations like PAQS, Bosch, OIZOM etc, who are dealing with environment sensors in India. Many of them are providing the sensors for implementation in the Smart city applications. OIZOM Instruments has implemented the air quality monitoring solutions in some smart cities like Surat Smart City, Varanasi Smart City, Palava Smart City. It is in process of implementation in Kakinada Smart City and Gandhinagar Smart City. Oizom Instruments have also been able to amalgate the design process from their experience with installations in UK,Turkey,Saudi Arabia and the US.

POLLUDRONE, ODOSENSE AND OIZOM TERMINAL

Polludrone, a product of an Environment Monitoring Solutions Company, Oizom in India is an advanced ambient air quality monitor which works on IoT Technology. Polludrone has a retrofit design, can be easily integrated into existing infrastructure and is fully solar powered. Polludrone measures various parameters of ambient air like dust particles, hazardous gases, ambient noise etc and can be seen in Figure 5.

An odor monitoring solution called 'Odosense' or 'enose' is designed such that it can detect and monitor smell and its intensity around locations that has odour. Generally odour needs to be monitored around waste treatment plants, landfill sites and industries where odours occur. The Oizom Odosense monitoring solution, consists of a network of devices that are kept on the periphery of the site. Odour dispersion is affected by meteorological parameters, thus, Odosense has an inbuilt weather station. It can detect the presence of gases like Ammonia (NH3), Hydrogen Sulfide (H2S) and measure their concentration also. Volatile Organic Compounds (TVOCs), Methane (CH4) and Methyl Mercaptan are also monitored by Odosense.

As odours are episodic and wind direction makes it difficult to track, Odosense is a proactive approach and helps to detect olfactory thresholds as they occur in real time. It is a solution with wireless data transfer capabilities and is fully solar powered, thus becomes an ideal choice for places like landfill sites, soil-treatment sites, wastewater treatment facilities and industries like fertilizers & paper-pulp.

The data from the Polludrone and Odosense is harvested in Oizom Terminal, a cloud based environmental analysis software. It is a Web-application to visualize and analyze the data from the Oizom Environmental Solution products as shown in Figure 6. It provides a Real time Air Quality Index and each parameters' data. The Polludrone devices are geo-mapped on the Device Location Map with Device Info for a quick look. Oizom Terminal Analytics enables you to review comprehensive analysis of environmental data. Quick reports on a periodic basis and configurable automatic alerts are also available.

Figure 5. Polludrone

Figure 6. Polludrone, Oziom terminal, and visualization

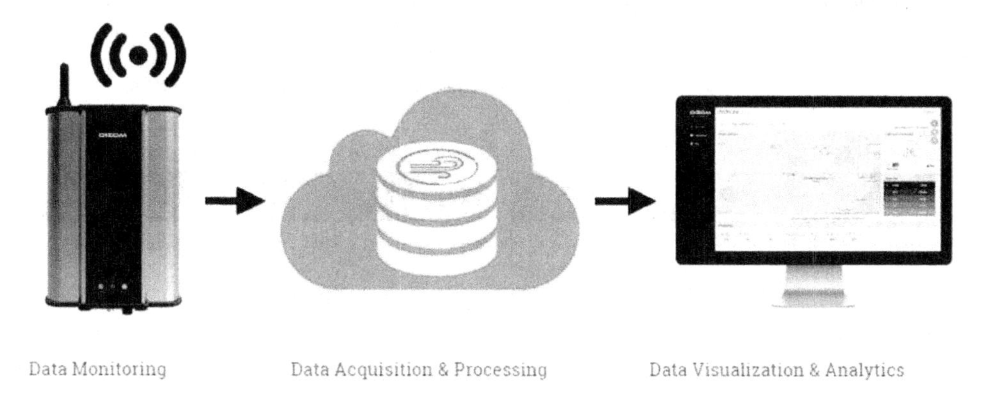

SURAT SMART CITY

In the state of Gujarat, on the western part of India is located a city Surat which was selected for the Smart City Mission in the first phase itself. Surat Municipal Corporation (SMC) is instrumental in the development of Surat as a Smart City. In order to implement projects of the Smart city, a Special Purpose Vehicle (SPV) called the Surat Smart City Development Limited (SSCDL) has been formed on 31st March 2016. Air Quality and Water Quality Monitoring projects are a part of the Surat Smart City area based plan in the Smart City Area (Surat Smart City, 2018).

Currently the two Air Quality Monitoring System installed viz. Limbayat and Varachha, assists in establishing the priorities for the control and reduction of pollutant levels. The installed IoT enabled Air Quality Monitors at two locations are of Oizom make. In order to know the real time air quality information of these areas weblink/page has been created on the Surat Municipal Corporation website. Figure 7 shows the air quality parameter indices at each location. System shows the AQI value and indicates

Figure 7. Air Quality details at Limbayat and Varachha
Source: https://www.suratmunicipal.gov.in/Home/AirQualityInfo

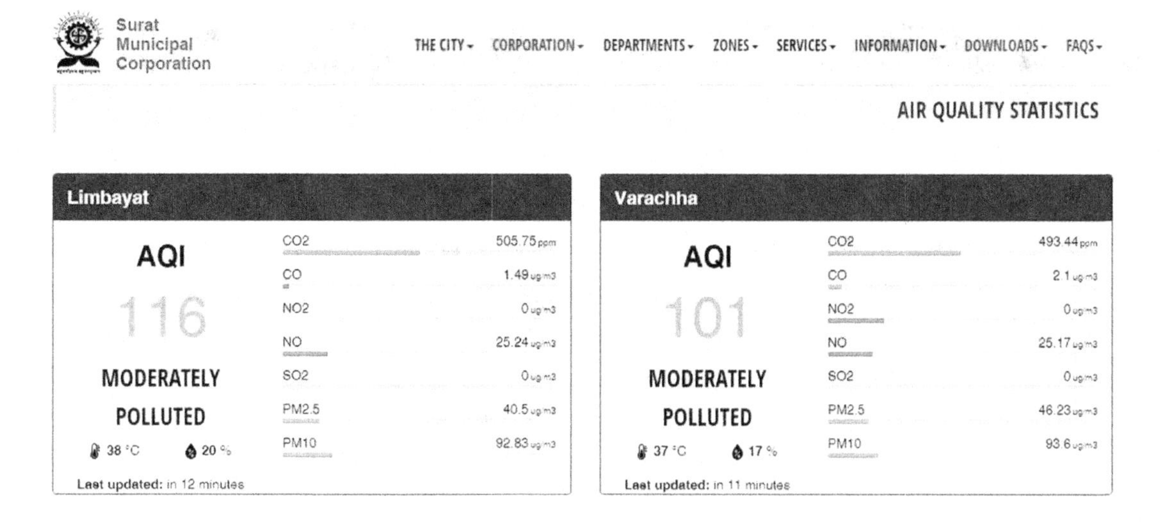

Figure 8. Last 24 hours AQI at Limbayat and Varachha
Source: https://www.suratmunicipal.gov.in/Home/AirQualityInfo

THE CITY▾ CORPORATION▾ DEPARTMENTS▾ ZONES▾ SERVICES▾ INFORMATION▾ DOWNLOADS▾ FAQS▾

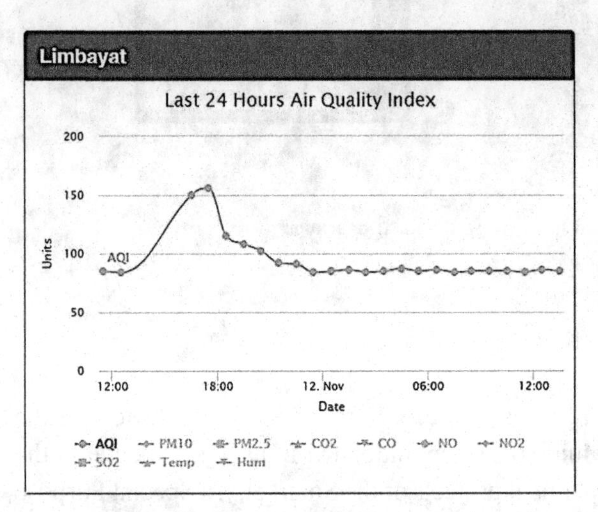

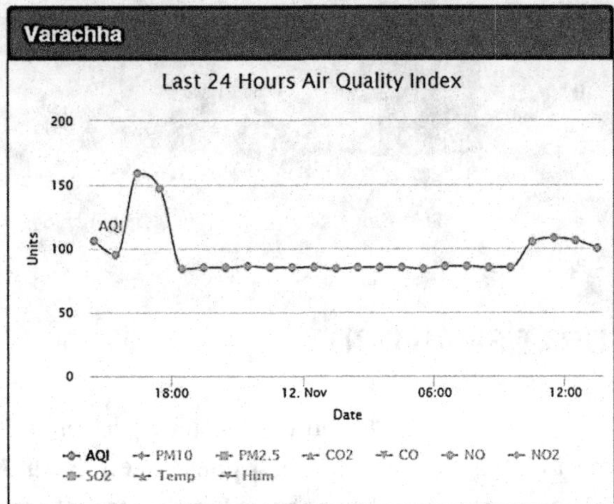

the result according to AQI Index, Figure 8 shows the air quality index over the last 24 hours. The device locations are also available on map as shown in Figure 9, future installation locations shall also be available on website and map view to check the air quality in the specific area (Surat Smart City, 2018)

Figure 9. Device Locations
Source: https://www.suratmunicipal.gov.in/Home/AirQualityInfo

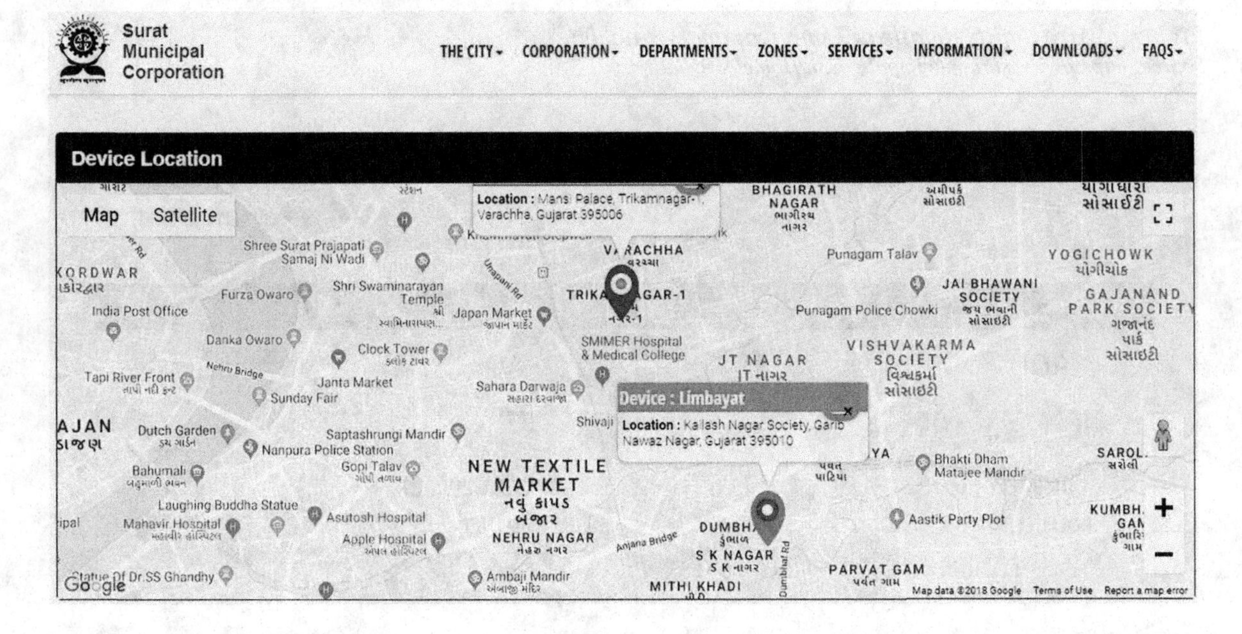

PALAVA SMART CITY ENVIRONMENTAL MONITORING

Thane, Navi Mumbai and Kalyan form an economic triangle. Palava is located at the centre of the triangle formed by Navi Mumbai, Thane and Kalyan. It is a home to over 29,000 families. In order to manage the city, the citizens and city administrators have formed the Palava City Management Association (PCMA). PCMA has been dedicated to monitor and maintain the air quality of the city. Spreading awareness, measuring air quality of Palava city and informing the residents about the air quality are one of the objectives of PCMA. Thus, based on a detailed survey conducted with guidance from experts, PCMA decided to install sensors to monitor the air quality. Polludrone Smart, which is an Air Quality Solution and also has an odour monitor was installed. This device measures particulate matter like PM2.5, PM10; Carbon monoxide (CO), Carbon dioxide (CO_2), Nitrogen Dioxide (NO_2) concentrations, Sulphur Dioxide (SO_2), Ozone (O_3) concentrations, Noise and Light Pollution.

The Polludrone Smart was installed at defined locations as shown in Figure 10. The PCMA administration team using the Oizom Terminal can view the data. Oizom Terminal is a cloud based application which helps visualise in a user friendly manner, the air quality data (AQI) on an hourly or a monthly basis. Intervals for automatic report generation and alerts in real time can be generated depending on PCMA requirements.

Odosense is developed particularly for the measurement of components in the atmosphere that have an odour or odourful gaseous contaminants. The same was deployed in Palava for continuous monitoring. Odosense Smart measures Sulphur Dioxide (SO_2), Nitrogen Dioxide (NO_2) concentrations, Carbon monoxide (CO), Hydrogen Sulphide (H_2S), Anhydrous Ammonia (NH_3), Methane (CH_4), Formaldehyde (CH_2O) and Volatile Organic Compounds (VOC).

Figure 10. A polludrone installation at Palava, Mumbai

Figure 11. The outdoor LED display at Palava, Mumbai

In Palava, a 2×3 feet outdoor LED display showcases the real time and average data (24 hour) of the air quality. This also creates awareness about the air quality in the city. This LED display can be seen in Figure 11 (Oizom Instruments Pvt Ltd, 2018).

KAKINADA SMART CITY PROJECT

In the state of Andhra Pradesh, the city of Kakinada was selected to be one of the Smart Cities in the Smart City Mission's first round or phase. Kakinada has seen a gradual an increase in urbanization and development. Kakinada belongs to the Special Economic Zone (SEZ) and it has a Petroleum, Chemical and Petrochemical Investment Region (PCPIR) which is proposed. Thus, in order to address the environmental conditions and bring about an awareness in the citizens of Kakinada city, Kakinada Smart City Corporation decided to setup smart poles equipped with environmental sensors.

A solution which monitors parameters like air quality, noise and disaster monitoring (flood and rainfall monitoring) was the requirement identified after an extensive research. A city-wide wireless sensor network, based on the LoRa Technology was what the Master System Integrator had envisioned. A Polludrone which was compact in size, easy to mount, had a rugged design to withstand extreme weather conditions was specially developed for Kakinada Smart City. The compact size made it easy to mount on the Smart Poles which were to be deployed in the city.

The Kakinada Smart City Project is executed in phases. Environment Sensors are in the phase 1 and are deployed on the Smart Poles. Installation of smart poles and environmental sensors on them has already started (Oizom Instruments Pvt Ltd, 2018)

GANDHINAGAR SMART CITY PROJECT

Gandhinagar the capital city of the state of Gujarat, India and a Municipal Corporation was chosen in the third phase of the Smart City Mission. As a part of its area based plan, installation of 10 environment sensors in different parts of the city is proposed. Along with this, 10 display boards would create awareness and have a constant air quality check. These sensors are currently under installation.

CONCLUSION

As a result of air pollution, it is said by World Health Organization (WHO) that approximately 5.5 million people die. In large cities, exhausts from automobiles, factories fills the air with harmful and hazardous particles. Many cities, have thus, starting to measure air pollution using a network of connected sensors, namely the Internet of Things (IoT) which gather and send data. With this data, it is possible from cities to have a tracking of changes in pollution levels over time, a mapping of areas where the pollution is high, identify pollutants and suggest interventions (Bousquet, 2017).

In an attempt to create awareness about air quality, availability of air quality data in near real time with technologies like IoT, many of the 100 identified smart cities in India are implementing solutions with a hope to take remedial action and improve the quality of life of citizens.

A distributed, multiple vendor and multiple platform framework that is evolving is Internet of Things (IoT). IoT is about being able to provide insights and knowledge about objects, the environment, social and human activities, that could be recorded by these devices in the physical environment. It does not mean only collecting and publishing data but enables systems to take actions that can be based on the knowledge obtained. 'Citizen Sensing' or human sensors is a growing as a result of social media platforms usage and crowd sourcing. Citizen's use or social media tools and smart devices to report the observations they make. To be able to discover, integrate and interpret the various data from multiple sources like social, physical and cyber streams; to be able to provide insights that are accurate, timely are some of the major challenges faced (Barnaghi & Sheth, 2014).

In February 2010, unique IPv4 addresses were over. Although the general public is not really affected by this, the progress of IoT would be slowed down. This is because every new IoT sensor requires a unique IP address. The auto configuration capability, enhanced security features of IPv6 makes management of networks easier thus promoting the migration to IPv6. In addition to IoT sensors becoming self-sustaining, efforts in the area of standards, privacy security, communications and architectures need to be addressed for IoT to reach its full potential (Evans, 2011).

Internet of Everything (IoE) is made up of interconnected networks of which Smart Cities can be considered the microcosms. The true value of IoT can be realized by using smart cities as a fertile ground where it can be implemented. In order that this may occur in our future cities and communities, leaders in the city must understand how the different components of the Internet of Things like process, people, data and things would be working together and what roles each would be playing (Mitchell S., Villa, Stewart-Weeks, & Lange, 2013).

ACKNOWLEDGMENT

This research received no specific grant from any funding agency in the public, commercial, or not-for-profit sectors.

REFERENCES

Barnaghi, P., & Sheth, A. (2014, September 9). *The Internet of Things- The Story so far*. Retrieved from http://iot.ieee.org/newsletter/september-2014/the-internet-of-things-the-story-so-far.html

Bhavani, D. S., & Reddy, R. R. (2015, December). Identification and Characterization of Particulate Matterin Hyderabad City. *International Journal of Innovative Research in Science*, *4*(11). doi:10.15680/IJIRSET.2015.0412160

Bousquet, C. (2017, April 19). *How Cities Are Using the Internet of Things to Map Air Quality*. Retrieved November 11, 2018, from https://datasmart.ash.harvard.edu/news/article/how-cities-are-using-the-internet-of-things-to-map-air-quality-1025

Central Pollution Control Board, Ministry of Environment, Forests and Climate Change. (2014). *National Air Quality Index*. Retrieved June 2018, from http://www.indiaenvironmentportal.org.in/files/file/Air%20Quality%20Index.pdf

CITI-SENSE. (2016, September). *Home Page*. Retrieved November 2018, from CITI-SENSE: http://www.citi-sense.eu/

Evans, D. (2011). *The Internet of things - How the next evolution of the intenet is changing everything*. Cisco Internet Business Solution Group.

Giffinger, R., Fertner, C., Kramar, H., Kalasek, R., Pichler-Milanović, N., & Meijers, E. (2007). Smart Cities: Ranking of European Medium-Sized Cities. Vienna, Austria: Academic Press.

Goldman Sachs. (2014). *The Internet of Things: Making sense of the next mega-trend.* Author.

Goldman Sachs Global Investment Research. (2014, September). *The Internet of Things: Making sense of the next mega-trend.* Retrieved June 2018, from https://www.goldmansachs.com/our-thinking/outlook/internet-of-things/iot-report.pdf

Harbour Research and Postscapes. (2013). *What exactly is the "Internet of Things"?* Retrieved from https://www.google.com/url?sa=t&rct=j&q=&esrc=s&source=web&cd=2&cad=rja&uact=8&ved=0CCMQFjAB&url=http%3A%2F%2Fharborresearch.com%2Fwp-content%2Fuploads%2F2014%2F03%2FHarbor-Postscapes-Infographic_March-2014.pdf&ei=uyxiVIWFDNGGuASE2ICgCA&usg=AFQjCNEWPdt47D

ISO 37120. (2014). *Sustainable Development in Communities: International Standard on city indicators*. Retrieved June 2018, from https://www.iso.org/files/live/sites/isoorg/files/archive/pdf/en/37120_briefing_note.pdf

ISO 37120:2018. (2018). *Sustainable Cities and Communities -Indicators for city services and quality of life*. Retrieved November 2018, from https://www.iso.org/obp/ui/#iso:std:iso:37120:ed-2:v1:en

Ministry of Housing and Urban Affairs, Government of India. (2017). *What is a Smart City*. Retrieved November 2018, from Smart Cities Mission: http://smartcities.gov.in/upload/uploadfiles/files/What%20is%20Smart%20City.pdf

Mitchell, S., Villa, N., Stewart-Weeks, M., & Lange, A. (2013). *Point of View:The Internet of Everything for Cities*. Cisco. doi:10.4324/9780203716687

Mitchell, S., Villa, N., Stewarts-Weeks, M., & Lange, A. (2013). Point of View: The Internet of Everything for Cities. *CISCO*. Retrieved June 2018, from https://www.cisco.com/c/dam/en_us/solutions/industries/docs/gov/everything-for-cities.pdf

Oizom Instruments Pvt Ltd. (2018a). *Kakinada Smart City Project*. Retrieved June 2018, from Oizom Redfining Resources: https://oizom.com/kakinada-smart-city-project/

Oizom Instruments Pvt Ltd. (2018b). *Palava Smart City Environmental Monitoring*. Retrieved June 2018, from Oizom Redefining Resources: https://oizom.com/palava-smart-city-environmental-monitoring/

Rohde, R. (2018a, November 5). *New Delhi - PM2.5 - Last 14 days*. Retrieved November 9, 2018, from https://twitter.com/RARohde/status/1059323321265852422

Rohde, R. (2018b, October 31). *Particulate Air Pollution in New Delhi, India for 2016, 2017, 2018*. Retrieved November 9, 2018, from https://twitter.com/RARohde/status/1057532316170551296

Surat Municipal Corporation. (2018, November). *Air Quality Statistics*. Retrieved November 11, 2018, from https://www.suratmunicipal.gov.in/Home/AirQualityInfo

Surat Smart City. (2018, March). *Air Quality and Water Quality Monitoring in Smart City Area*. Retrieved November 2018, from http://www.suratsmartcity.com/Documents/Projects/ABD/13.pdf

Times of India. (2018, November 11). *Air pollution level and air quality index in Delhi today*. Retrieved November 11, 2018, from https://timesofindia.indiatimes.com/city/delhi/air-pollution-level-and-air-quality-index-in-delhi-today/articleshow/66470391.cms

Twentyman, J. (2017, September). *Athens International Airport turns to IoT for environmental monitoring*. Retrieved from https://internetofbusiness.com/athens-international-airport-turns-to-iot-for-environmental-monitoring/

Vox Media. (2018, October 31). *Why India's air pollution is so horrendous*. Retrieved from https://www.vox.com/2018/5/8/17316978/india-pollution-levels-air-delhi-health

Wendling, Z. A., Emerson, J. W., Esty, D. C., Levy, M. A., & de Sherbinin, A. (2018). *2018 Environmental Performance Index*. New Haven, CT: Yale Centre for Environmental Law & Policy. Retrieved November 2018, from https://epi.yale.edu/

World Health Organization (WHO) India. (2018). *Air Pollution*. Retrieved November 2018, from http://www.searo.who.int/india/topics/air_pollution/en/

Chapter 20
Less Human Intervention (Automated) Waste Management System Using IoT for Next Gen Urbanization

M. Kavitha Margret
Sri Krishna College of Technology, India

D. Vijayanandh
Hindusthan College of Engineering and Technology, India

ABSTRACT

Next generation waste management in urbanization is the real and unpredictable challenge in the modern era. Government of modern city will face big challenges in handling and decomposing of waste. Suggestions were given by the researchers about handling of waste management with IoT-enabled devices. Smart bins were introduced that will focus on the capacity and type of the waste collected from different stakeholders, industries, and citizens. Smart bins were monitored by the municipality periodically. Government will focus on less human interventional (automated) waste management system that will lead to happy living of citizens in the nation. Due to tremendous growth of industry in cities, people migrated from village to cities. Handling this population and cleanliness of city is a very big focus of the government. Authors propose less human interventional (automated) waste management system for the next generation urbanization using smart IoT-enabled devices. Authors propose standard architecture model for tracking of smart bins in various region using self-efficient organization of wireless sensor networks (WSN) and grouping of those sensors in case of any malfunctioning or damage of sensors. Handling of large volume of data, cost of the underlying network topology, merging of devices, and speed of the data connectivity are focused to reduce human interventional waste management system that will organize the sensor group wisely. With the above architecture, the dream of smart city will come true in the future.

DOI: 10.4018/978-1-5225-9199-3.ch020

INTRODUCTION

Now a days nearly 0.1 tonnes of municipal solid waste generated every day. India is the world third largest garbage generator, Nearly 53% of the world population live in urban areas . By 2050 this ratio will expected to increase to 80% . Planning mega city is both scary and fascinating. Managing large volume of waste is the one of the key challenge of the government in urbanization . Managing includes collection of the waste from different sources like household, industry, organization or biological and converting into renewable energy . Waste can be of different type like solid, liquid, organic, recyclable and so on. Improper waste disposal will lead a threat to human body. When waste increases diseases also get increases . Stop looking waste is waste look around waste is a recourse. The question is how the waste to be transformed in to resource ? with the help of internet of things (IoT) efficient architecture model needed for disposal, recycling and converting waste into resource, with the new architecture transforming waste into resource is possible .

Smart city development can be achieved by proper waste management techniques, Smart city development and proper waste management are integral aspect of city management. Information and communication technology(ICT) leads a major role in Smart City development and waste management. Proper waste management and Continuous development in smart city needs novel technology for waste management and waste transformation . The waste collection is being understood through the use of sensors and real-time systems (Prajakta, Kalyani, Snehal,2015) . Smart bins are used to Collect and sending status of waste over a days or location. Supplementary waste forms like organic/inorganic, agricultural, biomedical, electronic, chemical, mineral, and radioactive are Considered by specific group points. Finding the level of waste from solid-waste-bins meets many difficulties due to the several indiscretions of the waste-bin filling process. Irregular disposal and the form of identifying process materials are the challenges exist for the smart waste management . Effective data aggregation from a large number of bins and the variety of the involved materials are more challenges in waste collection . Sensors are also be used to find the environmental conditions like humidity, temperature, and dust can suggestively affect the sensor accuracy and reliability in collected data amount and conditions constitute parameters that should also taken into interpretation for a complete waste management process, static route planning with static scheduling are followed in earlier waste collection system. Earlier systems indicates different areas that need to be developed continuously and encourages improvement in automation tools, recycling services, Privacy and Security Heterogeneity.

Authors interest is to ideate and encourage the use of the Internet of Things (IoT) to address the problems in waste management, starting with Challenges in design and implementation of Less Human Intervention (automated) Waste Management System, Architecture model for waste management, Types and Waste management techniques. Well suitable Software Technology Needed to Handle Waste Management in the development of urbanization, Innovative techniques were followed in different countries for handling Waste Clearance. Authors proposes less Human Intervention (automated) Waste Management System that will handle efficient transfer of waste into resource, on-site collection and proposes wise idea for monitoring and management system which involves a Wireless Sensor Network (WSN) for garbage bins in different regions .

Contribution of Internet of Things in the Development of Smart City

People, industries, government and educational system everything in the world are directly or indirectly rely on internet . IoT refers list of devices which connect each other for communication in terms of transferring data through internet . Every thing including cities are becoming "smarter," because of IoT. Very big and unsolvable issues in the urbanization are handled with care with the help of this IoT . Nowadays people wanted to live in pollution free cities, the proposed model can achieve this task with the help of IoT for waste management in smart city development, IoT Devices will link the breach between real world with smart devices, IoT devices will connect with society and industries for better performance . Real World Applications of IoT starts from Smart Home, wearable gadgets, Connected vehicles, Business or trade Internet, Smart Cities, waste management, agriculture, Smart Retail, Energy Engagement, IoT in Healthcare, future store, IoT in Poultry and Farming, IoT in construction.

IoT design with smart home and smart appliances, wearable gadgets, smart city, waste management are trending in this new era .

Contribution of IoT in Waste Management

Smart Cities

One of the powerful application of IoT in the world is smart city . Environmental monitoring, automated transportation, Smart surveillance, water distribution, urban security, pollution monitoring, traffic congestion control, automation of parking of vehicles and shortage of energy are examples of IoT in the development of smart city

Figure 1. Waste management using IoT

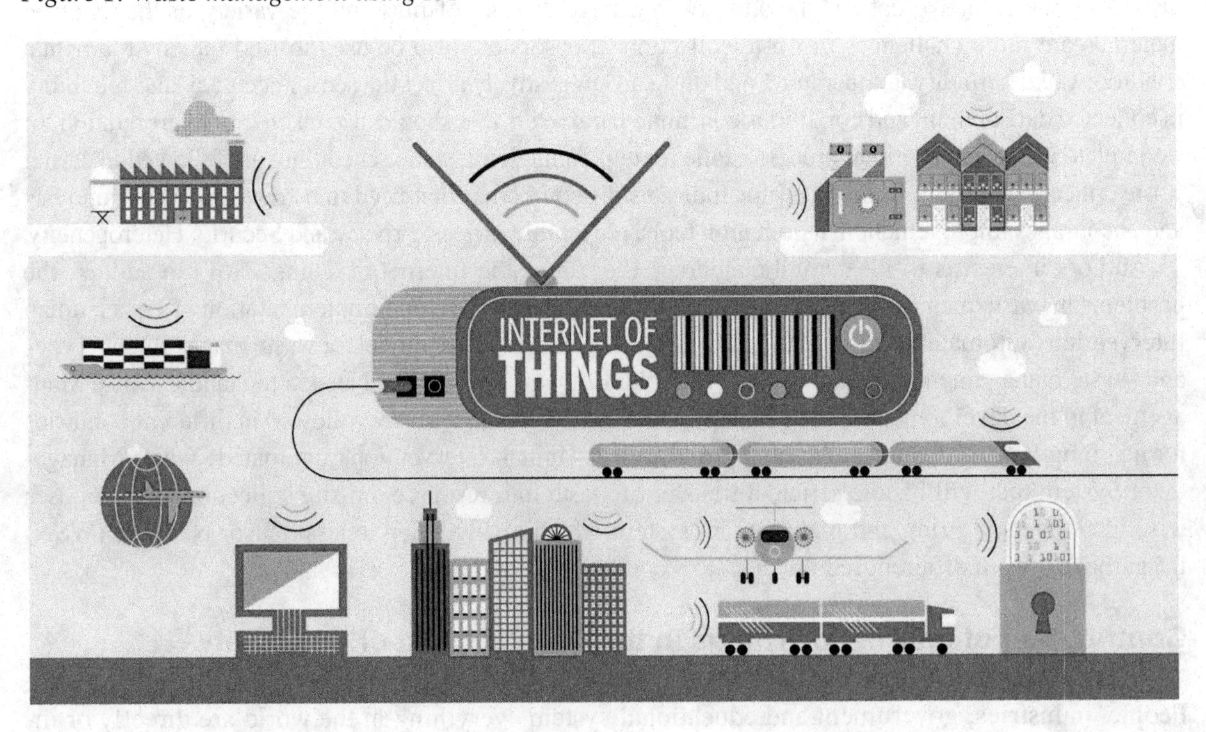

Smart Home

In home automation IoT plays an major role which includes security system, monitoring, lightening, energy saving, entertainment. Many manufacturers like Apple frequently releases their home products and accessories, those are being controlled by an application ios(iphone OS) . IoT provides assistive technology for elders in homecare, these systems are used to handle user specific disabilities and elders in home. Like Apple, different smart home product platforms are available in the market, for instance Amazon provides Echo, Samsung's provides SmartThings Hub

Smart Waste Bin

Smart waste bins which uses IoT in waste management helps real time monitoring system which integrates IoT technologies, sensors and wireless communication technologies. The objective of smart bin is to provide an effective and cost-effective waste collection and management system that will results in providing hygienic, healthy and green environment. Different new framework were suggested by researchers that enables real time monitoring of solid waste collection bins using Wi-Fi connection to support the waste management process .

Waste Management

Waste Management is a step by step process of collection of data from sensor, processing a data, managing and recycling of waste materials (Gaur, Scotney, Parr, Mcclean, 2015). . Smart waste bins were equipped with different sensors to detect the information about garbage level, types of garbage, conveying information to centralized server(cloud server) . IoT system encourages automation of waste management, reduces cost and time significantly.

Next Gen Urbanization

In the current decades (industrial Internet) traditional devices like PDAs and laptops, IP-based network platform, integrated buildings, connecting cars are becoming more popular using 'Internet of things'. To handle these integrated data the network becoming more challenge for countries, cities, organization and public services. IoT plays a major role in smart city development, enhancing productivity among employees, improving accessibility to public services to generate profits. By using network connectivity in urban planning, smart agriculture, telecommunication, cities around the world can change the way in exchanging services and manage the flow of data . In a smart city development highly secure data transfer, intelligent transportation, multimodal waste management system are needed to foster real-time IoT communications .It ensures security in global transaction, connecting and bridging Travel services, airlines, and hotels can all be united on a single IP-based platform. Wireless technologies (Wi-Fi, WiMAX, and 3G/4G/5G) can be enabled in any kind of transport. With the advanced technology workplaces and residential buildings can be integrate onto a single platform(Barton,Dalley,Patel,1996). Automate detection and analysis to access personal information in case of emergency also possible and level of security to personnel information are enhanced with the help of IOT enabled devices. The future dream of urbanization will become true by connecting all devices with IoT enabled platforms.

Smart Education

By adopting smart education systems with class room teaching using web based application, learning provided by other organizations and institutions are linked together to empower students to compete globally . IOT based smart education providing the high-quality learning opportunities for all education system. The education system will be common from rural to urban students to acquire a next generation education to foster the knowledge and to compete globally .

The Smart education system provides a common platform for sharing the knowledge. smart class room teaching, video conferencing, online certification courses are growing fast to increase the standard in teaching learning process. Institution are need to be accredited with national and international accreditation policy that will ensure the standard in modern teaching .

Present Available IoT Based Architecture for Handling Waste Management

An architecture (Sam Aleyadeh, 2018) based on IoT devices monitored the waste volume and the content in the bin . The architecture detailed an information about the surroundings . The algorithm detailed a dynamic scheduling of waste collection truck and routing based on the information collected from the smart bins. The architecture consists of four components starting from smart waste bin, mobile application, truck driver module, cloud store . In smart waste bin every bin equipped with multiple sensors for sensing the type and volume of the bin and its surroundings . Arduino Yun or a lattepanda board microcontrollers were used to manage several connected sensors. The implementation included Gps and microcontroller, 5m proximity sensor, high capacity load cell,humidity sensor,Lever activated switch . Proximity sensor detailed the information about the bin surrounding in case of any difficulty in accessing the level of the bin for example if the bin was surrounded by vehicles. The load cell determined the weight and passed the information to the controller. The dryness level of the bin was measured by humidity sensor, The lever activated switch was used to detect the event made on the bin .The position of the waste bin is identified by GPS module where the microcontroller was used for collection, aggregation and transmission of information to the cloud . Mobile application was used to connect to waste collecting vehicle driver to identify the route that was a function point to connect to cloud .

In The driver module sensory data information's were collected by OBDII and send via Bluetooth dongle to the driver mobile app .Finally route planning and shortest path were found out for better results .

An IGOE architecture for waste collection and optimization algorithm(Shashika Lokuliyana, Anuradha Jeyakody, 2017) resulted in automated waste management system . The proposed architecture consist of data gathering layer, data processing layer, data optimization layer and so on . Data gathering layer included different waste collection methods, waste disposal methods (Recycling, composting, Landfill),identifying waste disposal areas . Data processing Layer (DPL) included cloud server and the data base handling data received by the sensor and the individual. These data's were analysed in server and the decision was taken by the server by collecting the optimized location based on the decision alert, The alert was given to the supervisor, workers and waste collecting truck drivers. Data optimization layer runs an optimization algorithm for choosing best location for collecting waste

An architecture model (Wahab, Kadir, Tomari, Jabbar, 2014) based on crowd sensing to encourage public in handling and reusing waste and increasing awareness in recycling . The framework employed with RF readers attached to smart-bin to identify individual users based on RFID cards and reward points were given based on the weight of deposited items. The proposed model dealt with cloud based

Figure 2. IoT Based Waste Collection Architecture

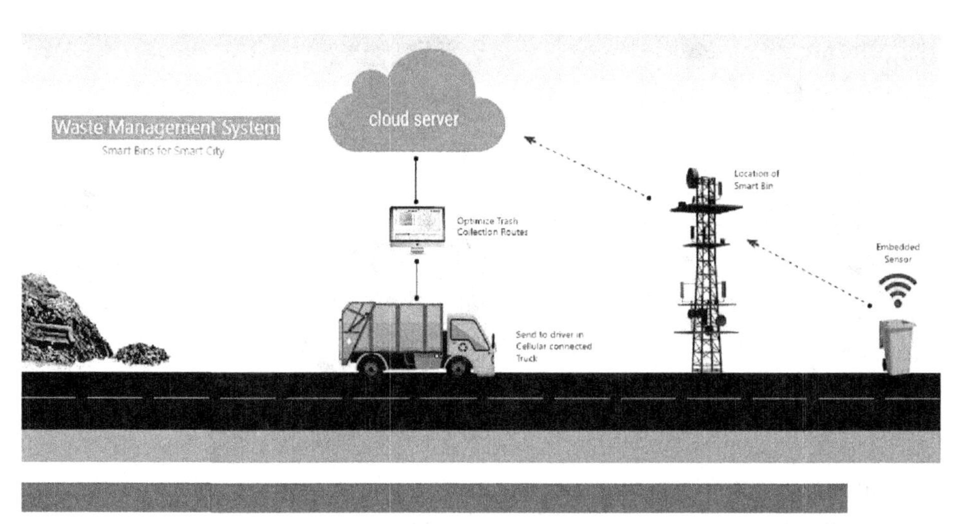

processing and aggregation model that enables data sharing, flexibility in design and scalability of devices, better resource utilization and optimal planning of waste collection allows the normal users and Municipal Corporation users to monitor the status of bins.

Figure 3. Shashika Lokuliyana IGOE Architecture for Automated Waste Collection Process

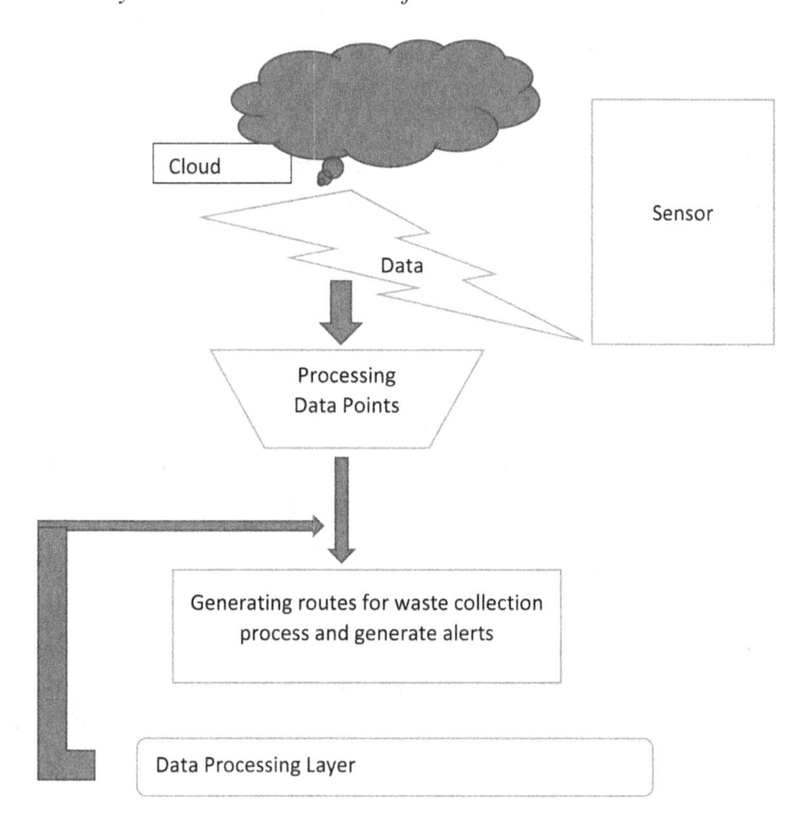

Figure 4. Planning of Waste Collection, Encourage Public in Handling and Reusing Waste

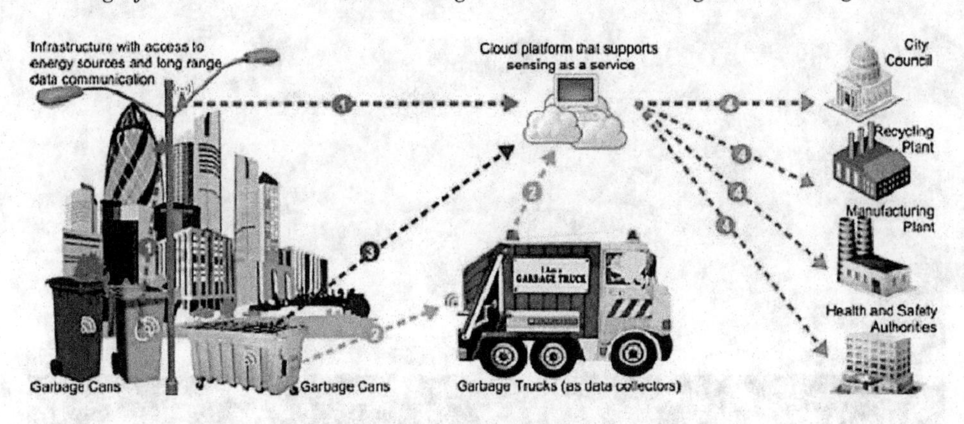

Proposed solution to handle Waste Management using Server less IoT Architecture (Eyhab Al-Masri', 2018) for planning of waste collection is mentioned below

The proposed architecture named Recycle.io waste management system having equipped with smart organic bins (SOB) and smart recycling bins (SRB) . These bins were furnished with a Raspberry Pi kit . This kit was a combination of ultrasonic sensor and camera. SOB and SRB implemented with edge computing that can detect abuses and take decisions at the edge of the IoT . This approach reduces costs associated with Source and targeting bearable consumption of production patterns . Grouped information from camera module and Sensors were sent to analytic unit that determines whether this waste substances were capable for recycling or any violation occurred, for example disposal of plastic material were not suitable for recycling. These information's were grouped and the summarized information were given to cloud based application that will handle further . The architecture was server less IoT based architecture results in identifying individual human violation in disposal of materials and reduction in cost

Challenges in Design and Implementation of Less Human Intervention (Automated) Waste Management System

- Handling large volume of data
- Cost of the underlying network topology
- Merging of devices

This chapter gives 360^0 view on the solutions which are needed to address the above problem . These problems are focused in designing an architectural model in such a way that reduce human interventional waste management system that will organize sensors and group them wisely.

Handling Large Volume of Data

Nearly 53% of the world population living in urban areas, providing resources to the citizens like power supply, water, healthcare, cleanliness of the city, handling air pollution are very big Challenge to the government. Everyday municipal waste collection system will not provide adequate solution for disposal and segregation of wastes. Improper disposal of solid waste is a very big threat to the human body .

During the holidays or absence of the waste collecting person the waste will not be collected for longer hour that will generate very unpleasant environment and bad smell over the surrounding. So effective monitoring solutions are needed to track the level of Bins in common place. A novel solution needed to handle the above problem.

To monitor the waste level in the bins Wireless Sensor Network (WSN) have been used . WSN is a group of autonomous sensor nodes distributed over the large scale. Challenges related to collecting large data and managing, processing listed (Rida Khatoun, Sherali Zeadally, 2016) . WSN are designed to monitor quantity, pressure, environment temperature, humidity, etc, Collecting the data related to waste management sensors embedded with on board processing unit and storage units . There are variations in the sensor nodes based on their different data rates, correlation and fusion of surrounded and own nodes .

Important characteristics of a WSN are:

- Battery power consumption constrains
- Capability to deal with node failures
- Network data Communication failures
- Capability to bear severe environmental conditions

Network Management in Wireless Sensor Network for Handling Data

Two different techniques followed in the WSN for data transportation and optimization process in garbage bin filling,

The Techniques are

- Minimum distance path searching
- Greedy path searching

These techniques are used for accessing the bin in an effective manner . In a smart city many sensors are deployed in an environment to track the level of the bins those are located in street and are connected directly to the controllers .Data processing stations using local area networks handles the sensor information. A sensors in a region is communicated the sensed data directly to the Centralized supervisor system,

Figure 5. Centralized waste collection approach

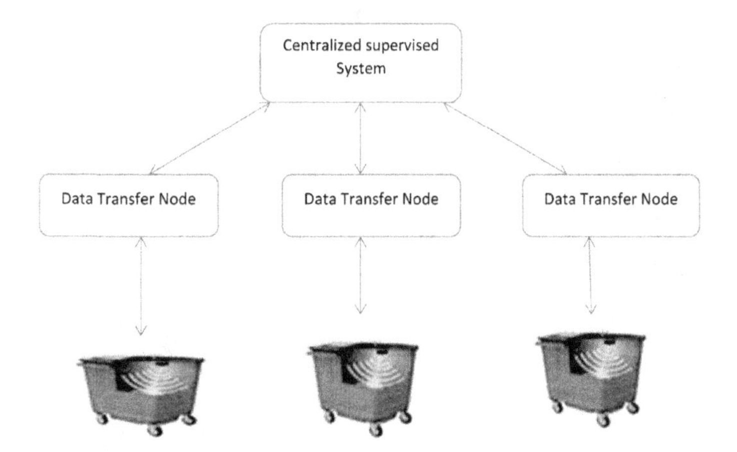

Collected data from the different bins are transmitted through RF technology to data transfer node through GSM/GPRS . Data transfer node forwards the groped data to the Centralized supervisor system which provide data storage and fusion.

- **Cost of Underlying Network Topology:** The major challenge in WSN (Rida Khatoun, Sherali Zeadally 2016) in deploying IoT devices at large scale in a city needs embedded OpenCPU based on GSM/GPRS module, It can be used in a software layer, running on an processor, which runs a support environment for handling and processing data in wireless node to node in a cost-effective manner . To handle data transmission between node to server two solutions have been suggested to reduce the Cost of the underlying network topology
    - Server Socket Module
    - User Application

First approach represents traditional model to send collected information from sensor to the network device like server through internet. TCP protocol is used to establish a communication over the network in order to establish multiple session, multi-threading also needed to establish multiple session . In the second approach user can get immediate update in their mobile phone using long range communication module without the use of the server . With the help of the second model the cost of the topology also gets reduced .Buffer Management in Sensor Networks is an another way to reduce cost of a network . Different type of congestion occurs when an area is densely populated with sensors, Radio collision and buffer overflow are two important types of congestion in a network. . To get neighbour bin information one-bit buffer state can be piggybacked and information's are periodically exchanged among neighbours using CSMA . CSMA with hidden ACK and 1/k-buffer solution prevents congestion from unknown source.

- **Merging of Devices:** In the case of failure in one device, the status of the smart bin are not delivered to the concerned authority . This can be monitored severely to prevent overloading of waste in the bins. *k* query region are fixed in handling of data in the database. When the query is generated from the bin about the indication of overflow or level of the bin dynamic routing is also possible using greedy technique .

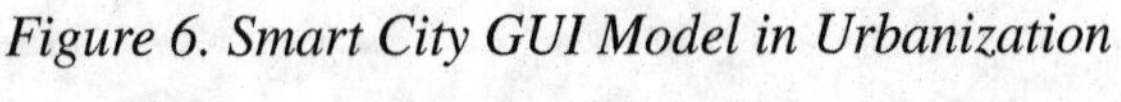

Figure 6. Smart City GUI Model in Urbanization

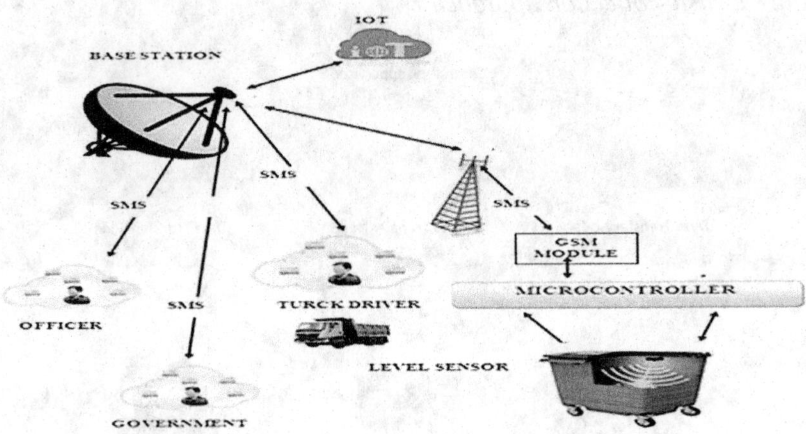

Graphical User Interface is implemented in Android app which incorporate Google Maps to choose dynamic route by the truck for collecting waste. In case of malfunctioning of sensor nodes identified nearest neighbour sensor with in the query region is selected for further communication.

Architecture Model for Waste Management

The proposed architecture shows tracking of smart bins in various region using self-efficient organization of wireless sensor networks (WSN) and grouping of those sensors in case of any malfunctioning or damage of sensors .There are four levels in the proposed architecture model starting from sensor layer and data collection layer, data processing correlation and fusion layer and application layer . unprocessed data's are collected from lower level to higher layer finally aggregated information are passed to final application layer . Intermediate layers uses technologies such as OpenCPU based GSM/GPRS module, k- query region can also be used for grouping the raw data then collected information's are integrate into standard format. Here, k-means clustering rules are used effectively to self organize (group) the data, in case of any malfunctioning or damage in the sensor region the information is passed to the centralized node for adjusting sensor query region .

- **Advantages of the Proposed System:**
  - Automated Real time sharing of information
  - Effective information about level of the dustbin.
  - Self organization of network in terms of failure .
  - Less Cost and improved optimization of resources .
  - Less time to ensure cleanliness of city
- **Types of MSW Municipal Solid Waste:**
  - Recyclable material- includes Bottles, Paper, cans, glass, certain plastics
  - Biodegradable waste - green waste including flowers, leaves vegetables

Figure 7. Architectural model for smart waste management

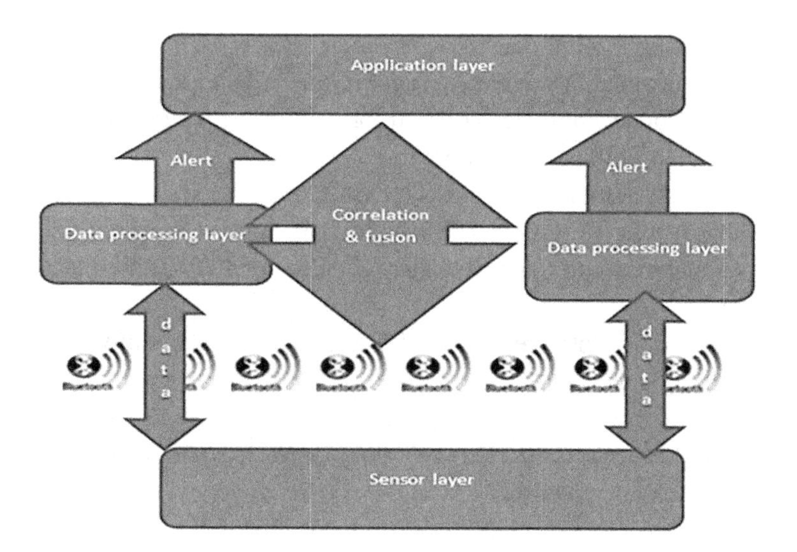

  - Composite waste – includes plastic, tetra packs, Waste clothing
  - Inert waste – includes Construction and demolition waste
  - Toxic waste & Domestic hazardous waste – fluorescent tubes, e-waste, medicines, chemicals, light bulbs, paints, fertilizer
- **Waste Management Technique:** Waste management technique listed from
  - Landfill
  - Composting
  - Incineration
  - Anaerobic Digestion
  - Pyrolysis And Gasification
- **Landfilling:**
  - Dumping of waste becoming a common practice in metropolitan cities this will lead to very serious threat to environment.
  - Landfilling is the process of removal of waste materials by burial. This is one of the oldest type of waste treatment . Landfills methods have been the most common and organized disposal of waste method around the world.
  - Advantages: It is a less cost-efficient way (Beigl, Salhofer, 2004) to dispose of waste many countries with large space can make use of this disposal method . Compared to other methods this technique requires less processing overhead and less cost . Landfill gas utilization is also possible in this technique. Leaving a specific location for disposal is an another advantage, where the special attention is given for monitoring and waste can be treated then recycled .
  - Compost: Compost is a decomposition of organic matter. Compost are used in gardens, landscaping, urban agriculture and organic farming. This process recycles various organic materials into rich nutrients.

This is beneficial for the land for fertilizer, soil conditioner, humic acids and also act as natural pesticide for soil. Compost is useful for erosion control. Wet organic matter such as food scraps, grass, leaves are broke down into earth for period of months with proper monitoring, watering and proper aeration are given regularly to form a compost. Earthworms and fungi are further broke up the material.

- Incineration: Incineration is a thermal waste treatment process which involves the combustion of organic substances contained in waste materials. In this technique waste materials are converted into energy . Gasification technologies are used for converting waste into energy another name of this technology is called as flue gas cleaning .
- Anaerobic Digestion: Anaerobic digestion is a digestion processes begins by decomposing input materials into biogas .First step is bacterial hydrolysis applied on input materials these bacteria converts waste material into renewable energy . Biogas directly used as a fuel by power gas engines
- Pyrolysis and Gasification: Both are advanced Thermal Technology in which biomass (waste) is heated in a vessel to produce a gas

Software Technology Needed to Handle Waste Management

Four technologies plays a major role in handling waste management

- IoT
- Cloud computing
- Geospatial Technology
- Artificial Intelligence Machine Learning

IoT

IoT refers List of devices which connect each other for communication in terms of transferring data through internet . Every thing including cities are becoming "smarter," because of IoT. Very big and unsolvable issues in the urbanization are handled with care with the help of this IoT . Nowadays people wanted to live in pollution free cities, the proposed model can achieve this task with the help of IoT for waste management in smart city development, IoT Devices will link the breach between real world with smart devices, IoT devices will connect with society and industries for better performance .

Cloud Computing

Cost

Cloud computing reduces the cost of setting up, running data stores and reduces cost of buying hardware, software and servers, speed of computing resources are provided on demand so massive amounts of computing resources can be given for businesses or industries Global scale Ability to scale around the country with increasing processing power, bandwidth, storage

Performance

Computing devices upgraded with efficient and latest version that reduced network latency for applications and increases financial prudence of scale.

Security

set of policies, controls are given by cloud providers that strengthen data security, apps and infrastructure security from potential extortions.

Cloud Computing Types

Different models of cloud types and packages have evolved to help to get the right solution.

Cloud types are public, private and hybrid:

- Public cloud: third-party cloud service are known as Public clouds . over the Internet servers and storage are delivered . Microsoft Azure is a public cloud this will provide supporting infrastructure and hardware, software component .
- Private cloud: A private cloud offers services to single business or organisation. on-site datacentre are provided privately in this type of sevice.
- Hybrid cloud: public and private services are combine together by technology . Data and applications can be shared in this type of services provides greater flexibility in infrastructure.

Cloud Services: IaaS, PaaS, serverless and SaaS

Cloud synchronizes IoT devices while collecting the data and helps to offer better decision towards optimized solution . Waste management deals with lot of data shared between IoT devices which requires large storage.Cloud computing provides storage for data processing, special networking and data analytics . Cloud provides back up data store, stream audio and video, delivers software on demand,analysis of gathered data, Implant intelligence.

Geospatial Technology

Geographic information systems (GIS) is its ability to assemble the geospatial data these data includes information on its exact location on the surface of geospatial area .

Types of Geospatial Technologies

- **Remote Sensing:** Images collected from space camera and sensor platforms appropriate for monitoring some commercial satellite image providers now offer images showing details of images, making these images appropriate for monitoring civilians needs and citizen right abuse.
- **Geographic Information Systems (GIS):** A collection of software tools for analysing data which is specific to location on the surface . Geographic patterns can be detected by GIS, such as pollutants, clusters resulting from given pattern.
- **Global Positioning System (GPS):** A network gives accurate coordinate locations to citizen and soldierly users . uses of GPS are Records a location point,it deals data management, Agents can easily go back to the same site for research, permitting public to access and verify results.
- **Internet Mapping Technologies:** Google Earth, Microsoft Virtual Earth details the changes in the geospatial. User interface apps are developed using such technologies making available to audience will solve the problems related to changes in the location .

Artificial Intelligence and Machine Learning

Artificial Intelligence (AI) technologies that dominate recent years, Recent years simulating computers that will think of its own and can easily learn and they can be extremely intelligent.

Machine learning (ML) techniques that allow computers to learn. APIs (application programming interface), training tools, big data, ML algorithms helps to achieve machine to think . Google, Amazon, Fractal Analytics,, Microsoft, Adext SAS,and Skytree these companies selling ML platforms for prediction and classification, Adext AI is the audience management tool applies real AI and machine learning to find the demographic group

Innovative Method to Waste Clearance

Concrete Roads with Plastic (India)

Paving the street with plastic. Plastic is the main ingredient in blacktop used for road construction professor Rajagopalan Vasudevan suggested a way to transform common plastic into a substitute for tar . Due to India's fast economy increased stages of plastic waste being produced by environment, to solve the issues related to environment idealist opinions are needed to handle the mass mess .This idea gives a solution to environmental overhead and saves money, 15 percent of money can be saved with the above idea

Trading Garbage (Indonesia)

55,000 tons of garbage grows yearly in every city in Indonesia, Dr. Gamala Albinsaid formed the Garbage Clinical Insurance which converts garbage into money in his healthcare network . The converted money is then retuned back as an investment in his network . The network is based on medical services and medicines .

Garbage into Useable Energy (Sweden)

In Sweden, garbage to energy system is developed to provide heating households waste to electricity to 260,000 homes. Sweden recycles its garbage and sorts its trash effectively through landfills.

Park Made from Garbage (Uganda)

An amusement park for kids from rejected materials, An environmentalist Ruganzu Bruno is transporting ecological art to the slums to promoting environmental awareness. Swings games and life-size board games were made from plastic bottles and some of the fascinations at the amusement park. The creativity improves the community and empowering kids effectively for managing waste and recycling.

Environmental Protection Agency (United States)

Environmental Protection Agency (EPA) in United States normalizes all unused substance under Resource Conservation and Recovery Act (RCRA). Solid waste like garbage, sludge from wastewater and unwanted materials from industrial may use this provisions on the disposal of solid and unsafe waste materials. This encourages individual states to manage municipal wastes. The western region in the country currently has large volume of landfills those are obey with federal regulations in preventing infection. It provides monitoring systems for groundwater pollution and landfill vapours. The government provides recycling waste and environmental protection of Municipal solid waste that has been recovered consistently over the several years.

Challenges in Waste Management

Automation Tools

To manage critical data's effective automation tools are needed to calculate invoice amounts for automated audits and optimization analysis for technology-assistance .

Recycling Services

To achieving zero waste and green economy the principal challenges is quality of Recyclers . Quality of Recyclers is referred by the amount of raw material is converted into other non-recyclable material. In the case of poor quality it is to be down-cycled or sent to other recovery options .For example in -manufacturing of clear glass the re-melt process products are to be followed for colored glass.

Privacy and Security

Internet of Things become a major key element in large-scale environment . Trust in security functions and standards are to be addressed to check integrity and reliability of data's.

New challenges are

- **Quality-of-Information:** In a heterogeneous environment giving trust in quality-of-information is to be ensured by security algorithms to re-use across many applications.
- **Providing Security:** Providing Security in exchange of data between heterogeneous IoT devices and handling their information.
- Providing protection tools for susceptible devices.

Heterogeneity

Waste management in a smart city development includes set of devices, software's, updating technologies, platform independence nature are different for every IoT system. Cloud stores need to handle all related issues. These issues will affect integration and device communication of IoT devices . Standards are not the only solution to manage Heterogeneity in IoT platforms.

Strength of this Chapter

This chapter details automated real time sharing of information and grouping of sensors in an effective manner, it details self-organization of sensors in case of any malfunctioning of sensors, how to handle large volume of data, how to reduce cost and improved optimization of resources when sensors deployed over a large space. This chapter details different ways in waste collection and management techniques, possible disposal methods, software technology needed to handle waste, challenges in waste management .

Future Enhancement

Due to mass growth in smart city development, the contribution of IoT leads an important role . Day by day the utilization of things related to internet growing exponentially,. Internet becomes the one of the important part in human life . The possible future needs in planning smart city are listed

- Best path selection base on through GSM/GPRS
- Time stamp can be added in the smart bins to analyse the maintenance of bins and audit log are prepared to ensure the cleanliness of bins
- Deep learning algorithms can be developed in future to form self-fault finding in the given query region in the sensor network

CONCLUSION

This chapter focuses on automated Waste Management System Using IoT . The proposed system helps to implement smart city for further development in urbanization. Tracking the bins and self-efficient network groups are framed to focuses on handling of large volume of data, cost of the underlying network topology, merging of devices, speed of the data connectivity . Architecture model are proposed to handle the above challenges .The proposed architecture model handles possible challenges in developing smart city and suggestions are given to overcome those challenges.

REFERENCES

Al-Masri. (2018). A Serverless IoT Architecture for Smart Waste Management Systems. *IEEE International Conference on Industrial Internet (ICII).*

Aleyadeh. (2018). *An Iot-Based Architecture For Waste Management.* Academic Press.

Barton, J. R., Dalley, D., & Patel, V. S. (1996). Life Cycle Assessment For WasteManagement. *Waste Management (New York, N.Y.), 16*(1-3), 35–50. doi:10.1016/S0956-053X(96)00057-8

Beigl, P., & Salhofer, S. (2004). Comparison Of Ecological Effects And Costs of Communal Waste Management System. Resources. *Conservation Andrecycling, 41*(2), 83–102. doi:10.1016/j.resconrec.2003.08.007

Gaur, A., Scotney, B., Parr, G., & Mcclean, S. (2015). Smart City Architecture And Its Applications Based On Iot. *Procedia Computer Science, 52*, 1089–1094. doi:10.1016/j.procs.2015.05.122

Khatoun, R., & Zeadally, S. (2016). Smart cities. *Communications of the ACM, 59*(8), 46–57. doi:10.1145/2858789

Shashika Lokuliyana & Jeyakody. (2017). *IGOE Iot Framework For Waste Collection And Optimization.* NCTM.

Prajakta, G., Kalyani, J., & And, M. (2015). Smart Garbage CollectionSystem In Residential Area. *International Journal of Research In Engineering And Technology, 4*(3), 122–124. doi:10.15623/ijret.2015.0403021

Wahab, K. Tomari, & Jabbar. (2014). Smart Recycle Bin: A Conceptual Approach Of Smart Waste Management With Integrated Web Based System. Academic Press.

Chapter 21
Smart Microbial Sources Management for Treatment:
Palm Oil Mill Effluent and Landfill Leachate

Hossein Farraji
University of Canterbury, New Zealand

Amin Mojiri
University of Hyroshima, Japan

Mohd Suffian Yusoff
Universiti Sains Malaysia, Malaysia

ABSTRACT

Overpopulation and industrialization are the major sources of wastewater in human society and water resources. Food production industries and municipal solid waste are the root origin of wastewaters containing palm oil mill effluent and municipal landfill leachate. Traditional treatment method for such highly polluted wastewaters cannot meet environmental discharge. Finding an advanced and smart de-contamination process for these types of polluted wastewater could be considered as a capable method for suitable adaptation with overpopulation in current condition and future coming decades. This chapter illustrates critical points through the application of traditional treatment techniques such as acclimatization in palm oil mill effluent and municipal landfill leachate as the most straightly polluted agro-industrial effluent.

INTRODUCTION

Waste and wastewater management is one of the most critical requirements of current life style of human-kind (Farraji et al., 2015b). Municipal Landfill Leachate (MLL) is known as highly polluted and toxic wastewater in human civilization. This liquid effluent creates when rainwater percolate into the landfill of waste disposal and through the process of degradation of waste materials. Several factors contribute to the amount of leachate generated in a municipal landfill including evapotranspiration, precipitation,

DOI: 10.4018/978-1-5225-9199-3.ch021

infiltration, surface runoff, the level of waste compaction as well as the trespass of ground water into the landfill (Costa et al., 2019). The characteristics of landfill leachate are defined by unpleasant odor, chemical oxygen demand (COD), heavy metals, biological oxygen demand (BOD), ammonium nitrogen (NH_4+- N), sulfur components and other toxic materials. As a result, landfill leachate can be considered as seriously polluted waste water that would impose major risks on the environment provided that they are treated accurately (Meky et al., 2017; Costa et al., 2019).

Based on the age of landfill, there are two sets of leachate: 1) Fresh leachate, in five years, with high concentration of COD and BOD and better biodegradability. 2) Aging leachate, over five years, with low concentration of COD and BOD as well as high degree of ammonia and also less biodegradability (Liu et al., 2017).

On the other hand, growing widespread demand of palm oil has caused critical waste rooted in the process of degradation of palm fruits. Palm oil mill effluent (POME) has a high potential of contamination with low pH and high concentration of BOD and COD. The characteristics of this liquid waste are defined by its viscous attribute, brownish color, the components of palm fruits which are water- soluble, a large amount of carbohydrates and lipids ranging from hemicellulose to simple sugars and phenolic materials and also nitrogenous components varying from amino acids to proteins as well as pathogenic bacteria (Wong et al., 2018). POME is markedly polluted which can pose detrimental effects on water bodies and terrestrial lands discharged directly from a mill. When a river or stream receive this kind of sewage, its color turns to brown, slimy and smelly and also brings about de-oxygenation which have harmful effects on aquatic species. Moreover, POME has undesirable effects on physicochemical and nutritional properties of soil and leads to increase in salinity and decline in pH (Islam et al., 2018). Consequently, MLL and POME can be considered as serious resources of contamination which need to treat by proper technologies.

Wastewater treatment even by a specific polishing post treatment method such as constructed wetland depends on the characteristics of wastewater and concentration of pollutants (Vymazal, 2009). Concentrated landfill leachate which is effluent of membrane treatment process through the landfill leachate treatment techniques of wastewater treatment highly depends on the types and concentration of pollutants in target in wastewater treatment plants(Zhang et al., 2013), required several high-tech treatment skills to providing suitable decontamination such as Electro-Ozonation (Mojiri et al., 2019) or combining with adsorbent in augmented sequencing batch reactor for achieving higher efficiencies (Mojiri et al., 2017; Zhang et al., 2013). In case of municipal landfill leachate (MLL) as a common industrial wastewater, there is no single method of treatment to meet environmental discharge regulation (Mojiri, 2014). Meanwhile, adsorbent application is a common treatment for MLL (Farraji et al., 2015c; Foo and Hameed, 2009). Most of the common treatment methods in MLL treatment contain mixing two or more spectacular and adaptive treatment methods such as soil infiltration (Liang and Liu, 2008), combined electro-Fenton oxidation and Sequencing Batch Reactor (Lin and Chang, 2000), biological method attached with growing biomass (Loukidou and Zouboulis, 2001), phytoremediation and adsorbent in microbiological method (Aziz H.A, Farraji, 2017; Mojiri et al., 2016). On the other hand, high concentration of ammonia, toxic compounds, and metallic elements found in MLL make the sludge acclimatization a proper method to treatment.

These methods would be costly and need to use chemical substances through the process of treatment. As a result, a cost-effective and eco-friendly treatment method is required (Islam et al., 2018). Microbiological treatment would be an easiest and a cheapest method with less environmental perils than the other alternatives for POME treatment. By considering the smart methods for treatment of wastewater, this book chapter as a comparing study, the main concerns will be on traditional sludge acclimatization which is an aerobic treatment method for MLL and applicability of this method for POME. Through the last half century, there was no innovative improvement with scaling up aspect presented for POME treatment. Even after this long period of time, current ponding system which is steps of cooling, acidification, anaerobic and aerobic treatment process, has not been changed significantly. In this study, capability of specific modified sludge acclimatization will present as a smart treatment method for POME decontamination.

BACKGROUND

POME as the effluent of vegetable oil industry reaches to 200 million tones per year in the world. Indonesia and Malaysia are the main producer of this effluent. None of current treatment method can't meet discharge standards of Malaysia. Consequently finding a short, cost effective and influenced treatment method is the main gap of knowledge for POME industry. Based on the biological treatment there should be a source of degradable compounds as feeding (organic and inorganic pollutants and nutrients), microbial community (could be artificially added) in an optimum growing circumstances (pH, temperature, absence of toxic and hazardous material. Acclimatization can be defined as a biological treatment process of adapting to a new environment. The process engaged this adaptation is called biodegradation which can be considered as the ability to decline the complexity of chemical and organic compounds (Stephenes, 1970). The research on the procedure of acclimatization is important in that it introduces the microorganisms to an unknown wastewater environment. In such circumstances, microorganisms can effectively enhance the level of degradability of pollutants (Zawani et al., 2013). Biodegradation would be carried out by microorganisms so that they can degrade organic compounds to carbon dioxide and sludge under aerobic conditions and to biogas (including mainly CO_2 and CH_4) under anaerobic conditions (Renou et al., 2008). It has been confirmed that biological processes is a very efficient method to remove organic and nitrogenous components from immature leachates when the ratio of BOD/COD, known as biodegradability index (BI), is a high value (>0.5). In old age leachates, biological treatment methods tend to be less effective due to the presence of refractory compounds mostly humic and fulvic acids (Renou et al., 2008). Before acclimatization, the seeding process would be conducted to grow up and develop the microorganisms' population in a bioreactor (Kahar et al., 2017). In this way, it is feasible to acclimatize bacterial population for the treatment of landfill leachate (Bohdziewicz et al., 2008).

Landfill leachate and also POME is a complex liquid that contains enormous concentrations of biodegradable and non-biodegradable substances which can lead to surface and ground water as well as soil contamination. Recently, several types of treatment technologies have been introduced to decline the threats of such waste water imposed on the environment. Nevertheless, some of the treatment methods remain complicated, expensive and necessitate acclimatization over the process (Kamaruddin et al., 2015; Neczaj et al., 2008). Among various treatment methods, biological treatments, due to its low cost and applicability, are useful and widely are employed to treat leachates and POME with excessive amounts of organic compounds. This process applies microbes to degrade the organic substances in these kind of waste water either under aerobic or anaerobic system(Kurniawan et al., 2010). Owing to

Table 1. Comparing anaerobic and aerobic treatment

Characteristics	Aerobic	Anaerobic
Organic removal	High	High
Effluent quality	Excellent	Moderate to poor
Organic loading rate	Moderate	High
Hydraulic retention time	Short	Long
Required area	Low	High
Sludge production	High	Low
Nutrient requirement	High	low
Alkalinity requirement	Low	High for certain industrial WW
Energy requirement	Low	Low to moderate
Temperature sensitivity	Low	High
Startup time	2-4 weeks	2-4 months
Odor	Less opportunity for odor	Potential odor problems
Bioenergy and nutrient recovery	No	Yes
Mode of treatment	Total (depending on feedstock)	Essentially pretreatment
Application for ponding POME treatment	Yes	Yes
Greenhouse gases emission	High (except in low concentration)	Low
Advantages	Short time can give high efficiency specially for toxic compounds	Effluent is suitable for land application
Disadvantages	High energy consume (aeration) Low inactivation of pathogen Effluent unsuitable for land use	Large area requirement, Pond renewal required Slow start up
Advised an applied in commercial POME treatment	Anaerobic - aerobic (ponding) 48% of Malaysian oil palm mills	Biogas production through the anaerobic digestion 52% of mills

(Borja et al., 1996; Cakir & Stenstrom, 2005; Chan et al., 2010; Chin et al., 2013; Doble & Kumar, 2005; Grady et al., 2011; Ma & Ong, 1985; Metcalf, 2014).

the cut down in the number of biodegradable organic substances in landfill leachate with the increase in the landfill age, the efficiency of the treatment will decrease. To improve the effectiveness of biological treatment, combined technical solutions could be considered (Torretta et al., 2016). While POME with high biodegradable organic compounds, compared to landfill leachate, would be treated more efficiently and effectively through biological treatment methods such as sludge acclimatization.

Based on the literature, anaerobic and aerobic treatment methods contain specific advantages and disadvantages. Meanwhile, selection of cost effective technology highly limited by artificial aeration which is electricity consumer method. Finding a smart treatment system with capability to shortening the treatment time period, will be an advanced improvement through the aerobic de-contamination process. Table 1 presents comparing anaerobic and aerobic treatment methods for wastewaters in literature.

Table 2. The characteristics of landfill leachate produced from decomposition of municipal solid waste in developing countries

Parameter	Range of Values (mg/L)	Parameter	Range of Values (mg/L)
pH	4.5 to 9	Organic N	10 to 4,250
Alkalinity (CaCO3)	300 to 11,500	Ammonia NH3	30 to 3,000
BOD_5	20 to 40,000	Nitrite nitrogen $NO2^-$	0 to 25
COD	500 to 60,000	Nitrate nitrogen $NO3^-$	0.1 to 50
Calcium	10 to 250	Total nitrogen	50 to 5,000
Chloride (Cl^-)	100 to 5,000	Total phosphate	0.1 to 30
Potassium	10 to 2,500	Sulphate ($SO4^{2-}$)	20 to 1,750
Sodium	50 to 4,000	Manganese	0.03 to 65
Magnesium	40 to 1,150	Total iron	3 to 2,100
TDS	0 to 42,300	Copper	4 to 1,400
Total SS	6 to 2,700	Lead	8 to 1,020
Hardness	0 to 22,800	Zinc	0.03 to 120

Source:(UNEP, 2005)

Landfill Leachate Characteristics

Sanitary landfill is widely applied for the disposal and elimination of municipal solid waste (MSW) in myriad countries (Renou et al., 2008). After landfilling, solid waste will experience physicochemical and biological changes. As a consequence, degradation of the waste organic compound along with percolation rainwater result in the generation of a strength polluted liquid called "leachate".(Guo et al., 2010). The characteristics of landfill leachate depend on the stage of decomposition in the landfill, the type of MSW being dumped, landfill age, pattern of the site hydrology, weather condition and moisture content (Guo et al., 2010; Mojiri et al., 2012).

Doubtless, landfill leachate as one of the outcomes of overpopulation and industrialization could be considered as environmental pollutant which need to be decontaminated by cost effective and influential smart treatment methods (Zolfaghari et al., 2016). Leachates carry various forms of pollutants such as biochemical oxygen demand (COD), ammonia–nitrogen (NH_3–N), biological oxygen demand (BOD_5), color, heavy metals and suspended solids (Mojiri, 2014) which have potential treat to soil, surface and groundwater which consequently pose a risk to public health (Aziz et al., 2013). The characteristics of landfill leachate produced from decomposition of MSW in developing countries are shown in Table 2.

The management of leachate is one of the most crucial parts of planning, design, operation, and management of municipal solid waste (MSW) landfills or municipal landfill leachate(MLL) (Halim et al., 2010). Leachate composition is a contributing factor to applying the leachate treatment methods. Highly biodegradable substances, young leachate for instance, will be treated via biological method and activated sludge process is widely applied. Heavy metals are commonly removed using chemical or physical treatments. Based on the literatures, a single method is not efficient and appropriate for all kinds of leachates to date and therefore treatment technologies need to occupy by smart and innovative methods (Mojiri, 2014).

Biodegradability Index

The ratio of biological oxygen demand (BOD) to chemical oxygen demand (COD) in wastewater engineering defined as biodegradability index (BI=BOD/COD). When BI in wastewater reaches to 0.5, it could be considered as a capable source for biological treatment methods(Metcalf and Eddy, 2014). Collection of BI in MLL is a common characteristics of wastewater engineering. Usually MLL is known as low BI source of wastewater and parts of treatment process concern on increasing the BI. The characteristics of some considerable pollutant in municipal landfill leachate along with its biological treatment technologies, BI before and after treatment are presented in Table 3.

As it clearly could be seen, Table 3 summarizes the final performance of the most applicable microbiological treatment technologies for the removal of COD, BOD and NH_3–N from MLL and other wastewaters. Based on the data in Table 3, the biological treatments demonstrate their effectiveness for the removal of organic and inorganic compounds along with improvement in the rate of biodegradability. The growth in the ratio of biodegradability index clarifies that the capability of wastewater for more biological treatment will be increased.

Although most biological processes can be appropriate for MLL treatment, it is obvious that none of the biological methods presented here is highly efficient or worldwide applicable for the removal of all kinds of pollutant. As a result, biological treatment can be considered as a pre-treatment process to degrade the organic compounds. The pre-treatment outcomes and its removal efficiency to eliminate COD and BOD_5 and also the rate of biodegradability achieved after treatment can set the stage for the selection of a suitable post-treatment method for MLL which would yield desirable results in terms of removal of overwhelmingly organic and chemical substances.

Color of Highly Polluted Wastewater

Color in some types of wastewater is a major or even critical pollution. Blocking the sunlight is the most deleterious effect of color as pollutant in wastewater which influenced as hinders of photosynthesis. Color pigments abstract light so oxygen mass transferring in aquatic media will be reduced (Silveira et al., 2009). Color of wastewater is the first recognized decontamination furthermore the aesthetic problem, obstruct light penetration and oxygen transfer in bodies of water (Banat et al., 1996; Patel and Suresh, 2008). In numbers of specific wastewater such as molasses- based industries (Satyawali and Balakrishnan, 2008), textile wastewater (Vakili, et al., 2016), pulp and paper mill (Mahesh et al., 2006), municipal landfill leachate especially concentrate landfill leachate (Mojiri et al., 2017), palm oil mill effluent,

A: POME

As an agricultural colloidal wastewater POME is one of the highly polluted industrial wastewater ever. Organic characteristics of the pollutants in POME causes to limitation of capable treatment method for this dark brown effluent as well as its straightly high concentration of non-degradable or low-degradable organic pollutants in very wide range of sizes (Farraji et al., 2015a). POME contain lignin, conventional separation methods are intractable about lignin (Mohan and Karthikeyan 1997). Lignin to cellulose ratio in POME is higher than other agro industrial wastewater (Jackson 1977) and it causes hardness of decoloring. Not only current commercial treatment method contain ponding and anaerobic digestion cannot meet discharge standard as well as color removal but, also most of lab scale treatment methods

Table 3. The characteristics of some critical pollutants in landfill leachate and effect of treatment technologies on biodegradability

Reference	**Biodegradability BOD_5/COD	**Concentration in Leachate after Treatment mg/L		Removal Efficiency	Treatment Techniques	Ammonia NH_3-N mg/L	BOD_5/COD	Initial Concentration in Leachate mg/L		Sample and Location
		BOD_5	COD					BOD_5	COD	
(Aghamohammadi et al., 2007)	-	-	2024.9	NH_3-N:12%- 74% COD: 4.5% -29.2% BOD_5: *NA	NPAC[1]-activated sludge process	1400	0.17	377	2860	Landfill leachate in Malaysia
	-	-	1544.4	NH_3-N:15- 78% COD:14.6% - 46% BOD_5: NA	PAC[2]-activated sludge process					
(Aziz et al., 2012a)	-	-	-	NH_3-N:NA COD:NA BOD_5: NA	PAC-SBR[3]	562	0.20	285	1295	landfill leachate in Malaysia
(Mojiri et al., 2014)	-	-	353.35	NH_3-N: 99.01% COD: 72.84% BOD_5: NA	PZ-SBR[4]	532.0	0.20	269	1301	landfill leachate in Malaysia
	-	-	703.58	NH_3-N:98.27% COD: 45.92% BOD_5: NA	SBR[5]					
(Aziz et al., 2012b)	0.005	4.2	739.44	NH_3-N: 89.40% COD: 42.90% BOD_5: 98.5%	NPAC-SBR[6]	562	0.20	285	1295	landfill leachate in Malaysia
	0.002	1.71	941.46	NH_3-N: 89.90% COD: 27.30% BOD_5: 99.40%	PAC-SBR					
(Aziz et al., 2012b)	0.036	33.99	929.50	NH_3-N:71.90% COD: 48.90% BOD_5: 81.22%	NPAC-SBR	1627	0.10	181	1819	landfill leachate in Malaysia
	0.002	1.50	753	NH_3-N:65.90% COD: 58.60% BOD_5:99.17%	PAC-SBR					
(Kahar et al., 2017)	0.22	424.2	1943.6	NH_3-N: NA COD: 58.93% BOD_5: 52.37%	Anaerobic Bioreactor	-	0.17	810.02	4732.56	Landfill leachate in Indonesia
(Loukidou and Zouboulis, 2001)	0.05	100	1750	NH_3-N: 60% COD: 65% BOD_5: 90%	SBR + polyurethane as biofilm-carriers	1800	0.2	1000	5000	Landfill leachate in Greece
	0.09	90	950	NH_3-N: 85% COD: 81% BOD_5: 91%	SBR using GAC[7] as biofilm-carriers					
(Guo et al., 2010)	0.15-0.21	353.4-778.6	2367-3550	NH_3-N: 96.6% COD: 21.1% BOD_5: 5.5%	air-stripping	1000-1750	0.09-0.22	374-824	3000-4500	Landfill leachate in China
	0.26-0.39	316.7-697.9	1176-1764	NH_3-N: 97.4% COD: 60.8% BOD_5: 15.3%	fenton					
	0.12-0.16	64.3-141.7	507-841.5	NH_3-N: 97.9% COD: 83.1% BOD_5: 82.8%	SBR					
	0.28-0.42	57.9-127.7	201-301.5	NH_3-N: 98.3% COD: 93.3% BOD_5: 84.5%	coagulation					

continued on following page

Table 3. Continued

Reference	**Biodegradability BOD_5/COD	**Concentration in Leachate after Treatment mg/L		Removal Efficiency	Treatment Techniques	Ammonia NH_3-N mg/L	BOD_5/COD	Initial Concentration in Leachate mg/L		Sample and Location
		BOD_5	COD					BOD_5	COD	
(Zawani et al., 2013)	-	-	805	NH_3-N: NA COD: 65% BOD_5: NA	SBR return activated sludge	-	0.008	20	2300	Kenaf-Retting Wastewater
(Li et al., 2009)	-	-	720	NH_3-N: 99% COD: 76% BOD_5: NA	SBR(air compressors, mechanical agitation &sludge)	1200	0.21	650	3000	Landfill leachate in China
(Neczaj et al., 2008)	0.17-0.19	11.6	60.8-68	NH_3-N: 99% COD: 98.4% BOD_5: 97.3%	SBR (co-treatment with sludge)	750-800	0.1	<430	3800-4250	Landfill leachate in Poland
	1.12	108-135	96-120			-	0.66	4000-5000	6000-7500	Industrial wastewater
(Liyan et al., 2009)	0.067	30.09	447.3	NH_3-N: 96.9% COD: 94.8% BOD_5: 97.8%	ARB[8]	861	0.15	1368	8603	Landfill leachate in China
(Zolfaghari et al., 2016)	0.008	4.44	567.3	NH_3-N: 98.2% COD: 63.4% BOD_5: 96.5%	MBR[9]	288	0.08	127	1550	landfill leachate in Canada summer))
	0.01	9.68	993	NH_3-N: 99.2% COD: 53.2% BOD_5: 96%	MBR	667	0.11	242	2122	landfill leachate in Canada winter))
(Kurniahan et al., 2010)	-	-	956.92	NH_3-N: 98% COD: 53% BOD_5: NA	activated sludge	115	0.17	**346	2036	Landfill leachate in Indonesia
	-	-	294.6	NH_3-N: 99.8% COD: 83% BOD_5: NA	AL[10]	104	0.56	**970	1733	Landfill leachate in UK
	-	-	800	NH_3-N: 75% COD: 80% BOD_5: NA	UASB[11]	160	NA	NA	4000	Landfill leachate in Finland
	-	-	64.4	NH_3-N: 99% COD: 99% BOD_5: NA	GAC+ coagulation & aerobic	1153	0.70	**4508	6440	Landfill leachate in Germany
(Yalmaz., & Öztürk, 2001)	-	-	780	NH_3-N: 99% COD: 97% BOD_5: NA	SBR	1000	0.58	**15080	26000	Landfill leachate in Turkey

1. None Powdered Activated Carbon
2. Powdered Activated Carbon
3. Powdered Activated Carbon Sequencing Batch Reactor
4. Powdered ZELIAC augmented Sequencing Batch Reactors
5. Sequencing Batch Reactors
6. Non-Powdered Activated Carbon Sequencing Batch Reactors
7. Granular Activated Carbon
8. Aged Refused Bioreactor
9. Membrane Bioreactor
10. Aerated Lagoons
11. Up flow Anaerobic Sludge Blanket

such as membrane bioreactor (Facta et al., 2014), adsorbent such as activated carbon could not remove color (Facta et al., 2014). Numerus lab scale decoloring researches by different adsorbents carried out for POME with high efficiency. Based on the literature, banana peel activated carbon (Mohammed and Chong, 2014), Suspended activated sludge and activated sludge-granular activated carbon (Zahrim et al., 2009), Microwave incinerated rice husk ash (Kutty and Ngatenah, 2011), Palm kernel shell (Mohammed, 2013), Zeolite (Farraji et al., 2017a; Farraji et al., 2017b) and TiO2 Nanoparticle (Tan et al., 2017) have been used for decoloring of POME. These methods are not cost effective, very time consumer, applicable only after traditional aerobic-anaerobic treatment process and rarely usable on industrial scale. POME treatment by activated carbon could not be considered as a perfect method for de coloring because of their short life time and continues replacing requirement (Facta et al., 2014). The source of color in POME is suspended solid.

B: MLL

As the most visible factor, color of MLL is a consideration issue in visual assessment. MLL contain true color and determined by American Dye Manufacturers Institute (ADMI) or (Pt. Co). Color in MLL mostly originated from spectacular functional and groups unsaturated bonds. Based on the Fourier Transform Infrared Spectroscopy (FTIR) analysis which intensively used for functional group analysis in organic compounds, hydroxyl group (-OH) will be increase after treatment with ozone oxidation. Furthermore, a capable coagulation pre-treatment could be considered as an enhancer of the efficiency of ozone oxidation. Meanwhile, positive effect of ozone oxidation on enhancing biodegradability of MLL in parallel with increasing the efficiency of decolorizing process. The ozone based treatment will show higher performance in addition with UV process in color removal as well as biodegradability index increasing in MLL. In fact, the most important part of ozone based treatment process for MLL is breaking large organic compounds (non or low biodegradable pollutants) to small organic ones which are more biodegradable than previous materials (Wu et al., 2004). Final results of Wu et al, (2004) indicate that 90% of color removal achieved after ozone based treatment process for MLL. Positive correlation between concentrations of UV/H2O2 process with maximum 91% decolorizing in two hours has been reported by (Shu et al., 2006) as a suitable pre-treatment of MLL. Higher performance of decolorizing (100% removal) reported by Xiao et al, (2013) with addition of UV 50 mA cm^{-2} to electrochemical process by RuO2/Ti as anode. Results of decolorizing process in MLL in presence of UV treatment were far better than single ozone based, H2O2, and electrochemical treatment techniques. To put in chest nut, large organic compounds which are the origin of color pollutants in wastewater should be breaking down by suitable technique (UV treatment) and then the biodegradation process by other capable treatment system could be carried out well.

COD of POME

Chemical oxygen demand (COD) defined as the volume of oxygen required for oxidizing the organic pollutants in a contaminated media without assistance of microbial communities. In other word, it could be defined as chemical decomposition of organic and inorganic contaminants. Agricultural wastewaters often contain high concentration of COD due to containing high volume of organic matter. Lignin, cellulose grease and oil, starch, and carbon hydrides are the most major types of organic contaminants in agro industrial wastewaters. The value of COD in agro industrial wastewater such as Sweet potato collected in range 73260-7940 mg/L (Tantipaibulvut et al., 2015), in Coffee wastewater which studied by Devi et al, (2008) reaches to 22000 mg/L meanwhile, Olive oil which is another agro industrial effluent differs from 15528 to 17343 mg/L (Paraskeva and Diamadop, 2006). Potato industry produces an effluent with 6760±970 mg/L (Manhokwe et al., 2015) and Sugar-beet as another main sources of agro industrial effluent has been collected by Alkaya and Demirer, (2011) with 6621±113 mg/L COD. Based on the review of Chin et al, (2013) POME has Chemical Oxygen Demand (COD) in range of 44300-102696 mg/L which is the highest collected COD in industrial wastewater. COD, BOD, TSS, and color are the main pollutants in POME (Liew et al., 2015). Based on the aforementioned values of COD in several agro industrial wastewater it could be clearly seen that POME is high strength polluted wastewater for microbiological treatment. Concentration of COD in a polluted wastewater is a critical influenced factor for selection of treatment method. Similarity of COD and BOD in volumes will lead treatment methods to microbiological systems.

BOD of POME

BOD_5 is a type of oxygen (dissolved oxygen) required for degrading organic matter by microorganisms at a certain time (frequently five days) and specific temperature (20 °C) in a dark place (Jouanneau et al., 2014). BOD_5 has been selected as an important characteristic of pollutants for river pollution in the United Kingdom in 1908 and has been adopted by the American Public Health Association (APHA) in 1936. BOD_5 is the amount of oxygen divided by the total volume of the treatment system that is processed in the respiration period of microorganisms. Aeration equipment will be sized in wastewater treatment on the basis of the value of BOD_5. Repeatability and inhabitation by wastewater toxic compounds and ions are the major drawbacks of BOD_5 measurement (Dubber and Gray, 2010). This characteristic of POME depends on discharge standards (100 mg/L) and has a strength limitation of 20 mg/L in certain areas of Malaysia (Sabah and Sarawak). POME with $BOD_5 < 5,000$ mg/L can be used for land application (Corley and Tinker, 2008). Since of nontoxic characteristic and organic base of POME its pollutants may biodegraded in soil even with $BOD_5 < 5,000$ mg/L.

Biodegradability Index in POME

There are limited studies about POME biodegradability. Based on the (Chin et al., 2013; Hansen et al., 2015) POME could be considered as a source of polluted agricultural effluent with capability for biologically treatment. Biodegradability of POME has not been concerned in the most of resent overviews of scientific literature(Abu Bakar et al., 2018; Choong et al., 2018; Liew et al., 2015; Ohimain and Izah, 2017). For giving a clear picture of POME biodegradability, characteristics of numerus studies have been collected in Table 4 and their related BI have been calculated by the authors. Varying range of BI in POME indicates that the common assumption about the biodegradability of this agricultural colloidal effluent could not be limited to such a narrow BI. Varying factors including, sampling point (raw or anaerobically treated POME), process of treatment, and BOD collecting errors, all play a part in the BI of POME. The mean BI in 29 collected research equals to 0.4 which is commonly considerable as non-biodegradable wastewater (Metcalf and Eddy, 2014). Nevertheless, in most of research community POME known as a biodegradable effluent. There is more requirement research about POME BI.

By comparing BI in MLL (Table 3) and POME (Table 4), it could be clearly seen that biodegradability of POME is several time higher than MLL. Furthermore, absent or very low concentration of hazardous heavy metal in POME in parallel with low volume of ammonia nitrogen (Rupani et al., 2010), makes it as non-toxic effluent. Consequently nontoxic characteristics of POME is significant different between POME and MLL in treatment process. Sludge acclimatization mostly conducted for adapting the microbial communities of augmented municipal wastewater in target wastewater so in POME because of aforementioned characteristics, municipal wastewater can be directly added as microbial communities to POME for enhancing biological degradation. High removal efficiency in direct application of municipal wastewater (Farraji et al., 2017a; Farraji et al., 2017b) is an applicable document for promising efficiency through the microbiological treatment.

Applications of POME

POME includes high concentrations of nitrogenous, protein, lipids, carbohydrate compounds and minerals that could be turned into value- added materials by microbial process. One of the most valuable by- products of palm oil mill effluent would be biodiesel which can be an alternative fuel to petro- diesel (Zahan and Kan., 2018). In addition, through the treatment of POME by means of anaerobic methods, biogas is produced which would be a major source of renewable energy and can contribute to the supplying of energy for communities and sustainable development (Loh et al., 2017). In fact, high biodegradability of POME makes it as a suitable source for Hydrogen and Methane production through the anaerobic digestion by anaerobic contact filter (Vijayaraghavan, and Ahmad 2006; Ismail et al., 2010; Taifor et al., 2017; Cahn and Chang, 2019).

Moreover, POME can be defined as a renewable sources of fertilizers and carotene. This effluent can be considered as a unique nutrient compound which has potential to decompose the soluble phosphate (or insoluble organic phosphate) and improve the physical and microbial soil sanitation, set the stage for increasing crops yield as well as declining the usage of chemical fertilizers (Foo and Hameed., 2010)

FUTURE RESEARCH DIRECTIONS

This book chapter presented some parts of specific characteristics of POME as colloidal agricultural effluent which should be considered on treatment process by traditional methods. Well understanding of a pollutant is the most critical and key factor through the selection of decontamination process. Smart thinking and innovative observation need to concerning on the characteristics of problem at the first and then seeking influenced techniques and solution methods. The first point about POME should be narrowed down in researches is non-hazardous specification and absence of toxic compounds in POME. It should be lighten that an extremely high concentration of organic compounds (BOD, COD, TSS, and color) is the main target in POME treatment process. Biodegradation is selected and innovative smart method for POME treatment in fast, easy, cheap, and adaptive system for future researches which may illustrated by following research objectives: a) Applicability of direct application of MWW for raw POME b) Inoculation of MWW at the first section of traditional treatment (ponding system) or in second part of ponding system (acidification) c) Effect of acidity through the MWW augmented treatment process d) The minimum volume of MWW as augmentation factor e) Adaption method for industrialization in current ponding system f) Applicability of post-treatment methods such as phytoremediation and constructed wetland for polishing of the final effluent. Since direct application of MWW in POME treatment is just in early process of researches so, wide range of studies required for future of this emerging agro- industry.

Table 4. POME characteristics in literatures

PH	BOD (mg/L)	COD (mg/L)	$\frac{BOD^*}{COD}$	Oil & Grease (mg/L)	Suspended Solids * (mg/L)	Nitrogen (mg/L)	Reference
4.5-5.5	10000-25000	30000-60000	0.33-0.41	2000-3000	3000- 5000	350-600	(Borja et al., 1996)
4.2-4.4	26000-30000	63000-73000	0.41	-	27000-36500	880-1100	(Chin et al., 1996)
4.7	25000	50000	0.5	4000	18000	750	(Ahmad et al., 2003)
3.8-4.4	23000-26000	42500-55700	0.54-0.46	4900-5700	16500-19500	500-700	(Najafpour et al., 2006)
4.05	22700	44300	0.51	4850	19780	780	(Zinatizadeh et al., 2006)
-	25000	50000	0.5	4000-6000	18000	750	(Ahmad et al., 2006)
3.5	24710	59300	0.41	-	17260	692	(Vijayaraghavan and Ahmad, 2006)
4.24-4.66	62500-69215	99465-112023	0.62-0.61	8845-10052	44680-47140	1305-1493	(Choorit and Wisarnwan, 2007)
-	22000-54300	75200-96300	0.29-0.56	8300-10600	8500-12000	830-920	(O-Thong et al., 2008)
4.18-4.7	19100-46700	65900-85300	0.28-0.54	8700-13800	24200-34300	900-1050	(Chan et al., 2010)
4.19-5.30	11730-37500	22660-73500	0.51-0.51	2500-16100	7100-24500	750-1200	(Poh and Chong, 2010)
4.15-4.45	21500-28500	45500-65000	0.45	1077-7582	15660-23560	500-800	(Wu et al., 2017)
4.5	30000	50000	0.6	-	59350	-	(Fadzil et al., 2013)
8.4	4700	1350	0.28	-	1800	-	(Mohammed and Chong, 2014)
4.69-4.75	33500-36400	69500-88150	0.48-0.41	6830-7610	23180-29100	1003-1053	(Khemkhao et al., 2015)
7.85	233	884	0.26	-	539	167	(Tee et al., 2016)
8.4	1350	4700	0.28	-	1800	-	(Kaman et al., 2017)
7.5	350	750	0.46	15	790	450	(Darajeh et al., 2016)
4.3	27000	75000	0.36	-	50000	-	(Tabassum et al., 2015)
7.85	233	884	0.26	-	539	167.4	(Tee et al., 2016)
5.1	3500	56500	0.06	109900	8300	960	(Krishnan et al., 2016)
4-5	25000	50000	0.50	4000	40500	-	(Kuppusamy et al., 2017)
5.8	1000	5000	0.2	-	-	-	(Najafpour et al., 2006)
4.5-5	27000	51000	0.52	4200	19000	580	(Abdullah et al., 2015)
4.5-5.2	27766	61994	0.44	-	7860	-	(Saleh et al., 2012)
4.92	9495	74730	0.12	3488	30390	-	(Tan et al., 2018)
4.5	30100	70000	0.43	10540	28900	980	(Chan et al., 2017)
4.74	34393	75900	0.45	191	14467	-	(Bala et al., 2015)
4.2	25000	51000	0.49	60000	18000	750	(Lam and Lee, 2011)

CONCLUSION

POME treatment through the last 50 years background concentrated on anaerobic-aerobic ponding system. In last two decades anaerobic digestion for methane production is going to improve treatment method to sustainable energy (methane) production. The most important part of POME treatment which make it as a capable agricultural effluent for the fastest, the easiest and the cheapest is high BI in this colloidal effluent for microbiological treatment. Absence or very low concentration of toxic compound is the main reason of independent of POME to sludge acclimatization. Direct application of municipal wastewater as microbial resources through the aerobic process could be a smart method for meeting environmental discharges in the shortest time ever.

ACKNOWLEDGMENT

The authors would like to recognize Dr. Mitra Ghasemi, who participated in the researching and writing of this chapter.

REFERENCES

Abdullah, M. A., Afzaal, M., Ismail, Z., Ahmad, A., Nazir, M. S., & Bhat, A. H. (2015). Comparative study on structural modification of Ceiba pentandra for oil sorption and palm oil mill effluent treatment. Desalin. *Water Treat.*, *54*(11), 3044–3053. doi:10.1080/19443994.2014.906326

Abu Bakar, S. N. H., Abu Hasan, H., Mohammad, A. W., Sheikh Abdullah, S. R., Haan, T. Y., Ngteni, R., & Yusof, K. M. M. (2018). A review of moving-bed biofilm reactor technology for palm oil mill effluent treatment. *Journal of Cleaner Production*, *171*, 1532–1545. doi:10.1016/j.jclepro.2017.10.100

Aghamohammadi, N., Aziz, H., Isa, M., & Zinatizadeh, A. (2007). Powdered activated carbon augmented activated sludge process for treatment of semi-aerobic landfill leachate using response surface methodology. *Bioresource Technology*, *98*(18), 3570–3578. doi:10.1016/j.biortech.2006.11.037 PMID:17280831

Ahmad, A. L., Ismail, S., Ibrahim, N., & Bhatia, S. (2003). Removal of suspended solids and residual oil from palm oil mill effluent. *Journal of Chemical Technology and Biotechnology (Oxford, Oxfordshire)*, *78*(9), 971–978. doi:10.1002/jctb.892

Ahmad, L., Sumathi, S., & Hameed, B. H. (2006). Coagulation of residue oil and suspended solid in palm oil mill effluent by chitosan, alum and PAC. *Chemical Engineering Journal*, *118*(1-2), 99–105. doi:10.1016/j.cej.2006.02.001

Alkaya, E., & Demirer, G. N. (2011). Anaerobic acidification of sugar-beet processing wastes: Effect of operational parameters. *Biomass and Bioenergy*, *35*(1), 32–39. doi:10.1016/j.biombioe.2010.08.002

Aziz, H. A., & Farraji, H. (2017). Phytoremediation of Landfill Leachate in Constructed Wetland Toxicological characteristics and Potential Impacts on the Environment. In *Biological Treatment* (pp. 143–161). Toxicological Characteristics and Potential Impacts on the Environment.

Aziz, H. A., Feng, C. T., & Bashir, M. J. K. (2013). Advanced treatment of landfill leachate effluent using membrane filtration. *J. Sci. Res. Environ. Sci.*, *1*, 36–43.

Aziz, S. Q., Aziz, D. H. A., Yusoff, M. S., Mojiri, A., & Amr, S. S. A. (2012). Adsorption isotherms in landfill leachate treatment using powdered activated carbon augmented sequencing batch reactor technique: Statistical analysis by response surface methodology. *International Journal of Chemical Reactor Engineering*, *10*(1). doi:10.1515/1542-6580.3112

Aziz, S. Q., Aziz, H. A., Yusoff, M. S., & Mohajeri, S. (2012). Removal of phenols and other pollutants from different landfill leachates using powdered activated carbon supplemented SBR technology. *Environmental Monitoring and Assessment*, *148*(10), 6147–6158. doi:10.100710661-011-2409-8 PMID:22068314

Bala, J. D., Lalung, J., & Ismail, N. (2015). Studies on the reduction of organic load from palm oil mill effluent (POME) by bacterial strains. *International Journal of Recycling of Organic Waste in Agriculture*, *4*(1), 1–10. doi:10.100740093-014-0079-6

Banat, I. M., Nigam, P., Singh, D., & Marchant, R. (1996). Microbial decolorization of textile-dyecontaining effluents: A review. *Bioresource Technology*, *58*(3), 217–227. doi:10.1016/S0960-8524(96)00113-7

Bohdziewicz, J., Neczaj, E., & Kwarciak, A. (2008). Landfill leachate treatment by means of anaerobic membrane bioreactor. *Desalination*, *221*(1-3), 559–565. doi:10.1016/j.desal.2007.01.117

Borja, R., Banks, C. J., & Sánchez, E. (1996). Anaerobic treatment of palm oil mill effluent in a two-stage up-flow anaerobic sludge blanket (UASB) system. *Journal of Biotechnology*, *45*(2), 125–135. doi:10.1016/0168-1656(95)00154-9

Cakir, F. Y., & Stenstrom, M. K. (2005). Greenhouse gas production: A comparison between aerobic and anaerobic wastewater treatment technology. *Water Research*, *39*(17), 4197–4203. doi:10.1016/j.watres.2005.07.042 PMID:16188289

Chan, Y. J., & Chong, M. F. (2019). Palm Oil Mill Effluent (POME) Treatment—Current Technologies, Biogas Capture and Challenges. In *Green Technologies for the Oil Palm Industry* (pp. 71–92). Singapore: Springer. doi:10.1007/978-981-13-2236-5_4

Chan, Y. J., Chong, M. F., & Law, C. L. (2010). Effects of Temperature on Aerobic Treatment of Anaerobically Digested Palm Oil Mill Effluent (POME). *Industrial & Engineering Chemistry Research*, *49*(15), 7093–7101. doi:10.1021/ie901952m

Chan, Y. J., Chong, M. F., & Law, C. L. (2017). Performance and kinetic evaluation of an integrated anaerobic–aerobic bioreactor in the treatment of palm oil mill effluent. *Environmental Technology*, *38*(8), 1005–1021. doi:10.1080/09593330.2016.1217053 PMID:27532518

Chin, K. K., Lee, S. W., & Mohammad, H. H. (1996). A study of palm oil mill effluent treatment using a pond system. *Water Science and Technology*, *34*(11), 119–123. doi:10.2166/wst.1996.0270

Chin, M. J., Poh, P. E., Tey, B. T., Chan, E. S., & Chin, K. L. (2013). Biogas from palm oil mill effluent (POME): Opportunities and challenges from Malaysia's perspective. *Renewable & Sustainable Energy Reviews*, *26*, 717–726. doi:10.1016/j.rser.2013.06.008

Choong, Y. Y., Chou, K. W., & Norli, I. (2018). Strategies for improving biogas production of palm oil mill effluent (POME) anaerobic digestion: A critical review. *Renewable & Sustainable Energy Reviews*, *82*, 2993–3006. doi:10.1016/j.rser.2017.10.036

Choorit, W., & Wisarnwan, P. (2007). Effect of temperature on the anaerobic digestion of palm oil mill effluent. *Electronic Journal of Biotechnology*, *10*(3), 376–385. doi:10.2225/vol10-issue3-fulltext-7

Costa, A. M., Alfaia, R. G. S. M., & Campos, J. G. (2019). Landfill leachate treatment in Brazil – An overview. *Journal of Environmental Management*, *232*, 110–116. doi:10.1016/j.jenvman.2018.11.006 PMID:30471544

Darajeh, N., Idris, A., Fard Masoumi, H. R., Nourani, A., Truong, P., & Sairi, N. A. (2016). Modeling BOD and COD removal from Palm Oil Mill Secondary Effluent in floating wetland by Chrysopogon zizanioides (L.) using response surface methodology. *Journal of Environmental Management*, *181*, 343–352. doi:10.1016/j.jenvman.2016.06.060 PMID:27393941

Doble, M., & Kumar, A. (2005). *Biotreatment of industrial effluents*. Elsevier.

Dubber, D., & Gray, N. F. (2010). Replacement of chemical oxygen demand (COD) with total organic carbon (TOC) for monitoring wastewater treatment performance to minimize disposal of toxic analytical waste. *J. Environ. Sci. Heal. Part A*, *45*(12), 1595–1600. doi:10.1080/10934529.2010.506116 PMID:20721800

Fadzil, N., Zainal, Z., & Abdullah, A. (2013). COD removal for palm oil mill secondary effluent by using UV/ferrioxalate/TiO2/O3 system. *Int. J. Emerg.*, *3*, 237–243.

Farraji, H., Qamaruz, Z. N., Hamidi, A., Aqeel, A.M., Amin, M., & Parsa, M. (2015a). Enhancing BOD/COD Ratio of POME Treatment in SBR System. *Appl. Mech. Mater.*, *802*, 437–442. Retrieved from www.scientific.net/AMM.802.437

Farraji, H., Qamaruz Zaman, N., & Mohajeri, P. (2015b). Waste Disposal:Sustainable Waste Treatments and Facility Siting Concerns. In H. Aziz & S. Amr (Eds.), *Control and Treatment of Landfill Leachate for Sanitary Waste Disposal* (pp. 43–74). IGI Global; doi:10.4018/978-1-4666-9610-5.ch003

Farraji, H., Zaman, N.Q., Abdul Aziz, H., Ashraf, M.A., Mojiri, A., & Mohajeri, P. (2015c). Landfill Leachate Treatment by Bentonite Augmented Sequencing Batch Reactor (SBR) System. *Appl. Mech. Mater.*, *802*, 466–471. Retrieved from www.scientific.net/AMM.802.466

Farraji, H., Zaman, N. Q., Aziz, H. A., & Sa'at, S. K. M. (2017a). Palm oil mill effluent treatment: Influence of zeolite, municipal wastewater and combined aerobic SBR system. AIP Conference Proceedings. AIP 1892, 1–7. doi:10.1063/1.5005693

Farraji, H., Zaman, N. Q., Aziz, H. A., & Sa'at, S. K. M. (2017b). Palm oil mill effluent and municipal wastewater co-treatment by zeolite augmented sequencing batch reactors: Turbidity removal. In AIP Conference Proceedings (Vol. 1892, No. 1, p. 040013). AIP Publishing. doi:10.1063/1.5005694

Foo, K. Y., & Hameed, B. H. (2009). An overview of landfill leachate treatment via activated carbon adsorption process. *Journal of Hazardous Materials*, *171*(1-3), 54–60. doi:10.1016/j.jhazmat.2009.06.038 PMID:19577363

Foo, K. Y., & Hameed, B. H. (2010). Insight into the applications of palm oil mill effluent: A renewable utilization of the industrial agricultural waste. *Renewable & Sustainable Energy Reviews*, *14*(5), 1445–1452. doi:10.1016/j.rser.2010.01.015

Grady, C. L. Jr, Daigger, G. T., Love, N. G., & Filipe, C. D. (2011). *Biological wastewater treatment*. CRC Press.

Guo, J.-S., Abbas, A. A., Chen, Y.-P., Liu, Z.-P., Fang, F., & Chen, P. (2010). Treatment of landfill leachate using a combined stripping, Fenton, SBR, and coagulation process. *Journal of Hazardous Materials*, *178*(1-3), 699–705. doi:10.1016/j.jhazmat.2010.01.144 PMID:20188464

Halim, A. A., Aziz, H. A., Johari, M. A. M., & Ariffin, K. S. (2010). Comparison study of ammonia and COD adsorption on zeolite, activated carbon and composite materials in landfill leachate treatment. *Desalination*, *262*(1-3), 31–35. doi:10.1016/j.desal.2010.05.036

Hansen, S. B., Padfield, R., Syayuti, K., Evers, S., Zakariah, Z., & Mastura, S. (2015). Trends in global palm oil sustainability research. *Journal of Cleaner Production*, *100*, 140–149. doi:10.1016/j.jclepro.2015.03.051

Islam, M. A., Yousuf, A., Karim, A., Pirozzi, M., Rahman Khan, M., & Wahid, Z. A. (2018). Bioremediation of palm oil mill effluent and lipid production by Lipomyces starkeyi: A combined approach. *Journal of Cleaner Production*, *172*, 1779–1787. doi:10.1016/j.jclepro.2017.12.012

Ismail, I., Hassan, M. A., Rahman, N. A. A., & Soon, C. S. (2010). Thermophilic biohydrogen production from palm oil mill effluent (POME) using suspended mixed culture. *Biomass and Bioenergy*, *34*(1), 42–47. doi:10.1016/j.biombioe.2009.09.009

Jouanneau, S., Recoules, L., Durand, M. J., Boukabache, A., Picot, V., Primault, Y., ... Thouand, G. (2014). Methods for assessing biochemical oxygen demand (BOD): A review. *Water Research*, *49*, 62–82. doi:10.1016/j.watres.2013.10.066 PMID:24316182

Kahar, A., Heryadi, E., Malik, L., Widarti, B. N., & Cahayanti, I. M. (2017). The study of seeding and acclimatization from leachate treatment in anaerobic bioreactor. *Journal of Engineering and Applied Sciences (Asian Research Publishing Network)*, *12*, 2610–2614.

Kamaruddin, M. A., Yusoff, M. S., Aziz, H. A., & Hung, Y. T. (2015). Sustainable treatment of landfill leachate. *Applied Water Science*, *5*(2), 113–126. doi:10.100713201-014-0177-7

Khemkhao, M., Techkarnjanaruk, S., & Phalakornkule, C. (2015). Simultaneous treatment of raw palm oil mill effluent and biodegradation of palm fiber in a high-rate CSTR. *Bioresource Technology*, *177*, 17–27. doi:10.1016/j.biortech.2014.11.052 PMID:25479389

Krishnan, S., Singh, L., Sakinah, M., Thakur, S., Wahid, Z. A., & Alkasrawi, M. (2016). Process enhancement of hydrogen and methane production from palm oil mill effluent using two-stage thermophilic and mesophilic fermentation. *International Journal of Hydrogen Energy*, *41*(30), 12888–12898. doi:10.1016/j.ijhydene.2016.05.037

Kuppusamy, P., Ilavenil, S., Srigopalram, S., Maniam, G. P., Yusoff, M. M., Govindan, N., & Choi, K. C. (2017). Treating of palm oil mill effluent using Commelina nudiflora mediated copper nanoparticles as a novel bio-control agent. *Journal of Cleaner Production*, *141*, 1023–1029. doi:10.1016/j.jclepro.2016.09.176

Kurniawan, T. A., Lo, W., Chan, G., & Sillanpää, M. E. (2010). Biological processes for treatment of landfill leachate. *Journal of Environmental Monitoring*, *12*(11), 2032–2047. doi:10.1039/c0em00076k PMID:20848046

Kutty, S., & Ngatenah, S. (2011). Removal of Zn (II), Cu (II), chemical oxygen demand (COD) and colour from anaerobically treated palm oil mill effluent (POME) using microwave incinerated rice. Int Conf Env. *Industry and Innovation*.

Lam, M. K., & Lee, K. T. (2011). Renewable and sustainable bioenergies production from palm oil mill effluent (POME): Win–win strategies toward better environmental protection. *Biotechnology Advances*, *29*(1), 124–141. doi:10.1016/j.biotechadv.2010.10.001 PMID:20940036

Li, H., Zhou, S., Sun, Y., Feng, P., & Li, J. (2009). Advanced treatment of landfill leachate by a new combination process in a full-scale plant. *Journal of Hazardous Materials*, *172*(1), 408–415. doi:10.1016/j.jhazmat.2009.07.034 PMID:19660862

Liang, Z., & Liu, J. (2008). Landfill leachate treatment with a novel process: Anaerobic ammonium oxidation (Anammox) combined with soil infiltration system. *Journal of Hazardous Materials*, *151*(1), 202–212. doi:10.1016/j.jhazmat.2007.05.068 PMID:17606322

Liew, W. L., Kassim, M. A., Muda, K., Loh, S. K., & Affam, A. C. (2015). Conventional methods and emerging wastewater polishing technologies for palm oil mill effluent treatment: A review. *Journal of Environmental Management*, *149*, 222–235. doi:10.1016/j.jenvman.2014.10.016 PMID:25463585

Lin, S. H., & Chang, C. C. (2000). Treatment of landfill leachate by combined electro-Fenton oxidation and sequencing batch reactor method. *Water Research*, *34*(17), 4243–4249. doi:10.1016/S0043-1354(00)00185-8

Liu, J., Zhang, P., Li, H., Tian, Y., Wang, S., Song, Y., ... Tian, Z. (2017). Denitrification of landfill leachate under different hydraulic retention time in a two-stage anoxic/oxic combined membrane bioreactor process: Performances and bacterial community. *Bioresource Technology*, *250*, 110–116. doi:10.1016/j.biortech.2017.11.026 PMID:29161569

Liyan, S., Youcai, Z., Weimin, S., & Ziyang, L. (2009). Hydrophobic organic chemicals (HOCs) removal from biologically treated landfill leachate by powder-activated carbon (PAC), granular-activated carbon (GAC) and biomimetic fat cell (BFC). *Journal of Hazardous Materials*, *163*(2-3), 1084–1089. doi:10.1016/j.jhazmat.2008.07.075 PMID:18752890

Loh, S. K., Nasrin, A. B., Azri, S. M., Adela, B. N., Muzzammil, N., Jay, T. D., ... Kaltschmitt, M. (2017). First report on Malaysia's experiences and development in biogas capture and utilization from palm oil mill effluent under the Economic Transformation Programme: Current and future perspectives. *Renewable & Sustainable Energy Reviews*, *74*, 1257–1274. doi:10.1016/j.rser.2017.02.066

Loukidou, M. X., & Zouboulis, A. I. (2001). Comparison of two biological treatment processes using attached-growth biomass for sanitary landfill leachate treatment. *Environ. Pollut.*, *111*(2), 273–281. doi:10.1016/S0269-7491(00)00069-5 PMID:11202731

Ma, A. N., & Ong, A. S. (1985). Pollution control in palm oil mills in Malaysia. *Journal of the American Oil Chemists' Society*, *62*(2), 261–266. doi:10.1007/BF02541389

Mahesh, S., Prasad, B., Mall, I. D., & Mishra, I. M. (2006). Electrochemical degradation of pulp and paper mill wastewater. Part 1. COD and color removal. *Industrial & Engineering Chemistry Research*, *45*(8), 2830–2839. doi:10.1021/ie0514096

Meky, N., Fujii, M., & Tawfik, A. (2017). Treatment of hypersaline hazardous landfill leachate using affled constructed wetland system: Effect of granular packing media and vegetation. *Environmental Technology*. doi:10.1080/09593330.2017.1397764 PMID:29073833

Metcalf, E., & Eddy, M. (2014). *Wastewater engineering: treatment and reuse* (5th ed.). New York: McGraw-Hill Education; doi:10.1016/0309-1708(80)90067-6

Mohammed, R. (2013). *Decolorisation of biologically treated palm oil mill effluent (POME) using adsorption technique. Int Ref. J Eng Sci.*

Mohammed, R., & Chong, F. (2014). Treatment and decolorization of biologically treated Palm Oil Mill Effluent (POME) using banana peel as novel biosorbent. *Journal of Environmental Management*, *132*, 237–249. doi:10.1016/j.jenvman.2013.11.031 PMID:24321284

Mojiri, A. (2014). *Co-treatment of landfill leachate and settled domestic wastewater using composite adsorbent in sequencing batch reactor* (Doctoral dissertation). Universiti Sains Malaysia.

Mojiri, A., Aziz, H. A., Aziz, S. Q., & Zaman, N. Q. (2012). Review on municipal landfill leachate and sequencing batch reactor (SBR) technique. *Arch. Des Sci.*, *65*, 22–31.

Mojiri, A., Aziz, H. A., Zaman, N. Q., Aziz, S. Q., & Zahed, M. A. (2014). Powdered ZELIAC augmented sequencing batch reactors (SBR) process for co-treatment of landfill leachate and domestic wastewater. *Journal of Environmental Management*, *139*, 1–14. doi:10.1016/j.jenvman.2014.02.017 PMID:24662109

Mojiri, A., Ziyang, L., Hui, W., Ahmad, Z., Tajuddin, R. M., Abu Amr, S. S., ... Farraji, H. (2017). Concentrated landfill leachate treatment with a combined system including electro-ozonation and composite adsorbent augmented sequencing batch reactor process. *Process Safety and Environmental Protection*, *111*, 253–262. doi:10.1016/j.psep.2017.07.013

Mojiri, A., Ziyang, L., Hui, W., & Gholami, A. (2019). Concentrated Landfill Leachate Treatment by Electro-Ozonation. In *Advanced Oxidation Processes (AOPs) in Water and Wastewater Treatment* (pp. 150–170). IGI Global. doi:10.4018/978-1-5225-5766-1.ch007

Mojiri, A., Ziyang, L., Tajuddin, R. M., Farraji, H., & Alifar, N. (2016). Co-treatment of landfill leachate and municipal wastewater using the ZELIAC/zeolite constructed wetland system. *Journal of Environmental Management*, *166*, 124–130. doi:10.1016/j.jenvman.2015.10.020 PMID:26496842

Najafpour, G. D., Zinatizadeh, A. A. L., Mohamed, A. R., Hasnain Isa, M., & Nasrollahzadeh, H. (2006). High-rate anaerobic digestion of palm oil mill effluent in an upflow anaerobic sludge-fixed film bioreactor. *Process Biochemistry (Barking, London, England)*, *41*(2), 370–379. doi:10.1016/j.procbio.2005.06.031

Neczaj, E., Kacprzak, M., Kamizela, T., Lach, J., & Okoniewska, E. (2008). Sequencing batch reactor system for the co-treatment of landfill leachate and dairy wastewater. *Desalination*, *222*(1-3), 404–409. doi:10.1016/j.desal.2007.01.133

O-Thong, S., Prasertsan, P., Intrasungkha, N., Dhamwichukorn, S., & Birkeland, N.-K. Å. (2008). Optimization of simultaneous thermophilic fermentative hydrogen production and COD reduction from palm oil mill effluent by Thermoanaerobacterium-rich sludge. *International Journal of Hydrogen Energy*, *33*(4), 1221–1231. doi:10.1016/j.ijhydene.2007.12.017

Ohimain, E. I., & Izah, S. C. (2017). A review of biogas production from palm oil mill effluents using different configurations of bioreactors. *Renewable & Sustainable Energy Reviews*, *70*, 242–253. doi:10.1016/j.rser.2016.11.221

Patel, R., & Suresh, S. (2008). Kinetic and equilibrium studies on the biosorption of reactive black 5 dye by Aspergillus foetidus. *Bioresource Technology*, *99*(1), 51–58. doi:10.1016/j.biortech.2006.12.003 PMID:17251011

Poh, P., & Chong, M. (2010). Biomethanation of Palm Oil Mill Effluent (POME) with a thermophilic mixed culture cultivated using POME as a substrate. *Chemical Engineering Journal*, *164*(1), 146–154. doi:10.1016/j.cej.2010.08.044

Renou, S., Givaudan, J. G., Poulain, S., Dirassouyan, F., & Moulin, P. (2008). Landfill leachate treatment: Review and opportunity. *Journal of Hazardous Materials*, *150*(3), 468–493. doi:10.1016/j.jhazmat.2007.09.077 PMID:17997033

Rupani, P., Singh, R., & Ibrahim, M. (2010). *Review of current palm oil mill effluent (POME) treatment methods: vermicomposting as a sustainable practice*. World Appl. Sci.

Saleh, A. F., Kamarudin, E., Yaacob, A. B., Yussof, A. W., & Abdullah, M. A. (2012). Optimization of biomethane production by anaerobic digestion of palm oil mill effluent using response surface methodology. Asia-Pacific. *Chemical Engineering Journal*, *7*, 353–360.

Satyawali, Y., & Balakrishnan, M. (2008). Wastewater treatment in molasses-based alcohol distilleries for COD and color removal: A review. *Journal of Environmental Management*, *86*(3), 481–497. doi:10.1016/j.jenvman.2006.12.024 PMID:17293023

Stephenes, A. O. (1970). *Acclimation of activated sludge to industrial wastes* (Unpublished master dissertation). McMaster University, Canada.

Tabassum, S., Zhang, Y., & Zhang, Z. (2015). An integrated method for palm oil mill effluent (POME) treatment for achieving zero liquid discharge – A pilot study. *Journal of Cleaner Production*, *95*, 148–155. doi:10.1016/j.jclepro.2015.02.056

Taifor, A. F., Zakaria, M. R., Yusoff, M. Z. M., Toshinari, M., Hassan, M. A., & Shirai, Y. (2017). Elucidating substrate utilization in biohydrogen production from palm oil mill effluent by Escherichia coli. *International Journal of Hydrogen Energy*, *42*(9), 5812–5819. doi:10.1016/j.ijhydene.2016.11.188

Tan, H. M., Poh, P. E., & Gouwanda, D. (2018). Resolving stability issue of thermophilic high-rate anaerobic palm oil mill effluent treatment via adaptive neuro-fuzzy inference system predictive model. *Journal of Cleaner Production*, *198*, 797–805. doi:10.1016/j.jclepro.2018.07.027

Tan, Y. H., Goh, P. S., Ismail, A. F., Ng, B. C., & Lai, G. S. (2017). Decolourization of aerobically treated palm oil mill effluent (AT-POME) using polyvinylidene fluoride (PVDF) ultrafiltration membrane incorporated with coupled zinc-iron oxide nanoparticles. *Chemical Engineering Journal*, *308*, 359–369. doi:10.1016/j.cej.2016.09.092

Tee, P.-F., Abdullah, M. O., Tan, I. A. W., Mohamed Amin, M. A., Nolasco-Hipolito, C., & Bujang, K. (2016). Performance evaluation of a hybrid system for efficient palm oil mill effluent treatment via an air-cathode, tubular upflow microbial fuel cell coupled with a granular activated carbon adsorption. *Bioresource Technology*, *216*, 478–485. doi:10.1016/j.biortech.2016.05.112 PMID:27268432

Torretta, V., Ferronato, N., Katsoyiannis, I., Tolkou, A., & Airoldi, M. (2016). Novel and Conventional Technologies for Landfill Leachates Treatment: A Review. *Sustainability*, *9*(1), 9. doi:10.3390u9010009

UNEP. (2005). *Solid Waste Management*. Retrieved from www.unep.or.jp/ietc/publications/spc/solid_waste_management/Vol_I/Binder1.pdf

Vakili, M., Rafatullah, M., Gholami, Z., & Farraji, H. (2016). Treatment of reactive dyes from water and wastewater through chitosan and its derivatives. In B. A. K. Mishra (Ed.), *Smart Materials for Waste Water Applications* (pp. 347–377). Wiley. doi:10.1002/9781119041214.ch14

Vijayaraghavan, K., & Ahmad, D. (2006). Biohydrogen generation from palm oil mill effluent using anaerobic contact filter. *International Journal of Hydrogen Energy*, *31*(10), 1284–1291. doi:10.1016/j.ijhydene.2005.12.002

Vymazal, J. (2009). The use constructed wetlands with horizontal sub-surface flow for various types of wastewater. *Ecological Engineering*, *35*(1), 1–17. doi:10.1016/j.ecoleng.2008.08.016

Wong, K. A., Lam, S. M., & Sin, J. C. (2018). Wet chemically synthesized ZnO structures for photodegradation of pre-treated palm oil mill effluent and antibacterial activity. *Ceramics International*. doi:10.1016/j.ceramint.2018.10.078

Yalmaz, G., & Öztürk, I. (2001). Biological ammonia removal from anaerobically pre-treated landfill leachate in sequencing batch reactors (SBR). *Water Science and Technology*, *43*(3), 307–314. doi:10.2166/wst.2001.0151 PMID:11381921

Zahan, K. A., & Kano, M. (2018). Biodiesel Production from Palm Oil, Its By-Products and Mill Effluent: A Review. *Energies*, *11*(8), 2132. doi:10.3390/en11082132

Zahrim, A., Rachel, F., Menaka, S., Su, S., & Melvin, F. (2009). Decolourisation of anaerobic palm oil mill effluent via activated sludge-granular activated carbon. *World Applied Sciences Journal*, 126–129.

Zawani, Z., Chuah-Abdullah, L., Ahmadun, F. R., & Abdan, K. (2013). Acclimatization process of microorganisms from activated sludge in Kenaf-Retting Wastewater. In Developments in Sustainable Chemical and Bioprocess Technology (pp. 59-64). Springer. doi:10.1007/978-1-4614-6208-8_8

Zhang, Q. Q., Tian, B. H., Zhang, X., Ghulam, A., Fang, C. R., & He, R. (2013). Investigation on characteristics of leachate and concentrated leachate in three landfill leachate treatment plants. *Waste Management (New York, N.Y.)*, *33*(11), 2277–2286. doi:10.1016/j.wasman.2013.07.021 PMID:23948053

Zinatizadeh, A. A. L., Mohamed, A. R., Abdullah, A. Z., Mashitah, M. D., Hasnain Isa, M., & Najafpour, G. D. (2006). Process modeling and analysis of palm oil mill effluent treatment in an up-flow anaerobic sludge fixed film bioreactor using response surface methodology (RSM). *Water Research*, *40*(17), 3193–3208. doi:10.1016/j.watres.2006.07.005 PMID:16949124

Zolfaghari, M., Jardak, K., Drogui, P., Brar, S. K., Buelna, G., & Dubé, R. (2016). Landfill leachate treatment by sequential membrane bioreactor and electro-oxidation processes. *Journal of Environmental Management*, *184*, 318–326. doi:10.1016/j.jenvman.2016.10.010 PMID:27733297

Compilation of References

Abbasi, J. S., Javaid, N., Gull, S., Islam, S., Imran, M., Hassan, N., & Nasr, K. (2017). Balanced Energy Efficient Rectangular Routing Protocol for Underwater Wireless Sensor Networks. *IEEE Access: Practical Innovations, Open Solutions.*

Abdulkader, O., Bamhdi, A. M., Thayananthan, V., Jambi, K., & Alrasheedi, M. (2018, February). A novel and secure smart parking management system (SPMS) based on integration of WSN, RFID, and IoT. In *2018 15th Learning and Technology Conference (L&T)* (pp. 102-106). IEEE.

Abdullah, A., Al Enazi, S., & Damaj, I. (2016). AgriSys: A Smart and Ubiquitous Controlled-Environment Agriculture System. *International Conference on Big Data and Smart City.* 10.1109/ICBDSC.2016.7460386

Abdullahi, H. S., Sheriff, R. E., & Mahieddine, F. (2017). Convolution neural network in precision agriculture for plant image recognition and classification. *2017 Seventh International Conference on Innovative Computing Technology (INTECH),* 1–3. 10.1109/INTECH.2017.8102436

Abdullah, M. A., Afzaal, M., Ismail, Z., Ahmad, A., Nazir, M. S., & Bhat, A. H. (2015). Comparative study on structural modification of Ceiba pentandra for oil sorption and palm oil mill effluent treatment. Desalin. *Water Treat., 54*(11), 3044–3053. doi:10.1080/19443994.2014.906326

Abu Bakar, S. N. H., Abu Hasan, H., Mohammad, A. W., Sheikh Abdullah, S. R., Haan, T. Y., Ngteni, R., & Yusof, K. M. M. (2018). A review of moving-bed biofilm reactor technology for palm oil mill effluent treatment. *Journal of Cleaner Production, 171,* 1532–1545. doi:10.1016/j.jclepro.2017.10.100

Aburukba, R., Al-Ali, A. R., Kandil, N., & AbuDamis, D. (2016). Configurable ZigBee - based control system for people with multiple disabilities in smart homes. IEEE.

Accenture. (2016). *Industrial Internet Insights Report.* Author.

Adler, L. (2016). *How smart city Barcelona brought the internet of things to life.* Retrieved July 3, 2017, from http://datasmart.ash.harvard.edu/news/article/how-smart-city-barcelona-brought-the-internet-of-things-to-life-789

Aghamohammadi, N., Aziz, H., Isa, M., & Zinatizadeh, A. (2007). Powdered activated carbon augmented activated sludge process for treatment of semi-aerobic landfill leachate using response surface methodology. *Bioresource Technology, 98*(18), 3570–3578. doi:10.1016/j.biortech.2006.11.037 PMID:17280831

Ahmad, A. L., Ismail, S., Ibrahim, N., & Bhatia, S. (2003). Removal of suspended solids and residual oil from palm oil mill effluent. *Journal of Chemical Technology and Biotechnology (Oxford, Oxfordshire), 78*(9), 971–978. doi:10.1002/jctb.892

Ahmadi, P., Technology, A., & Agriculture, F. (2017). *Early Detection of Ganoderma Basal Stem Rot of Oil Palms Using Artificial Neural Network Spectral Analysis.* Academic Press.

Ahmad, L., Sumathi, S., & Hameed, B. H. (2006). Coagulation of residue oil and suspended solid in palm oil mill effluent by chitosan, alum and PAC. *Chemical Engineering Journal*, *118*(1-2), 99–105. doi:10.1016/j.cej.2006.02.001

Ahvenniemi, H., Huovila, A., Pinto-Seppa, I., & Airaksinen, M. (2017). What are the differences between sustainable and smart cities? *Cities (London, England)*, *60*, 234–245. doi:10.1016/j.cities.2016.09.009

Aina, Y. A. (2017). Achieving smart sustainable cities with GeoICT support: The Saudi evolving smart cities. *Cities (London, England)*, *71*, 49–58. doi:10.1016/j.cities.2017.07.007

Akkaya, K., Guvenc, I., Aygun, R., Pala, N., & Kadri, A. (2015, March). IoT-based occupancy monitoring techniques for energy-efficient smart buildings. In Wireless Communications and Networking Conference Workshops (WCNCW), 2015 IEEE (pp. 58-63). IEEE. doi:10.1109/WCNCW.2015.7122529

Akyildiz, I. F. (2005). Underwater Acoustic Sensor Networks (UW-ASN). *BWN Laboratory*. Retrieved from http://bwn.ece.gatech.edu/Underwater/index.html

Akyildiz, I. F., Pompili, D., & Melodia, T. (2005). Underwater acoustic sensor networks: research challenges. Ad Hoc Networks, 257–279.

Alahakoon, D., & Yu, X. (2016). Smart electricity meter data intelligence for future energy systems: A survey. *IEEE Transactions on Industrial Informatics*, *12*(1), 425–436. doi:10.1109/TII.2015.2414355

Alavi, A. H., Jiaob, P., Buttlar, W. G., & Lajnef, N. (2018). Internet of Things-Enabled Smart Cities: State-of-the-Art and Future Trends. *Measurement*, *129*, 589–606. doi:10.1016/j.measurement.2018.07.067

Albino, V., Berardi, U., & Dangelico, R. M. (2015). Smart cities: Definitions, dimensions, performance, and initiatives. *Journal of Urban Technology*, *22*(1), 3–21. doi:10.1080/10630732.2014.942092

Aleyadeh. (2018). *An Iot-Based Architecture For Waste Management*. Academic Press.

Al-Fuqaha, Guizani, Mohammadi, Aledhari, & Ayyash. (2015). Internet of Things: Architectures, Protocols, and Applications. IEEE Communication Survey and Tutorials, 17(4).

Al-Hamadi & Chen. (2017). *Trust-Based Decision Making for Health IoT Systems. IEEE Internet of Things Journal, 4(5).*

Ali, T., Jung, L. T., & Faye, I. (2014). Diagonal and Vertical Routing Protocol for Underwater Wireless Sensor Network. *International Conference on Innovation, Management and Technology Research*, 372 – 379. 10.1016/j.sbspro.2014.03.690

Alkaya, E., & Demirer, G. N. (2011). Anaerobic acidification of sugar-beet processing wastes: Effect of operational parameters. *Biomass and Bioenergy*, *35*(1), 32–39. doi:10.1016/j.biombioe.2010.08.002

Alleven, M. (2014). Sigfox launches IoT network in 10 UK cities. *Fierce Wireless Tech*. Retrieved from https://www.fiercewireless.com/tech/sigfox-launches-iot-network-10-uk-cities

Al-Masri. (2018). A Serverless IoT Architecture for Smart Waste Management Systems. *IEEE International Conference on Industrial Internet (ICII).*

Almeida, V. A. F., Doneda, D., & Moreira Da Costa, E. (2018). Humane smart cities: The need for governance. *IEEE Internet Computing*, *22*(2), 91–95. doi:10.1109/MIC.2018.022021671

Almotiri, S. H., Khan, M. A., & Alghamdi, M. A. (2016). Mobile Health (m-Health) System in the Context of IoT. *2016 IEEE 4th International Conference on Future Internet of Things and Cloud Workshops (FiCloudW)*, 39-42.

Alphonsa, A., & Ravi, G. (2016). Earthquake early warning system by IoT using Wireless sensor networks. *2016 International Conference on Wireless Communications, Signal Processing and Networking (WiSPNET)*.10.1109/WiSPNET.2016.7566327

Alphonse, A. S., & Dharma, D. (2018). Novel directional patterns and a Generalized Supervised Dimension Reduction System (GSDRS) for facial emotion recognition. *Multimedia Tools and Applications*, *77*(8), 9455–9488. doi:10.100711042-017-5141-8

Ammar, M., Russello, G., & Crispo, B. (2018). Internet of Things: A survey on the security of IoT frameworks. *Journal of Information Security and Applications*, *38*, 8–27. doi:10.1016/j.jisa.2017.11.002

Anagnostopoulos, T., Zaslavsky, A., & Medvedev, A. (2015). Robust waste collection exploiting cost efficiency of IoT potentiality in smart cities. In *Recent Advances in Internet of Things (RIoT), 2015 International Conference on* (pp. 1-6). IEEE.

Anagnostopoulos, T., Zaslavsky, A., Kolomvatsos, K., Medvedev, A., Amirian, P., Morley, J., & Hadjieftymiades, S. (2017). Challenges and Opportunities of Waste Management in IoT-Enabled Smart Cities: A Survey. *IEEE Transactions on Sustainable Computing*, *2*(3), 275–289. doi:10.1109/TSUSC.2017.2691049

Anand, S., & Routray, S. K. (2017, March). Issues and challenges in healthcare narrowband IoT. In *Inventive Communication and Computational Technologies (ICICCT), 2017 International Conference on* (pp. 486-489). IEEE. 10.1109/ICICCT.2017.7975247

Anggoro, A. G. P. (2018). *Monitoring server room temperature remotely in real time using raspberry pi and firebase* (Doctoral dissertation). Universitas muhammadiyah surakarta.

Ankitha, S., & Balajee, M. (2016). Security And Privacy Issues in IoT. *SCIREA Journal of Agriculture*, *1*(2), 135–142.

Anthopoulos, L. (2017). Smart utopia vs smart reality: Learning by experience from 10 smart city cases. *Cities (London, England)*, *63*, 128–148. doi:10.1016/j.cities.2016.10.005

Antonelli, G., & Cappiello, G. (Eds.). (2016). *Smart Development in Smart Communities*. Taylor & Francis. doi:10.4324/9781315641850

Ardito, L., Ferraris, A., Petruzzelli, A. M., Bresciani, S., & Del Giudice, M. (2018). The role of universities in the knowledge management of smart city projects. *Technological Forecasting and Social Change*. doi:10.1016/j.techfore.2018.07.030

Arebey, M., Hannan, M. A., Basri, H., Begum, R. A., & Abdullah, H. (2010, June). Solid waste monitoring system integration based on RFID, GPS and camera. In *Intelligent and Advanced Systems (ICIAS), 2010 International Conference on* (pp. 1-5). IEEE.

Arora, R. U. (2018). Financial sector development and smart cities: The Indian case. *Sustainable Cities and Society*, *42*, 52–58. doi:10.1016/j.scs.2018.06.013

Atlam, H. F., Alenezi, A., Alassafi, M. O., & Wills, G. (2018). Blockchain with Internet of Things: Benefits, challenges, and future directions. *International Journal of Intelligent Systems and Applications*, *10*(6), 40–48. doi:10.5815/ijisa.2018.06.05

Atzori, L., Iera, A., & Morabito, G. (2010). The internet of things: A survey. *Computer Networks*, *54*(15), 2787–2805. doi:10.1016/j.comnet.2010.05.010

Ayaz, M., Abdullah, A., & Jung, L.T. (2010). Temporary cluster based routing for underwater wireless sensor networks. *Informat Tech (ITSim) Internat Symposin*, 1009–14.

Ayoub, W., Samhat, A. E., Nouvel, F., Mroue, M., & Prévotet, J. C. (2018). Internet of Mobile Things: Overview of LoRaWAN, DASH7, and NB-IoT in LPWANs standards and Supported Mobility. *IEEE Communications Surveys & Tutorials*.

Aziz, H. A., & Farraji, H. (2017). Phytoremediation of Landfill Leachate in Constructed Wetland Toxicological characteristics and Potential Impacts on the Environment. In *Biological Treatment* (pp. 143–161). Toxicological Characteristics and Potential Impacts on the Environment.

Aziz, H. A., Feng, C. T., & Bashir, M. J. K. (2013). Advanced treatment of landfill leachate effluent using membrane filtration. *J. Sci. Res. Environ. Sci.*, *1*, 36–43.

Aziz, S. Q., Aziz, D. H. A., Yusoff, M. S., Mojiri, A., & Amr, S. S. A. (2012). Adsorption isotherms in landfill leachate treatment using powdered activated carbon augmented sequencing batch reactor technique: Statistical analysis by response surface methodology. *International Journal of Chemical Reactor Engineering*, *10*(1). doi:10.1515/1542-6580.3112

Aziz, S. Q., Aziz, H. A., Yusoff, M. S., & Mohajeri, S. (2012). Removal of phenols and other pollutants from different landfill leachates using powdered activated carbon supplemented SBR technology. *Environmental Monitoring and Assessment*, *148*(10), 6147–6158. doi:10.100710661-011-2409-8 PMID:22068314

Babu, Naidu, &Meenakshi. (2018). Earthquake Detection and Alerting Using IoT. *International Journal of Engineering Science Invention*, *7*(5), 14-18. Retrieved from www.ijesi.org

Baimel, D., Tapuchi, S., & Baimel, N. (2016). Smart grid communication technologies- overview, research challenges and opportunities. *International Symposium on Power Electronics, Electrical Drives, Automation and Motion (SPEEDAM)*. 10.1109/SPEEDAM.2016.7526014

Baimel, D., Tapuchi, S., & Baimel, N. (2016). *Smart Grid Communication Technologies*. Scientific Research Publishing. doi:10.4236/jpee.2016.48001

Bala, J. D., Lalung, J., & Ismail, N. (2015). Studies on the reduction of organic load from palm oil mill effluent (POME) by bacterial strains. *International Journal of Recycling of Organic Waste in Agriculture*, *4*(1), 1–10. doi:10.100740093-014-0079-6

Banat, I. M., Nigam, P., Singh, D., & Marchant, R. (1996). Microbial decolorization of textile-dyecontaining effluents: A review. *Bioresource Technology*, *58*(3), 217–227. doi:10.1016/S0960-8524(96)00113-7

Banu, J. R., Raj, E., Kaliappan, S., Beck, D., & Yeom, I. T. (2007). Solid state biomethanation of fruit wastes. *Journal of Environmental Biology*, *28*(4), 741–745. PMID:18405106

Banyan Water's Data-Driven Water Management Solutions Save Businesses And Organizations 1.75 Billion Gallons Of Water To Date. (2015, September 10). Retrieved from https://www.prnewswire.com/news-releases/banyan-waters-data-driven-water-management-solutions-save-businesses-and-organizations-175-billion-gallons-of-water-to-date-300139938.html

Barnaghi, P., & Sheth, A. (2014, September 9). *The Internet of Things- The Story so far*. Retrieved from http://iot.ieee.org/newsletter/september-2014/the-internet-of-things-the-story-so-far.html

Barton, J. R., Dalley, D., & Patel, V. S. (1996). Life Cycle Assessment For WasteManagement. *Waste Management (New York, N.Y.)*, *16*(1-3), 35–50. doi:10.1016/S0956-053X(96)00057-8

Battarra, R., Gargiulo, C., Tremiterra, M. R., & Zucar, F. (2018). Smart mobility in Italian metropolitan cities: A comparative analysis through indicators and actions. *Sustainable Cities and Society*, *41*, 556–567. doi:10.1016/j.scs.2018.06.006

Bayindiretal, R. (2016). Renewable and Sustainable Energy Reviews.*Smart grid technologies and applications. Renewable & Sustainable Energy Reviews*, *66*, pp499–pp516. doi:10.1016/j.rser.2016.08.002

Beigl, P., & Salhofer, S. (2004). Comparison Of Ecological Effects And Costs of Communal Waste Management System. Resources. *Conservation Andrecycling*, *41*(2), 83–102. doi:10.1016/j.resconrec.2003.08.007

Bélissent, J. (2010). *Getting clever about smart cities: New opportunities require new business models*. Retrieved July 10, 2017, from http://193.40.244.77/iot/wp-content/uploads/2014/02/getting_clever_about_smart_cities_new_opportunities.pdf

Bellavista, P., Cardone, G., Corradi, A., & Foschini, L. (2013). Convergence of MANET and WSN in IoT urban scenarios. *IEEE Sensors Journal*, *13*(10), 3558–3567. doi:10.1109/JSEN.2013.2272099

Bello, H., Jian, X., Wei, Y., & Chen, M. (2018). Energy-Delay Evaluation and Optimization for NB-IoT PSM with Periodic Uplink Reporting. *IEEE Access: Practical Innovations, Open Solutions.*

Benbrahim, M., Daoudi, A., Benjelloun, K., & Ibenbrahim, A. (2007). Discrimination of Seismic Signals Using Artificial Neural Networks. *International Journal of Computer, Electrical, Automation, Control and Information Engineering*, *1*(4).

Benghanem, M. (2010). *RETRACTED: A low cost wireless data acquisition system for weather station monitoring*. Elsevier.

Benini, L. (2013). Designing next-generation smart sensor hubs for the Internet-of-Things. *5th IEEE International Workshop on Advances in Sensors and Interfaces IWASI.* 10.1109/IWASI.2013.6576075

Bersch, S. D., Azzi, D., Khusainov, R., Achumba, I. E., & Ries, J. (2014). Sensor data acquisition and processing parameters for human activity classification. *Sensors (Basel)*, *14*(3), 4239–4270. doi:10.3390140304239 PMID:24599189

Berst, J., Enbysk, L., & Williams, C. (2013). Smart cities readiness guide: The planning manual for building tomorrow's cities today. *Smart Cities Council.* Retrieved from http://www.corviale.com/wp-content/uploads/2013/12/guida-per-le-smart-city.pdf

Bhalerao, Ghosh, Mhatre, Vadgaonkar, Wajge, & Shinde. (2017). IoT Based Smart Waste Management System. *IJCTA*, *10*(8), 607–611.

Bharadwaj, A. S., Rego, R., & Chowdhury, A. (2016, December). IoT based solid waste management system: A conceptual approach with an architectural solution as a smart city application. In *India Conference (INDICON), 2016 IEEE Annual* (pp. 1-6). IEEE. 10.1109/INDICON.2016.7839147

Bhardwaj, R., Sharma, A. L., & Kumar, A. (2016). Multi – parameter algorithm for Earthquake Early Warning. *Geomatics, Natural Hazards & Risk*, *7*(4), 1242–1264. doi:10.1080/19475705.2015.1069409

Bhargava, N., Katiyar, V. K., Sharma, M. L., & Pradhan, P. (2009). Earthquake Prediction through Animal Behavior: A Review. *Indian Journal of Biomechanics,* 7 – 8.

Bhavani, D. S., & Reddy, R. R. (2015, December). Identification and Characterization of Particulate Matterin Hyderabad City. *International Journal of Innovative Research in Science*, *4*(11). doi:10.15680/IJIRSET.2015.0412160

Bifulco, F., Tregua, M., Amitrano, C. C., & D'Auria, A. (2016). ICT and sustainability in smart cities management. *International Journal of Public Sector Management*, *29*(2), 132–147. doi:10.1108/IJPSM-07-2015-0132

Biswas, K., & Muthukkumarasamy, V. (2016). Securing smart cities using blockchain technology. In *High Performance Computing and Communications; IEEE 14th International Conference on Smart City; IEEE 2nd International Conference on Data Science and Systems (HPCC/SmartCity/DSS), 2016 IEEE 18th International Conference on* (pp. 1392-1393). IEEE. 10.1109/HPCC-SmartCity-DSS.2016.0198

Bohdziewicz, J., Neczaj, E., & Kwarciak, A. (2008). Landfill leachate treatment by means of anaerobic membrane bioreactor. *Desalination*, *221*(1-3), 559–565. doi:10.1016/j.desal.2007.01.117

Bohli, Kurpatov, & Schmidt. (2015). *Selective Decryption of Outsourced IoT Data*. IEEE.

Boman, J., Taylor, J., & Ngu, A. H. (2014, October). Flexible IoT middleware for integration of things and applications. In *Collaborative Computing: Networking, Applications and Worksharing (CollaborateCom), 2014 International Conference on* (pp. 481-488). IEEE. 10.4108/icst.collaboratecom.2014.257533

Borgohain, T., Kumar, U., & Sanyal, S. (2015). *Survey of security and privacy issues of internet of things*. arXiv preprint arXiv:1501.02211

Borja, R., Banks, C. J., & Sánchez, E. (1996). Anaerobic treatment of palm oil mill effluent in a two-stage up-flow anaerobic sludge blanket (UASB) system. *Journal of Biotechnology*, *45*(2), 125–135. doi:10.1016/0168-1656(95)00154-9

Borsekova, K., Korónya, S., Vaňovàb, A., & Vitálišov, K. (2018). Functionality between the size and indicators of smart cities: A research challenge with policy implications. *Cities (London, England)*, *78*, 17–26. doi:10.1016/j.cities.2018.03.010

Bouskela, M., Casseb, M., Bassi, S., Luca, C. D., & Facchina, M. (2016). The road toward smart cities: Migrating from traditional city management to the smart city. *Inter-American Development Bank (IDB)*. Retrieved June 5, 2017, from https://publications.iadb.org/bitstream/handle/11319/7743/The-Road-towards-Smart-Cities-Migrating-from-Traditional-City-Management-to-the-Smart-City.pdf?sequence=3

Bousquet, C. (2017, April 19). *How Cities Are Using the Internet of Things to Map Air Quality*. Retrieved November 11, 2018, from https://datasmart.ash.harvard.edu/news/article/how-cities-are-using-the-internet-of-things-to-map-air-quality-1025

Braun, T., Fung, B. C. M., Iqbal, F., & Shah, B. (2018). Security and privacy challenges in smart cities. *Sustainable Cities and Society*, *39*, 499–507. doi:10.1016/j.scs.2018.02.039

Breiman, L. (2001). Random forests. *Machine Learning*, *45*(1), 5–32. doi:10.1023/A:1010933404324

Brown, E. (2016). *21 Open Source Projects for IoT*. Retrieved from Linux.com

Buchanan R. C., Newell, D. K., Evans, S. C., & Miller, D. R. (2014). Induced Seismicity: The Potential for Triggered Earthquakes in Kansas. *Kansas Geological Survey, Public Information Circular, 36.*

Burgess, T. (2016). Water: At what cost? The state of the world's water 2016. *WaterAid*. Retrieved July 4, 2017, from https://www.wateraid.org/uk/~/media/Publications/Water--At-What-Cost--The-State-of-the-Worlds-Water-2016.pdf?la=en-GB

Business Insider. (2015). *The Enterprise Internet of Things Market*. Retrieved from https://www.businessinsider.in/The-Corporate-Internet-Of-Things-Will-Encompass-More-Devices-Than-The-Smartphone-And-Tablet-Markets-Combined/articleshow/45483725.cms

Byun, J. H., Kim, S. Y., Sa, J. H., Shin, Y. T., Kim, S. P., & Kim, J. B. (2016). Smart city implementation models based on IoT(Internet of Things) technology. *Proceedings of Advanced Science and Technology Letters*, *129*, 209–212. doi:10.14257/astl.2016.129.41

Caesarendra, W., & Tjahjowidodo, T. (2017). A review of feature extraction methods in vibration-based condition monitoring and its application for degradation trend estimation of low-speed slew bearing. *Machines*, *5*(4), 21. doi:10.3390/machines5040021

Cakir, F. Y., & Stenstrom, M. K. (2005). Greenhouse gas production: A comparison between aerobic and anaerobic wastewater treatment technology. *Water Research*, *39*(17), 4197–4203. doi:10.1016/j.watres.2005.07.042 PMID:16188289

Cao, X., & Li, Y. (2018, January). Data Collection and Network Architecture Analysis in Internet of Vehicles Based on NB-IoT. In *2018 International Conference on Intelligent Transportation, Big Data & Smart City (ICITBS)* (pp. 157-160). IEEE. 10.1109/ICITBS.2018.00048

Carreiro, Antunes, & Jorge. (2012). Energy Smart House Architecture for a Smart Grid. *IEEE International Symposium on Sustainable Systems and Technology (ISSST).*

Central Pollution Control Board, Ministry of Environment, Forests and Climate Change. (2014). *National Air Quality Index*. Retrieved June 2018, from http://www.indiaenvironmentportal.org.in/files/file/Air%20Quality%20Index.pdf

Chahuara, P., Fleury, A., Portet, F., & Vacher, M. (2016). On-line human activity recognition from audio and home automation sensors: Comparison of sequential and non-sequential models in realistic Smart Homes 1. *Journal of Ambient Intelligence and Smart Environments*, *8*(4), 399–422. doi:10.3233/AIS-160386

Chandu, Kumar, Prabhukhanolkar, Anish, & Rawal. (2017). *Design and Implementation of Hybrid Encryption for Security of IOT Data*. IEEE.

Chan, Y. J., & Chong, M. F. (2019). Palm Oil Mill Effluent (POME) Treatment—Current Technologies, Biogas Capture and Challenges. In *Green Technologies for the Oil Palm Industry* (pp. 71–92). Singapore: Springer. doi:10.1007/978-981-13-2236-5_4

Chan, Y. J., Chong, M. F., & Law, C. L. (2010). Effects of Temperature on Aerobic Treatment of Anaerobically Digested Palm Oil Mill Effluent (POME). *Industrial & Engineering Chemistry Research*, *49*(15), 7093–7101. doi:10.1021/ie901952m

Chan, Y. J., Chong, M. F., & Law, C. L. (2017). Performance and kinetic evaluation of an integrated anaerobic–aerobic bioreactor in the treatment of palm oil mill effluent. *Environmental Technology*, *38*(8), 1005–1021. doi:10.1080/09593330.2016.1217053 PMID:27532518

Chatterji, T. (2018). Digital urbanism in a transitional economy– A review of India's municipal e-governance policy. *Journal of Asian Public Policy*, *11*(3), 334–349. doi:10.1080/17516234.2017.1332458

Chatzigiannakis, I., Vitaletti, A., & Pyrgelis, A. (2016). A privacy-preserving smart parking system using an IoT elliptic curve based security platform. *Computer Communications*, *89*, 165–177. doi:10.1016/j.comcom.2016.03.014

Chen, C.C. (2010). A performance evaluation of MSW management practice in Taiwan. *Resour. Conserv. Recycl.*, *54*(12).

Chen, Yeh, Hsieh, & Chang. (2010). Communication Infrastructure of Smart Grid. *Proceedings* of *the 4th International Symposium* on *Communications, Control* and *Signal Processing.*

Chin, K. K., Lee, S. W., & Mohammad, H. H. (1996). A study of palm oil mill effluent treatment using a pond system. *Water Science and Technology*, *34*(11), 119–123. doi:10.2166/wst.1996.0270

Chin, M. J., Poh, P. E., Tey, B. T., Chan, E. S., & Chin, K. L. (2013). Biogas from palm oil mill effluent (POME): Opportunities and challenges from Malaysia's perspective. *Renewable & Sustainable Energy Reviews*, *26*, 717–726. doi:10.1016/j.rser.2013.06.008

Choong, Y. Y., Chou, K. W., & Norli, I. (2018). Strategies for improving biogas production of palm oil mill effluent (POME) anaerobic digestion: A critical review. *Renewable & Sustainable Energy Reviews*, *82*, 2993–3006. doi:10.1016/j.rser.2017.10.036

Choorit, W., & Wisarnwan, P. (2007). Effect of temperature on the anaerobic digestion of palm oil mill effluent. *Electronic Journal of Biotechnology*, *10*(3), 376–385. doi:10.2225/vol10-issue3-fulltext-7

Chourabi, H., Nam, T., Walker, S., Gil-Garcia, J. R., Mellouli, S., Nahon, K., ... Scholl, H. J. (2012). Understanding smart cities: An integrative framework. In *Proceedings of 45th Hawaii International Conference on System Sciences*. Maui, HI, USA: IEEE.

Cho, Y. W., Kim, J. M., & Park, Y. Y. (2016). Design and Implementation of Marine Elevator Safety Monitoring System based on Machine Learning. *Indian Journal of Science and Technology*, *9*(S1).

Chui, M., Loffler, M., & Roberts, R. (2014). The Internet of Things. *McKinsey Quarterly*. Retrieved from https://www.mckinsey.com/industries/high-tech/our-insights/the-internet-of-things

CitiesS. (n.d.). Retrieved from https://www.gsma.com/iot/smart-cities/

CITI-SENSE. (2016, September). *Home Page*. Retrieved November 2018, from CITI-SENSE: http://www.citi-sense.eu/

Climent, S., Sanchez, A., Capella, J. V., Meratnia, N., & Serrano, J. J. (2014). Underwater Acoustic Wireless Sensor Networks: Advances and Future Trends in Physical, MAC and Routing Layers. *Sensors (Basel)*, *14*(1), 795–833. doi:10.3390140100795 PMID:24399155

Cocchia, A. (2014). Smart and digital city: A systematic literature review. In R. P. Dameri & C. R. Sabroux (Eds.), *Smart city: How to create public and economic value with high technology in urban space* (pp. 13–43). Springer International Publishing.

Coconuts Singapore. (2014). *Western Singapore becomes test-bed for smart city solutions*. Retrieved from https://coconuts.co/singapore/news/western-singapore-becomes-test-bed-smart-city-solutions

Cominola, A., Giuliani, M., Piga, D., Castelletti, A., & Rizzoli, A. E. (2015). Benefits and challenges of using smart meters for advancing residential water demand modeling and management: A review. *Environmental Modelling & Software*, *72*, 198–214. doi:10.1016/j.envsoft.2015.07.012

Costa, A. M., Alfaia, R. G. S. M., & Campos, J. G. (2019). Landfill leachate treatment in Brazil – An overview. *Journal of Environmental Management*, *232*, 110–116. doi:10.1016/j.jenvman.2018.11.006 PMID:30471544

Cretu, N., & Pop, M. (2008). Higher order statistics in signal processing and nanometric size analysis. *Journal of Optoelectronics and Advanced Materials*, *10*(12), 3292–3299.

Cui, Xie, Qu, Gao, & Yang. (2018). Security and Privacy in Smart Cities: Challenges and Opportunities. *IEEE Access*.

DaCosta, C. A., Pasluosta, C. F., Eskofier, B., DaSilva, D. B., & DaRosaRighi, R. (2018). Internet of Health Things: Toward intelligent vital signs monitoring in hospital wards. *Artificial Intelligence in Medicine, 89*, 61 - 69.

Darajeh, N., Idris, A., Fard Masoumi, H. R., Nourani, A., Truong, P., & Sairi, N. A. (2016). Modeling BOD and COD removal from Palm Oil Mill Secondary Effluent in floating wetland by Chrysopogon zizanioides (L.) using response surface methodology. *Journal of Environmental Management*, *181*, 343–352. doi:10.1016/j.jenvman.2016.06.060 PMID:27393941

Daugherty, P., Negm, W., Banerjee, P., & Alter, A. (2016). *Driving Unconventional Growth through the Industrial Internet of Things*. Accenture.

Davies, N. (2015). *How the Internet of Things will enable 'smart buildings*. Extreme Tech.

Davis, A., & Chang, H. (2012). Underwater Wireless Sensor Networks. *Oceans*.

De Angelis, E., Ciribini, A. L. C., Tagliabue, L. C., & Paneroni, M. (2015). The Brescia Smart Campus Demonstrator. Renovation toward a zero energy classroom building. *Procedia Engineering*, *118*, 735–743. doi:10.1016/j.proeng.2015.08.508

Deakin, M., Diamantini, D., & Borrelli, N. (2018). The governance of a smart city food system: The 2015 Milan World Expo Mark. *City, Culture and Society*.

Deka, K., & Goswami, K. (2018). IoT-Based Monitoring and Smart Planning of Urban Solid Waste Management. In *Advances in Communication, Devices and Networking* (pp. 895–905). Singapore: Springer. doi:10.1007/978-981-10-7901-6_96

Demiris, G., & Hensel, K. (2008). Technologies for an Aging Society: A Systematic Review of 'Smart Home' Applications. *IMIA Yearbook of Medical Informatics*, *2008*, 33–40.

Desai, P., Sheth, A., & Anantharam, P. (2015). Semantic gateway as a service architecture for iot interoperability. *Mobile Services (MS), IEEE International Conference on*, 313–319.

Deshmukh, Surendran, & Sardey. (2017). Air and Sound Pollution Monitoring System using IoT. *International Journal on Recent and Innovation Trends in Computing and Communication*, *5*(6), 175–178.

Dhanalaxmi, B., & Naidu, G. A. (2017). A survey on design and analysis of robust IoT architecture. *2017 International Conference on Innovative Mechanisms for Industry Applications (ICIMIA)*, 375-378. 10.1109/ICIMIA.2017.7975639

Dilip, K. P., Dnyandeo, J. M., Changdev, J. P., & Lavhate, S. S. (2018). IoT based solid waste management for the smart city. *International Journal of Advance Research, Ideas And Innovations in Technology*, *4*(2).

Doble, M., & Kumar, A. (2005). *Biotreatment of industrial effluents*. Elsevier.

Dohler, M. (2011). Smart cities: An action plan. *Proc. Barcelona Smart Cities Congress*.

Dong, S., Duan, S., Yang, Q., Zhang, J., Li, G., & Tao, R. (2017). Mems-based smart gas metering for internet of things. *IEEE Internet of Things Journal*, *4*(5), 1296–1303. doi:10.1109/JIOT.2017.2676678

Dorothy, Kumar, & Sharmila. (2016). *IoT based Home Security through Digital Image Processing Algorithms*. IEEE. DOI doi:10.1109/WCCCT.2016.15

Dorri, A., Kanhere, S. S., & Jurdak, R. (2016). *Blockchain in internet of things: challenges and solutions*. arXiv preprint arXiv:1608.05187

Dorri, A., Kanhere, S. S., Jurdak, R., & Gauravaram, P. (2017). Blockchain for IoT security and privacy: The case study of a smart home. In *Pervasive Computing and Communications Workshops (PerCom Workshops), 2017 IEEE International Conference on*, (pp. 618-623). IEEE. 10.1109/PERCOMW.2017.7917634

Dubber, D., & Gray, N. F. (2010). Replacement of chemical oxygen demand (COD) with total organic carbon (TOC) for monitoring wastewater treatment performance to minimize disposal of toxic analytical waste. *J. Environ. Sci. Heal. Part A*, *45*(12), 1595–1600. doi:10.1080/10934529.2010.506116 PMID:20721800

Earthquakes – Technical Hazard Sheet – Natural Disaster Profile, Humanitarian Health Action, World Health Organization. (2018). *Earthquakes - Why do we have earthquakes: British Geological Survey*. Retrieved from http://earthquakes.bgs.ac.uk

El Moulat., Debauche, Mahmoudi, Ait Brahim, Manneback, & Lebeau. (2018). Monitoring system using Internet of Things for potential landslides. *Proceedings of the 15th International Conference on Mobile Systems and Pervasive Computing (MobiSPC 2018)*.

Elder, C. Y. (2006). Mean-Squared Error Sampling and Reconstruction in the Presence of Noise. *IEEE Transactions on Signal Processing*, *54*(12).

Elmaghraby, A. S., & Losavio, M. M. (2014). Cyber security challenges in smart cities: Safety, security and privacy. *Journal of Advanced Research*, *5*(4), 491–497. doi:10.1016/j.jare.2014.02.006 PMID:25685517

Engage Mobile Blog. (2016). *Goldman Sachs Report: How the Internet of Things Can Save the American Healthcare System $305 Billion Annually*. Engage Mobile Solutions, LLC. Retrieved https://www.engagemobile.com/goldman-sachs-report-how-the-internet-of-things-can-save-the-american-healthcare-system-305-billion-annually

Ersue, M., Romascanu, D., Schoenwaelder, J., & Sehgal, A. (2014). *Management of Networks with Constrained Devices: Use Cases*. IETF Internet Draft.

Escolar, S., Villanueva, F. J., Santofimia, M. J., Villa, D., del Toro, X., & Lopez Carlos, J. A multiple-attribute decision making-based approach for smart city rankings design. *Technological Forecasting and Social Change*. doi:10.1016/j.techfore.2018.07.024

Esteban, E., Salgado, O., Iturrospe, A., & Isasa, I. (2016). Model-based approach for elevator performance estimation. *Mechanical Systems and Signal Processing*, *68*, 125–137. doi:10.1016/j.ymssp.2015.07.005

Evans, D. (2011). *The Internet of things - How the next evolution of the intenet is changing everything*. Cisco Internet Business Solution Group.

Fadzil, N., Zainal, Z., & Abdullah, A. (2013). COD removal for palm oil mill secondary effluent by using UV/ferrioxalate/TiO2/O3 system. *Int. J. Emerg.*, *3*, 237–243.

Farraji, H., Qamaruz, Z. N., Hamidi, A., Aqeel, A.M., Amin, M., & Parsa, M. (2015a). Enhancing BOD/COD Ratio of POME Treatment in SBR System. *Appl. Mech. Mater.*, *802*, 437–442. Retrieved from www.scientific.net/AMM.802.437

Farraji, H., Zaman, N. Q., Aziz, H. A., & Sa'at, S. K. M. (2017a). Palm oil mill effluent treatment: Influence of zeolite, municipal wastewater and combined aerobic SBR system. AIP Conference Proceedings. AIP 1892, 1–7. doi:10.1063/1.5005693

Farraji, H., Zaman, N. Q., Aziz, H. A., & Sa'at, S. K. M. (2017b). Palm oil mill effluent and municipal wastewater co-treatment by zeolite augmented sequencing batch reactors: Turbidity removal. In AIP Conference Proceedings (Vol. 1892, No. 1, p. 040013). AIP Publishing. doi:10.1063/1.5005694

Farraji, H., Zaman, N.Q., Abdul Aziz, H., Ashraf, M.A., Mojiri, A., & Mohajeri, P. (2015c). Landfill Leachate Treatment by Bentonite Augmented Sequencing Batch Reactor (SBR) System. *Appl. Mech. Mater.*, *802*, 466–471. Retrieved from www.scientific.net/AMM.802.466

Farraji, H., Qamaruz Zaman, N., & Mohajeri, P. (2015b). Waste Disposal:Sustainable Waste Treatments and Facility Siting Concerns. In H. Aziz & S. Amr (Eds.), *Control and Treatment of Landfill Leachate for Sanitary Waste Disposal* (pp. 43–74). IGI Global; doi:10.4018/978-1-4666-9610-5.ch003

Fattah, H. (2018). *5G LTE Narrowband Internet of Things (NB-IoT)*. CRC Press. doi:10.1201/9780429455056

Feltrin, L., Tsoukaneri, G., Condoluci, M., Buratti, C., Mahmoodi, T., Dohler, M., & Verdone, R. (2018). *NarrowBand-IoT: A Survey on Downlink and Uplink Perspectives*. IEEE Wireless Communication Network, ToAppear.

Fernandez-Aneza, V., Fernández-Güell, J. M., & Giffing, R. (2018). Smart City implementation and discourses: An integrated conceptual model. The case of Vienna. *Cities (London, England)*, *78*, 4–16. doi:10.1016/j.cities.2017.12.004

Fernández-Caramés, T. M., & Fraga-Lamas, P. (2018). A Review on the Use of Blockchain for the Internet of Things. *IEEE Access: Practical Innovations, Open Solutions*, *6*, 32979–33001. doi:10.1109/ACCESS.2018.2842685

Fernando, T. S., & Rafael, T. M. (2012). A cooperative approach to traffic congestion detection with complex event processing and Vanet. *IEEE Transactions on Intelligent Transportation Systems*, *13*(2).

FIT French Project. (2014). *Use case: Sensitive wildlife monitoring*. Author.

Fitchard, K. (2014). Sigfox brings its internet of things network to San Francisco. *Gigaom*. Retrieved from https://gigaom.com/2014/05/20/sigfox-brings-its-internet-of-things-network-to-san-francisco

Flores, A. Q., Carvalho, J. B., & Cardoso, A. J. M. (2008, September). Mechanical fault detection in an elevator by remote monitoring. In *Electrical Machines, 2008. ICEM 2008. 18th International Conference on* (pp. 1-5). IEEE. 10.1109/ICELMACH.2008.4800064

Fluke corporation. (2018). *An introduction to machinery vibration*. Retrieved from https://www.reliableplant.com/Read/24117/introduction-machinery-vibration

Folianto, F., Low, Y. S., & Yeow, W. L. (2015). Smartbin: Smart Waste Management System. *IEEE Tenth International Conference on Intelligent Sensors, Sensor Networks and Information Processing*.

Foo, K. Y., & Hameed, B. H. (2009). An overview of landfill leachate treatment via activated carbon adsorption process. *Journal of Hazardous Materials*, *171*(1-3), 54–60. doi:10.1016/j.jhazmat.2009.06.038 PMID:19577363

Foo, K. Y., & Hameed, B. H. (2010). Insight into the applications of palm oil mill effluent: A renewable utilization of the industrial agricultural waste. *Renewable & Sustainable Energy Reviews*, *14*(5), 1445–1452. doi:10.1016/j.rser.2010.01.015

Forte, V. J. Jr. (2010). *Smart Grid at National Grid*. IEEE. doi:10.1109/ISGT.2010.5434729

Frank, R., Giordano, E., & Gerla, M. (2010). TrafRoute: A different approach routing in vehicular networks. *Proc. VECON*. 10.1109/WIMOB.2010.5645018

Fujdiak, R., Masek, P., Mlynek, P., Misurec, J., & Olshannikova, E. (2016, July). Using genetic algorithm for advanced municipal waste collection in smart city. In *Communication Systems, Networks and Digital Signal Processing (CSNDSP), 2016 10th International Symposium on* (pp. 1-6). IEEE. 10.1109/CSNDSP.2016.7574016

Gandhi, B. K., & Rao, M. K. (2016). A prototype for IoT based car parking management system for smart cities. *Indian Journal of Science and Technology*, *9*(17).

Gauer, A., Scotney, B., Parr, G., & McClean, S. (2015). Smart city architecture and its applications based on IoT. *Procedia Computer Science*, *52*, 1089–1094. doi:10.1016/j.procs.2015.05.122

Geology Universe. (2018). Retrieved from: https://geologyuniverse.com/3-major-seismic-belts-of-the-earth/

Giagopoulos, D., Chatziparasidis, I., & Sapidis, N. S. (2018). Dynamic and structural integrity analysis of a complete elevator system through a Mixed Computational-Experimental Finite Element Methodology. *Engineering Structures*, *160*, 473–487. doi:10.1016/j.engstruct.2018.01.018

Giffinger, R., Fertner, C., Kramar, H., Kalasek, R., Pichler-Milanović, N., & Meijers, E. (2007). Smart Cities: Ranking of European Medium-Sized Cities. Vienna, Austria: Academic Press.

Giffinger, R., Fertner, C., Kramar, H., Kalasek, R., Pichler-Milanović, N., & Meijers, E. (2007). *Smart Cities: Ranking of European Medium-Sized Cities. Centre of Regional Science (SRF)*. Vienna, Austria: Vienna University of Technology.

Giri & Pippal. (2017). Use of Linear Interpolation for Automated Irrigation System in Agriculture using Wireless Sensor Network. *International Conference on Energy, Communication, Data Analytics and Soft Computing*. 10.1109/ICECDS.2017.8389716

Goel, Bush, & Bakken. (2013). *IEEE Vision for Smart Grid Communications: 2030 and Beyond*. IEEE Communication Society.

Gohar, M., Muzammal, M., & Rahman, A. U. (2018). SMART TSS: Defining transportation system behavior using big data analytics in smart cities. *Sustainable Cities and Society*, *41*, 114–119. doi:10.1016/j.scs.2018.05.008

Göksenli, A., & Eryürek, I. B. (2009). Failure analysis of an elevator drive shaft. *Engineering Failure Analysis*, *16*(4), 1011–1019. doi:10.1016/j.engfailanal.2008.05.014

Goldman Sachs Global Investment Research. (2014, September). *The Internet of Things: Making sense of the next mega-trend.* Retrieved June 2018, from https://www.goldmansachs.com/our-thinking/outlook/internet-of-things/iot-report.pdf

Goldman Sachs. (2014). *The Internet of Things: Making sense of the next mega-trend.* Author.

Government of India. (2017). *Economic Survey 2016-17.* Retrieved June 6, 2017, from http://indiabudget.nic.in/es2016-17/echapter.pdf

Grady, C. L. Jr, Daigger, G. T., Love, N. G., & Filipe, C. D. (2011). *Biological wastewater treatment.* CRC Press.

Granda-Cantu˜na, J., Molina-Colcha, C., Hidalgo-Lupera, S.-E., & Valarezo-Varela, C.-D. (2018). Design and Implementation of a Wireless Sensor Network for Precision Agriculture Operating in API Mode. *International Conference on eDemocracy & eGovernment.* 10.1109/ICEDEG.2018.8372346

Gubbi, J., Buyya, R., Marusic, S., & Palaniswami, M. (2013). Internet of Things (IoT): A vision, architectural elements, and future directions. *Future Generation Computer Systems, 29*(7), 1645 - 1660.

Guo, J.-S., Abbas, A. A., Chen, Y.-P., Liu, Z.-P., Fang, F., & Chen, P. (2010). Treatment of landfill leachate using a combined stripping, Fenton, SBR, and coagulation process. *Journal of Hazardous Materials*, *178*(1-3), 699–705. doi:10.1016/j.jhazmat.2010.01.144 PMID:20188464

Gutierrez, J. M., Jensen, M., Henius, M., & Riaz, T. (2015). Smart waste collection system based on location intelligence. *Procedia Computer Science*, *61*, 120–127. doi:10.1016/j.procs.2015.09.170

Haase, J., Alahmad, M., Nishi, H., Ploennigs, J., & Tsang, K. F. (2016). The IOT mediated built environment: A brief survey. *IEEE 14th International Conference on Industrial Informatics (INDIN),* 1065 - 1068. 10.1109/INDIN.2016.7819322

Halim, A. A., Aziz, H. A., Johari, M. A. M., & Ariffin, K. S. (2010). Comparison study of ammonia and COD adsorption on zeolite, activated carbon and composite materials in landfill leachate treatment. *Desalination*, *262*(1-3), 31–35. doi:10.1016/j.desal.2010.05.036

Hamouda, Y. E. M. (2017). Smart Irrigation Decision Support based on Fuzzy Logic using Wireless Sensor Network. *International Conference on Innovation, Management and Technology Research*, 361-368. 10.1109/ICPET.2017.26

Hannan, M. A., Arebey, M., Begum, R. A., & Basri, H. (2011). Radio Frequency Identification (RFID) and communication technologies for solid waste bin and truck monitoring system. *Waste Management (New York, N.Y.)*, *31*(12), 2406–2413. doi:10.1016/j.wasman.2011.07.022 PMID:21871788

Hansen, S. B., Padfield, R., Syayuti, K., Evers, S., Zakariah, Z., & Mastura, S. (2015). Trends in global palm oil sustainability research. *Journal of Cleaner Production*, *100*, 140–149. doi:10.1016/j.jclepro.2015.03.051

Harbour Research and Postscapes. (2013). *What exactly is the "Internet of Things"?* Retrieved from https://www.google.com/url?sa=t&rct=j&q=&esrc=s&source=web&cd=2&cad=rja&uact=8&ved=0CCMQFjAB&url=http%3A%2F%2Fharborresearch.com%2Fwp-content%2Fuploads%2F2014%2F03%2FHarbor-Postscapes-Infographic_March-2014.pdf&ei=uyxiVIWFDNGGuASE2ICgCA&usg=AFQjCNEWPdt47D

Harsha, S. S., Reddy, S. C., & Mary, S. P. (2017, February). Enhanced home automation system using internet of things. In *I-SMAC (IoT in Social, Mobile, Analytics and Cloud)(I-SMAC), 2017 International Conference on* (pp. 89-93). IEEE. 10.1109/I-SMAC.2017.8058302

Hart, J. K., & Martinez, K. (2015). Toward an environmental Internet of Things. *Earth & Space Science, 2*(5), 194 - 200.

Hearst, M. A., Dumais, S. T., Osuna, E., Platt, J., & Scholkopf, B. (1998). Support vector machines. *IEEE Intelligent Systems & their Applications, 13*(4), 18–28. doi:10.1109/5254.708428

Heather, K. (2010). *This elevator could shape the cities of future.* Retrieved from https://money.cnn.com/2016/05/03/technology/maglev-elevator-smart-city/index.html

Hendricks, D. (2015). *The Trouble with the Internet of Things. London Data store*. Greater London Authority. Retrieved https://data.london.gov.uk/blog/the-trouble-with-the-internet-of-things

Heryandi, A. (2018, August). Developing Application Programming Interface (API) for Student Academic Activity Monitoring using Firebase Cloud Messaging (FCM). *IOP Conference Series. Materials Science and Engineering, 407*(1), 012149. doi:10.1088/1757-899X/407/1/012149

Hoang, V., Julien, N., Berruet, P., Detection, A. F., & Fdi, I. (2013). *On-line self-diagnosis based on power measurement for a wireless sensor node*. Academic Press.

Hoglund, A., Bergman, J., Lin, X., Liberg, O., Ratilainen, A., Razaghi, H. S., ... Yavuz, E. A. (2018). Overview of 3GPP Release 14 Further Enhanced MTC. *IEEE Communications Standards Magazine, 2*(2), 84–89. doi:10.1109/MCOM-STD.2018.1700050

Hong, I., Park, S., Lee, B., Lee, J., Jeong, D., & Park, S. (2014). IoT-based smart garbage system for efficient food waste management. *The Scientific World Journal*. PMID:25258730

Hsu, C. L., & Lin, J. C. C. (2016). An empirical examination of consumer adoption of Internet of Things services: Network externalities and concern for information privacy perspectives. *Computers in Human Behavior, 62*, 516–527. doi:10.1016/j.chb.2016.04.023

Huang, G. B., & Siew, C. K. (2005). Extreme learning machine with randomly assigned RBF kernels. *International Journal of Information Technology, 11*(1), 16–24.

Huang, G. B., Zhou, H., Ding, X., & Zhang, R. (2012). Extreme learning machine for regression and multiclass classification. *IEEE Transactions on Systems, Man, and Cybernetics. Part B, Cybernetics, 42*(2), 513–529. doi:10.1109/TSMCB.2011.2168604 PMID:21984515

Huang, Y., & Yu, W. (2016). Elevator Safety Monitoring and Early Warning System Based on Directional antenna transmission technology. *Electronics (Basel), 19*(2), 101–104.

Hussain & Abdullah. (2018). *Review of Different Encryption and Decryption Techniques Used for Security and Privacy of IoT in Different Applications*. IEEE.

IBM. (2013a). *IBM harnesses power of big data to improve Dutch flood control and water management systems*. Retrieved June 29, 2017, from https://www-03.ibm.com/press/us/en/pressrelease/41385.wss

IBM. (2013b). *Miami-Dade police department: New patterns offer breakthroughs for cold cases.* Retrieved June 21, 2017, from http://smartcitiescouncil.com/system/tdf/public_resources/Miami_Dade%20police.pdf?file=1&type=node&id=200

IJSMI. (2018). Overview of recent advances in Health care technology and its impact on health care delivery. *International Journal of Statistics and Medical Informatics, 7*, 1 - 6.

IMS Center, . (2016). *Center for Intelligent Maintenance Systems. Author.*

Intel Newsroom. (2014). *San Jose Implements Intel Technology for a Smarter City*. Retrieved from https://newsroom.intel.com/news-releases/san-jose-implements-intel-technology-for-a-smarter-city/#gs.6AIlWLiP

Islam, M. S., Arebey, M., Hannan, M. A., & Basri, H. (2012). Overview for solid waste bin monitoring and collection system. In *Innovation Management and Technology Research (ICIMTR), 2012 International Conference on* (pp. 258-262). IEEE. 10.1109/ICIMTR.2012.6236399

Islam, N. S., & Wasi-ur-Rahman, M. (2009). *An intelligent SMS-based remote water metering system*. Paper presented at the 2009 12th International Conference on Computers and Information Technology.

Islam, M. A., Yousuf, A., Karim, A., Pirozzi, M., Rahman Khan, M., & Wahid, Z. A. (2018). Bioremediation of palm oil mill effluent and lipid production by Lipomyces starkeyi: A combined approach. *Journal of Cleaner Production*, *172*, 1779–1787. doi:10.1016/j.jclepro.2017.12.012

Ismail, I., Hassan, M. A., Rahman, N. A. A., & Soon, C. S. (2010). Thermophilic biohydrogen production from palm oil mill effluent (POME) using suspended mixed culture. *Biomass and Bioenergy*, *34*(1), 42–47. doi:10.1016/j.biombioe.2009.09.009

ISO 37120. (2014). *Sustainable Development in Communities: International Standard on city indicators*. Retrieved June 2018, from https://www.iso.org/files/live/sites/isoorg/files/archive/pdf/en/37120_briefing_note.pdf

ISO 37120:2018. (2018). *Sustainable Cities and Communities -Indicators for city services and quality of life*. Retrieved November 2018, from https://www.iso.org/obp/ui/#iso:std:iso:37120:ed-2:v1:en

Istepanian, R., Hu, S., Philip, N., & Sungoor, A. (2011). The potential of Internet of m-health Things "m - IoT" for non - invasive glucose level sensing. *Annual International Conference of the IEEE Engineering in Medicine and Biology Society (EMBC)*, 5264 - 6.

ITU. (2019). *Internet of Things Global Standards Initiative*. Retrieved from https://www.itu.int/en/ITU-T/gsi/iot/Pages/default.aspx

Jaag, C., & Bach, C. (2017). Blockchain Technology and Cryptocurrencies: Opportunities for Postal Financial Services. In M. Crew, P. Parcu, & T. Brennan (Eds.), *The Changing Postal and Delivery Sector. Topics in Regulatory Economics and Policy*. Springer. doi:10.1007/978-3-319-46046-8_13

Jain, M., Kumar, P., Bhansali, I., Liao, Q. V., Truong, K., & Patel, S. (2018). *FarmChat : A Conversational Agent to Answer Farmer Queries*. Academic Press.

Jain, S., Garg, R., Bhosle, V., & Sah, L. (2017, August). Smart university-student information management system. In *Smart Technologies For Smart Nation (SmartTechCon), 2017 International Conference On* (pp. 1183-1188). IEEE.

Javaid, N., Shah, M., Ahmad, A., Imran, M., Khan, M. I., & Vasilakos, A. V. (2016). An Enhanced Energy Balanced Data Transmission Protocol for Underwater Acoustic Sensor Networks. *Sensors (Basel)*, *16*(4), 487. doi:10.339016040487 PMID:27070605

Ji, Z., & Ganchev, I. (2015). A Cloud-Based Car Parking Middleware for IoTBased Smart Cities; Design and Implementation. *Sensors (Basel)*, *14*, 22372–22393. doi:10.3390141222372 PMID:25429416

Johnsen, F. T., & … . (2018). Application of IoT in military operations in a smartcity. *2018 IEEE International Conference on Military Communications and Information Systems (ICMCIS)*, 1-8.

Jolliffe, I. (2011). Principal component analysis. In *International encyclopedia of statistical science* (pp. 1094–1096). Berlin: Springer. doi:10.1007/978-3-642-04898-2_455

Jouanneau, S., Recoules, L., Durand, M. J., Boukabache, A., Picot, V., Primault, Y., ... Thouand, G. (2014). Methods for assessing biochemical oxygen demand (BOD): A review. *Water Research*, *49*, 62–82. doi:10.1016/j.watres.2013.10.066 PMID:24316182

Juan, M., Nunez, V., Faruk Fonthal, R., Yasmin, M., & Quezada, L. (2017). Design and implementation of WSN for precision agriculture in white cabbage crops. *IEEE International Conference on Electronics, Electrical Engineering and Computing (INTERCON).*

Jukaria, Singh, & Kumar. (2017). *A Comprehensive Review on Smart Meter Communication Systems in Smart Grid for Indian Scenario*. Academic Press.

Jung, D., Zhang, Z., & Winslett, M. (2017, April). Vibration Analysis for IoT Enabled Predictive Maintenance. In *Data Engineering (ICDE), 2017 IEEE 33rd International Conference on* (pp. 1271-1282). IEEE. 10.1109/ICDE.2017.170

Jurek, A., Nugent, C., Bi, Y., & Wu, S. (2014). Clustering-based ensemble learning for activity recognition in smart homes. *Sensors (Basel)*, *14*(7), 12285–12304. doi:10.3390140712285 PMID:25014095

Kahar, A., Heryadi, E., Malik, L., Widarti, B. N., & Cahayanti, I. M. (2017). The study of seeding and acclimatization from leachate treatment in anaerobic bioreactor. *Journal of Engineering and Applied Sciences (Asian Research Publishing Network)*, *12*, 2610–2614.

Kamaruddin, M. A., Yusoff, M. S., Aziz, H. A., & Hung, Y. T. (2015). Sustainable treatment of landfill leachate. *Applied Water Science*, *5*(2), 113–126. doi:10.100713201-014-0177-7

Kanamori, H., & Wu, Y.-M. (2005, June). Rapid Assessment of Damage Potential of Earthquakes in Taiwan from the Beginning of P Waves. *Bulletin of the Seismological Society of America*, *95*(3), 1181–1185. doi:10.1785/0120040193

Kang, W. M., Moon, S. Y., & Park, J. H. (2017). An enhanced security framework for home appliances in smart home. *Human-Centric Computing and Information Sciences, 7*(6).

Karaci, A. (2018). IOT-Based Earthquake Warning System Development And Evaluation. *Mugula Journal of Science and Technology, 4,* 156-161. doi:10.22531/muglajsci.442492

Karlgren, J., Fahlén, L., Wallberg, A., Hansson, P., Ståhl, O., Söderberg, J., & Åkesson, K. P. (2008). Socially Intelligent Interfaces for Increased Energy Awareness in the Home. The Internet of Things. Lecture Notes in Computer Science, 4952, 263 - 275. doi:10.1007/978-3-540-78731-0_17

Kaur, S., Pandey, S., & Goel, S. (2018). Semi-automatic leaf disease detection and classification system for soybean culture. *IET Image Processing*, *12*(6), 1038–1048. doi:10.1049/iet-ipr.2017.0822

Kavitha, S., Banu, J. R., Priya, A. A., Uan, D. K., & Yeom, I. T. (2017). Liquefaction of food waste and its impacts on anaerobic biodegradability, energy ratio and economic feasibility. *Applied Energy*, *208*, 228–238. doi:10.1016/j.apenergy.2017.10.049

Khanna, A., & Anand, R. (2016, January). IoT based smart parking system. In *2016 International Conference on Internet of Things and Applications (IOTA)* (pp. 266-270). IEEE. 10.1109/IOTA.2016.7562735

Khan, S. A., & Kim, J. M. (2016). Rotational speed invariant fault diagnosis in bearings using vibration signal imaging and local binary patterns. *The Journal of the Acoustical Society of America*, *139*(4), EL100–EL104. doi:10.1121/1.4945818 PMID:27106344

Khatoun, R., & Zeadally, S. (2016). Smart cities. *Communications of the ACM*, *59*(8), 46–57. doi:10.1145/2858789

Khemkhao, M., Techkarnjanaruk, S., & Phalakornkule, C. (2015). Simultaneous treatment of raw palm oil mill effluent and biodegradation of palm fiber in a high-rate CSTR. *Bioresource Technology*, *177*, 17–27. doi:10.1016/j.biortech.2014.11.052 PMID:25479389

Kim, T., Ramos, C., & Mohammed, S. (2017). Smart City and IoT. *Future Generation Computer Systems*, *76*, 159–162. doi:10.1016/j.future.2017.03.034

Kirsch, R. A. (1971). Computer determination of the constituent structure of biological images. *Computers and Biomedical Research, an International Journal*, *4*(3), 315–328. doi:10.1016/0010-4809(71)90034-6 PMID:5562571

Kishore Kumar Reddy, N. G., & Rajeshwari, K. (2017). Interactive clothes based on IOT using NFC and Mobile Application. *2017 IEEE 7th Annual Computing and Communication Workshop and Conference (CCWC)*, 1-4.

Kiss, P., Reale, A., Ferrari, C. J., & Istenes, Z. (2018, January). Deployment of IoT applications on 5G edge. In *Future IoT Technologies (Future IoT), 2018 IEEE International Conference on* (pp. 1-9). IEEE.

Köhn, R. (2018). *Corporations are joining forces against hackers*. Retrieved from https://www.faz.net/aktuell/wirtschaft/diginomics/grosse-internationale-allianz-gegen-cyber-attacken-15451953-p2.html?printPagedArticle=true#pageIndex_1

Komninos, N. (2018). Connected Intelligence in Smart Cities Shared, engagement and awareness spaces 4 innovation. URENIO Research, Aristotle University.

Kramers, A., Höjer, M., Lövehagen, N., & Wangel, J. (2014). Smart sustainable cities–exploring ICT solutions for reduced energy use in cities. *Environmental Modelling & Software*, *56*, 52–62. doi:10.1016/j.envsoft.2013.12.019

Krishnan, N. C., & Cook, D. J. (2014). Activity recognition on streaming sensor data. *Pervasive and Mobile Computing*, *10*, 138–154. doi:10.1016/j.pmcj.2012.07.003 PMID:24729780

Krishnan, S., Singh, L., Sakinah, M., Thakur, S., Wahid, Z. A., & Alkasrawi, M. (2016). Process enhancement of hydrogen and methane production from palm oil mill effluent using two-stage thermophilic and mesophilic fermentation. *International Journal of Hydrogen Energy*, *41*(30), 12888–12898. doi:10.1016/j.ijhydene.2016.05.037

Kumar Sharma, P., & Park, J. H. (2018). Blockchain based hybrid network architecture for the smart city. *Future Generation Computer Systems*, *86*, 650–655. doi:10.1016/j.future.2018.04.060

Kumar, J. S., & Patel, D. R. (2014). A survey on internet of things: Security and privacy issues. *International Journal of Computer Applications, 90*(11).

Kumar, M., Sabale, K., Mini, S., & Panigrahi, T. (2018, January). Priority based deployment of IoT devices. In *Information Networking (ICOIN), 2018 International Conference on* (pp. 760-764). IEEE. 10.1109/ICOIN.2018.8343220

Kumar, N. S., Vuayalakshmi, B., Prarthana, R. J., & Shankar, A. (2016, November). IOT based smart garbage alert system using Arduino UNO. In Region 10 Conference (TENCON), 2016 IEEE (pp. 1028-1034). IEEE.

Kumar, H., Singh, M. K. S., Gupta, M. P., & Madaan, J. (2018). Moving towards smart cities: Solutions that lead to the Smart City Transformation Framework. *Technological Forecasting and Social Change*. doi:10.1016/j.techfore.2018.04.024

Kuppusamy, P., Ilavenil, S., Srigopalram, S., Maniam, G. P., Yusoff, M. M., Govindan, N., & Choi, K. C. (2017). Treating of palm oil mill effluent using Commelina nudiflora mediated copper nanoparticles as a novel bio-control agent. *Journal of Cleaner Production*, *141*, 1023–1029. doi:10.1016/j.jclepro.2016.09.176

Kurniawan, T. A., Lo, W., Chan, G., & Sillanpää, M. E. (2010). Biological processes for treatment of landfill leachate. *Journal of Environmental Monitoring*, *12*(11), 2032–2047. doi:10.1039/c0em00076k PMID:20848046

Kurzon, I., Vernon, F. L., & Rosenberg, A. (2014). Real-Time Automatic P and S Waves Singular Value Decomposition. *Bulletin of the Seismological Society of America, 104*(4), 1696–1708. doi:10.1785/0120130295

Kutty, S., & Ngatenah, S. (2011). Removal of Zn (II), Cu (II), chemical oxygen demand (COD) and colour from anaerobically treated palm oil mill effluent (POME) using microwave incinerated rice. Int Conf Env. *Industry and Innovation.*

Lakshmi K. R., Nagesh, Y., & Krishna, V. M. (2014). Analysis on Predicting Earthquakes through an Abnormal Behavior of Animals. *International Journal of Scientific & Engineering Research, 5*(4).

Lakshminarasimhan. (2016). Advanced traffic management system using Internet of Things. *IEEE Journal,* (1), 1-9.

Lam, M. K., & Lee, K. T. (2011). Renewable and sustainable bioenergies production from palm oil mill effluent (POME): Win–win strategies toward better environmental protection. *Biotechnology Advances, 29*(1), 124–141. doi:10.1016/j.biotechadv.2010.10.001 PMID:20940036

Laursen, L. (2014). *Barcelona's smart city ecosystem.* Retrieved June 5, 2017, from https://www.technologyreview.com/s/532511/barcelonas-smart-city-ecosystem/

LeCun, Y., Bengio, Y., & Hinton, G. (2015). Deep learning. *Nature, 521*(7553), 436.

Lee, J. (2003). E - manufacturing - fundamental, tools and transformation. *Robotics and Computer - Integrated Manufacturing. Leadership of the Future in Manufacturing, 19*(6), 501 - 507.

Lee, J. (2014). Keynote Presentation: Recent Advances and Transformation Direction of PHM. *Road mapping Workshop on Measurement Science for Prognostics and Health Management of Smart Manufacturing Systems Agenda.*

Lee, J., Bagheri, B., & Kao, H. A. (2015). A cyber - physical systems architecture for industry 4.0 - based manufacturing systems. *Manufacturing Letters, 3*, 18 - 23.

Lee, J. (2015). *Industrial Big Data.* Mechanical Industry Press.

Li, S., Wang, H., Xu, T., & Zhou, G. (2011). Application Study on Internet of Things in Environment Protection Field. Lecture Notes in Electrical Engineering, 133, 99 - 106. doi:10.1007/978-3-642-25992-0_13

Li, X., & Zhao, D. (2017). Capacity Research in Cluster-Based Underwater Wireless Sensor Networks Based on Stochastic Geometry. IEEE, 14, 80-87.

Liang, X., Zhang, H., Tingting, L., & Gulliver, A. (2016). A Novel Time of Arrival Estimation Algorithm based on Skewness and Kurtosis. *International Journal of Signal Processing, Image Processing and Patten Recognition, 9*(3), 247–260. doi:10.14257/ijsip.2016.9.3.22

Liang, Z., & Liu, J. (2008). Landfill leachate treatment with a novel process: Anaerobic ammonium oxidation (Anammox) combined with soil infiltration system. *Journal of Hazardous Materials, 151*(1), 202–212. doi:10.1016/j.jhazmat.2007.05.068 PMID:17606322

Liew, W. L., Kassim, M. A., Muda, K., Loh, S. K., & Affam, A. C. (2015). Conventional methods and emerging wastewater polishing technologies for palm oil mill effluent treatment: A review. *Journal of Environmental Management, 149*, 222–235. doi:10.1016/j.jenvman.2014.10.016 PMID:25463585

Li, H., Zhou, S., Sun, Y., Feng, P., & Li, J. (2009). Advanced treatment of landfill leachate by a new combination process in a full-scale plant. *Journal of Hazardous Materials, 172*(1), 408–415. doi:10.1016/j.jhazmat.2009.07.034 PMID:19660862

Likotiko, E. D., Nyambo, D., & Mwangoka, J. (2017). *Multi-agent based IoT smart waste monitoring and collection architecture.* Academic Press.

Lin, S. H., & Chang, C. C. (2000). Treatment of landfill leachate by combined electro-Fenton oxidation and sequencing batch reactor method. *Water Research*, *34*(17), 4243–4249. doi:10.1016/S0043-1354(00)00185-8

Lipsky, J. (2015). IoT Clash Over 900 MHz Options. *EETimes*. Retrieved from https://www.eetimes.com/document.asp?doc_id=1326599

Liu, J., Zhang, P., Li, H., Tian, Y., Wang, S., Song, Y., ... Tian, Z. (2017). Denitrification of landfill leachate under different hydraulic retention time in a two-stage anoxic/oxic combined membrane bioreactor process: Performances and bacterial community. *Bioresource Technology*, *250*, 110–116. doi:10.1016/j.biortech.2017.11.026 PMID:29161569

Liu, X. K., Chen, Y., & Yu, H. N. (2014). Research on web-based elevator failure remote monitoring system. *Applied Mechanics and Materials*, *494*, 797–800. doi:10.4028/www.scientific.net/AMM.494-495.797

Li, Y., Cheng, X., Cao, Y., Wang, D., & Yang, L. (2018). Smart choice for the smart grid: Narrowband Internet of Things (NB-IoT). *IEEE Internet of Things Journal*, *5*(3), 1505–1515. doi:10.1109/JIOT.2017.2781251

Liyan, S., Youcai, Z., Weimin, S., & Ziyang, L. (2009). Hydrophobic organic chemicals (HOCs) removal from biologically treated landfill leachate by powder-activated carbon (PAC), granular-activated carbon (GAC) and biomimetic fat cell (BFC). *Journal of Hazardous Materials*, *163*(2-3), 1084–1089. doi:10.1016/j.jhazmat.2008.07.075 PMID:18752890

Li, Z., Meier, M. A., Hauksson, E., Zhan, Z., & Andrews, J. (2018). Machine Learning Seismic Wave Discrimination: Application to Earthquake Early Warning. *Geophysical Research Letters*, *45*(10), 4773–4779. doi:10.1029/2018GL077870

Llorca, D. F., & Sotelo, M. A. (2010). *Traffic Data Collection for Floating Car Data Enhancement in V2I Networks. EURASIP Journal on Advances in Signal Processing*. doi:10.1155/2010/719294

Lloret, J., Tomas, J., Canovas, A., & Parra, L. (2016). An integrated IoT architecture for smart metering. *IEEE Communications Magazine*, *54*(12), 50–57. doi:10.1109/MCOM.2016.1600647CM

Lockman, B. A. (2005). Single-Station Earthquake Characterization for Early Warning. *Bulletin of the Seismological Society of America*, *95*(6), 2029–2039. doi:10.1785/0120040241

Loh, S. K., Nasrin, A. B., Azri, S. M., Adela, B. N., Muzzammil, N., Jay, T. D., ... Kaltschmitt, M. (2017). First report on Malaysia's experiences and development in biogas capture and utilization from palm oil mill effluent under the Economic Transformation Programme: Current and future perspectives. *Renewable & Sustainable Energy Reviews*, *74*, 1257–1274. doi:10.1016/j.rser.2017.02.066

Loukidou, M. X., & Zouboulis, A. I. (2001). Comparison of two biological treatment processes using attached-growth biomass for sanitary landfill leachate treatment. *Environ. Pollut.*, *111*(2), 273–281. doi:10.1016/S0269-7491(00)00069-5 PMID:11202731

Ma, A. N., & Ong, A. S. (1985). Pollution control in palm oil mills in Malaysia. *Journal of the American Oil Chemists' Society*, *62*(2), 261–266. doi:10.1007/BF02541389

Madakam, Ramaswamy, & Tripathi. (2015). Internet of Things(IoT): A Literature Review. *Journal of Computer and Communications, 3*, 164-173. Retrieved from http://www.ieccr.net/comsoc/ijcis/

Mahesh, S., Prasad, B., Mall, I. D., & Mishra, I. M. (2006). Electrochemical degradation of pulp and paper mill wastewater. Part 1. COD and color removal. *Industrial & Engineering Chemistry Research*, *45*(8), 2830–2839. doi:10.1021/ie0514096

Mahizhnan, A. (1999). Smart cities: The Singapore case. *Cities (London, England)*, *16*(1), 13–18. doi:10.1016/S0264-2751(98)00050-X

Mahmud, K., Town, G. E., Morsalin, S., & Hossain, M. J. (2018). Integration of electric vehicles and management in the internet of energy. *Renewable and Sustainable Energy Reviews, 82*, 4179 - 4203.

Ma, J. (2014). Internet-of-Things: Technology evolution and challenges. *IEEE MTT-S International Microwave Symposium*, 1-4.

Malandrino, F., & Chiasserini, C. F. (2013). Optimal Content Downloading In Vehicular Networks. *IEEE Transactions on Mobile Computing, 12*(7), 1377–1391. doi:10.1109/TMC.2012.115

Manasi, H. K., & Smithkumar, B. S. (2016). A Novel approach to Garbage Management Using Internet of Things for smart cities. *International Journal of Current Trends in Engineering & Research*, 2(5), 348–353.

Maribor. (n.d.). *Free wireless internet access in the municipality of Maribor*. Retrieved May 26, 2017, from http://www.smartcitymaribor.si/en/Projects/Smart_Living_and_Urban_Planning/Free_wireless_internet_in_the_Municipality_of_Maribor/

Martin, C. J., Evans, J., & Karvonen, A. (2018). Smart and sustainable? Five tensions in the visions and practices of the smart-sustainable city in Europe and North America. *Technological Forecasting and Social Change, 133*, 269–278. doi:10.1016/j.techfore.2018.01.005

Masuo, T. (2011). *Deliverable on Smart Grid Architecture*. International Telecommunication Union.

Mavridou & Papa. (2012). *A Situational Awareness Architecture for the Smart Grid*. Institute for Computer Sciences, Social Informatics and Telecommunications Engineering.

Ma, Y.-W., & Chen, J.-L. (2018). Toward Intelligent Agriculture Service Platform with LoRa-based Wireless Sensor Network. *Proceedings of IEEE International Conference on Applied System Innovation*, 13-17. 10.1109/ICASI.2018.8394568

McDaniel, P., & Smith, S. W. (2009). *Security and Privacy Challenges in the Smart Grid*. IEEE Computer and Reliability Societies. doi:10.1109/MSP.2009.76

Mehra, S., & Verma, S. (2016). Smart transportation - Transforming Indian cities. *Grant Thornton India*. Retrieved May 24, 2017, from http://www.grantthornton.in/globalassets/1.-member-firms/india/assets/pdfs/smart-transportation-report.pdf

Mekki, K., Bajic, E., Chaxel, F., & Meyer, F. (2018). *A comparative study of LPWAN technologies for large-scale IoT deployment*. ICT Express.

Meky, N., Fujii, M., & Tawfik, A. (2017). Treatment of hypersaline hazardous landfill leachate using affled constructed wetland system: Effect of granular packing media and vegetation. *Environmental Technology*. doi:10.1080/09593330.2017.1397764 PMID:29073833

Meola, A. (2016a). How IoT & smart home automation will change the way we live. *Business Insider*. Retrieved from https://www.businessinsider.com/internet-of-things-smart-home-automation-2016-8?IR=T

Meola, A. (2016b). Why IoT, big data & smart farming are the future of agriculture. *Business Insider*. Retrieved from https://www.businessinsider.com/internet-of-things-smart-agriculture-2016-10?IR=T

Merritt, R. (2015). 13 Views of IoT World. *EETimes*. Retrieved from https://www.eetimes.com/document.asp?doc_id=1326596

Meseguer, J., & Quevedo, J. (2017). *Real-time monitoring and control in water systems. In Real-time monitoring and operational control of drinking-water systems* (pp. 1–19). Springer.

Metcalf, E., & Eddy, M. (2014). *Wastewater engineering: treatment and reuse* (5th ed.). New York: McGraw-Hill Education; doi:10.1016/0309-1708(80)90067-6

Microsoft Services. (2011). *London transport manages 2.3 million website hits a day with new data feed.* Retrieved May 26, 2017, from http://smartcitiescouncil.com/system/tdf/public_resources/London%20Transport%20and%20its%20very%20busy%20website.pdf?file=1&type=node&id=542

Ministry of Housing and Urban Affairs, Government of India. (2017). *What is a Smart City.* Retrieved November 2018, from Smart Cities Mission: http://smartcities.gov.in/upload/uploadfiles/files/What%20is%20Smart%20City.pdf

Ministry of Urban Development. (2015). *Smart cities: Mission statement & guidelines.* Retrieved May 10, 2017, from http://164.100.161.224/upload/uploadfiles/files/SmartCityGuidelines(1).pdf

Minoli. (2013). Layer 1/2 Connectivity: Wireless Technologies for the IoT. In *Building the Internet of Things with IPv6 and MIPv6 The Evolving World of M2M Communications*. Academic Press.

Misra, D., Das, G., Chakrabortty, T., & Das, D. (2018). An IoT-based waste management system monitored by cloud. *Journal of Material Cycles and Waste Management*, 1–9.

Mitchell, S., Villa, N., Stewarts-Weeks, M., & Lange, A. (2013). Point of View: The Internet of Everything for Cities. *CISCO*. Retrieved June 2018, from https://www.cisco.com/c/dam/en_us/solutions/industries/docs/gov/everything-for-cities.pdf

Mitchell, S., Villa, N., Stewart-Weeks, M., & Lange, A. (2013). *Point of View:The Internet of Everything for Cities.* Cisco. doi:10.4324/9780203716687

Mitton, Papavassiliou, Puliafito, & Trivedi. (2012). *Combining Cloud and sensors in a smart city environment.* Academic Press.

Moghadam, P., Ward, D., Goan, E., Jayawardena, S., Sikka, P., & Hernandez, E. (2017). *Plant Disease Detection using Hyperspectral Imaging*. Academic Press.

Mogili, U. M. R., & Deepak, B. B. V. L. (2018). ScienceDirect ScienceDirect Review on Application of Drone Systems in Precision Agriculture. *Procedia Computer Science*, *133*, 502–509. doi:10.1016/j.procs.2018.07.063

Mohammed, & Ahmed. (2017). Internet of Things Applications, Challenges and Related Future Technologies. *International Journal World Science News*, *67*(2), 126–148.

Mohammed, R. (2013). *Decolorisation of biologically treated palm oil mill effluent (POME) using adsorption technique. Int Ref. J Eng Sci.*

Mohammed, R., & Chong, F. (2014). Treatment and decolorization of biologically treated Palm Oil Mill Effluent (POME) using banana peel as novel biosorbent. *Journal of Environmental Management*, *132*, 237–249. doi:10.1016/j.jenvman.2013.11.031 PMID:24321284

Mojiri, A. (2014). *Co-treatment of landfill leachate and settled domestic wastewater using composite adsorbent in sequencing batch reactor* (Doctoral dissertation). Universiti Sains Malaysia.

Mojiri, A., Aziz, H. A., Aziz, S. Q., & Zaman, N. Q. (2012). Review on municipal landfill leachate and sequencing batch reactor (SBR) technique. *Arch. Des Sci.*, *65*, 22–31.

Mojiri, A., Aziz, H. A., Zaman, N. Q., Aziz, S. Q., & Zahed, M. A. (2014). Powdered ZELIAC augmented sequencing batch reactors (SBR) process for co-treatment of landfill leachate and domestic wastewater. *Journal of Environmental Management*, *139*, 1–14. doi:10.1016/j.jenvman.2014.02.017 PMID:24662109

Mojiri, A., Ziyang, L., Hui, W., Ahmad, Z., Tajuddin, R. M., Abu Amr, S. S., ... Farraji, H. (2017). Concentrated landfill leachate treatment with a combined system including electro-ozonation and composite adsorbent augmented sequencing batch reactor process. *Process Safety and Environmental Protection, 111*, 253–262. doi:10.1016/j.psep.2017.07.013

Mojiri, A., Ziyang, L., Hui, W., & Gholami, A. (2019). Concentrated Landfill Leachate Treatment by Electro-Ozonation. In *Advanced Oxidation Processes (AOPs) in Water and Wastewater Treatment* (pp. 150–170). IGI Global. doi:10.4018/978-1-5225-5766-1.ch007

Mojiri, A., Ziyang, L., Tajuddin, R. M., Farraji, H., & Alifar, N. (2016). Co-treatment of landfill leachate and municipal wastewater using the ZELIAC/zeolite constructed wetland system. *Journal of Environmental Management, 166*, 124–130. doi:10.1016/j.jenvman.2015.10.020 PMID:26496842

Mondal, A., Misra, I. S., & Bose, S. (2017). Building a Low Cost Solution using Wireless Sensor Network for Agriculture Application. *International Conference on Innovations in Electronics, Signal Processing and Communication.* 10.1109/IESPC.2017.8071865

Mooney, W. D., Prodehl, C., & Pavlenkova, N. I. (2002). Seismic Velocity Structure of the Continental Lithosphere from Controlled Source Data. International Handbook of Earthquake and Engineering Seismology, 81A.

Mora, H., Signes-Pont, M. T., Gil, D., & Johnsson, M. (2018). Collaborative Working Architecture for IoT-Based Applications. *Sensors (Basel), 18*(6), 1676. doi:10.339018061676 PMID:29882868

Morello, R., Mukhopadhyay, S. C., Liu, Z., Slomovitz, D., & Samantaray, S. R. (2017). Advances on Sensing Technologies for Smart Cities and Power Grids: A Review. *IEEE Sensors Journal*, 99.

Moriarty, P., & Honnery, D. (2015). Future cities. *Future, 66*, 45–53. doi:10.1016/j.futures.2014.12.009

Moses, N., & Chincholkar, Y. D. (2016). Smart parking system for monitoring vacant parking. *Int. J. Adv. Res. Comput. Commun. Eng, 5*(6), 717–720.

Mostaço, G. M., Costa de Souza, Í. R., Campos, L. B., & Cugnasca, C. E. (2018). *AgronomoBot : a smart answering Chatbot applied to agricultural sensor networks*. Academic Press.

Mostafa, A. E., Zhou, Y., & Wong, V. W. (2017, May). Connectivity maximization for narrowband IoT systems with NOMA. In *Communications (ICC), 2017 IEEE International Conference on* (pp. 1-6). IEEE. 10.1109/ICC.2017.7996362

Mukherjee, B. (2017). The internet of things (IOT): Revolutionized the way we live. *Postscapes*. Retrieved from http://www.bizmak.xyz/internet-of-things-iot-revolutionized-the-way-we-live

Mulligan & Olsson. (2013). Architectural implications of smart city business models: An evolutionary perspective. *IEEE Communications Magazine, 51*(6), 80-85.

Mulvenna, M., Hutton, A., Martin, S., Todd, S., Bond, R., & Moorhead, A. (2017). Views of Caregivers on the Ethics of Assistive Technology Used for Home Surveillance of People Living with Dementia. *Neuroethics, 10*(2), 255 - 266.

Mundhe, M., Pandagale, P., & Pathan, A. K. (2014). Smart water for Aurangabad city. *International Journal of Advanced Research in Computer Science and Software Engineering, 4*(10), 649–652.

Myint, Gopal, & Aung. (2017). Reconfigurable Smart Water Quality Monitoring System in IoT Environment. *Proceedings of the IEEE/ACIS 16th International Conference on Computer and Information Science (ICIS).*

Najafpour, G. D., Zinatizadeh, A. A. L., Mohamed, A. R., Hasnain Isa, M., & Nasrollahzadeh, H. (2006). High-rate anaerobic digestion of palm oil mill effluent in an upflow anaerobic sludge-fixed film bioreactor. *Process Biochemistry (Barking, London, England), 41*(2), 370–379. doi:10.1016/j.procbio.2005.06.031

Nam, T., & Pardo, T. A. (2011). Conceptualizing smart city with dimensions of technology, people, and institutions. In *Proceedings of the 12th Annual International Digital Government Research Conference: Digital Government Innovation in Challenging Times*. New York, NY: ACM. Retrieved April 6, 2018, from https://dl.acm.org/citation.cfm?id=2037602

Nandan, A., Yadav, B. P., Baksi, S., & Bose, D. (2017). Recent Scenario of Solid Waste Management in India. *World Scientific News*, (66), 56-74.

NASSCOM & Accenture. (2015). *Integrated ICT and geospatial technologies framework for 100 smart cities mission.* Retrieved May 28, 2017, from http://agiindia.com/pdf/Integrated-ICT-Geospatial-Technologies-2015%20(Nasscom-Accenture).pdf

National Common Mobility Card. (2016, November 23). *Press Information Bureau.* Retrieved May 24, 2017, from http://pib.nic.in/newsite/PrintRelease.aspx?relid=154158

National Crime Records Bureau. (2014). *Crime in India 2013 Statistics.* Ministry of Home Affairs. Retrieved June 27, 2017, from http://ncrb.nic.in/StatPublications/CII/CII2013/Statistics-2013.pdf

National Crime Records Bureau. (2017). *CCTNS Pragati Dashboard (31.05.2017).* Retrieved July 4, 2017, from http://www.ncrb.gov.in/BureauDivisions/CCTNS/CCTNS_Dashboard/PRGATI%20dashboard%2028.02.2017%20ver%20 9.0%20for%20MHA.pdf

National Crime Records Bureau. (n.d.). *Crime and Criminal Tracking Network & Systems (CCTNS).* Retrieved July 4, 2017, from http://www.ncrb.gov.in/BureauDivisions/CCTNS/cctns.htm

National Institute of Disaster Management. (n.d.). Retrieved from: http://nidm.gov.in/safety_earthquake.asp

Navarra, D. D. (2013). Perspectives on the evaluation of Geo-ICT for sustainable urban governance: Implications for e-government policy. *URISA Journal*, *25*(1), 19–29.

Navghane, S. S., Killedar, M. S., & Rohokale, D. V. (2016). IoT based smart garbage and waste collection bin. *Int. J. Adv. Res. Electron. Commun. Eng*, *5*(5), 1576–1578.

Navghane, S. S., Killedar, M. S., & Rohokale, V. M. (2016). IoT Based Smart Garbage and Waste Collection Bin. *International Journal of Advanced Research in Electronics and Communication Engineering*, *5*(5), 1576–1578.

Neczaj, E., Kacprzak, M., Kamizela, T., Lach, J., & Okoniewska, E. (2008). Sequencing batch reactor system for the co-treatment of landfill leachate and dairy wastewater. *Desalination*, *222*(1-3), 404–409. doi:10.1016/j.desal.2007.01.133

Neirotti, P., De Marco, A., Cagliano, A. C., Mangano, G., & Scorrano, F. (2014). Current trends in smart city initiatives: Some stylised facts. *Cities (London, England)*, *38*, 25–36. doi:10.1016/j.cities.2013.12.010

Nick, A. (2010). *Smart Elevators Bring You There Faster & More Efficiently.* Retrieved from https://www.triplepundit.com/2010/04/smart-elevators-bring-you-there-faster-more-efficiently

Nintanavongsa, P., & Pitimon, I. (2017). Impact of Sensor Mobility on UAV-based Smart Farm Communications. *International Electrical Engineering Congress.*

Nirde, K., Mulay, P. S., & Chaskar, U. M. (2017). IoT based Solid Waste Management System for Smart city. *International Conference on Intelligent Computing and Control Systems.*

Nordrum, A. (2016). Popular Internet of Things Forecast of 50 Billion Devices by 2020 Is Outdated. *IEEE.* Retrieved from https://spectrum.ieee.org/tech-talk/telecom/internet/popular-internet-of-things-forecast-of-50-billion-devices-by-2020-is-outdated

Nutter, R. S. (1983). Hazard Evaluation Methodology for Computer-Controlled Mine Monitoring/Control Systems. *IEEE Transactions on Industry Applications*, *IA-19*(3), 445–449. doi:10.1109/TIA.1983.4504222

O'Flynn, De Donno, Barrett, Robinson, & O'Riordan. (2017). Smart Microneedle Sensing Systems for Security in Agriculture, Food and the Environment (SAFE). *IEEE Sensors*.

Ochoa, L. H., Nino, L. F., & Vargas, C. A. (2017). *Fast magnitude determination using a single seismological station record implementing machine learning techniques*. Geodesy and Geodynamics. doi:10.1016/j.geog.2017.03.010

Odaka, T., Ashiya, K., Tsukada, S., Sato, S., Ohtake, K., & Nozaka, D. (2003). A New Method of Quickly Estimating Epicentral Distance and Magnitude from a Singe Seimic Record. *Bulletin of the Seismological Society of America, 93*(1), 526 – 532.

Ohimain, E. I., & Izah, S. C. (2017). A review of biogas production from palm oil mill effluents using different configurations of bioreactors. *Renewable & Sustainable Energy Reviews*, *70*, 242–253. doi:10.1016/j.rser.2016.11.221

Oizom Instruments Pvt Ltd. (2018a). *Kakinada Smart City Project*. Retrieved June 2018, from Oizom Redfining Resources: https://oizom.com/kakinada-smart-city-project/

Oizom Instruments Pvt Ltd. (2018b). *Palava Smart City Environmental Monitoring*. Retrieved June 2018, from Oizom Redefining Resources: https://oizom.com/palava-smart-city-environmental-monitoring/

Ojala, V.-M. (2017). *Addressing the interoperability challenge of combining heterogeneous data sources in data-driven solution* (Master's Thesis). Academic Press.

Ojo, A., Curry, E., & Janowski, T. (2014). *Designing Next Generation Smart City Initiatives Harnessing Findings and Lessons From a Study of Ten Smart City Programs*. Academic Press.

Olalere, I. O., Dewa, M., & Nleya, B. (2018). Remote Condition Monitoring of Elevator's Vibration and Acoustics Parameters for Optimised Maintenance Using IoT Technology. In *IEEE Canadian Conference on Electrical & Computer Engineering* (pp. 1-4). IEEE. 10.1109/CCECE.2018.8447771

Oshin, O., Owoniyi, A., Oni, O., & Idachaba, F. E. (2017). *Programming of NFC Chips: A University System Case Study*. Academic Press.

Oskin, B. (2015). What is a Subduction zone? *Planet Earth, Live Science.* Retrieved from https://www.livescience.com/43220-subduction-zone-definition.html

Osman, A. M. S. (2018). A novel big data analytics framework for smart cities. *Future Generation Computer Systems*. doi:10.1016/j.future.2018.06.046

O-Thong, S., Prasertsan, P., Intrasungkha, N., Dhamwichukorn, S., & Birkeland, N.-K. Å. (2008). Optimization of simultaneous thermophilic fermentative hydrogen production and COD reduction from palm oil mill effluent by Thermoanaerobacterium-rich sludge. *International Journal of Hydrogen Energy*, *33*(4), 1221–1231. doi:10.1016/j.ijhydene.2007.12.017

Otuoze, Mustafa, & Larik. (2018). Smart grids security challenges: Classification by sources of threats. *Journal of Electrical Systems and Information Technology*, 468–483.

Padode, P., Padode, F., Mali, D., Gawde, S., & Yadav, S. (2016). *Smart cities India readiness guide: The planning manual for building tomorrow's cities.* Retrieved June 7, 2017, from http://india.smartcitiescouncil.com/system/files/india/premium_resources/Smart_Cities_Council_India_Readiness_Guide_v2016-02.pdf?file=1&type=node&id=3330

Padwal & Kurde. (2016). Long-Term Environment Monitoring for IOT Applications using Wireless Sensor Network. *International Journal of Engineering Technology, Management and Applied Sciences*, *4*(2), 50–55.

Palomo-Navarro, A., & Navío-Marco, J. (2017). Smart city networks' governance: The Spanish smart city network case study. *Telecommunications Policy*, *xxx*, 1–9.

Parello, J., Claise, B., Schoening, B., & Quittek, J. (2014). *Energy Management Framework*. IETF Internet Draft <draft-ietf-eman-framework-19>.

Parikh, Kanabar, & Sidhu. (2010). Opportunities and challenges of wireless communication technologies for smart grid applications. *Proceedings of power energy soc. gen. meet. IEEE*, 1–7.

Parkash, P. V. (2016). IoT Based Waste Management for Smart City. *International Journal of Innovative Research in Computer and Communication Engineering*, *4*(2), 1267–1274.

Park, Y., & Rue, S. (2015). Analysis on Smart City service technology with IoT. *Technology Review*, *13*(2), 31–37.

Parvatha Reddy, P. (2016). *Sustainable Crop Protection under Protected Cultivation. Sustainable Crop Protection under Protected Cultivation*. Academic Press. doi:10.1007/978-981-287-952-3

Patel & Patel. (2016). Internet of Things – IOT: Definition, Characteristics, Architecture, Enabling Technologies, Application & Future Challenges. *IJESC, 6*(5).

Patel, K. K., & Patel, S. M. (2016). Internet of things-IOT: definition, characteristics, architecture, enabling technologies, application & future challenges. *International Journal of Engineering Science and Computing*, *6*(5).

Patel, R., & Suresh, S. (2008). Kinetic and equilibrium studies on the biosorption of reactive black 5 dye by Aspergillus foetidus. *Bioresource Technology*, *99*(1), 51–58. doi:10.1016/j.biortech.2006.12.003 PMID:17251011

Patil, V. C., Al-Gaadi, K. A., Biradar, D. P., & Rangaswamy, M. (2012). Internet of Things (IoT) and Cloud computing for Agriculture: An overview. *Proceedings of the third national conference on Agro-Informatics and Precision Agriculture.*

Patil, H. A., Baljekar, P. N., & Basu, T. K. (2012). Novel Temporal and Spectral Features Derived from TEO for Classification Normal and Dysphonic Voices. In S. Sambath & E. Zhu (Eds.), *Frontiers in Computer Education. Advances in Intelligent and Soft Computing* (Vol. 133). Berlin: Springer. doi:10.1007/978-3-642-27552-4_76

Pavitra & Janani. (2017). Low Delay and High Throughput Data Collection in Wireless Sensor Networks with Mobile Sinks. *International Journal of Scientific & Engineering Research*. doi:10.1109/IEECON.2017.8075822

Pedersen, T. H., Nielsen, K. U., & Petersen, S. (2017). Method for room occupancy detection based on trajectory of indoor climate sensor data. *Building and Environment*, *115*, 147–156. doi:10.1016/j.buildenv.2017.01.023

Peng, F., Wang, W., & Liu, H. (2014). Development of a reflective PPG signal sensor. *2014 7th International Conference on Biomedical Engineering and Informatics,* 612-616.

Perera, C., Liu, C. H., & Jayawardena, S. (2015). The Emerging Internet of Things Marketplace From an Industrial Perspective: A Survey. *IEEE Transactions on Emerging Topics in Computing, 3*(4), 585 - 598.

Perera, C., Zaslavsky, A., Christen, P., & Georgakopoulos, D. (2014). Sensing as a service model for smart cities supported by internet of things. *Transactions on Emerging Telecommunications Technologies*, *25*(1), 81–93. doi:10.1002/ett.2704

Petrov, V., Samuylov, A., Begishev, V., Moltchanov, D., Andreev, S., Samouylov, K., & Koucheryavy, Y. (2018). Vehicle-based relay assistance for opportunistic crowdsensing over narrowband IoT (NB-IoT). *IEEE Internet of Things Journal, 5*(5), 3710-3723.

Poh, P., & Chong, M. (2010). Biomethanation of Palm Oil Mill Effluent (POME) with a thermophilic mixed culture cultivated using POME as a substrate. *Chemical Engineering Journal, 164*(1), 146–154. doi:10.1016/j.cej.2010.08.044

Poon, L. (2018). Sleepy in Songdo, Korea's Smartest City. City Lab. *Atlantic Monthly Group*. Retrieved from https://www.citylab.com/life/2018/06/sleepy-in-songdo-koreas-smartest-city/561374

Popli, S., Jha, R. K., & Jain, S. (2018). A Survey on Energy Efficient Narrowband Internet of things (NBIoT): Architecture, Application and Challenges. *IEEE Access: Practical Innovations, Open Solutions.*

Popoola, S. I., Atayero, A. A., Okanlawon, T. T., Omopariola, B. I., & Takpor, O. A. (2018). Smart campus: Data on energy consumption in an ICT-driven university. *Data in Brief, 16*, 780–793. doi:10.1016/j.dib.2017.11.091 PMID:29276746

Postscapes. (2014). *Smart Trash*. Retrieved from https://www.postscapes.com/smart-trash

Pradhan, P., Naik, A., & Patel, P. (2014). Location Privacy in Ubiquitous Computing. *International Journal of Research in Science and Technology*, 54.

Prajakta, G., Kalyani, J., & And, M. (2015). Smart Garbage CollectionSystem In Residential Area. *International Journal of Research In Engineering And Technology*, *4*(3), 122–124. doi:10.15623/ijret.2015.0403021

Quach, K. (2018). *Google goes bilingual, Face book fleshes out translation and Tensor Flow is dope - And, Microsoft is assisting fish farmers in Japan*. Retrieved from https://www.theregister.co.uk/2018/09/01/ai_roundup_310818

Ramcharan, A., Baranowski, K., Mccloskey, P., Ahmed, B., Legg, J., & Hughes, D. P. (2017). *Deep Learning for Image-Based Cassava Disease Detection*. Academic Press. doi:10.3389/fpls.2017.01852

Ramesh. (2018). IoT based smart water distribution system. *International Conference on Applied Soft Computing Techniques ICASCT-18.*

Ramnath, S., Javali, A., Narang, B., Mishra, P., & Routray, S. K. (2017, May). IoT based localization and tracking. In *IoT and Application (ICIOT), 2017 International Conference on* (pp. 1-4). IEEE. 10.1109/ICIOTA.2017.8073629

Randhawa, G., & Sidhu, A. S. (2011). Status of public health care services during the process of liberalization: A study of Punjab. *Pravara Management Review*, *10*(2), 10–15.

Ratasuk, R., Mangalvedhe, N., Zhang, Y., Robert, M., & Koskinen, J. P. (2016, October). Overview of narrowband IoT in LTE Rel-13. In *Standards for Communications and Networking (CSCN), 2016 IEEE Conference on* (pp. 1-7). IEEE. 10.1109/CSCN.2016.7785170

Ratasuk, R., Vejlgaard, B., Mangalvedhe, N., & Ghosh, A. (2016, April). NB-IoT system for M2M communication. In *Wireless Communications and Networking Conference (WCNC)*, 2016 *IEEE* (pp. 1-5). IEEE.

Rathore, M. M., Ahmad, A., Paul, A., & Rho, S. (2016). Urban planning and building smart cities based on the internet of things using big data analytics. *Computer Networks, 101*, 63–80. doi:10.1016/j.comnet.2015.12.023

Rawidean, M., Kassim, M., & Harun, A. N. (2017). Wireless Sensor Networks and Cloud Computing Integrated Architecture for Agricultural Environment Applications. *International Conference on Sensing Technology.*

Raya, M., & Hubaux, J. P. (2007). Securing vehicular ad hoc networks. *Journal of Computer Security, 15*(1), 39–68. doi:10.3233/JCS-2007-15103

Regmi, J. (2017). Rupture Dynamics and Seismological Variables in Earthquake. The Himalayan Physics, 6-7, 96-99.

Rehman, S. U., Khan, I. U., Moiz, M., & Hasan, S. (2016). Security and privacy issues in IoT. *International Journal of Communication Networks and Information Security*, *8*(3), 147.

Renou, S., Givaudan, J. G., Poulain, S., Dirassouyan, F., & Moulin, P. (2008). Landfill leachate treatment: Review and opportunity. *Journal of Hazardous Materials*, *150*(3), 468–493. doi:10.1016/j.jhazmat.2007.09.077 PMID:17997033

Reshma, Mundhe, & Dabhade. (2016). *Environmental Monitoring For IoT Applications Based On Wireless Sensor Network*. Academic Press.

Riazul Islam, S. M., & Daehan Kwak, Md. (2015). *The Internet of Things for Health Care: A Comprehensive Survey. IEEE Access, 3*. doi:10.1109/ACCESS.2015.2437951

Rico, J. (2014). *Going beyond monitoring and actuating in large scale smart cities*. NFC & Proximity Solutions - WIMA Monaco.

Rohde, R. (2018a, November 5). *New Delhi - PM2.5 - Last 14 days*. Retrieved November 9, 2018, from https://twitter.com/RARohde/status/1059323321265852422

Rohde, R. (2018b, October 31). *Particulate Air Pollution in New Delhi, India for 2016, 2017, 2018*. Retrieved November 9, 2018, from https://twitter.com/RARohde/status/1057532316170551296

Roman, D. H., & Conlee, K. D. (2015). *The Digital Revolution Comes to US Healthcare*. Goldman Sachs.

Roman, R., Zhou, J., & Lopez, J. (2013). On the features and challenges of security and privacy in distributed internet of things. *Computer Networks*, *57*(10), 2266–2279. doi:10.1016/j.comnet.2012.12.018

Roueff, A., Chanussot, J., & Mars, I. J. (2006). Estimation of polarization parameters using time-frequency representations and its application to wave separation. *Signal Processing*, *86*(12), 3714–3731. doi:10.1016/j.sigpro.2006.03.019

Routray, S. K., & Sharmila, K. P. (2017, February). Green initiatives in IoT. In *Advances in Electrical, Electronics, Information, Communication and Bio-Informatics (AEEICB), 2017 Third International Conference on* (pp. 454-457). IEEE. 10.1109/AEEICB.2017.7972353

Routray, S. K., Jha, M. K., Sharma, L., Nyamangoudar, R., Javali, A., & Sarkar, S. (2017, May). Quantum cryptography for IoT: A Perspective. In *IoT and Application (ICIOT), 2017 International Conference on* (pp. 1-4). IEEE. 10.1109/ICIOTA.2017.8073638

Rupa, Ms., Rajni Kumari, Ms., Nisha Bhagchandani, Ms., & Ashish Madhur, Mr. (2018, May). Smart Garbage Management System Using Internet of Things (IoT) For Urban Areas. *IOSR Journal of Engineering*, *08*(5), 78–84.

Rupani, P., Singh, R., & Ibrahim, M. (2010). *Review of current palm oil mill effluent (POME) treatment methods: vermicomposting as a sustainable practice*. World Appl. Sci.

Sadhukhan, P. (2017, September). An IoT-based E-parking system for smart cities. In *2017 International Conference on Advances in Computing, Communications and Informatics (ICACCI)* (pp. 1062-1066). IEEE. 10.1109/ICACCI.2017.8125982

Sahitya, Balaji, Naidu, & Abinaya. (2017). Designing a Wireless Sensor Network for Precision Agriculture using Zigbee. *IEEE International Advance Computing Conference*, 287-291. 10.1109/IACC.2017.0069

Sahitya, G. (2016). Wireless Sensor Network for Smart Agriculture. Academic Press.

Sahoo, S., & Das, J. K. (2018). Bearing Fault Detection and Classification Using ANC-Based Filtered Vibration Signal. In *International Conference on Communications and Cyber Physical Engineering* (pp. 325-334). Academic Press.

Sai Ram, K. S., & Gupta, A. N. P. S. (2016). IoT based Data Logger System for weather monitoring using Wireless sensor networks. *International Journal of Engineering Trends and Technology*, *32*(2), 71–75. doi:10.14445/22315381/IJETT-V32P213

Saleh, A. F., Kamarudin, E., Yaacob, A. B., Yussof, A. W., & Abdullah, M. A. (2012). Optimization of biomethane production by anaerobic digestion of palm oil mill effluent using response surface methodology. Asia-Pacific. *Chemical Engineering Journal*, *7*, 353–360.

Salman, T., & Jain, R. (2017). Advanced. *Computer Communications*, *1*(1).

Saravanan, K., Julie, E. G., & Robinson, Y. H. (2019). Smart Cities & IoT: Evolution of Applications, Architectures & Technologies, Present Scenarios & Future Dream. In *Internet of Things and Big Data Analytics for Smart Generation* (pp. 135–151). Cham: Springer. doi:10.1007/978-3-030-04203-5_7

Saravanan, K., & Srinivasan, P. (2018). Examining IoT's applications using cloud services. In *Examining cloud computing technologies through the Internet of Things* (pp. 147–163). IGI Global. doi:10.4018/978-1-5225-3445-7.ch008

Satyawali, Y., & Balakrishnan, M. (2008). Wastewater treatment in molasses-based alcohol distilleries for COD and color removal: A review. *Journal of Environmental Management*, *86*(3), 481–497. doi:10.1016/j.jenvman.2006.12.024 PMID:17293023

Schaffers, H., Komninos, N., Pallot, M., Trousse, B., Nilsson, M., & Oliveira, A. (2011). Smart cities and the future internet: Towards cooperation frameworks for open innovation. In *The future internet assembly* (pp. 431–446). Berlin: Springer. doi:10.1007/978-3-642-20898-0_31

Schlienz, J., & Raddino, D. (2016). *Narrowband internet of things whitepaper.* White Paper, Rohde&Schwarz.

Securityinfowatch.com. (2012). *STE Security Innovation Awards Honorable Mention: The End of the Disconnect.* Retrieved from https://www.securityinfowatch.com/video-surveillance/video-transmission-equipment/article/10840006/innovative-wireless-network-connects-new-york-waterways-ferries-for-safety-security-roi-and-more

Seebo Blog. (2018). *What's So Smart About Intelligent Maintenance Systems.* Retrieved from https://blog.seebo.com/whats-so-smart-about-intelligent-maintenance-systems

Sen, R., & Raman, B. (2012). *Intelligent transport systems for Indian cities.* Paper presented at 6th USENIX/ACM Workshop on Networked Systems for Developing Regions.

Sensors for Smart Cities. (2015). *ESRI India*, 9. Retrieved May 24, 2017, from http://www.esri.in/~/media/esri-india/files/pdfs/news/arcindianews/Vol9/sensors-for-smart-cities.pdf?la=en

Sertyesilisik, B., & Sertyesilisik, E. (2015). Sustainability Leaders for Sustainable Cities. In Leadership and Sustainability in the Built Environment. Spon Research, Routledge.

Sertyesilisik, B., & Sertyesilisik, E. (2016). Eco industrial Development: As a Way of Enhancing Sustainable Development. *Journal of Economic Development Environment and People*, *5*(1), 6–27. doi:10.26458/jedep.v5i1.133

Severi, S., Abreu, G., Sottile, F., Pastrone, C., Spirito, M., & Berens, F. (2014). M2M Technologies: Enablers for a Pervasive Internet of Things. *The European Conference on Networks and Communications (EUCNC2014).* 10.1109/EuCNC.2014.6882661

SF Park Pilot Evaluation. (n.d.). Retrieved July 5, 2017, from http://sfpark.org/about-the-project/pilot-evaluation/

Shah & Venkatesan. (n.d.). Authentication of IoT Device and IoT Server Using Vaults. In *Computing And Communications 12th IEEE International Conference On Big Data Science And Engineering*. IEEE. DOI 10.1109/TrustCom/BigDataSE.2018.00117

Shaheryar, A., Xu-Cheng, Y., & Ramay, W. Y. (2017). Robust Feature Extraction on Vibration Data under Deep-Learning Framework: An Application for Fault Identification in Rotary Machines. *International Journal of Computers and Applications*, *167*(4).

Shamshiri, R. R., Kalantari, F., Ting, K. C., Thorp, K. R., Hameed, I. A., Weltzien, C., … Shad, Z. M. (2018). *Advances in greenhouse automation and controlled environment agriculture : A transition to plant factories and urban agriculture*. Academic Press. doi:10.25165/j.ijabe.20181101.3210

Shanthi, M., Banu, J. R., & Sivashanmugam, P. (2018). Effect of surfactant assisted sonic pretreatment on liquefaction of fruits and vegetable residue: Characterization, acidogenesis, biomethane yield and energy ratio. *Bioresource Technology*, *264*, 35–41. doi:10.1016/j.biortech.2018.05.054 PMID:29783129

Sharif-Yazd, M., Khosravi, M. R., & Moghimi, M. K. (2017). A Survey on Underwater Acoustic Sensor Networks: Perspectives on Protocol Design for Signaling, MAC and Routing. *Journal of Computer and Communications*, *5*(05), 12–23. doi:10.4236/jcc.2017.55002

Sharma, L., Javali, A., Nyamangoudar, R., Priya, R., Mishra, P., & Routray, S. K. (2017, July). An update on location based services: Current state and future prospects. In *Computing Methodologies and Communication (ICCMC), 2017 International Conference on* (pp. 220-224). IEEE.

Sharma, P. K., Moon, S. Y., & Park, J. H. (2017). Block-VN: A Distributed Blockchain Based Vehicular Network Architecture in Smart City. *JIPS*, *13*(1), 184–195.

Shashika Lokuliyana & Jeyakody. (2017). *IGOE Iot Framework For Waste Collection And Optimization*. NCTM.

Sher, Javaid, Azam, Ahmad, Abdul, Ghouzali, … Khan. (2017). Monitoring square and circular fields with sensors using energy-efficient cluster-based routing for underwater wireless sensor networks. *International Journal of Distributed Sensor Networks*, 13.

Shi, D., & Xu, B. (2018, June). Intelligent elevator control and safety monitoring system. *IOP Conference Series. Materials Science and Engineering*, *366*(1), 012076. doi:10.1088/1757-899X/366/1/012076

Shin, J. H., & Jun, H. B. (2014). A study on smart parking guidance algorithm. *Transportation Research Part C, Emerging Technologies*, *44*, 299–317. doi:10.1016/j.trc.2014.04.010

Shyam, G. K., Manvi, S. S., & Bharti, P. (2017, February). Smart waste management using Internet-of-Things (IoT). In *Computing and Communications Technologies (ICCCT), 2017 2nd International Conference on* (pp. 199-203). IEEE.

Sicari, S., Rizzardi, A., Grieco, L. A., & Coen-Porisini, A. (2015). Security, privacy and trust in Internet of Things: The road ahead. *Computer Networks*, *76*, 146–164. doi:10.1016/j.comnet.2014.11.008

Silva, B. N., Khan, M., & Ha, K. (2018). Towards sustainable smart cities: A review of trends, architectures, components, and open challenges in smart cities. *Sustainable Cities and Society*, *38*, 697–713. doi:10.1016/j.scs.2018.01.053

Singapore Water Story. (n.d.). Retrieved July 4, 2017, from https://www.pub.gov.sg/watersupply/singaporewaterstory

Singh, A., Aggarwal, P., & Arora, R. (2016, September). IoT based waste collection system using infrared sensors. In *Reliability, Infocom Technologies and Optimization (Trends and Future Directions), 2016 5th International Conference on* (pp. 505-509). IEEE. 10.1109/ICRITO.2016.7785008

Singhvi, A., Saget, B., & Lee, C. J. (2018). What went wrong with Indonesia's Tsunami warning system. *Asia Pacific, The New York Times*. Retrieve from https://www.nytimes.com/interactive/2018/10/02/world/asia/indonesia-tsunami-early-warning-system.html

Sino - Singapore Guangzhou Knowledge City. (2014). *A vision for a city today, a city of vision tomorrow*. Retrieved from http://www.ssgkc.com/index.asp

Smart Cities Mission Integrated Command and Control Centre. (2017). In *National workshop for project management consultants of smart cities*. Retrieved http://smartcities.gov.in/upload/uploadfiles/files/ICCC.pdf

Smart Education. (n.d.). Retrieved June 24, 2017, from http://www.ibigroup.com/new-smart-cities-landing-page/education-smart-cities/

Snehal, Shinde, Karode, & Suralkar. (2017). Review on - IOT based environment monitoring system. *International Journal of Electronics and Communication Engineering and Technology*, *8*(2), 103–108.

Solid Waste Management Rules. (2016). *Ministry of Environment*. Forest and Climate Change Government of India.

Srivastava, H. N. (Ed.). (1983). *Earthquakes Forecasting & Mitigation*. National Book Trust.

Stein, S., & Wysession, M. (2003). *An Introduction to Seismology, Earthquakes and Earth structure*. Blackwell Science, Inc.

Stephenes, A. O. (1970). *Acclimation of activated sludge to industrial wastes* (Unpublished master dissertation). McMaster University, Canada.

Suárez, A. D., Parra, O. J. S., & Forero, J. H. D. (2018). Design of an Elevator Monitoring Application using Internet of Things. *International Journal of Applied Engineering Research*, *13*(6), 4195–4202.

Surapaneni, P., Maguluri, L. P., & Symala, M. (2018). Solid Waste Management in Smart Cities using IoT. *International Journal of Pure and Applied Mathematics*, *118*(7), 635–640.

Surat Municipal Corporation. (2018, November). *Air Quality Statistics*. Retrieved November 11, 2018, from https://www.suratmunicipal.gov.in/Home/AirQualityInfo

Surat Smart City. (2018, March). *Air Quality and Water Quality Monitoring in Smart City Area*. Retrieved November 2018, from http://www.suratsmartcity.com/Documents/Projects/ABD/13.pdf

Swan, M. (2012). Sensor Mania! The Internet of Things, Wearable Computing, Objective Metrics, and the Quantified Self 2.0. *Sensor and Actuator Networks, 1*(3), 217 - 253.

Tabassum, S., Zhang, Y., & Zhang, Z. (2015). An integrated method for palm oil mill effluent (POME) treatment for achieving zero liquid discharge – A pilot study. *Journal of Cleaner Production*, *95*, 148–155. doi:10.1016/j.jclepro.2015.02.056

Taifor, A. F., Zakaria, M. R., Yusoff, M. Z. M., Toshinari, M., Hassan, M. A., & Shirai, Y. (2017). Elucidating substrate utilization in biohydrogen production from palm oil mill effluent by Escherichia coli. *International Journal of Hydrogen Energy*, *42*(9), 5812–5819. doi:10.1016/j.ijhydene.2016.11.188

Talari, S., Shafie-khah, M., Siano, P., Loia, V., Tommasetti, A., & Catalão, J. P. (2017). A review of smart cities based on the internet of things concept. *Energies*, *10*(4), 421. doi:10.3390/en10040421

Tallapragada, V. V. S., Rao, N. A., & Kanapala, S. (2017). EMOMETRIC: An IOT Integrated Big Data Analytic System for Real Time Retail Customer's Emotion Tracking and Analysis. *International Journal of Computational Intelligence Research*, *13*(5), 673–695.

Tallapragada, V. V. S., Rao, N. A., & Kanapala, S. (2017). Leaf Disease Detection Using Combined Feature of Texture, Colour and Wavelet Transform. *International Journal of Control Theory and Applications*, *10*(21), 159–167.

Tanappagol & Kondikopp. (2017). IoT Based Energy Efficient Environmental Monitoring Alerting and Controlling System. *International Journal of Latest Technology in Engineering Management & Applied Science*, *6*(7), 83–86.

Tan, H. M., Poh, P. E., & Gouwanda, D. (2018). Resolving stability issue of thermophilic high-rate anaerobic palm oil mill effluent treatment via adaptive neuro-fuzzy inference system predictive model. *Journal of Cleaner Production, 198*, 797–805. doi:10.1016/j.jclepro.2018.07.027

Tan, L., & Wang, N. (2010). Future Internet: The Internet of Things. *3rd International Conference on Advanced Computer Theory and Engineering (ICACTE)*, 5, 376 - 380.

Tan, Y. H., Goh, P. S., Ismail, A. F., Ng, B. C., & Lai, G. S. (2017). Decolourization of aerobically treated palm oil mill effluent (AT-POME) using polyvinylidene fluoride (PVDF) ultrafiltration membrane incorporated with coupled zinc-iron oxide nanoparticles. *Chemical Engineering Journal, 308*, 359–369. doi:10.1016/j.cej.2016.09.092

Tapashetti, S., & Shobha, K. R. (2018). Precision Agriculture using LoRa. *International Journal of Scientific & Engineering Research, 9*(5), 2023–2028.

Tapia, E. M., Intille, S. S., & Larson, K. (2004, April). Activity recognition in the home using simple and ubiquitous sensors. In *International conference on pervasive computing* (pp. 158-175). Springer. 10.1007/978-3-540-24646-6_10

Tee, P.-F., Abdullah, M. O., Tan, I. A. W., Mohamed Amin, M. A., Nolasco-Hipolito, C., & Bujang, K. (2016). Performance evaluation of a hybrid system for efficient palm oil mill effluent treatment via an air-cathode, tubular upflow microbial fuel cell coupled with a granular activated carbon adsorption. *Bioresource Technology, 216*, 478–485. doi:10.1016/j.biortech.2016.05.112 PMID:27268432

Tel Aviv Smart City. (n.d.). Retrieved May 27, 2017, from https://www.tel-aviv.gov.il/en/WorkAndStudy/Documents/Tel-Aviv%20Smart%20City%20(pdf%20booklet).pdf

Tendulkar, Sonawane, Vakte, Pujari, & Dhomase. (2016). A Review of Traffic Management System Using IoT. *International Journal of Modern Trends in Engineering and Research, 3*(4), 247–249.

Thangam, E. C., Mohan, M., Ganesh, J., & Sukesh, C. V. (2018). Internet of Things (IoT) based Smart Parking Reservation System using Raspberry-pi. *International Journal of Applied Engineering Research, 13*(8), 5759–5765.

The Economist Intelligence Unit. (2018). The Global Liveability Index 2018 A free overview. *The Economist.*

Theopoulos, A., Boursianis, A., Koukounaras, A., & Samaras, T. (2018). Prototype wireless sensor network for real-time measurements in hydroponics cultivation. *International Conference on Modern Circuits and Systems Technologies.* 10.1109/MOCAST.2018.8376576

Times of India. (2018, November 11). *Air pollution level and air quality index in Delhi today.* Retrieved November 11, 2018, from https://timesofindia.indiatimes.com/city/delhi/air-pollution-level-and-air-quality-index-in-delhi-today/articleshow/66470391.cms

Tipping, M. (2003). *U.S. Patent No. 6,633,857.* Washington, DC: U.S. Patent and Trademark Office.

Tissera, M. D., & McDonnell, M. D. (2016). Deep extreme learning machines: Supervised autoencoding architecture for classification. *Neurocomputing, 174*, 42–49. doi:10.1016/j.neucom.2015.03.110

Torretta, V., Ferronato, N., Katsoyiannis, I., Tolkou, A., & Airoldi, M. (2016). Novel and Conventional Technologies for Landfill Leachates Treatment: A Review. *Sustainability, 9*(1), 9. doi:10.3390u9010009

Traffic cops mull over intelligent system to decongest city. (2018, September 10). *The Times of India.* Retrieved February 27, 2019 from https://timesofindia.indiatimes.com/city/visakhapatnam/traffic-cops-mull-over-intelligent-system-to-decongest-city-roads/articleshow/65746292.cms

Traffic Index Rate. (2017). Retrieved March 21, 2017, from https://www.numbeo.com/traffic/rankings_current.jsp

Trak.in. (2016). *How IoT's are Changing the Fundamentals of "Retailing"*. Retrieved from https://trak.in/tags/business/2016/08/30/internet-of-things-iot-changing-fundamentals-of-retailing

Transport for London. (2014). *Tfl to launch world-leading trials of intelligent pedestrian technology to make crossing the road easier and safer*. Retrieved May 24, 2017, from https://tfl.gov.uk/info-for/media/press-releases/2014/march/tfl-to-launch-worldleading-trials-of-intelligent-pedestrian-technology-to-make-crossing-the-road-easier-and-safer

Trencher, G. (2018). Towards the smart city 2.0: Empirical evidence of using smartness as a tool for tackling social challenges. *Technological Forecasting and Social Change*. doi:10.1016/j.techfore.2018.07.033

Twentyman, J. (2017, September). *Athens International Airport turns to IoT for environmental monitoring*. Retrieved from https://internetofbusiness.com/athens-international-airport-turns-to-iot-for-environmental-monitoring/

Uddin, J., Kang, M., Nguyen, D. V., & Kim, J. M. (2014). Reliable fault classification of induction motors using texture feature extraction and a multiclass support vector machine. *Mathematical Problems in Engineering*.

UNEP. (2005). *Solid Waste Management*. Retrieved from www.unep.or.jp/ietc/publications/spc/solid_waste_management/Vol_I/Binder1.pdf

United Nations ESCWA. (2015). *Smart cities: Regional perspectives*. Retrieved April 5, 2017, from https://worldgovernmentsummit.org/api/publications/document/d1d75ec4-e97c-6578-b2f8-ff0000a7ddb6

United Nations. (2014). *The future we want: Sustainable cities*. Retrieved from http://www.un.org/en/sustainablefuture/cities.shtml#facts

Urban Transport. (2014). Retrieved March 24, 2017, from http://www.worldbank.org/en/country/india/brief/urban-transport

Uribe-Pérez, N., Hernández, L., de la Vega, D., & Angulo, I. (2016). State of the art and trends review of smart metering in electricity grids. *Applied Sciences*, *6*(3), 68. doi:10.3390/app6030068

Uusitalo, J. (2018). *Novel Sensor Solutions with Applications to Monitoring of Elevator Systems* (Master of Science thesis).

Vaisali, G., Sai Bhargavi, K., Kumar, S., & Satyanarayana, S. (2017, December). Smart solid waste management system by IOT. *International Journal of Mechanical Engineering and Technology*, *8*(12), 841–846.

Vakili, M., Rafatullah, M., Gholami, Z., & Farraji, H. (2016). Treatment of reactive dyes from water and wastewater through chitosan and its derivatives. In B. A. K. Mishra (Ed.), *Smart Materials for Waste Water Applications* (pp. 347–377). Wiley. doi:10.1002/9781119041214.ch14

Van Kooten, O., Heuvelink, E., & Stanghellini, C. (2008). New developments in greenhouse technology can mitigate the water shortage problem of the 21st century. *Acta Horticulturae*, (767), 45–52. doi:10.17660/ActaHortic.2008.767.2

Van Merode, D., Tabunshchyk, G., Patrakhalko, K., & Yuriy, G. (2016, February). Flexible technologies for smart campus. In *Remote Engineering and Virtual Instrumentation (REV), 2016 13th International Conference on* (pp. 64-68). IEEE. 10.1109/REV.2016.7444441

Vasagade, T. S., Tamboli, S. S., & Shinde, A. D. (2017). Dynamic solid waste collection and management system based on sensors, elevator and GSM. In *Inventive Communication and Computational Technologies, 2017 International Conference on* (pp. 263-267). IEEE. 10.1109/ICICCT.2017.7975200

Venticinque, S., & Amato, A. (2018). A methodology for deployment of IoT application in fog. *Journal of Ambient Intelligence and Humanized Computing*, 1–22.

Veroustraete, F. (2020). The Rise of the Drones in Agriculture. *Cronicon, 2*(2015), 325–327.

Vijayaraghavan, K., & Ahmad, D. (2006). Biohydrogen generation from palm oil mill effluent using anaerobic contact filter. *International Journal of Hydrogen Energy*, *31*(10), 1284–1291. doi:10.1016/j.ijhydene.2005.12.002

Viola, P., & Jones, M. (2002). Fast and robust classification using asymmetric adaboost and a detector cascade. In Advances in neural information processing systems (pp. 1311-1318). Academic Press.

Vongsingthong, S., & Smanchat, S. (2014). *Internet of Things: A review of applications & technologies. Suranaree Journal of Science and Technology.*

Vox Media. (2018, October 31). *Why India's air pollution is so horrendous*. Retrieved from https://www.vox.com/2018/5/8/17316978/india-pollution-levels-air-delhi-health

Vyas, Bhatt, & Jha. (2016). IoT: Trends, Challenges and Future Scope. *IJCSC, 7*(1), 186-197.

Vymazal, J. (2009). The use constructed wetlands with horizontal sub-surface flow for various types of wastewater. *Ecological Engineering*, *35*(1), 1–17. doi:10.1016/j.ecoleng.2008.08.016

Wahab, K. Tomari, & Jabbar. (2014). Smart Recycle Bin: A Conceptual Approach Of Smart Waste Management With Integrated Web Based System. Academic Press.

Wang & Liu. (2014). The application of internet of things in agricultural means of production supply chain management. *Journal of Chemical and Pharmaceutical Research*, *6*(7), 2304–2310.

Wang, X., Ge, H., Zhang, W., & Li, Y. (2015, September). Design of elevator running parameters remote monitoring system based on Internet of Things. In *Software Engineering and Service Science (ICSESS), 2015 6th IEEE International Conference on* (pp. 549-555). IEEE. 10.1109/ICSESS.2015.7339118

Wang, F., Cheng, Z., Reisner, A., & Liu, Y. (2018). Compliance with household solid waste management in rural villages in developing countries. *Journal of Cleaner Production*, *202*, 293–298. doi:10.1016/j.jclepro.2018.08.135

Wang, Y. P. E., Lin, X., Adhikary, A., Grovlen, A., Sui, Y., Blankenship, Y., ... Razaghi, H. S. (2017). A primer on 3GPP narrowband Internet of Things. *IEEE Communications Magazine*, *55*(3), 117–123. doi:10.1109/MCOM.2017.1600510CM

Wang, Z., & Zhao, B. (2017). Automatic event detection and picking of P, S seismic phase for earthquake early warning and application for the 2008 Wenchuan earthquake. *Soil Dynamics and Earthquake Engineering*, *97*, 172–181. doi:10.1016/j.soildyn.2017.03.017

Wan, Z., Yi, S., Li, K., Tao, R., Gou, M., Li, X., & Guo, S. (2015). Diagnosis of elevator faults with LS-SVM based on optimization by K-CV. *Journal of Electrical and Computer Engineering*, *2015*, 70. doi:10.1155/2015/935038

Weiser, M., Gold, R., & Brown, J. S. (1999). The origins of ubiquitous computing research at PARC in the late 1980s. *IBM Systems Journal*, *38*(4), 693–696. doi:10.1147j.384.0693

Wendling, Z. A., Emerson, J. W., Esty, D. C., Levy, M. A., & de Sherbinin, A. (2018). *2018 Environmental Performance Index*. New Haven, CT: Yale Centre for Environmental Law & Policy. Retrieved November 2018, from https://epi.yale.edu/

Whetton, R. L., Hassall, K. L., Waine, T. W., & Mouazen, A. M. (2017). ScienceDirect Hyperspectral measurements of yellow rust and fusarium head blight in cereal crops : Part 1 : Laboratory study. *Biosystems Engineering*, *166*, 101–115. doi:10.1016/j.biosystemseng.2017.11.008

Wiemer, S. (2015). Earthquake Statistics and Earthquake Prediction Research. Institute of Geophysics, Zurich, Switzerland.

Williams, M. (2016). What is an Earthquake? *Universe Today, Space and astronomy news*. Retrieved from https://www.universetoday.com/47813/what-is-an-earthquake

Wipro Limited. (2015). *Mobility in Urban India*. Retrieved May 24, 2017, from http://www.wipro.org/earthian/pdf/mobility-in-urban-India.pdf

Wong, K. A., Lam, S. M., & Sin, J. C. (2018). Wet chemically synthesized ZnO structures for photodegradation of pre-treated palm oil mill effluent and antibacterial activity. *Ceramics International*. doi:10.1016/j.ceramint.2018.10.078

World Economic Forum. (2016). *Inspiring future cities & urban services*. Retrieved March 21, 2017, from http://www3.weforum.org/docs/WEF_Urban-Services.pdf

World Health Organization (WHO) India. (2018). *Air Pollution*. Retrieved November 2018, from http://www.searo.who.int/india/topics/air_pollution/en/

Wowk, V. (2005). *A Brief Tutorial on Machine Vibration*. Machine Dynamics, Inc.

Wrona, K. (2015). Securing the Internet of Things a military perspective. *Proceedings of the IEEE 2nd World Forum on Internet of Things (WF-IoT)*, 502–507. 10.1109/WF-IoT.2015.7389105

Wu, Chiang, Chang, & Chang. (2017). An Interactive Telecare System Enhanced with IoT Technology. In *Pervasive Computing*. IEEE.

Wynn, R. B., Huvenne, V. A. I., Le Bas, T. P., Murton, B. J., Connelly, D. P., Bett, B. J., ... Hunt, J. E. (2014). Autonomous Underwater Vehicles (AUVs): Their past, present and future contributions to the advancement of marine geosciences. *Marine Geology*, *352*, 451–468. doi:10.1016/j.margeo.2014.03.012

Xie, X. F. (2016). *Key Applications of the Smart IoT to Transform Transportation*. Retrieved from http://www.wiomax.com/what-can-the-smart-iot-transform-transportation-and-smart-cities

Xie, X. F., & Wang, Z. J. (2017). *Integrated in - vehicle decision support system for driving at signalized intersections: A prototype of smart IoT in transportation*. Transportation Research Board (TRB) *Annual Meeting*, Washington, DC.

Xu, J., Riccioli, C., & Sun, D. (2016). Comparison of Vis or NIR Hyperspectral Imaging and Computer Vision for Automatic Differentiation of Organically and Conventionally Farmed Salmon. *Journal of Food Engineering*. doi:10.1016/j.jfoodeng.2016.10.021

Yalmaz, G., & Öztürk, I. (2001). Biological ammonia removal from anaerobically pre-treated landfill leachate in sequencing batch reactors (SBR). *Water Science and Technology*, *43*(3), 307–314. doi:10.2166/wst.2001.0151 PMID:11381921

Yamaski, E. (2012). What We Can Learn from Japan's Early Earthquake Warning System. *Momentum*, *1*(1), 2.

Yang, C., Shen, W., & Wang, X. (2018). The Internet of Things in Manufacturing: Key Issues and Potential Applications. *IEEE Systems, Man, and Cybernetics Magazine*, *4*(1), 6 - 15.

Yan, G., Yang, W., Rawat, D. B., & Olariu, S. (2011). SmartParking: A secure and intelligent parking system. *IEEE Intelligent Transportation Systems Magazine*, *3*(1), 18–30. doi:10.1109/MITS.2011.940473

Yang, D. H., Kim, K. Y., Kwak, M. K., & Lee, S. (2017). Dynamic modeling and experiments on the coupled vibrations of building and elevator ropes. *Journal of Sound and Vibration*, *390*, 164–191. doi:10.1016/j.jsv.2016.10.045

Yao, Z., Wan, J., Li, X., Shi, L., & Qian, J. (2011). The Design of Elevator Failure Monitoring System. In *Advances in Automation and Robotics* (Vol. 1, pp. 437–442). Berlin: Springer. doi:10.1007/978-3-642-25553-3_54

Yasin, M., Tekeste, T., Saleh, H., Mohammad, B., Sinanoglu, O., & Ismail, M. (2017). Ultra-Low Power, Secure IoT Platform for Predicting Cardiovascular Diseases. *IEEE Transactions on Circuits and Systems. I, Regular Papers*, *64*(9), 2624–2637. doi:10.1109/TCSI.2017.2694968

Yeh. (2016). A Secure IoT-Based Healthcare System With Body Sensor Networks. *IEEE Access, 4.* . doi:10.1109/ACCESS.2016.2638038

Yeom, I. T., Sharmila, V. G., Banu, J. R., Kannah, R. Y., & Sivashanmugham, P. (2018). *Municipal waste management. In municipal and industrial waste: source, management practices and future challenges.* Nova Science Publisher.

Yi, X. (2016, December). Design of Elevator Monitoring Platform on Big Data. In *Industrial Informatics-Computing Technology, Intelligent Technology, Industrial Information Integration (ICIICII), 2016 International Conference on* (pp. 40-43). IEEE. 10.1109/ICIICII.2016.0021

Yigitcanlar, T., & Kamruzzaman, M. (2018). Does smart city policy lead to sustainability of cities? *Land Use Policy, 73*, 49–58. doi:10.1016/j.landusepol.2018.01.034

Youtube website. (n.d.a). *Anakata Wind Power Resources - Innovative Wind Turbine Technology.* Retrieved from https://www.youtube.com/watch?v=Rhh1zWM6SiQ

Youtube website. (n.d.b). *Baker Turbo-Vortex Wind Turbine Turbina.* Retrieved from https://www.youtube.com/watch?v=wTeiSHpbFt4

Youtube website. (n.d.c). *4 Most Popular Vertical Wind Turbines.* Retrieved from https://www.youtube.com/watch?v=a3n-VBpcqzM

Youtube website. (n.d.d). *Funnel wind turbine: radical new design harnesses 600% more electricity from wind – TomoNews.* Retrieved from https://www.youtube.com/watch?v=im8W4z4og-8

Youtube website. (n.d.e). *Future of Wind Energy - new Vertical axis Wind Turbine invention.* Retrieved from https://www.youtube.com/watch?v=Bmz_YZOWdV8

Youtube website. (n.d.f). *Floating wind turbine takes to the sky - BBC Click.* Retrieved from https://www.youtube.com/watch?v=RzCK9Ht0SWk

Youtube website. (n.d.g). *Heppolt Wind Turbine Progress Report - Wind Turbine New 2015.* Retrieved from https://www.youtube.com/watch?v=IaplRH7ldzQ

Youtube website. (n.d.h). *Wind Tulip in Jerusalem.* Retrieved from https://www.youtube.com/watch?v=28ok_7bSFHc

Youtube website. (n.d.i). *Introducing the Altaeros BAT: The Next Generation of Wind Power.* Retrieved from https://www.youtube.com/watch?v=kldA4nWANA8

Youtube website. (n.d.j). *New Wind Power System: Polish engineers develop more efficient wind turbine system.* Retrieved from https://www.youtube.com/watch?v=61Ekas-xbfU

Youtube website. (n.d.k). *New Wind Turbine FloDesign.* Retrieved from https://www.youtube.com/watch?v=WB5CawKfE2M

Youtube website. (n.d.l). *SeaTwirl puts a new spin on offshore wind turbines.* Retrieved from https://www.youtube.com/watch?v=Ccs3RP9LxIY

Youtube website. (n.d.m). *Sky Wolf Wind Turbine Animation 2014.* Retrieved from (https://www.youtube.com/watch?v=jGTO886FKMA)

Youtube website. (n.d.n). *Urban Power USA new 10 KW vertical wind turbine.* Retrieved from https://www.youtube.com/watch?v=oIAq3wvVOBA

Youtube website. (n.d.o). *Windjuicer a new wind power technology.* Retrieved from https://www.youtube.com/watch?v=NgOjbNy2HFk

Youtube website. (n.d.p). *WindTamer Turbines WindTamerTurbines.com.* Retrieved from https://www.youtube.com/watch?v=mpUPlHx_2gw

Youtube website. (n.d.q). *windtrap - a new wind turbine*. Retrieved from https://www.youtube.com/watch?v=ldGafXZdvlo

Youtube website. (n.d.r). *Wind turbines of the future*. Retrieved from https://www.youtube.com/watch?v=18ogee_Gj7k

Yushi, L., Fei, J., & Hui, Y. (2012). Study on application modes of military internet of things (miot). *IEEE International Conference on Computer Science and Automation Engineering (CSAE)*, 630–634. 10.1109/CSAE.2012.6273031

Zahan, K. A., & Kano, M. (2018). Biodiesel Production from Palm Oil, Its By-Products and Mill Effluent: A Review. *Energies*, *11*(8), 2132. doi:10.3390/en11082132

Zahrim, A., Rachel, F., Menaka, S., Su, S., & Melvin, F. (2009). Decolourisation of anaerobic palm oil mill effluent via activated sludge-granular activated carbon. *World Applied Sciences Journal*, 126–129.

Zambrano, M., Perez, I., Palau, C., & Esteve, M. (2016). Technologies of Internet of Things applied to an Earthquake Early Warning System. *Future Generation Computer Systems*. doi:10.1016/j.future.2016.10.009

Zanella, A. (2014). *Internet of Things for Smart Cities. IEEE Internet of Things Journal, 1(1).*

Zavare, S., Parashare, R., Patil, S., Rathod, P., & Babanne, V. (2017). Smart City Waste Management System Using GSM. *Int. J. Comput. Sci. Trends Technol.*, *5*, 74–78.

Zawani, Z., Chuah-Abdullah, L., Ahmadun, F. R., & Abdan, K. (2013). Acclimatization process of microorganisms from activated sludge in Kenaf-Retting Wastewater. In Developments in Sustainable Chemical and Bioprocess Technology (pp. 59-64). Springer. doi:10.1007/978-1-4614-6208-8_8

Zayas, A. D., & Merino, P. (2017, May). The 3GPP NB-IoT system architecture for the Internet of Things. In *Communications Workshops (ICC Workshops), 2017 IEEE International Conference on* (pp. 277-282). IEEE.

Zhang, K., Ni, J., Yang, K., Liang, X., Ren, J., & Shen, X. S. (2017). Security and privacy in smart city applications: Challenges and solutions. *IEEE Communications Magazine*, *55*(1), 122–129. doi:10.1109/MCOM.2017.1600267CM

Zhang, Q. (2015). *Precision Agriculture Technology for Crop Farming*. CRC Press. doi:10.1201/b19336

Zhang, Q. Q., Tian, B. H., Zhang, X., Ghulam, A., Fang, C. R., & He, R. (2013). Investigation on characteristics of leachate and concentrated leachate in three landfill leachate treatment plants. *Waste Management (New York, N.Y.)*, *33*(11), 2277–2286. doi:10.1016/j.wasman.2013.07.021 PMID:23948053

Zhang, Y., Sun, X., Zhao, X., & Su, W. (2018). Elevator ride comfort monitoring and evaluation using smartphones. *Mechanical Systems and Signal Processing*, *105*, 377–390. doi:10.1016/j.ymssp.2017.12.005

Zhao, J., & Cao, G. (2008). VADD: Vehicle-assisted data delivery in vehicular adhoc networks. *IEEE Transactions on Vehicular Technology*, *57*(3), 1910–1922. doi:10.1109/TVT.2007.901869

Zhou, Y., Wang, K., & Liu, H. (2018). An Elevator Monitoring System Based On The Internet Of Things. *Procedia Computer Science*, *131*, 541–544. doi:10.1016/j.procs.2018.04.262

Zinatizadeh, A. A. L., Mohamed, A. R., Abdullah, A. Z., Mashitah, M. D., Hasnain Isa, M., & Najafpour, G. D. (2006). Process modeling and analysis of palm oil mill effluent treatment in an up-flow anaerobic sludge fixed film bioreactor using response surface methodology (RSM). *Water Research*, *40*(17), 3193–3208. doi:10.1016/j.watres.2006.07.005 PMID:16949124

Zolfaghari, M., Jardak, K., Drogui, P., Brar, S. K., Buelna, G., & Dubé, R. (2016). Landfill leachate treatment by sequential membrane bioreactor and electro-oxidation processes. *Journal of Environmental Management*, *184*, 318–326. doi:10.1016/j.jenvman.2016.10.010 PMID:27733297

Zubair, Szewczyk, Valli, Rabadia, Hannay, Chernyshev, … Peacock. (2017). Future challenges for smart cities: Cyber-security and digital forensics. Security Research Institute & School of Science, Edith Cowan University.

About the Contributors

Krishnan Saravanan is working as a Senior Assistant professor, Department of Computer Science & Engineering at Anna University, Regional Campus, Tirunelveli, Tamilnadu. He has published papers in 12 international conferences and 23 international journals. He has also written 6 book chapters and one edited book with international publishers. He is an active researcher and academician. Also, he is reviewer for many reputed journals in Elsevier, IEEE, etc.

Golden Julie received her B.E degree in Computer Science and Engg in 2005 from Anna University Chennai and ME degree in Computer Science and Engineering in 2008 from Anna University Chennai. He finished PhD in Anna University, Chennai in 2016.Presently she is working as assistant professor in Regional centre Anna university, Tirunelveli, India She has published many research papers in various fields. Her research area includes Wireless Sensor Adhoc Networks and Image Processing. She is a member of ISTE.

* * *

Rajesh Banu, Assistant Professor, Department of Civil Engineering, Anna University Regional Campus-Tirunelveli was formerly Post-Doctoral Fellow and Lecturer in Department of Civil and Environmental Engineering, Sungkyunkwan University, South Korea. He also had an experience of working in an environmental consultancy for one year. He has been the reviewer of many national and international journals. He is the author of more than 180 scientific papers. His publications achieved 2900 citation and 31 H-Index. He has guided 12 Ph.D students and 55 master degree students. Dr. Rajesh banu's research centers on biological wastewater treatment, Nutrient removal, Membrane Bioreactor mesophilic and thermophilic high rate anaerobic treatment of wastewater, Microbial Fuel cells, Energy recovery from waste and Sludge reduction. His prime area of research is generation of energy from solid waste.

Payal Patel is an experienced professional with a demonstrated history of working in the environmental services industry. Payal is currently leading Business Development for Oizom Instruments. She is passionate about supporting the UN Sustainable Development Goals and intend to work and spread awareness about attaining Sustainable Environment through Environment Data Acquisitions and Analytics.

Thilagavathi C. is currently working as an Assistant Professor in IT Department at M. Kumarasamy College of Engineering.

Rani Chellasamy is currently working as an Assistant Professor in the Department of Computer Science and Engineering, Government College of Engineering, Salem. She received his B.Tech, M.Tech. and PhD in Information and Communication Engineering from Madurai Kamaraj University, Anna University Chennai, Anna University Chennai in 2003, 2008 and 2012 respectively. Her research interest includes development of intelligent optimization algorithms for data mining.

Vincenzo Cimino completed his Master's Degree in Computer Science at University of Palermo, Italy, in 2018. Vincenzo's favorite subjects are Big Data and Machine Learning and he has a great interest in open source development, Internet of Things and web design. Vincenzo began programming in C during high school and has since expanded his languages to C++, Java, Javascript, Python and Php. In his spare time Vincenzo enjoys photography, robotics and graphic design.

Deepa Devassy is currently working as an Assistant Professor in CSE at Sahrdaya College of Engineering and Technology, Thrissur.

Usha Devi is presently working as Head of Electronics and Communication Engineering Department in Regional Campus of Anna University at Tirunelveli, Tamilnadu, India since 2008.

Gayatri Doctor with a background of an experience in the IT Industry has now been in Academics for the past two decades. She is currently an Associate Professor at the Faculty of Management, CEPT University and is engaged in the field of Urban Management from a technology perspective. Her interests include exploring various technologies, their applications in the urban context, analyzing the user acceptance, challenges of using these technologies, understanding the use of Internet of Things (IoT), Open Data in future cities. She teaches studio courses like Ward Management, Urban Services Management and courses like e-governance, IT in Urban, Future Cities and Technology to share the concepts and applications of technology usage to improve services, quality of life in the Urban environment. She has a number of publications in International Journals, Conferences and edited books. Her areas of expertise - Smart Cities, E-Governance, Technologies in Cities, Enterprise Resource Planning, Digital Repositories, Knowledge Management.

Hossein Farraji is a Ph.D in Civil (Environmental) Engineering from Universiti Sains Malaysia (USM). Author completed his Master of (Horticulture)Engineering at Azad University of Tehran (AUT), central unit of science and researches in 1999. Author has completed his Bachelor of (Horticulture) Engineering in 1997. He has 11 years governmental job experiences in Institute of Standard and Industrial Researches of Iran (ISIRI). His research area is application green technology (phytoremediation and constructed wetland) for wastewater treatment. Three years of working experiences as Graduate Assistant in USM followed by Postdoctoral researches on Medicinal Plant production under LED light in plant Factory system at Eman Biodiscovery (USM).

Ramesh Kesavan is working as an Assistant Professor in Department of Computer Applications, Anna University Regional Campus, Tirunelveli, India. He received his B.Sc in Maths from ManonmaniamSundaranar University in 1998, MCA from ManonmaniamSundaranar University in 2001, M.Phil in Computer Science from Periyar University in 2008 and Ph.D from Anna University, Chennai in 2014. He has published more than 20 papers in well renowned International and National Journals and Conferences and guiding 10 research scholars. His areas of interests are Data Mining, Prediction, Forecasting, Time-series analysis, Classification, Clustering and Big Data Analytics.

M. Dinesh Kumar is currently a research scholar in Department of Civil Engineering, Anna University Regional Campus, Tirunelveli.

Sathish Kumar received his B.Tech, M.E, and Ph.D in Information Technology, Computer Science Engineering and Information and Communication Engineering from Anna University Coimbatore and Anna University Chennai, Anna University Chennai in 2011, 2013, and 2019 respectively. He is currently working as Assistant Professor in Computer Science and Engineering department at Presidency University Bangalore in Karnataka. He received best paper award in IEEE sponsored International Conference on Advanced Computing & Communication System, India. His research area includes Big Data Analytics, IoT and Machine Learning Techniques.

Venkatanaresh M. works as assistant professor in the Department of ECE at Sree Vidyanikethan Engg. College.

Starvin M. S., B.E (Mechanical Engineering) in Dr. Sivanthi Aditanar College Of Engineering, Tiruchendure, Tamilnadu, India. M.E. (Manufacturing Engineering) in Government College of Technology, Coimbatore, Tamilnadu, India. He got Two years Industrial experience and worked as a Senior Research Fellow under the BRNS funded project for two years. He completed his PhD in the Department of Mechanical Engineering in 2015. Ten years Engineering college teaching experience. He is working as an Assistant Professor in University College of Engineering Nagercoil, Tamilnadu, India. He Published 12 International Journal papers. He also published 11 International conferences and 3 National Conference papers. He has undertaken and completed a sponsored project worth of 9.57 lacks from DRDO, India. Dr.M.S.Starvin is a life member in ISTE. He is expertise in the area of Finite Element Analysis, Numerical simulation using ABAQUS, ANSYS. He handled the Finite Element Analysis subject to the Under Graduate as well as Post Graduate Students. He also conducted various workshops and delivered expert talk on Finite Element Analysis and Composite Materials.

Kavitha Margret Mansingh works as an assistant professor in the department of Computer scienec and Engineering, Sri Krishna college of Technology, Coimbatore. she had received her UG Degree from Madurai Kamaraaj University and her PG with Anna University . her area of interest is wireless sensor networks in which she had presented many number of papers in international and national journals . she presently working on waste management in smartcity with recent tools.

Amin Mojiri received his PhD in 2015 from University Science Malaysia. He was working as Post-doctoral Research Fellow in Shanghai Jiao Tong University- China and University Technology Mara-Malaysia. He is currently working at Hiroshima University as JSPS Research Fellow. He could publish more than 40 papers in International Journals.

Thangaraj Muthuraman is an Associate Professor at the Department of Computer Science, Madurai Kamaraj University, Madurai. His research interests are in the area of Data Semantics, Networks and Wireless Sensor Networks. At MKU additionally, he is also a Resource Executive of many academic councils. He has worked on many Governments sponsored UGC projects that involve innovation and Domain specific benefits. He has over two decades of experience in the field of Information Technology.

Mahima Nanda received her B.Tech Degree in Computer Science Engineering in 2013 and then did her MBA in Human Resource Management from Panjab University, Chandigarh. After that, she worked in a private company for one and half year and then went on to pursue her Ph.D in Management from Guru Nanak Dev University, Amritsar in 2016. She has keen interest in studying about human resource management as well as labour and social welfare issues.

P. J. Beslin Pajila received the B.E degree from the Computer Science and Engineering Department, Karunya University, Coimbatore, India, in 2009, and the M.E degree from the Computer science and Engineering Department, Einstein College of Engineering, Tirunelveli, India, in 2011. Currently working as an assistant professor with the Computer Science and Engineering at Francis Xavier Engineering College, Tirunelveli and currently pursuing the Ph.D. degree with Anna University, Chennai.

Punitha Ponmalar Pichaiah is now an Associate Professor in Sri Meenakshi Government College, Madurai. She received her post-graduate degree in Computer Science from Alagappa University, Karaikudi and M.Phil. degree in Computer Science from Mother Terasa University, Madurai. Her research interests include Wireless Sensor Networks, QOS in sensor networks and Adhoc networks.

S. Suja Priyadharsini received her Bachelor Degree in 2001 from Manonmaiam Sundaranar University, Tirunelveli, Tamilnadu. She completed her ME in Applied Electronics from Anna University, Chennai, Tamilnadu in 2006. She was conferred her PhD in Information and communication Engineering by Anna University, Chennai, Tamilnadu in 2015. She has been working as an Assistant Professor in Anna University Regional Campus, Tirunelveli, Tamilnadu since 2008. Her area of interests includes signal processing and soft computing.

Saravanan Radhakrishnan holds a Masters degree in Computer Science and is currently a Research Scholar in Vellore Institute of Technology. He has 22 years of experience in Software Development and is now a Founder and Chief Technology Officer of a Technology Startup focusing on Artificial Intelligence, Internet of Things and Robotics for varied domains including Farm Automation. His research interests are in the areas of Deep Learning, Vision Computing and Internet of Things. His passion lies in creatively combining technology and nature; his fully automated roof-top garden and many of the automations in his house are the results of this passion.

Gurpreet Randhawa is Assistant Professor, University Business School (UBS), Guru Nanak Dev University, Amritsar, Punjab. Dr. Randhawa has received her B. Tech degree from Punjab Agricultural University, Ludhiana, and MBA and PhD from Kurukshetra University, Kurukshetra. She has more than sixteen years of teaching and research experience and has published many research papers in the journals of national and international repute. Her research interests include job attitudes, employee behaviour and public health care system. She also worked as co-investigator in UGC Major Research Project on Status of Public Health Care Services in Punjab: A Case Study of Punjab Health Systems Corporation (PHSC).

Simona E. Rombo is Assistant Professor at Department of Mathematics and Computer Science (DMI) of the University of Palermo. Her main research interests are on Bioinformatics, Algorithms and Data Structures, Data Mining. Her main research contributions concern the design and development of efficient algorithms for the analysis of biological networks, sequences and structures. In particular, she contributed to the proposal of approaches for local/global alignment and querying of biological networks, clustering of protein-protein interaction networks, pattern discovery from biological sequences and 2D arrays, prediction and classification of protein structures and functions, data compression. More recently, she started working on Epigenomics, proposing a kmer based analysis for the study of chromatin organization in eukaryotes.

Sudhir K. Routray has got a BE degree in Electrical Engineering from Utkal University, India; MSc degree in Data Communications from University of Sheffield, UK; and PhD degree in Communication Engineering from University of Aveiro, Portugal. He has more than fourteen years of experience in teaching, research and industry. Currently, Dr Sudhir K. Routray works as an associate professor of Electrical Engineering at Addis Ababa Science and Technology University, Addis Ababa. He is a senior member of IEEE and volunteers for IEEE and several other non-profit organizations. He has more than sixty research articles in the form of journals, book chapters and conferences.

Anuradha S. has around 12 years of software industry experience in Microsoft, . Net, J2EE technologies that involve in analysis, development and integration of Web and Windows based applications, enterprise business servers. She is very strong in business process modelling and design.

Dynisha S. is a graduate student of Environmental Engineering at Regional campus of Anna University Tirunelveli.

Ramalakshmi S. received her bachelor degree in Electronics and Communication Engineering from Anna University, Chennai in 2016. She completed her master degree in Applied Electronics from Anna University Chennai in 2018. Her main area of interest is signal and image processing.

Begum Sertyesilisik is an Assoc. Prof. at the Department of Architecture in the Istanbul Technical University.

Rajeswari Shanmugam is currently working as an Associate Professor in the Department of Computer Science and Engineering in Sahrdaya College of Engineering and Technology, Thrissur, Kerala. She has received Ph.D degree in Information and Communication Engineering from Anna University,

Chennai in the year 2016. Completed M.E degree in Computer Science and Engineering in Nandha Engineering College, Tamilnadu in the year 2008 and did her B.Tech – Information Technology in Maharaja Engineering College, Tamilnadu. She is having more than 12 years of experience in teaching. She has published more than 22 papers in various International Journals and presented more than 14 papers in both national and International Conferences. She has written a chapter in a book named as Recent Development in Wireless Sensor and Ad-hoc Networks by Springer Publication. She is acting as a Reviewer of Computational Intelligence and Neuroscience journal by Hindawi Publication and International Journal of Wireless Personal Communication by Springer Publication. She has given a Guest Lecturer in various subjects such as Computer Architecture, System Software, Theory of Computation, Formal Languages and Automata Theory and Operating Systems in Premier Institutions. She has also given an invited talk in Anna University Sponsored FDP on Programming and Data Structures – II. She has acted as a jury in a National level project fair, Symposium and Various International Conferences. She has been awarded as a best faculty in the year 2010-2011 in Angel College of Engineering and Technology, Tirupur, Tamilnadu. She has attended various Seminars, Workshops and Faculty Development Programmes to enhance the knowledge of student's community. She is also an active life time member in Indian Society of Technical Education.

A. Sherly Alphonse has received her B.E degree from Manonmaniam Sundaranar university, India and M.E from Anna University, India. She has five years teaching experience in various prestigious institutions. She completed her full time Ph.D at Anna University, Chennai, India. Her research interest includes image processing and facial expression analysis.

V. V. Satyanarayana Tallapragada is working as Associate Professor in the Department of ECE, Sree Vidyanikethan Engineering College, Tirupati, Andhra Pradesh, India. He completed his Ph.D from Jawaharlal Nehru Technological University, Hyderabad in 2015. He obtained in Master's Degree from Acharya Nagarjuna University in 2007 in Communications and Radar Systems. He has 12 years of teaching experience. His research interests include Signal and Image Processing, Internet of Things, Communication Systems.

Jaganathan Thirumal works as Assistant Professor in Regional Campus of Anna University at Tirunelveli, Tamilnadu, India since 2008.

Vijayanandh V. works as an assistant professor in the department of Electrical and Electronics Engineering, Hindusthan College of Engineering and Technology Coimbatore. He had received his UG Degree from Madras University and his PG from Annamalai University . He is currently pursuing his Ph.D in Anna University .His area of interest is wireless sensor networks in which he had presented many number of papers in international and national journals . He presently working on waste management in smart city with recent tools.

Vijayarajan V. is currently working as an Associate Professor in School of Computer Science and Engineering, Department of Software Systems, Vellore Institute of Technology. He has 18 years of teaching experience and has published more than 30 peer reviewed publications. His research interests are theory of computation, operating systems, machine learning, deep learning, vision computing and medical analysis and disease prediction.

Rahul Verma, MCom (International Business) & PGDBA (Finance & Marketing), is a Lecturer in management with the Department of Training and Technical Education, India, for the last seven years and is also pursuing a Ph. D. in Commerce from Mewar University, India. His research interests include international business, marketing, and finance. He has attended and presented more than 19 research papers at several national and international conferences and seminars. He has published more than 10 research papers and 3 books with prestigious publishers like IGI Global & Apple Academic Press. He is also on editorial boards of several peer - reviewed journals.

Index

T

U

V

W

Printed in the United States
By Bookmasters